TURBULENCE AND COMBUSTION

Combustion: An International Series
Norman Chigier, *Editor*

Lefebvre, Atomization and Sprays
Lefebvre, Gas Turbine Combustion
Kuznetsov and Sabel'nikov, Turbulence and Combustion

Forthcoming titles
Chigier, Stevenson, and Hirleman, Flow Velocity and Particle Size Measurement

TURBULENCE AND COMBUSTION

Revised and Augmented for the English Edition

V. R. Kuznetsov
*Baranov Central Institute of
Aviation Motor Designing*

V. A. Sabel'nikov
*Zhukovsky Central Institute of
Aerohydrodynamics*

English-Edition Editor

P. A. Libby
*Department of Applied Mechanics and
Engineering Sciences
University of California, San Diego*

◉HEMISPHERE PUBLISHING CORPORATION
A member of the Taylor & Francis Group
New York   Washington   Philadelphia   London

TURBULENCE AND COMBUSTION. Revised and Augmented for the English Edition

Copyright © 1990 by Hemisphere Publishing Corporation. All rights reserved. Printed in the United States of America. Except as permitted under the United States Copyright Act of 1976, no part of this publication may be reproduced or distributed in any form or by any means, or stored in a data base or retrieval system, without the prior written permission of the publisher. Originally published as Turbulentnost' i goreniye by Nauka, Moscow, 1986.
Translated by Jamil I. Ghojel.

1 2 3 4 5 6 7 8 9 0   B R B R   8 9 8 7 6 5 4 3 2 1 0 9

This book was set by Allen Computype.
The editor was Janine Ludlam.
Cover design by Sharon Martin DePass.
Braun—Brumfield, inc. was the printer and binder.
A CIP Catalog record for this book is available from the British Library.

Library of Congress Cataloging-in-Publication Data

Kuznetsov, V. R. (Vadim Rostislavovich)
 [Turbulentnost' i gorenie. English]
 Turbulence and combustion / V. R. Kuznetsov, V. A. Sabel'nikov ;
English-edition editor P. A. Libby. — Rev. and augm. for English ed.
   p.   cm.
 Translation of: Turbulentnost' i gorenie.
 Includes bibliographical references.
  1. Combustion engineering. 2. Turbulence. I. Sabel'nikov, V. A.
(Vladimir Anatol'evich) II. Libby, Paul A. III. Title.
TJ254.5.K8813   1990
 621.402'3—dc20   89-19832
                                                           CIP

ISBN 0-89116-873-7
ISSN 1040-2756

CONTENTS

		PREFACE	ix
		INTRODUCTION	xiii
CHAPTER ONE		**INTERMITTENCY AND THE QUALITATIVE FORM OF PROBABILITY DENSITY FLUCTUATIONS IN TURBULENT FLOWS**	1
	1.1	Intermittency in turbulent flows	1
	1.2	Qualitative structure of two-point and three-point PDFs of velocity differences in locally homogeneous turbulence	21
	1.3	Qualitative form of PDF of the concentration of a passive contaminant in turbulent flows	24
	1.4	Qualitative form of concentration PDF upon combustion of a homogeneous mixture	35
CHAPTER TWO		**EQUATIONS FOR PROBABILITY DENSITY FLUCTUATION (PDF)**	41
	2.1	Equation for concentration PDF	42
	2.2	Equation for two-point velocity difference PDF in locally homogeneous turbulence	48
	2.3	Closure hypotheses of the equations for PDFs	51

CHAPTER THREE		**PASSIVE CONTAMINANT CONCENTRATION PDF**	57
	3.1	Equation for the conditional concentration PDF in turbulent fluids. Boundary conditions	58
	3.2	Closure hypothesis for the component describing the mixing to the molecular level	61
	3.3	Closure hypothesis for the convective component	67
	3.4	Concentration PDF in homogeneous turbulence	75
	3.5	Approximate description of concentration PDF in free turbulent flows	81
	3.6	Mathematical properties of the equation for concentration PDF in free turbulent flows. Statement of the boundary-value problem	92
	3.7	Numerical solution of the boundary-value problem	117
	3.8	Structure of isoscalar surfaces in turbulent flows	122
CHAPTER FOUR		**STATISTICAL CHARACTERISTICS OF SMALL-SCALE TURBULENCE**	129
	4.1	Similarity hypothesis	132
	4.2	Relation of the characteristics of turbulent and nonturbulent fluids	136
	4.3	Interaction between turbulent and nonturbulent fluids	139
	4.4	Effect of viscosity on the structure of small-scale pulsations	146
	4.5	Experimental investigations of intermittency and the structure of the fine-scale part of the turbulence spectra	149
	4.6	Theory of inertial interval and the problem of turbulence modeling	183
CHAPTER FIVE		**TURBULENT DIFFUSION COMBUSTION**	189
	5.1	Gas dynamic effects in turbulent diffusion combustion	194
	5.2	Effect of temperature and concentration fluctuations on the mean reaction rate	202
	5.3	Effect of radiation on the characteristics of turbulent diffusion combustion	205
	5.4	Effect of the rate of chemical reactions on turbulent diffusion combustion	210
	5.5	Formation of nitrogen oxides in turbulent diffusion combustion	224
CHAPTER SIX		**TURBULENT COMBUSTION OF A HOMOGENEOUS MIXTURE**	233
	6.1	Main problems	233
	6.2	Mechanism of turbulent combustion	242
	6.3	Effect of differences in the coefficients of molecular transport on the flame structure at the leading points	247

	6.4	Spectral representation of the velocity of flame propagation	251
	6.5	Effect of flame instability on the turbulent combustion of a homogeneous mixture	253
	6.6	Criterial description of turbulent combustion of a homogeneous mixture	262
	6.7	Maximum possible intensity of the flow in turbulent combustion of a homogeneous mixture	267
	6.8	Flame stabilization by blunt bodies	275
CHAPTER SEVEN		**TURBULENT COMBUSTION OF PARTIALLY PREMIXED GASES**	285
	7.1	Effect of turbulence on the combustion of atomized liquid fuel	286
	7.2	Qualitative structure of the zone of chemical reactions during the turbulent combustion of partially premixed gases	295
	7.3	Effect of chemical reactions on the combustion of partially premixed gases	302
	7.4	After-burning of the unburned fuel penetrating into the lean part of the flame	306
		CONCLUSION	313
		REFERENCES	321
		INDEX	357

PREFACE

A significant part of the energy currently used is produced by burning gases or vaporized liquid fuel in turbulent flows. Seemingly, this method of obtaining energy will remain dominant for a long time to come, particularly in transportation. Therefore, the study of turbulent combustion is of considerable practical interest.

Investigations of turbulent combustion reveal a noticeable gap between fundamental and applied research which thwarts the development of the technology, since fuel combustion devices have reached high levels of development and further improvement of their performance is not feasible without thorough analysis of the fluid dynamic features of combustion chambers, in particular, the characteristics of turbulence. In the meantime, the theory of turbulence has lately progressed considerably. Naturally, an adequate quantitative description of all turbulent flows is not presently feasible. However, the available qualitative understanding of many features of turbulence, and the accumulated experimental data and dimensional considerations allow sufficiently accurate estimates to be made of the characteristics of turbulence in a wide class of flows. Furthermore, many features of the combustion of gases in laminar flow are currently clear. Therefore, these prerequisites for the creation of the theory of turbulent combustion exist.

The proposed monograph sets out to expound, from a single standpoint, the principles of this theory which have been formulated to date. The selection of the scope of problems on which the principal concepts, ideas, and the techniques of the theory have been largely dictated by the authors' own investigations. The very first stages of these investigations showed that the construction of the theory of turbulent combustion is impossible without the development of the theory of turbulence proper and, in particular, those parts which were less studied (intermittency, probability distributions of various fluid dynamic parameters). As is often the case, the investigation of the indicated problems acquired independent significance and it is hoped that the obtained results will affect the development of the theory of turbulence. These considerations determined the title of the book and the selection of the material. Despite its definite weakness and tendentiousness, such an approach apparently enables the fulfillment of the set task. The theory of turbulent combustion is still in the developmental stage. The authors realize that the proposed monograph is only a small (albeit necessary) step toward the solution of the problems of turbulence and turbulent combustion.

The monograph is comprised of an introduction, seven chapters, and conclusion. With the exception of §1.4, the first four chapters are devoted to the investigation of flows without chemical reactions. This investigation was conducted on the basis of the equations for probability distribution densities of various turbulence characteristics.

Chapter 1 covers intermittency of turbulent flows and its effect on the qualitative form of velocity and concentration probability distribution densities. In Chapter 2, equations for the probability distribution densities of various fluid dynamic quantities are derived and a review of the known closure methods of these equations is conducted.

In Chapter 3, an equation for the probability distribution density of a passive contaminant is analyzed. The solution of this equation is obtained, analyzed, and compared with the experiment. Investigated in Chapter 4 is the equation for the probability density of velocity difference at two points separated by a distance in the inertial range of the turbulent spectrum.

The results obtained in the first four chapters are utilized for the analysis of diffusion combustion (Chapter 5) and combustion of a homogeneous mixture (Chapter 6). The principal result of Chapter 5 is a method of computation of the main characteristics of turbulent diffusion combustion. The method accounts for the chain character of the chemical reactions. It is shown that the effect of turbulence on the conditions of progress of chemical reactions is determined by scalar dissipation. A method for the calculation of the concentration of nitrogen oxides is given.

In Chapter 6, a qualitative picture is constructed providing the framework for the effect of flame instability and the differences in the molecular transport coefficients on the progress of combustion of a homogeneous mixture. A number of nontrivial criteria have been obtained characterizing flame propagation. An estimate of the ultimate heat release of the combustion process is given on the basis of the theory of locally homogeneous turbulence, and it is shown that this heat release is substantially lower than that in normal flames if the integral turbulence scale is greater than the thickness of the normal flame front.

The development of the theory of turbulence and turbulent combustion, like the development of any physical theory, is impossible without a close and continuous link with experiment. Therefore, great attention is given to the selection and analysis of experimental data illustrating the adopted hypotheses and conclusions. The methods developed in the monograph can be used in scientific and applied problems that are associated with the investigation of the effect of turbulence on the progress of chemical reactions (for example, in chemical technology, in gas dynamic lasers and so on). The authors are hopeful that the present monograph will facilitate further mutual unravelling and enrichment of the methods of the theory of turbulence and the theory of turbulent combustion, and stimulate new investigations at the interface of these two theories.

All the main physical ideas and mathematical techniques of the theory of turbulence and theory of combustion are given in the text as deemed necessary for direct use in the theory of turbulent combustion. Additional information on the theory of turbulence, theory of combustion, and also the widely used (in the monograph) similarity theory can be found in the special literature (see Batchelor [1953], Townsend [1956], Hinze [1959], Monin and Yaglom [1965, 1967], Frank-Kamenetskii [1967], Shchetinkov [1965], Williams [1965], Sedov [1977], and other books which are referred to in the text of the monograph).

The development of the general plan of the monograph and the discussion of the separate chapters were conducted with the combined efforts of both authors; therefore, they are equally responsible for possible shortcomings and omissions.

The authors would like to express their gratitude to Academician O. M. Belotserkovskii and corresponding members of the AS USSR V. M. Ievlev and V. V. Sychev for reviewing several sections of the manuscript of this monograph. The authors are deeply grateful to the editor of this book, A. N. Sekundov, whose remarks substantially increased the quality of the book, and Yu. Ya. Buriko, with whose cooperation a number of results have been obtained.

V. R. Kuznetsov
V. A. Sabel'nikov

INTRODUCTION

To construct the theory of turbulent combustion, one must combine, in a single unit, the methods and concepts that have been formed in two substantially differing branches of science—the theory of turbulence and the kinetics of chemical reactions. In order to do so, one must first of all establish which turbulence characteristics are of principal interest for the theory of combustion. The answer is furnished by the theory of laminar combustion which proceeds from the fact that the thickness of the zone where the processes of chemical conversion take place is much less than the characteristic dimension of the problem. For example, in laminar combustion of a homogeneous stoichiometric propane-air mixture under normal conditions the thickness of the reaction zone is less than 0.5 mm. Furthermore, during combustion strong temperature changes take place and the rates of chemical reactions are strongly dependent on the temperature. This circumstance also leads to the chemical reactions being localized in thin zones.

On the basis of these indicated features, methods have been developed in the theory of laminar combustion which enable the description of the phenomenon to be significantly simplified. Indeed, the zone of chemical reactions can be regarded as a boundary layer. Then, the solution of this internal problem (i.e., distribution of concentrations and temperature in the reaction zone) is found with the aid of relatively simple methods, since heat and mass transfer along the flame front in the equations of diffusion and heat transfer is insignificant and, hence, it suffices to integrate a system of ordinary differential equations. When solving the external problem, chemical reactions can be ignored, and the combination of the internal and external solutions enables the location of the flame front to be determined.

To illustrate this, we can cite two characteristic examples—laminar combustion of nonpremixed gases (diffusion combustion) and combustion of a homogeneous mixture. In the first case, an elegant technique, proposed by Burke and Schumann [1928], is used; namely, the so-called reduced concentration of the fuel is introduced $z = (\text{St } c_f - c_o + 1)/(1 + \text{St})$, where St is the stoichiometric factor (indicating the number of grams of the oxidant needed for the complete combustion of one gram of fuel), c is the mass concentration, and subscripts f and o refer to the fuel and oxidant, respectively. The quantity z can be interpreted as the concentration of the atoms in all the formed chemical compounds. Therefore, chemical reactions do not evidently affect its distribution, and the latter is found from the solution of the equation of diffusion without sources. Since chemical reactions occur rapidly, the concentration of the fuel and the oxidant at the flame front are simultaneously close to zero, i.e., the flame front, as a first approximation, is a surface on which condition $z = z_s = 1/(1 + \text{St})$ is fulfilled. The investigation of the internal structure of the flame front in diffusion combustion was conducted by Zel'dovich [1949], on the assumption that the chemical reaction is a single step, and an irreversible process. This study is reduced to the solution of one ordinary differential equation with a nonlinear source term.

As another example, we can cite laminar combustion of a homogeneous mixture. The solution of this problem was obtained by Zel'dovich and Frank-Kamenetskii [1938 a, b] who analyzed propagation of the normal (plane) flame front. Two zones are identified in the flame. In the first (the preheat zone) chemical reactions are insignificant. In it, the mixture is heated as a result of convection and heat conduction. In the second zone, (zone of chemical reactions) chemical conversion of the reactants takes place. Convection in this zone is insignificant, and heat transfer is determined only by conduction. It is important that the thickness of the reaction zone is several times less than the thickness of the preheat zone. Therefore, the reaction zone can be regarded as a surface on which specific boundary conditions apply. The first condition is obvious—the temperature is equal to the temperature of the equilibrium products of combustion. The second condition links the jump of the temperature derivative normal to the reaction zone with the rate of the chemical reaction and the coefficients of molecular transport (the existence of such a jump follows from the fact that heat release is concentrated on the surface).

The method of Zel'dovich and Frank-Kamenetskii can also be used to describe the combustion of a homogeneous mixture in general flows. To this end, the heat transfer equation without sources and with two boundary conditions must be solved: 1) on the surface of the flame front the temperature is equal to the temperature of the combustion products and 2) at an infinite distance from the flame front the temperature is equal to the temperature of the fresh mixture. One more boundary condition for the jump of the temperature derivative along the normal to the reaction zone determines the location of the flame front.

The solution, in a number of cases, can be simplified further, since not only the thickness of the reaction zone, but also the thermal flame front thickness is frequently small compared with the characteristic dimension of the problem (as already mentioned, upon combustion of a homogeneous mixture of propane with air in normal conditions the main change of temperature occurs in a distance less than a millimeter). Therefore, the flame front can be regarded as a surface on which discontinuous changes of velocity, density, concentrations, and temperature take place. The velocity of this surface relative to the fresh mixture is a physicochemical constant which is often referred to as the normal propagation velocity u_n. Thus, the determination of the position of the flame front is reduced to the solution of a purely kinematic problem: to finding the surface which moves with velocity $u + u_n n$ (u is the velocity of the fresh mixture and n is the unit normal to the flame front).

It follows from the above that in the laminar combustion of both nonpremixed and homogeneous gases, the position of the flame front can be determined without consideration of the details of chemical kinetics.

The methods indicated above can be used for the analysis of the turbulent combustion of a homogeneous mixture (Damköhler [1940], Shchelkin [1943] and of turbulent diffusion combustion (Hawthorne, Wedell, and Hottel [1949]). There is little doubt at the present time that the reaction zone in turbulent diffusion combustion can be assumed to be thin (henceforth, the reaction zone is defined by the nonaveraged distributions of temperature and concentrations). As regards turbulent combustion of a homogeneous mixture, the known experimental data (though presently very limited) also indicate that the thickness of the reaction zone in the majority of cases is small compared with the characteristic dimension of the problem.

Thus, the investigation of turbulent combustion can be conducted in two stages. In the first stage, the internal structure of the reaction zone is studied. This structure is determined only by the local characteristics of turbulence. For example, if the reaction zone thickness is much less than the minimum spatial scale of velocity fluctuations, then it can be assumed that the velocity of the medium and the reduced concentration of the fuel in the vicinity of the reaction zone vary linearly. Such an approach has already been employed in the theory of combustion of a homogeneous mixture (Klimov [1963]). In the second stage, large-scale oscillations of the reaction zone, which lead to its mixing as a whole, are investigated. In this case, the amplitude characteristics of the oscillations are of main interest, since they determine the mean length of the combustion zone. In this stage, the details of chemical kinetics do not have particular significance.

In order to formulate which turbulence parameters and which methods of investigation ought to be used in each of the stages, let us consider the specific features of turbulent flows and the directions resulting from these features relative to the theory of turbulence. Two main approaches can be currently identified in this theory. The purpose of the first approach (traditional) is to find various statistical characteristics. The second appeared relatively recently with the development of powerful computers and is based on the numerical integration of the Navier-Stokes equations, i.e., on the elucidations of a more or less detailed picture of the flow. The investigation of the so-called coherent structures pertains to this direction, i.e., nonrandom or completely random large-scale velocity fluctuations.

The selection of the correct approach must be based on the principal feature of turbulence which is determined by the condition of large-scale flow characteristics.

To clarify, it should be remembered that any complex spatial velocity distribution can be represented in the form of superposition of harmonic oscillations. The wavelengths of large-scale oscillations in turbulent flows are comparable with the characteristic linear dimension of the flow. The wavelength of the small-scale oscillations is much less than the characteristic dimension of the flow and, most importantly, decreases with increasing Reynolds number. Therefore, in the superposition under consideration (i.e., in the turbulence spectrum) a large number of oscillations are considered with a larger range of wavelengths. Large-scale oscillations determine the energy of turbulence, and small-scale oscillations determine its dissipation, which is significant at all Reynolds numbers (i.e., at any viscosity, however small).

The last circumstance deserves particular attention. It is evident from the general considerations that large-scale velocity fluctuations are practically independent of viscosity, since the Reynolds numbers for them are usually very high. Such fluctuations are, however, unstable, as a result of which fluctuations with less spatial scale and somewhat lower Reynolds number are formed. This process continues until the fluctuations are so small in spatial scale that their Reynolds number becomes of the order of unity. These fluctuations are stable due to the strong effect of viscous dissipation. Therefore, with decreasing viscosity the scale of the smallest spatial fluctuation drops, leading to an increase in the velocity gradient and, hence, the energy dissipation stays, on the average, unchanged.

It follows from the described picture that the multiscale state of the processes of turbulent transport leads to self-similarity of turbulent flows with respect to Reynolds numbers. Expressed more precisely, this means that the mean values of all quantities, determined by large-scale velocity fluctuations, are independent of the Reynolds number if this number tends to infinity. To such quantities belong, for example, velocity, pressure or concentration of the inert contaminant and also the various intensities of these quantities. The principle of self-similarity with respect to Reynolds number is not, generally speaking, applicable to the description of the gradients of fluid dynamic parameters, since these gradients are determined by small-scale velocity fluctuations. The validity of the principle under consideration is well supported experimentally and does not presently cause particular doubts.

A number of important conclusions are drawn from this argument. First of all, it is evident that the description of the energy carrier and large-scale velocity fluctuations alone cannot be closed. Indeed, the evolution of such fluctuations is determined by viscous dissipation which depends on small-scale fluctuations. Furthermore, since the energy spectrum of the fluctuations is continuous, large- and small-scale fluctuations cannot be considered in isolation as in the theory of the laminar flow of a continuous medium when microscopic and molecular motions are considered. Therefore, only two paths are possible for constructing the theory of turbulence. In the first, all the characteristics of the fluctuation of all scales are considered. Moreover, viscous effects must be accounted for; consequently, the coefficient of kinematic viscosity must appear in such a theory. This path, however, involves superfluous information, since the main features of turbulence are independent of Reynolds numbers.

Therefore, the second path is more natural and is based on the search for universal links between the characteristics of small- and large-scale fluctuations. As is evident from the theory of Kolmogorov [1941] and Obukhov [1941], similar links do indeed exist, if the characteristic scales of the fluctuations determining the energy of turbulence and its dissipation are significantly different. These links are regarded as consequences of the principle of turbulence self-similarity with respect to Reynolds number.

Strictly speaking, such an approach presumes consideration of the limit of the Navier-Stokes equations when Re → ∞. In this case, difficulties arise, and in order to eliminate them we apply the known quantum mechanics reasoning which illustrates the need for statistical description of the problem. Pursuing the ideas of quantum mechanics, we consider whether it is possible to measure any quantity (including energy dissipation) at Re → ∞. The term "measurement" is naturally understood to mean not only the measurement proper with the aid of some physical instrument, but also the numerical solution of the Navier-Stokes equations.

Obviously, in any experiment or numerical solution of the Navier-Stokes equations only quantities averaged over some space–time domain are determined (such quantities are convenient to refer to as partially averaged). If the Reynolds number tends to infinity, then the problem of measurement (or numerical solution) becomes particularly important, since the spatial scales of velocity fluctuations determining dissipation tend to zero. Evidently, the theory has an objective value only if we consider the quantities having a limit when the dimension of the domain, over which averaging takes place, tends to zero (otherwise, different measuring devices or different numerical algorithms will yield inconsistent results). Thus, it is necessary to analyze the double limit when, on the one hand, the Reynolds number tends to infinity, and, on the other, the dimension of the domain l, over which averaging takes place, approaches zero. From the practical standpoint, this means that a series of tests are carried out in which quantities l and Re are varied, and the measurement results are then extrapolated in the region $l = 0$, Re $= \infty$.

The existing experimental data indicate quite definitely that such extrapolation is impossible. This conclusion is based on numerous tests in which intermittency was investigated, i.e., extremely irregular distribution of velocity and concentration gradients in turbulent flows when the regions with extremely small gradient values (nonturbulent fluid) alternate irregularly with the regions in which the values of the gradients are very large (turbulent fluid)—Corrsin [1943], Batchelor and Townsend [1949], Townsend [1956] and others. The rate of turbulence energy dissipation

$$\epsilon = \frac{1}{2} \nu \left(\frac{\partial u_i}{\partial x_j} + \frac{\partial u_j}{\partial x_i} \right)^2$$

and the scalar dissipation of concentration nonhomogeneity

$$N = D \left(\frac{\partial z}{\partial x_k} \right)^2$$

(ν and D are the kinematic viscosity and molecular diffusion coefficient, respectively) are quadratic with respect to the gradients; therefore, the previous discussion is equally pertinent to the fields of ϵ and N, which play the most important role in the theory of turbulence and turbulent mixing.

In a nonturbulent fluid, energy dissipation and small-scale fluctuations are absent. In a turbulent fluid, dissipation and small-scale fluctuations always play an important role.

Therefore, if measurements are being conducted within turbulent fluid, and if l = const, there will always be a sufficiently large Reynolds number so that energy dissipation will change significantly within the measurement volume. Hence, the measured value of dissipation depends on the dimension and form of the domain over which averaging is effected. This means that there is apparently no objective means of finding partially averaged dissipation energy in a turbulent fluid. Thus, if Reynolds number tends to infinity, then only a statistical description of the flow in the turbulent fluid is possible (i.e., where $\epsilon > 0$). On the other hand, the deterministic description of flow in a nonturbulent fluid is possible, since small-scale velocity fluctuations are nonexistent and, hence, the determination of quantities ϵ and N is unnecessary.

Hence, the difficulties arising in the investigation of turbulence associated with the detailed picture of the flow are obvious. Indeed, the numerical integration of the Navier-Stokes equations is possible at not-so-high Reynolds numbers that are of principal interest in specific scientific and applied problems. To overcome these difficulties, a number of the so-called estimate turbulence models are proposed in which partially averaged flow characteristics are directly considered. The computation of these characteristics is based on equations of motion in which the effect of fluctuations with wavelengths which are less than the averaging scale is described by means of the coefficient of turbulent (more precisely, microturbulent) viscosity. However, it is evident that not all problems are solved with the aid of estimate models, since, as already mentioned, intermittency leads to the fact that there is no objective method for finding partially averaged energy dissipation. For the same reason, one cannot give a closed description of coherent structures.

Considerable difficulties also exist in the statistical approach. Three directions can be identified in this approach: 1) investigation of the formalism of the moments linked by the infinite chain of the Keller-Fridman equation [1924]; 2) the functional approach to the theory of turbulence based on the consideration of the characteristic functional introduced by Kolmogorov [1935] for which a linear equation in variational derivations was obtained by Hopf [1952]; and 3) formalism of finite-dimensional probability distributions, introduced relatively recently in the works of Monin [1967], Lundgren [1967], Novikov [1967], Ulinich and Lyubimov [1968], and Kuznetsov [1967].

Only the functional approach to the theory of turbulence is closed. However, the lack of a mathematical theory of the equations in variational derivatives and, equally important, the lack of clarity in the additional limitations which enable the identification of the number of functionals that are of interest for the theory of turbulence, do not allow, as yet, any firm results in this direction to be obtained.*
Furthermore, the characteristic functional describes all the statistical properties of the velocity field, including the properties of small-scale fluctuations that are dependent

---

*Detailed discussion of the problem and presentation of rigorous results which relate to Hopf's equation is included in a book by Vishik and Fursikov [1980].

on the Reynolds number. Consequently, such an approach is linked, in a sense, with the processing of superfluous information.

The other directions, including turbulence models, in which the main efforts are directed toward seeking the first two single-point moments of the velocity and concentration field, are based on accurate (though not closed) relations resulting from the Navier-Stokes equations. To close these relations, information gained from experience is introduced. Despite some success, this path did not lead, however, to the creation of a universal theory that is capable of describing all turbulent flows with sufficient practical accuracy.

Thus, considerable difficulties arise both upon using statistical methods and attempting to elucidate the details of the flow. Apparently, only the combination of both approaches, statistical and deterministic, permits the solution of the problem of describing turbulence. Such a combination, as will become clear now, is also important in the theory of turbulent combustion.

It was indicated earlier that the reaction zone thickness in the majority of cases is small. Therefore, for the approximate description of combustion a certain surface can be introduced close to which the reactions are localized. This surface is strongly distorted by turbulent fluctuations of different scales and, if it can thus be expressed, has the shape of a strongly crumpled piece of paper with multiple internal voids of widely varied sizes. Despite such a complex structure, two problems can be separately considered: 1) what is the internal structure of the reaction zone, and 2) what is the structure of the surface near to which it is localized?

Let us consider the first problem. Since the thickness of the chemical reaction zone is small, its structure is determined by the local characteristics of turbulence. On the basis of the theory of Kolmogorov [1941] and Obukhov [1941, 1949], it can be assumed that these characteristics are the coefficients of molecular transport, energy dissipation ϵ, and, if we are considering the combustion of nonpremixed gases, scalar dissipation N.

When investigating the internal structure of the reaction zone, statistical description of the process is, in a sense, superfluous. Indeed, if the reaction zone is locally planar, then all the characteristics of the process are described in a coordinate system, linked with some isotherm in this zone, by one-dimensional, quasi-stationary equations of diffusion and heat. These equations contain the components of the velocity of the medium, the normal to the reaction zone and, if we are considering combustion of nonpremixed gases, the reduced concentration of the fuel. In view of the fact that the reaction zone thickness is small, it can be assumed that these quantities are linearly dependent on the coordinate normal to the reaction zone. Therefore, the solution of the diffusion and heat equations inside the reaction zone can be obtained by means of simple methods. The solution obviously features direct velocity gradients and reduced concentration of the fuel which must be regarded as random parameters. Within the framework of such an approach, statistical methods are required only to determine the mean characteristics of the process inside the reaction zone.

It is quite clear that the solution of this problem is very closely associated with the investigation of statistical characteristics of the small-scale part of the turbulence spectrum. Hence, it is obvious that energy dissipation and scalar dissipations are of fundamental importance not only in the theory of turbulence (Kolmogorov [1941], Obukhov [1941, 1949]), but also in the theory of turbulent combustion.

Let us consider the second problem. The solution must be based on the analysis of probability distributions of various fluid dynamic parameters. Indeed, it is quite clear from geometrical considerations that the probability density of temperature (or concentration) can be linked with the volume enclosed between two adjacent isotherms—in particular, between those where the main conversion of the substance takes place. The latter volume is proportional to the surface near which chemical reactions are localized. This circumstance stipulates the particular role of probability densities in the theory of turbulent combustion. Formally, this role is evident in the fact that upon solution of the equations describing the behavior of the reacting gas, the rates of chemical reactions which are nonlinearly dependent on temperature and concentration are averaged.

At the stage of investigation under consideration, the details of chemical kinetics are of little importance. For example, in the combustion of nonpremixed gases the reaction zone is located near a surface on which the reduced concentration of fuel, i.e., concentration of the contaminant not taking part in the reaction, is constant. This means that the statistical characteristics of the flame front oscillations are directly dependent on the rate of chemical reaction, i.e., the problem is reduced to the investigation of probability distribution of the inert (nonreacting) contaminant.

In the combustion of homogeneous mixtures, the reaction zone is located near the surface on which the temperature is constant and close to the temperature of the combustion products. The description of large-scale fluctuations of the flame front is, thereby, reduced to the study of temperature probability distribution. The details of chemical kinetics do not have significance in this case either, since it can often be assumed that the velocity of the reaction zone relative to the medium is close to u_n and, thus, the specific features of the chemical reactions affect only the normal velocity of flame propagation.

When using these considerations in the investigation of turbulent combustion, a whole series of nontrivial effects must be taken into account. One of them is related to the effect of gas density change on the fluid dynamic structure of the flow as a result of which the mean velocity of the initial fuel components and combustion products changes. This process, first of all, acts on the large-scale part of the turbulence spectrum. Consequently, the characteristics of the latter depend weakly on the kinetics of chemical reaction and are mainly determined by the ratio of the densities of the initial fuel components and combustion products. In the combustion of a homogeneous mixture, in addition to the noted factors, a substantial role is played by the fluid dynamic (thermal) instability of the laminar flame which can, under specific conditions, lead to additional flow turbulence. From the linear theory of stability (Landau [1944]) it is known that the increment in the amplitude of the harmonic flame fluctuation increases with a decreasing scale of this fluctuation or with an increasing normal velocity of flame propagation. Therefore, small-scale turbulence must first increase. The characteristics of this turbulence are dependent on u_n, i.e., are determined by the rate of chemical reactions.

Another group of effects is associated with the influence of the processes of molecular transport on the structure of the chemical reaction zone. It should be emphasized, in this regard, that since the reaction zone thickness is small, its structure is determined by the small-scale part of the turbulence spectrum. The principle of self-similarity with respect to Reynolds number is not applicable to the latter. It should also be noted that the differences between the coefficients of molecular transport, which lead to a change in the composition and temperature in the reaction zone, play an important role.

Finally, when analyzing the internal structure of the flame front, the chain character of chemical reactions must be taken into account; i.e., one must bear in mind that chemical conversion occurs in several stages in which many intermediate substances are formed (atoms and free radicals).

CHAPTER
ONE
INTERMITTENCY AND THE QUALITATIVE FORM OF PROBABILITY DENSITY FLUCTUATIONS IN TURBULENT FLOWS

In this chapter, the main concepts of turbulent motion at high Reynolds number which are needed for the analysis of the structure of turbulent flows and the mechanisms of chemical reaction in them are discussed. The scales of length and velocity, determining the Reynolds number Re, correspond to large-scale fluctuations in the flow, i.e., Re $= qL/\nu$, where q is the root-mean-square value of the fluctuating velocity, L is the integral turbulence scale, and ν is the kinematic molecular viscosity. Considered in this chapter are intermittency and the qualitative form of probability density fluctuations in turbulent flows. As was indicated in the Introduction, these characteristics are of paramount significance for the theory of turbulent combustion and the theory of turbulence proper. It has become apparent, owing to extensive experimental investigations, that the qualitative form of probability density fluctuations (PDF) are essentially determined by intermittency and local turbulence structure as a result of which these equations cannot be considered in isolation from each other.

Our discussion follows the viewpoint expressed in a paper by Kuznetsov and Sabel'nikov [1981a].

§1.1 INTERMITTENCY IN TURBULENT FLOWS

In the theory of turbulence, the assumption that the mean values of the rate of energy dissipation $\langle \epsilon \rangle$ and scalar dissipation $\langle N \rangle$ tend to finite, nonzero limits when Re $\to \infty$ is of fundamental importance (henceforth, the word "rate" in these expressions is omitted for the sake of brevity). Here

$$\epsilon = \frac{\nu}{2}\left(\frac{\partial u_i}{\partial x_j} + \frac{\partial u_j}{\partial x_i}\right)^2 = \nu|\vec{\omega}|^2 + 2\nu\frac{\partial^2 u_i u_j}{\partial x_i \partial x_j} \cdot N = D(\nabla z)^2$$

D, ν are the coefficients of diffusion and kinematic viscosity, respectively; u the velocity vector; x the radius vector of a point in rectangular Cartesian coordinate system, $\nabla = (\partial/\partial x_1, \partial/\partial x_2, \partial/\partial x_3)$ is the gradient operation (Hamilton's operation); $\vec{\omega} = \nabla x u$ is the vorticity vector; z is the concentration of inert contaminant (i.e., contaminant not taking part in the chemical reaction): summing is effected by the repeated subscripts, and the angular brackets $\langle\ \rangle$ denote unconditional averaging.

Just as quantity ϵ characterizes the decrease in energy of turbulence as a result of viscosity, scalar dissipation N describes the rate with which reductions in concentration inhomogeneities owing to molecular diffusion take place. Therefore, it can be said that scalar dissipation gives the rate of substance mixing at the molecular level.

The assumption that finite limits of the quantities $\langle\epsilon\rangle$ and $\langle N\rangle$ exist when Re $\to \infty$ is well confirmed experimentally. Its qualitative substantiation is usually based on the theory of fluid dynamic stability and in particular that when viscosity decreases due to flow instability, more small-scale motions are generated. This process automatically leads to an increase in the instantaneous gradients of velocity and concentration as to compensate on the average for the decrease of viscosity. These concepts lie at the basis of the theory of locally homogeneous and isotropic turbulence of Kolmogorov [1941], Obukhov [1941, 1949]. A detailed statement of this theory is given in the book by Monin and Yaglom [1967]. According to this theory, the spatial scale of the more small-scale motions, the so-called Kolmogorov or internal turbulent scale η, is equal to $\eta = \nu^{3/4}\langle\epsilon\rangle^{-1/4}$.

It is pertinent to note that scalar dissipation and energy dissipation are independent of the coefficients of molecular transport even in the laminar boundary layer at high Reynolds number. As an example we can cite the flow in the boundary layer with zero pressure gradient or in a mixing layer between two plane-parallel flows. The increase in Reynolds number in both cases leads to a decrease in the thickness of the boundary layer and to the corresponding increase in velocity and concentration gradients. As a result, quantities ϵ and N remain exactly unchanged as it can be easily verified from the Blasius equation (see, for example, Schlichting [1960]). This flow pattern is observed only inside a narrow boundary layer (layer thickness tends to zero as the Reynolds number increases) outside of which the processes of molecular transport are insignificant, i.e., $\epsilon = N = 0$, and the characteristics of flow are described by Euler's equation (in a number of cases, one can use the assumption of potential flow for the description of flow outside the boundary layer).

This flow structure is characteristic for many problems of fluid dynamics in the sense that at very high Reynolds numbers viscous forces are significant only in a very narrow region outside of which the processes of molecular transport are of no importance. This conclusion is a natural consequence of the Navier-Stokes equations in which a small parameter multiplying the highest derivatives is included at high Reynolds numbers.

An analogous structure is also observed in turbulent flows. In this case, the dissipation processes also take place only in narrow regions. The features of turbulent flow become apparent as these regions are chaotically displaced in space and the values of ϵ and N in them are, generally speaking, dependent on Reynolds number. The described phenomenon was first revealed by Corrsin [1943] and is usually referred to as intermittency. It is established now that this phenomenon is characteristic for all turbulent flows.

Two forms of intermittency are distinguished — external and internal. In order to explain the meaning of these terms we turn to an obvious example — discharge of a jet of smoke from the smoke stack of a power station. Observations of such a jet indicate that a sharp boundary exists beyond which smoke does not penetrate. This boundary is distorted and fluctuates in a non-stationary manner. Such a boundary is observed not only in jets, but also in wakes and in boundary layers. Measurements indicate that energy dissipation outside the boundary is equal to zero (Townsend [1956]). Thus, the spatial distribution of dissipation energy turns out to be very nonuniform: the regions in which $\epsilon > 0$ alternate with regions in which $\epsilon = 0$. The fluctuation of the boundaries of the wakes, jets, and boundary layers are usually referred to as external intermittency; moreover, the word "external" is often omitted.

It is generally known that inside the boundaries of these flows the distribution of energy dissipation and scalar dissipation are also very nonuniform: the regions in which intensive fluctuations of velocity and concentration gradient are observed alternate with the regions in which such fluctuations are practically nonexistent. This phenomenon was first revealed by Batchelor and Townsend [1949]. It is referred to as internal intermittency.

It is significant that as a result of pressure fluctuation, velocity fluctuations are observed in the entire flow. Therefore, the investigation of intermittency cannot be based on the consideration of the velocity field. On account of the aforementioned, the most widespread method of studying intermittency is associated with the analysis of the distribution of velocity gradients, i.e., energy dissipation. However, the problem of division of the internal and external intermittencies arises in this method. In order to emphasize the importance of this problem, let us analyze the results of investigations of both types of intermittency.

In accordance with the established terminology, we shall say that the region with small values of velocity or concentration gradients are filled with nonturbulent fluid, and the region with large values of the gradients are filled with turbulent fluid. In the literature, instead of the term "nonturbulent" fluid, the expression "potential" fluid is used, i.e., it is assumed that there exists a region in which the flow is potential. This assumption will be used in this book, although its validity is not completely clear at the present time (see, for example, Monin [1978]).

The characteristics of external intermittency have been investigated in detail in a number of works (Hedley and Keffer [1974a, b], Thomas [1973], Fabris [1979a, b], Paizis and Schwars [1974, 1975], Mobbs [1968], Murlis, Tsai, and Bradshaw [1982], Wood and Bradshaw [1982] and so on). They are based on the measurements of various conditionally averaged moments, i.e., on the averaging over such a time interval when turbulent fluid is observed at the point under study. Such averaging is henceforth denoted by the symbol $\langle \ \rangle_t$. It is usually assumed that the fluid is turbulent if the square of the velocity gradient exceeds some threshold value (Townsend [1956]). It is sometimes assumed that the velocity gradient must

4 TURBULENCE AND COMBUSTION

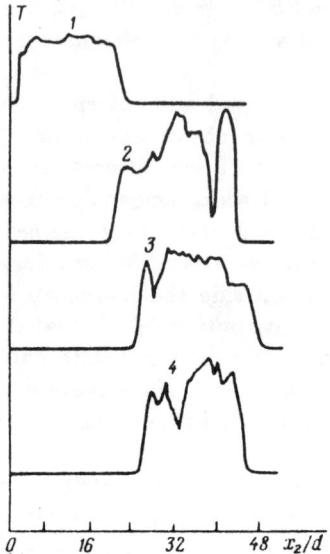

Figure 1.1 "Frozen" temperature distributions in plane submerged jet from data by Uberoi and Singh [1975]. Distribution obtained at section $x_1/d = 45$, $d = 3.18$ mm. Maximum overheat at the initial section equal to 50°C; maximum mean velocity at the section of measurement 0.305 m/sec; speed of travel of the resistance thermometer equal to 6.1 m/sec. Oscillograms 1–4 obtained at different moments of time. Measurement units along the ordinate axis are arbitrary.

exceed the threshold value during some interval of time whose duration is not less than some specified value (Hedley and Keffer [1974a, b]). In a number of cases, the concentration field is used as an indicator of turbulence, i.e., the fluid is assumed turbulent if $z > 0$. A detailed review of various experimental techniques of the selection of turbulence indicator upon measuring conditionally averaged characteristics in turbulent flows can be found in the works of Antonia [1981], Schon and Charnay [1977].

To illustrate the applicability of the field of concentration (or temperature) as a turbulent index, oscillograms which were obtained by Uberoi and Singh [1975]* are presented in Fig. 1.1. Experiments were conducted with a slightly heated plane jet discharging into a quiescent ambient. The resistance thermometer in these tests was moved perpendicular to the plane of symmetry at a speed 20 times higher than the maximum velocity of the jet at that section where measurements

---

*In this and in other figures the following notations are used: x_1 is the longitudinal coordinate measured along the main direction of flow; x_2 is the transverse coordinate for plane flows and radial coordinate for axially symmetric flows. The origin of the coordinates is located at: for a jet — in the plane of the outlet section of the nozzle on the axis or in plane of symmetry; for a wake behind a circular cylinder — on the axis of the cylinder; for a mixing layer — at a point where the contact of two flows takes place. Frequently, instead of x_1 and x_2, symbols x and y are used, respectively.

were taken. Hence, depicted in Fig. 1.1 are the "frozen" temperature distributions. It is evident that a sudden temperature change takes place at the boundary of the jet. Consequently, identification of turbulent fluid in this case is simple. Such a situation is apparently characteristic for moderate Reynolds numbers (in the present case Reynolds number calculated from the nonaveraged jet width and maximum mean velocity lies in the range $10^3 - 1.3 \times 10^3$). As it will be seen later, the region which is completely filled with fluctuations in Fig. 1.1 when Re → ∞ acquires a far more complex structure.

As a result of investigations of external intermittency, the following facts have been established. At those flow regions where intermittency is significant (i.e., the flow boundary is frequently observed), single-point probability distributions of velocity and concentration differ strongly from normal distributions. For example, in the works of La Rue and Libby [1974], Antonia, Prabhu, and Stephenson [1975], it is shown that the factors of asymmetry A and kurtosis E of concentration fluctuations can be greater than ten. Here

$$A = \frac{\langle(\xi - \langle\xi\rangle)^3\rangle}{\langle(\xi - \langle\xi\rangle)^2\rangle^{3/2}}, \qquad E = \frac{\langle(\xi - \langle\xi\rangle)^4\rangle}{\langle(\xi - \langle\xi\rangle)^2\rangle^2} - 3$$

where quantity ξ is understood to mean velocity or concentration. We shall further utilize quantities A_t and E_t, which are obtained if in the formulas for A and E averaging is effected with respect to the turbulent fluid.

In contrast to the unconditional PDFs, conditional PDF of velocity and concentration in turbulent fluid differs little from the normal distribution. For example, the measurements of La Rue and Libby [1974], Antonia, Prabhu, and Stephenson [1975] indicate that quantities A_t and E_t for the concentration field does not exceed several tenths. A similar statement applies also for the velocity field (Townsend [1956]).

It is also established that in jets and wakes the quantities averaged with respect to the turbulent fluid vary in the transverse direction much less than the unconditionally averaged quantity. This conclusion was first made by Townsend [1956]. As an example, presented in Fig. 1.2 are the results of measurements of energy dissipation in a wake of a cylinder which are given in his work (d is the diameter of the cylinder; u_0 is the velocity of incoming flow). Analogous data are also obtained for the concentration field in axisymmetric jets by Becker, Hottel, and Williams [1967], Antonia, Prabhu, and Stephenson [1975]. The results of the first work are given in Fig. 1.3, 1.4, and the second — in Fig. 1.5, 1.6 (d is the nozzle diameter; u_0 is the velocity of jet discharge; u_∞ is the velocity of the coflowing stream).

Figures 1.3 and 1.4 show that when x_2/x_1 changes in the range 0.14 – 0.26, the unconditionally averaged concentration changes by two orders of magnitude, whereas concentration averaged with respect to the turbulent fluid changes only by 2.5 times, and the intensity of concentration fluctuations in the turbulent fluid remains practically unchanged.

As the measurements of Fabris [1979a, b] in the wake of a heated circular cylinder show, the mean values of the squared temperature derivative $(\partial T/\partial t)^2$ have

6 TURBULENCE AND COMBUSTION

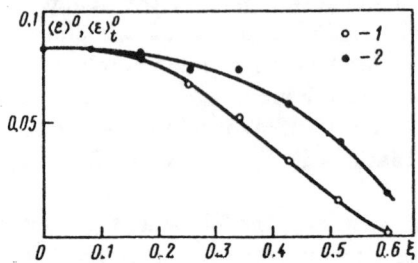

Figure 1.2 Profiles of unconditional and conditional averaged rate of turbulence energy dissipation in the wake of a circular cylinder from data by Townsend [1956].

the same character (Fig. 1.7). Using Taylor's hypotheses (i.e., the assumption that $\partial/\partial t = -\langle u_1 \rangle \partial/\partial x_1$) and the assumption of local turbulence isotropy, one can conclude that these remarks are also equally pertinent to scalar dissipation.

In contrast to the concentration of a contaminant, which does not penetrate into the nonturbulent fluid, the fluctuating energy in nonturbulent fluid is not equal to zero but is caused by pressure fluctuations. The latter, as is generally known, determines the nonlocal character of the transport processes in turbulent flows. As an illustration, we turn to Fig. 1.8 which depicts the results of

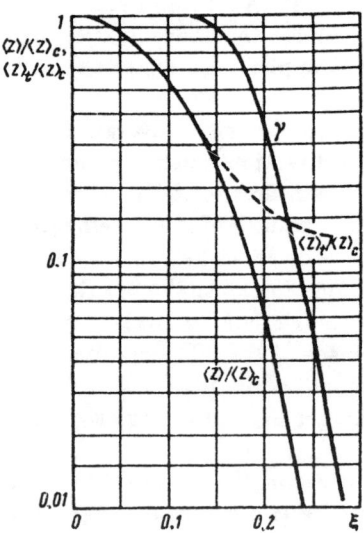

Figure 1.3 Profiles of unconditional and conditional averaging of concentration in submerged axisymmetric jet from data by Becker, Hottel, and Williams [1967]. $x_1/d = 20 - 36$, $\mathrm{Re}_d = u_0 d/\nu = 5.4 \cdot 10^4$, $d = 2.41$ cm, $u_0 = 130$ m/s, $\langle z \rangle_c$.

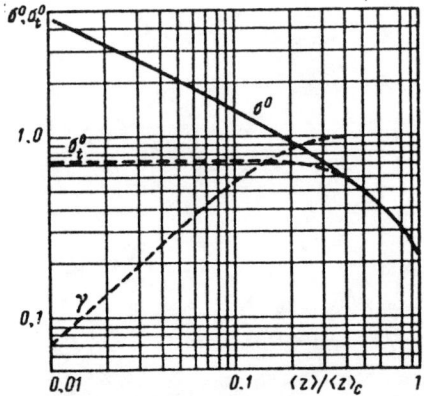

Figure 1.4 Unconditional and conditional root-mean-square concentrations in a submerged axisymmetric jet from data by Becker, Hottel, and Williams [1967]. $\sigma^0 = \langle(z-\langle z \rangle)^2\rangle^{1/2}/\langle z \rangle$, $\sigma_t^0 = \langle(z-\langle z \rangle_t)^2 \rangle_t^{1/2}/\langle z \rangle_t$. Conditions of the experiments and notations are as in Fig. 1.3.

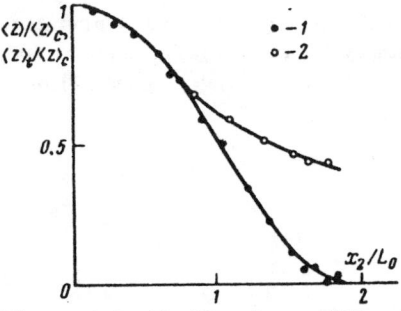

Figure 1.5 Profiles of unconditional and conditional averaged concentration in coflowing axisymmetric jet from data by Antonia, Prabhu, and Stephenson [1975]. $x_1/d = 59$, $Re_d = u_0 d/\nu = 4.3 \cdot 10^4$, $u_0/u_\infty = 6.6$, $d = 2.03$ cm, $u_0 = 32$ m/s; 1) $\langle z \rangle/\langle z \rangle_c$, 2) $\langle z \rangle_t/\langle z \rangle_c$, $\langle z \rangle_c$ is the mean concentration on jet axis; L_0 is the laminar scale, equal to the distance from axis of symmetry at which $\langle z \rangle = \langle z \rangle_c/2$.

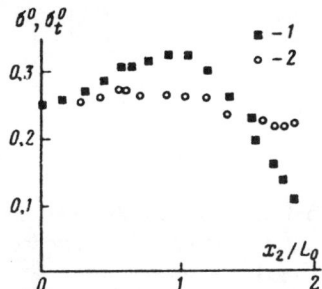

Figure 1.6 Profiles of unconditional and conditional root-mean-square concentrations in coflowing axisymmetric jet from data by Antonia, Prabhu, and Stephenson [1975]. 1) $\sigma^0 = \langle(z-\langle z \rangle)^2\rangle^{1/2}/\langle z \rangle_c$, 2) $\sigma_t^0 = \langle(z-\langle z \rangle_t)^2\rangle_t^{1/2}/\langle z \rangle_c$, $\langle z \rangle_c$ is the mean concentration on axis of jet. Conditions of the experiments as in Fig. 1.5.

8 TURBULENCE AND COMBUSTION

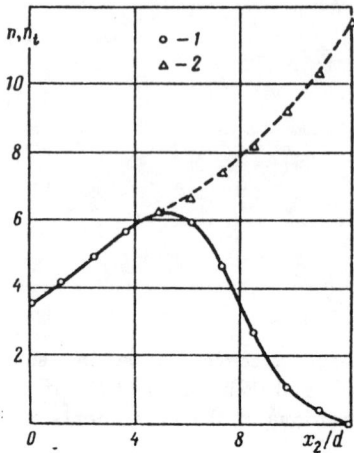

Figure 1.7 Profiles of unconditional and conditional averaged squared temperature derivative in a wake of a circular cylinder from data by Fabris [1979a, b]. $x_1/d = 400$, $\text{Re}_d = u_0 d/\nu = 2.7 \cdot 10^3$, $d = 6.25$ mm, $u_0 = 6.46$ m/s; *1)* $n = \langle(\partial z/\partial t)^2\rangle d^2/u_0^2$, *2)* $n_t = \langle(\partial z/\partial t)^2\rangle_t d^2/u_0^2$, $z = \Delta T/(\Delta T)_{\max}$, $\Delta T = T - T_\infty$.

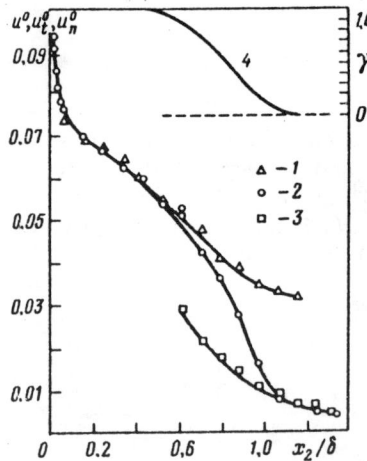

Figure 1.8 Profiles of unconditional and conditional root-mean-square fluctuations of longitudinal velocity in turbulent and nonturbulent fluids and the intermittency factor in the turbulent boundary layer on a plate from data by Kovasznay, Kibens, and Blackwelder [1970]. Measurements conducted in a section located at distance $x_1 = 9$ m from the plate tip, boundary layer thickness $\delta = 10$ cм, $R\delta = u_\infty \delta/\nu = 2.75 \cdot 10^4$, $u_\infty = 4.3$ m/s; *1)* $u_t^0 = \langle(u_1 - \langle u_1\rangle_t)^2\rangle_t^{1/2}/u_\infty$, *2)* $u^0 = \langle(u_1 - \langle u_1\rangle)^2\rangle^{1/2}/u_\infty$, *3)* $u_n^0 = \langle(u_1 - \langle u_1\rangle_n)^2\rangle_n^{1/2}/u_\infty$, *4)* - γ.

measurements of the conditional averaged moments $\langle(u_1 - \langle u_1\rangle_t)^2\rangle_t^{1/2}$ and $\langle(u_1 - \langle u_1\rangle_n)^2\rangle_n^{1/2}$ in the turbulent boundary layer on a plate (the subscript n corresponds to averaging over the nonturbulent fluid). These data are given by Kovasznay, Kibens, and Blackwelder [1970] (analogous results were obtained in a plane jet by Jenkins and Goldschmidt [1976], Oler, Jenkins, and Goldschmidt [1981]. It is evident that if the point under study is located not too far from the mean boundary of the boundary layer, then the fluctuating energy in both fluids is of the same order, i.e., pressure fluctuations effectively redistribute energy.

Let us now consider the results of the investigations of internal intermittency. It was established in the first work dedicated to this problem (Batchelor and Townsend [1949]) that the kutrosis of the velocity gradient fluctuations (and, hence, the factor of the kurtosis of energy dissipation fluctuations) is very large. It was also shown (which is more important) that these factors increase with increasing Reynolds number. The results of the experiments were later confirmed by Pond and Stewart [1965], Gurvich [1966, 1967], Stewart, Wilson, and Burling [1970], Wingard and Tennekes [1970], Gibson, Stegen, and Williams [1970], Gibson, Stegen, and McConnell [1970], Wygnanski and Fiedler [1970], Kuo and Corrsin [1971], Chen [1971], Gibson and Masiello [1971], Sheih, Tennekes, and Lumley [1971], Antonia [1973], Wingaard and Pao [1975], Frenkiel and Klebanoff [1975], Wingaard, Pao, and Wignanski [1976], Antonia and Danh [1977], and also by a number of other investigators.

As an example, presented in Figs. 1.9 and 1.10 are the dependences of the kurtosis of the fluctuations of one of the velocity gradient components $\partial u_1/\partial x_1$ on the number Re $\lambda = \sqrt{\langle(u_1 - \langle u_1\rangle)^2\rangle}\, \lambda/\nu \sim \text{Re}^{1/2}$, where $\lambda^2 = \langle(u_1 - \langle u_1\rangle)^2\rangle/\langle(\partial u_1/\partial x_1)^2\rangle$, λ is Taylor's microscale. Figure 1.9 is taken from the work of Van Atta and Antonia [1980], and Fig. 1.10 — from the work of Frenkiel and Klebanoff [1975]. The results of measurements of the kurtosis of concentration gradient is of an analogous character (Fig. 1.11 is borrowed from a work by Tavoularis and Corrsin [1981b]).

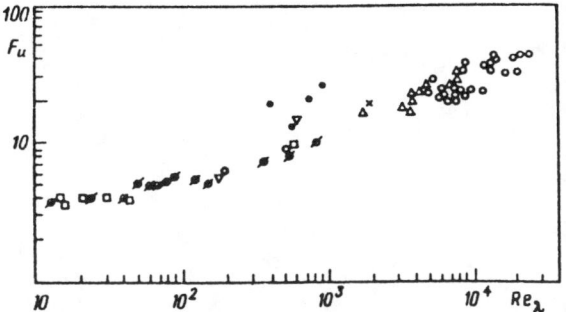

Figure 1.9 Dependence of the kurtosis of the velocity derivative on Reynolds number from data by different authors (the figure corresponds to tests in the boundary layer of the Earth's surface in laboratory conditions). $F_u = \langle(\partial u/\partial x_1)^4\rangle/\langle(\partial u/\partial x_1)^2\rangle^2 = 3+E_u$. The different symbols denote data by different authors.

10 TURBULENCE AND COMBUSTION

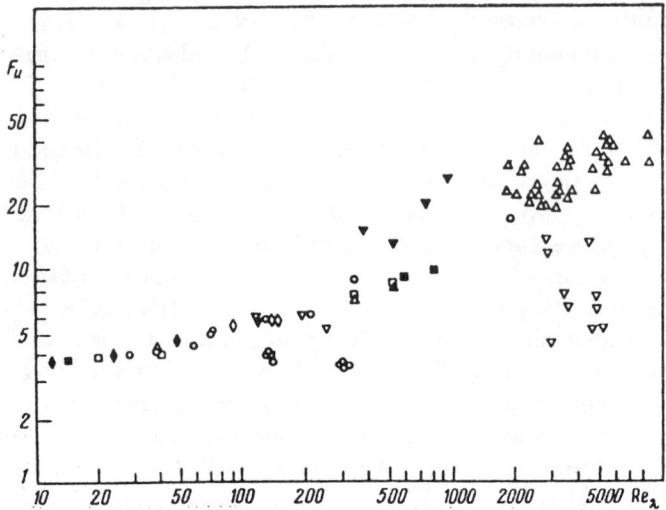

Figure 1.10 Dependence of the kurtosis of velocity derivative on Reynolds number in the boundary layer of the Earth's surface in laboratory conditions from data by different authors. $F_u = \langle(\partial u_1/\partial x_1)^4\rangle/\langle(\partial u_1/\partial x_1)^2\rangle^2 = 3 + E_u$. The different symbols denote data by different authors.

The results of the considered measurements are evidence that there are regions in the flow in which energy dissipation far exceeds the mean value. Since $\langle\epsilon\rangle$ is independent of Reynolds number and the kurtosis of energy dissipation fluctuations apparently increases in an unrestricted fashion with increasing Reynolds number, it then follows that energy dissipation takes place in a volume which tends to zero when $\text{Re} \to \infty$. To elucidate this conclusion we note that probability distribution of the random quantity ξ can be regarded as the ratio of that part

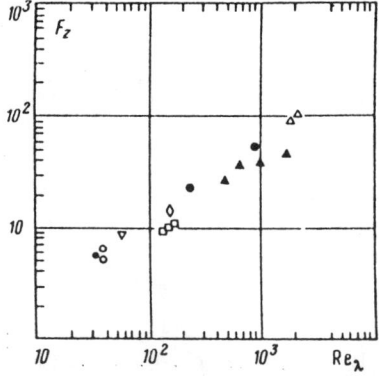

Figure 1.11 Dependence of the kurtosis of concentration derivative on Reynolds number from data by different authors. $F_z = \langle(\partial z/\partial x_1)^4\rangle/\langle(\partial z/\partial x_1)^2\rangle^2 = 3 + E_z$. The different symbols denote different authors.

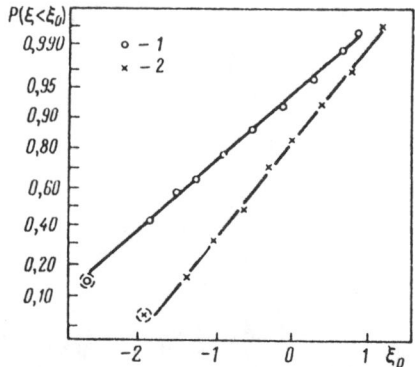

Figure 1.12 Probability distribution of the squared temperature difference at two close points in the boundary layer of the Earth's surface from data by Gurvich [1967]. The sensors were mounted at a height $H = 4$ m from the Earth's surface at a distance of $\ell = 2$ cm from each other in the direction of the mean wind velocity. *1)* $\text{Re}_H = \langle u_1 \rangle H / \nu = 1.8 \cdot 10^6$, $\langle u_1 \rangle = 7$ m/s, *2)* $\text{Re}_H = \langle u_1 \rangle H / \nu = 1.4 \cdot 10^6$, $\langle u_1 \rangle = 5$ m/s, $\xi_0 = 1$ lg $[(\Delta_\ell T)^2 / \langle (\Delta_\ell T)^2 \rangle]$. $P(\xi < \xi_0)$ is the probability of event $\xi < \xi_0$, $\langle u_1 \rangle$ is the mean wind velocity; dependences *1* and *2* correspond to two sections of recording $\Delta_\ell T$ obtained in different days; the scale along the ordinate axis is selected so that a straight line would correspond to the normal distribution law.

of the volume, in which condition $\xi < \xi_0$ is fulfilled, to the total volume of the system (for the sake of simplicity, it is assumed that distribution $\xi(x)$ is statistically homogeneous and the process is ergodic). In view of this, let us consider the question of probability distribution of energy dissipation and scalar dissipation.

Despite the fact that direct measurements of these distributions are also known (see the literature cited above), it is convenient to draw a number of conclusions here on the basis of measurements of velocity difference or temperature PDFs at two closely spaced points. From the theoretical works of Kolmogorov [1962a, b], Obukhov [1962], Yaglom [1966], Betchov [1974, 1975, 1976], Betchov and Lorenzen [1974], Betchov and Larsen [1981], it is evident that these distributions essentially differ from the normal distribution. Illustrated in Fig. 1.12 are the results of measurements in the boundary layer at the Earth's surface of the PDF of the squared temperature difference at two closely located points $(\Delta_\ell T)^2 = [T(x_1 + \ell) - T(x_1)]^2$ (axis x_1 is located in the direction of the mean wind). These data were obtained by Gurvich [1967]. Similar data for the squared velocity difference $(\Delta_\ell u_1)^2 = [u_1(x_1 + \ell) - u_1(x_1)]^2$ are presented in Fig. 1.13 from the experimental results of Gibson, Stegen, and Williams [1970] (see also Gurvich [1966]). In both cases the distance between the points ℓ satisfies the inequality $\eta \ll \ell \ll L$.

The presented data indicate that the distribution under consideration approach, as a first approximation, the lognormal distribution (i.e., the logarithm of the quantity is distributed by a normal law). An attempt to theoretically

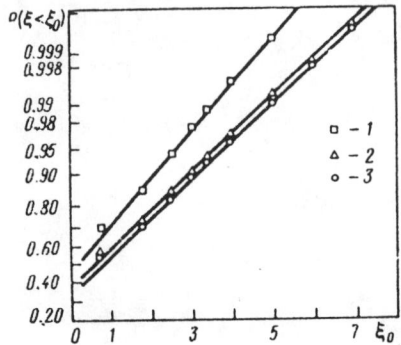

Figure 1.13 Probability distribution of the squared velocity difference at two close points in the boundary layer at the Earth's surface from data by Gibson, Stegen, and Williams [1970]. The sensors were mounted at height $H = 1.2$ and 7 m from the Earth's surface at distance $\ell = 1$ cm from each other in the direction of the mean wind velocity. 1) $H = 7$m, 2) $H = 2$ m, 3) $H = 1$ m, $\xi_0 = \ln{(\Delta_\ell u_1)^2} + \text{const}$, $P(\xi < \xi_0)$ is the probability of event $\xi < \xi_0$; the scale along the ordinate axis is selected the same way as in Fig. 1.12.

substantiate this law is included in the work by Yaglom [1966]. The energy dissipation field ϵ^ℓ, averaged over the region with small ℓ, is considered in this work. Based on a credible hypothesis, it is shown that the probability density of quantity ϵ^ℓ has the form

$$P(\epsilon^\ell) = \frac{1}{\sqrt{2\pi}\,\sigma_\epsilon \epsilon^\ell} \exp\left\{-\frac{(\ln(\epsilon^\ell/\langle\epsilon\rangle) + \sigma_\epsilon^2/2)^2}{2\sigma_\epsilon^2}\right\}$$

$$\sigma_\epsilon^2 = \mu \ln(L/\ell) = \langle(\ln \epsilon^\ell - \langle\ln \epsilon^\ell\rangle)^2\rangle$$

$$\langle \ln \epsilon^\ell \rangle = \ln\langle\epsilon\rangle - \frac{1}{2}\sigma_\epsilon^2 \tag{1.1}$$

where μ is a constant. It is often assumed that this formula yields the PDF of direct dissipation if $\ell \sim \eta$ (see, for example, Frenkiel and Klebanoff [1975]). From the above formula it follows that when $\ell = \eta$ the limit

$$\lim_{Re \to \infty} \int_{k\langle\epsilon\rangle}^{\infty} P(\epsilon^\ell)\,d\epsilon^\ell$$

is equal to zero at any value of k, however small it might be. Hence, it follows that when $Re \to \infty$, the dissipation energy is concentrated in a region whose volume tends to zero (Mandelbrot [1974, 1976]).

The question is how to interpret the above stated results. It should be recalled that for high Reynolds numbers the flow is described by a system of equations with perturbations of the highest derivatives. It is well known that the solution of such equations are distinguished, as a rule, by the fact that regions of two types appear. In the first type the highest derivatives (which in the present case describe the viscous forces) are inessential. In the second type of regions viscous forces are always dominant: the volume of these regions tends to zero with increasing Reynolds number. The theory of laminar boundary layer on a rigid surface well illustrates this flow picture.

Such boundary layers can, apparently, also arise in a free flow. This conclusion is particularly confirmed in the linear theory of stability (we should keep in mind the so-called critical layer inside which the viscous forces are always essential). From the standpoint of the ideal fluid theory, boundary layers of the type under consideration can be interpreted as tangential discontinuities.

This interpretation must be viewed with great caution, since it is possible that the development of instabilities in an ideal fluid leads to the appearance of a singularity of another type. The results obtained in the numerical computation of the flow corresponding to the Taylor-Green vortex [1937] confirm this, i.e., $u_1 = \cos x_1 \sin x_2 \cos x_3, u_2 = -\sin x_1 \cos x_2 \cos x_3, u_3 = 0$ when $t = 0$. Similar computations for an ideal fluid were conducted by Morf, Orszag, and Frisch [1980], and for viscous fluid — by Brachet et al. [1983]. It is established that when $t > 0$, the flow becomes three-dimensional, and vortex tubes are stretched at an increasing rate in time and vorticity becomes, as a result, infinitely large after a finite time interval. Of course, there is not yet complete confidence that the indicated result reflects essential properties of the equations of fluid dynamics and that it is not caused by the imperfection of the numerical algorithm. A more detailed discussion of the question of the emergence of singularities in the solutions of Euler and Navier-Stokes equations is given in the works of Ladyzhenskaya [1970], Saffman [1978, 1981], Saffman and Baker [1979], Marsden and McCracken [1976]. We shall only note here the most important results, namely that singularities might arise only in three-dimensional cases, since the theorem concerning the existence and uniqueness of the solution for both Euler and Navier-Stokes equations has been proved in a two-dimensional case for smooth initial conditions (see, for example, Ladyzhenskaya [1970]). Proving such a theorem in a three-dimensional case has not been successful and, therefore, a hypothesis has been put forward which states that a singularity arises in the solution of nonstationary Euler and Navier-Stokes equations with smooth initial conditions after a finite length of time (Leray [1934], Kraichnan [1975], Mandelbrot [1976]).

Let us now continue the discussion of the causes of the emergence of intermittency. Let us consider the subsequent "fate" of the tangential discontinuity. It is generally known that in the linear approximation this discontinuity is unstable, and that the incremental increase of the amplitude of the perturbation (surface) is inversely proportional to the wavelength of the perturbation (see, for example, Landau and Lifshits [1954], Batchelor [1967]). The last property is, evidently, of great significance, since under a sufficiently large initial amplitude of small scale perturbations, the surface area of the discontinuity increases sharply over a short period of time.

It can be assumed that the discontinuity surface is compact (i.e., its points do not extend beyond the bounds of the restricted spatial domain), and its area is infinitely large. Such a surface resembles, to a certain extent, a sponge, i.e., a formation with a large number of internal voids having a wide spectrum of characteristic dimensions and without distinctly defined external boundary (thin "channels" exist connecting the interior of the formation with the external space). From this viewpoint, the fluctuations of the external boundary of the "sponge" can be referred to as external intermittency, and the fluctuations of the internal channels as internal intermittency.

The geometrical intermittency systems which reflect the above pattern were developed by Novikov and Stewart [1964], Novikov [1965, 1966, 1969, 1971], Yaglom [1966], Mandelbrot [1976] (see also the review in the book by Monin and Yaglom [1967]). Attempts to describe intermittency based on the numerical integration of the hydrodynamic equations were made by Siggia and Patterson [1978], Siggia [1981], Pullin [1981], Chorin [1981], Brachet et al. [1983].

Apparently, the essence of the problem is most accurately and simply reflected in a model by Novikov and Stewart [1964] in which it is assumed that the flow can be divided into n^3 identical cubes so that all the energy dissipation would be concentrated in γn^3 of the cubes ($0 < \gamma < 1$) and is equal to zero in the remaining cubes. It is also supposed that each of the γn^3 cubes can be divided again in a similar way and the division process can continue indefinitely when Re $\to \infty$. Thus, nonturbulent fluid can be found at any point of the flow. The characteristic dimension of the domains, filled with this fluid, varies from zero to the integral turbulence scale. The domains that are occupied by the turbulent fluid are closely "intertwined" with the domains which contain nonturbulent fluid.

Similar hypotheses have been stated by Mandelbrot [1974, 1975, 1976, 1977], in whose works the concept of fractals is used to describe the structure of the turbulent fluid.

This concept has appeared in practical investigations connected with the determination of a shore line on the basis of cartographic measurements. This line is found to increase stepwise with increasing map scale (minute details are enlarged more and more). The principal quantitative characteristic of fractals is the Hausdorff-Bezikovich space which determines the measure of "entanglement" of the set. We shall limit ourselves to a simple but nonrigorous explanation of this space without citing its formal definition.

Let there be a set of points in a certain cube. We divide this cube into N^3 identical small cubes. We shall assume that there is a rule enabling us to find those cubes in which the points of the analyzed set are located. Let $M(N)$ be the number of such cubes. The Hausdorff-Bezikovich space can be defined by formula $d = \lim_{N \to \infty} \ln M(N)/\ln N$. It is readily seen that if the set of points has a familiar form, then d is an integer; $d = 0$ for a finite number of points; $d = 1$ for a line of finite length; $d = 2$ for a surface of finite area; $d = 3$ for a finite volume. Generally, d is not an integer. Thus, the quantity d yields a generalization of the concept of the number of dimensions of the space.

According to Mandelbrot [1976, 1977], when Re $\to \infty$, energy dissipation is concentrated at a set of points which, from the topographical standpoint, is a surface. This surface has an infinite area and is so "entangled" that it can be regarded as a fractal for which $2 < d < 3$.

Another point of view regarding modeling of intermittency can be based on the concept of strange attractors. A special collection of translated works *Strange Attractors* edited by Sinai and Shil'nikov [1981], and the reviews by Rabinovich [1978] and Monin [1978] are also devoted to the problem of strange attractors.

The main result attained in this field can be formulated as follows. The solutions of deterministic systems of ordinary nonlinear differential equations of the first order with the number of equations equal to or greater than three are often found to be badly "predictable" (and stochastic from the standpoint of the experimenter), even when the solution exists and is unique and, therefore, no singularities arise in this solution. This solution structure is caused by the fact that each phase trajectory is unstable (i.e., with time the distance between two initially close phase trajectories increases exponentially). The set of phase trajectories (strange attractors) are compact in the sense that all its points do not extend beyond the bounds of some finite phase volume. For nonconservative systems the phase volume (more precisely, Lebesque measure) is equal to zero, just as the volume of turbulent fluid is equal to zero when Re $\to \infty$. Phase point distributions also resemble the distribution of points which belong to the turbulent fluid in the physical space. A link between strange attractors and fractals is quite distinctly observed (Mandelbrot [1976]).

The experimental data considered earlier and the results of their theoretical analysis point to the fact that the quantitative determination of the characteristics of intermittency is associated with a number of essential difficulties. First, it is unclear how to determine the boundaries of the turbulent fluid (as already indicated, thin channels filled with nonturbulent fluid can "penetrate" inside the region which, on the face of it, is completely filled with turbulent fluid). Second, since the viscous effects when Re $\to \infty$ are apparently essential only in regions with zero volume, the definition of the intermittency factor which is usually understood to mean the relative value of the volume filled with turbulent fluid becomes unclear. Third, the question arises as which fluid dynamic characteristic is most appropriate for the determination of intermittency. Indeed, nonturbulent regions are always found next to turbulent regions. If the dimensions of both regions are of the same order, then the dissipation energy is also of the same order, which is evident from the works of Landau and Lifshits [1954], Phillips [1955] (pressure fluctuations lead to nonlocal energy transport). Therefore, both fluids might become indistinguishable in the character of their velocity fluctuations.

Let us consider how to overcome the indicated difficulties. It is evident that the last one can be eliminated if quantities that are defined by the more small-scale fluctuations are to be analyzed. One of these quantities is energy dissipation which will be considered henceforth.

We note now that energy dissipation at any Reynolds number is nonzero owing to viscous vorticity diffusion. The analogous statement is also valid for contaminant concentration, i.e., as a result of scalar dissipation the molecular diffusion is different from zero everywhere. Thus, intermittency, if it can be put this way, is a phenomenon which emerges only when Re = ∞ and cannot rigorously be determined at a finite Reynolds number. Nonetheless, as is quite often the case in physics in general and in turbulence theory in particular, the introduction of the concept of intermittency at a finite Reynolds number (even though nonrigorous and approximate) is found to be extremely useful[†]. In this case, however, for the quantitative determination of intermittency it is necessary to introduce some boundary level ϵ_0 assuming that the fluid is turbulent if $\epsilon > \epsilon_0$, and nonturbulent otherwise. Since there is not any satisfactory method of selecting ϵ_0, this definition, despite the fact that it is natural and is widely used in experimental investigations, is nonconstructive. One would think that this difficulty can be overcome if Reynolds number is made infinite and it is hence assumed that $\epsilon_0 \to 0$.

However, a new difficulty is encountered in this case, one associated with the fact that the volume of the turbulent fluid tends to zero and, hence, the characteristics of intermittency cannot be measured. Despite the indicated difficulty, the quantitative characteristics of intermittency can be presented on the basis of the scheme due to Obukhov [1962] (see also Yaglom [1966], Gurvich and Yaglom [1967]). Consider the dissipation field ϵ^ℓ averaged over regions with dimension ℓ, say, over spheres with radius ℓ and the center at point x. Consideration of field ϵ^ℓ is not only natural, but even obligatory if we take into account that in experiments quantities are always averaged over some finite region (the dimension of this region is obviously determined by the scale of the measuring device). We define $\epsilon_0 = \langle\epsilon\rangle/\text{Re}$ as the limiting energy dissipation level ϵ_0. Such a choice is based on intuitive considerations according to which it can be expected that $\lim_{\text{Re}\to\infty} \partial u_i/\partial x_j$ is finite in the nonturbulent fluid and, hence, energy dissipation in this fluid is inversely proportional to Reynolds number.

Let us introduce the function (index) of intermittency $\Gamma^\ell(x, \ell)$

$$\Gamma^\ell = 1, \quad \text{if} \quad \epsilon^\ell > \epsilon_0$$
$$\Gamma^\ell = 0, \quad \text{if} \quad \epsilon^\ell < \epsilon_0$$

For the moments of this function we use the following notation:

$$\langle\Gamma^\ell\rangle = \gamma^\ell$$

$$\langle[\Gamma^\ell(x^{(1)}) - \Gamma^\ell(x^{(2)})]^2\rangle = D^\ell_{\gamma\gamma}$$

$$\langle\Gamma^\ell(x^{(1)})\Gamma^\ell(x^{(2)})\Gamma^\ell(x^{(3)})\rangle = S^\ell_{\gamma\gamma\gamma} \qquad (1.2)$$

[†]The purpose of the introduction of such a concept is based on the desire to divide the flow region into sections in which molecular viscosity and dissipation are significant, and sections where viscous processes are unimportant. Such a division enables us to describe not only qualitatively, but also quantitatively many complex phenomena that are observed in turbulent flows.

In the case of locally homogeneous turbulence, we have

$$D^\ell_{\gamma\gamma} = D^\ell_{\gamma\gamma}(r), \qquad S^\ell_{\gamma\gamma\gamma} = S^\ell_{\gamma\gamma\gamma}(r, R)$$

$$r = x^{(2)} - x^{(1)}, \qquad R = x^{(3)} - x^{(1)} \tag{1.3}$$

Consider also the probability that there is turbulent fluid at one set of the space points and nonturbulent fluid at the other. Such an analysis is required to elucidate the structure of two- and three-point velocity PDFs (see Chapter 4). These probabilities will be denoted by symbol γ^ℓ with subscripts t and n (subscript t refers to turbulent fluid and n to nonturbulent fluid; the ordinal number of the subscript corresponds to the number of the point). For example, symbol γ^ℓ_{ntn} denotes the probability of an event when there is nonturbulent fluid at the first and third points and turbulent fluid at the second. These probabilities are linked with the intermittency function. Thus, for example,

$$\gamma^\ell_{tt} = \langle \Gamma^\ell(x^{(1)})\Gamma^\ell(x^{(2)}) \rangle, \qquad \gamma^\ell_{tnt} = \langle \Gamma^\ell(x^{(1)})[1 - \Gamma^\ell(x^{(2)})]\Gamma^\ell(x^{(3)}) \rangle$$

and so on. By using the definition of the structure function $D^\ell_{\gamma\gamma}$ and the relation linking $\gamma^\ell_{tt}, \ldots, \gamma^\ell_{nn}$ with the intermittency function for a locally homogenous turbulence, we find

$$\langle \Gamma^\ell(x^{(1)})\Gamma^\ell(x^{(2)}) \rangle = \gamma^\ell_{tt} = \gamma^\ell - \frac{1}{2}D^\ell_{\gamma\gamma}(r)$$

$$\langle \Gamma^\ell(x^{(1)})[1 - \Gamma^\ell(x^{(2)})] \rangle = \gamma^\ell_{tn} = \frac{1}{2}D^\ell_{\gamma\gamma}(r)$$

$$\langle [1 - \Gamma^\ell(x^{(1)})]\Gamma^\ell(x^{(2)}) \rangle = \gamma^\ell_{nt} = \frac{1}{2}D^\ell_{\gamma\gamma}(r)$$

$$\langle [1 - \Gamma^\ell(x^{(1)})][1 - \Gamma^\ell(x^{(2)})] \rangle = \gamma^\ell_{nn} = 1 - \gamma^\ell - \frac{1}{2}D^\ell_{\gamma\gamma}(r) \tag{1.4}$$

It is evident from general considerations that the rational determination of the characteristics of intermittency must be based on the double limit involving $Re \to \infty$ and $\ell \to 0$. The value of such a limit is, apparently, ambiguous. This conclusion is indirectly confirmed by the fact that a generally accepted algorithm for the measurement of intermittency has not been developed to date. Direct proofs of its validity are also known. In particular, a measurement algorithm was selected by Kuo and Corrsin [1971] which yielded unusual results. First, it was established that the intermittency factor decreases with increasing Reynolds number. Second, the measured values of γ were found to be substantially less than unity in regions where it is traditionally assumed that $\gamma = 1$ (for example, on the axis of a submerged jet).

Thus, it can be assumed that the moments of the intermittency function tend to some values which are substantially dependent on how ℓ tends to zero and Re to infinity. In particular, any value for the intermittency factor can be obtained in the range $0 \leq \gamma \leq \gamma_{max} \leq 1$. The lower estimate of γ here is chosen on the basis of the stated concepts and of experimental data; it follows that the volume occupied by the turbulent fluid tends to zero when Re $\rightarrow \infty$. It is evident that if ℓ tends to zero first and then Re to infinity, i.e., all the measurements are conducted accurately, then a zero value of γ will be obtained. Thereby, the characteristics of internal intermittency will be measured. Another variant of the limit transition is based on the fact that Re first tends to infinity, and then the dimension of the region ℓ tends to zero. In this case, the maximum smoothing of all small-scale details takes place and, hence, it can be assumed that the maximum possible value of γ is obtained in this variant.

Based on the analogy discussed earlier between the turbulent fluid and the sponge, it can be expected that such smoothing would "close" all internal channels that extend inwards from the external boundary. Moreover, multiply-connected nonturbulent regions, whose dimensions strongly vary (even on the order of the integral turbulence scale), can remain in the turbulent fluid. Thus, it seems that the second variant of the limit transition yields the characteristics of the external intermittency. Comparison of both variants of the limit process indicates that there are, in a strict sense, not two different types of intermittency, but only one, whose characteristics have so far been defined ambiguously.

It seems that the second variant of the limit transition $\left(\lim_{\ell \to 0} \lim_{Re \to \infty}\right)$ corresponds best with the algorithms which are used in investigations of the external intermittency.

The conclusion reached here is based on the following considerations. We turn to formula (1.1). Although there are presently serious doubts as to the validity of the lognormal law (see Chapter 4), this formula illustrates the main qualitative feature of dissipation distribution, namely that the amplitude of fluctuations of quantity ϵ varies over an exceptionally wide range of values. Therefore, when estimating the resolution of the measuring device, the dimension of the region over which averaging is effected must be compared not with the Kolmogorov scale determined from quantity $\langle \epsilon \rangle$, but with the Kolmogorov scale computed from the true (nonaveraged) value of ϵ. Since $\epsilon \gg \langle \epsilon \rangle$ in a turbulent fluid, the restriction on the resolution of the measuring device strongly increases. In particular, the estimates made in Chapter 4 show that in the majority of cases for accurate measurement of ϵ it is required that the dimension of the averaging region be less than $\nu^{3/4}/\langle \epsilon \rangle^{1/4}$. This condition, to the best of our knowledge, has not been fulfilled in any of the known investigations. Thus, considerably smoothed characteristics are measured in experiments and, hence, there is a certain analogy between the proposed theoretical algorithm and the algorithms used in the experimental investigations (it is readily seen that when Re $\rightarrow \infty$ and $\ell \neq 0$, smoothed turbulence characteristics are also considered). It is also pertinent to mention that in a number of investigations algorithms that are actually based on the consideration of smoothed characteristics are specially used; for example, the fluid is assumed whose duration is not less than a specified value (Hedley and Keffer [1974a, b]).

In accordance with the above mentioned analysis, quantities γ, $D_{\gamma\gamma}$, $S_{\gamma\gamma\gamma}$ are, henceforth, understood to mean such limits as γ^ℓ, $D^\ell_{\gamma\gamma}$, $S^\ell_{\gamma\gamma\gamma}$ when first Re $\to \infty$ at $\ell = $ const, and then $\ell \to 0$. Moreover, the superscript ℓ in formulas (1.2) – (1.4) will be omitted.

In concluding this section, let us dwell on the results of measurements of some intermittency characteristics which are required for the interpretation of the results obtained in Chapters 3 and 4. Let us start with the study of the behavior of the structure of function $D_{\gamma\gamma}$ which can be judged from the results of measurements of correlation spectrum of the intermittency function $\langle \Gamma(x^{(1)})\Gamma(x^{(2)}) \rangle$ in the work by La Rue and Libby [1976] (measurement of this correlation was also conducted by Kovasznay, Kibens, and Blackwelder [1970], Barsoum, Kawall, and Keffer [1978], Moum, Kawall, and Keffer [1979]; their results are analogous to the data by La Rue and Libby [1976]). Investigated in this work was the wake of a heated circular cylinder, and the boundaries of the turbulent fluid were determined from the temperature field (the applicability of this method of intermittency measurement was considered above; see also §1.3). Although there is not complete confidence in the adequacy of the above introduced definition of intermittency and the definition used in the experiments of La Rue and Libby [1976], it can still be anticipated that the qualitative character of function $D_{\gamma\gamma}(r)$ in the experiment is correctly established. According to the data by La Rue and Libby [1976] (Fig. 1.14), the structure function spectrum $D_{\gamma\gamma}$ is proportional to k^{-2} (k is the wave number) and, hence, $D_{\gamma\gamma}(r) \sim r$ (in the experiments the values of r lie within the range of $r = 0.01 - 0.1L_c$, L_c — wake width). From relation

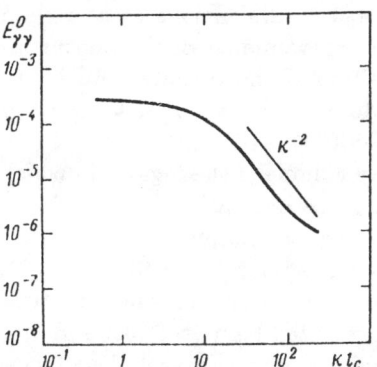

Figure 1.14 Normalized spectrum of intermittency function in a wake of circular cylinder from data by La Rue and Libby [1976] $x_1/d = 400$, $\text{Re}_d = u_0 d/\nu = 2.8 \cdot 10^3$, $d = 0.66$ cm, $u_0 = 7.6$ m/sec, $E^0_{\gamma\gamma} = E_{\gamma\gamma}/\ell_c \langle (\Gamma - \langle \Gamma \rangle)^2 \rangle$, $E_{\gamma\gamma}$ is the correlation spectrum $\langle \Gamma(x_1)\Gamma(x_1) + r\rangle)$; k is the wave number, $\ell_c = \sqrt{(x_1 - x_{10})d}$, $x_{10} = -40d$. Shown in the figure are data at various distances from the plane of wake symmetry in the range $x_2/\ell_c = 0.294 - 0.396$, $\gamma = 0.299 - 0.762$.

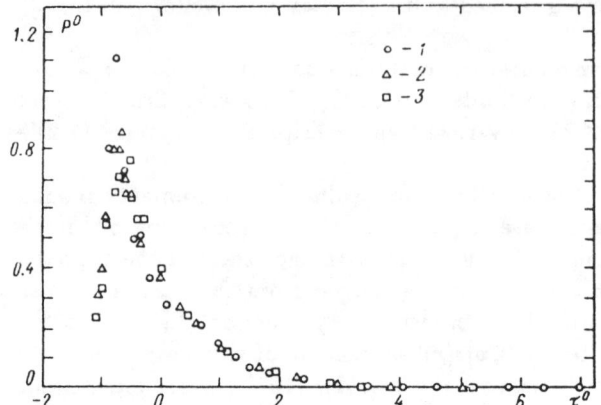

Figure 1.15 Probability distribution density of the duration of "turbulent" ranges from data by La Rue and Libby [1976]. *1)* $x_2/\ell_c = 0.294$, $\gamma = 0.762$, *2)* $x_2/\ell_c = 0.349$, $\gamma = 0.509$, *3)* $x_2/\ell_c = 0.39$, $\gamma = 0.299$, $P^0 = \tau' P$, $\tau^0 = (\tau - \langle \tau \rangle)/\tau'$, $\langle \tau \rangle$ is the mean duration of "turbulent" ranges, $\tau' = \langle (\tau - \langle \tau \rangle)^2 \rangle^{1/2}$ is the root-mean-square of the value of duration of "turbulent" ranges. Test condition the same as in Fig. 1.14.

(1.4) we then obtain that γ_{tt} increases with decreasing r. Since the quantity γ_{tt}, by its physical meaning, characterizes the relative volume of the regions occupied by the turbulent fluid and having a dimension of the order r, the emergence of intermittency is reduced upon transition to small scales.

Since energy is transferred from large-scale to small-scale fluctuations, it is natural to assume that the fluctuations with a dimension of the order r would appear as a result of flow instability in the region with dimensions greater than r. An increase of γ_{tt} when $r/L \to 0$ indicates that entrainment of the nonturbulent fluid takes place. The reverse process (transition of the turbulent fluid into nonturbulent) is impossible, since fluid particles do not leave the region occupied by vortex flow (see, for example, Landau and Lifshits [1954]).

Quantities γ_{nt} and γ_{tn} characterize the relative volume of the regions in which both fluids are simultaneously observed. Hence, these quantities can serve as a measure of the distortion of the boundary of the region occupied by the turbulent fluid. It follows from (1.4) that $\gamma_{tn} = \gamma_{nt} \to 0$ when $r/L \to 0 (D_{\gamma\gamma} \to 0)$, i.e., the boundaries of the turbulent fluid are not too distorted. The quantitative measure of these distortions is conveniently presented later (see §3.1 and 3.8) when discussing the question of the area of isoscalar surfaces in turbulent flows. The obtained result is confirmed by measurements of PDF $P(\tau)$ of the duration τ of "turbulent" ranges (intervals of time during which turbulent fluid is observed at a given point; similar time intervals can be linked with the spatial scales using Taylor's hypotheses on turbulence quenching). In the works by Corrsin and Kistler [1955], La Rue and Libby [1976, 1978], Kawall and Keffer [1979], Chevray and Tutu [1978], Barsoum, Kawall, and Keffer [1978], Moum, Kawall, and Keffer [1979], it is shown that $P(\tau) \to 0$ when $\tau \to 0$. As an example illustrating the previous argument Fig. 1.15 shows the experimental data by La Rue and Libby [1976] (measurements conducted in the wake of a heated circular cylinder). If the boundaries of the turbulent fluid were strongly distorted, then the PDF under consideration would possess very large values when $r \to 0$. Thus, the behavior of $P(r)$ as $r \to 0$, which is observed in experiments, coincides with the adopted definition of the intermittency factor.

§1.2 QUALITATIVE STRUCTURE OF TWO-POINT AND THREE-POINT PDFs OF VELOCITY DIFFERENCES IN LOCALLY HOMOGENEOUS TURBULENCE

It has already been noted in the last section that intermittency exerts significant influence on n-point ($n \geq 2$) velocity PDF in turbulent flows. This refers particularly to the case when the distance between any two points is small in comparison with the scale of turbulence $L = q^3/\langle \epsilon \rangle$, $q^2 = \langle (u - \langle u \rangle)^2 \rangle$. The purpose of this section is the qualitative analysis of the effect of intermittency on the form of the probability densities under consideration for $n = 2$ and $n = 3$.

We first analyze the structure of the two-point velocity difference PDF $v = u(x^2) - u(x^{(1)})$, $r = x^{(2)} - x^{(1)}$. We introduce four conditional two-point velocity difference PDFs: P_{tt}, P_{tn}, P_{nn} which correspond to probabilities γ_{tt}, γ_{tn}, γ_{nt}, γ_{nn}. According to Bayes formula for unconditional velocity difference probability density, we have

$$P(\mathbf{v}, r) = \gamma_{tt} P_{tt} + \gamma_{nt} P_{nt} + \gamma_{tn} P_{tn} + \gamma_{nn} P_{nn} \tag{1.5}$$

Consider the case when the distance between the points satisfies the inequality $r \ll L$. As already mentioned, pressure fluctuations arising in the turbulent fluid generate velocity fluctuations in the nonturbulent fluid. Therefore, it can be assumed that upon crossing through the boundary of this fluid, turbulence energy does not change too much.

This assumption agrees with measurements in jets, wakes, and boundary layers of various one-point moments obtained by conditional averaging over the turbulent and nonturbulent fluid (Kovasznay, Kibens, and Blackwelder [1970], Jenkins and Goldschmidt [1976], Oler, Jenkins, and Goldschmidt [1981] and so on). As a typical example, we present in Fig. 1.8 the results of measurements in the boundary layer on a plate.

Recall also the theoretical results of Landau and Lifshits [1954] and Phillips [1955] with respect to motion in nonturbulent fluid. They show that with increasing distance from the boundary of the turbulent fluid, the velocity fluctuations attenuate smoothly. Thus, it can be assumed that pressure fluctuations are so effective in redistributing the energy between both fluids that

$$P_{nt} = P_{tn} = P_{tt} \quad \text{at} \quad r/L \to 0 \tag{1.6}$$

These formulas are not valid at any fixed value of r/L, since it is known from the works of Landau and Lifshits and Phillips that with increasing penetration deep into the nonturbulent fluid, the amplitude of the fluctuations attenuates more rapidly, the larger the wave number.

Consider now the general properties of function P_{nn}. The two points can either lie close to the boundaries of the turbulent fluid or away from it. As established in §1.1, these boundaries are moderately distorted and, hence, when $r/L \to 0$, the probability of the first event can be ignored and only the second is considered, i.e., the largest contribution to the conditional probability density P_{nn} is made by events taking place at points located deep in the nonturbulent fluid (at distances much larger than r). With increasing penetration into the turbulent fluid, harmonic fluctuations of the pressure and velocity attenuate exponentially and the decrement of this attenuation is inversely proportional to the wave number (Landau and Lifshits [1954], Phillips [1955]). Hence, a number of important conclusions can be drawn. The intensity of small-scale fluctuations in a nonturbulent fluid is small, so that the velocity in the region with a characteristic dimension of the order $r(r/L \to 0)$ can be assumed to change by a linear law (this assumption is not trivial, since Re $= \infty$), i.e., $v_i = A_{ij} r_j$, where A_{ij} is some random tensor. Since mainly large-scale fluctuations penetrate into the nonturbulent fluid, tensor A_{ij} is determined by large-scale motions in the turbulent fluid. Consequently, velocity fluctuations in the nonturbulent fluid are locally homogeneous (v is dependent only on the distance between the points) but, generally speaking, not isotropic (tensor A_{ij} is dependent on large-scale fluctuations). Hence, it also follows that the structure function of velocity fluctuations $\langle v_i v_j \rangle_n$ in the nonturbulent fluid for a small distance between the points is determined by large-scale fluctuations.

Thus, the effect of turbulent fluid on the nonturbulent fluid is determined by direct interaction of large-scale and small-scale fluctuations. It also follows from this analysis that since $v_i = A_{ij} r_j$, the following estimate is valid

$$\langle v^k \rangle_n \approx A_k (r/L)^k \tag{1.7}$$

where A_k is independent of r.

It was established in §1.1 that $D_{\gamma\gamma}(r) \to 0$ when $r/L \to 0$. Using this result we obtain from (1.5) with the aid of (1.4), (1.6), (1.7), the following asymptotic representation of the unconditional two-point velocity difference probability density

$$P(\mathbf{v}, r) \to \gamma P_{tt} + (1-\gamma) P_{nn} \quad \text{at} \quad r/L \to 0 \tag{1.8}$$

Formula (1.8) shows that if $r/L \to 0$, then when seeking quantities $\gamma_{tt}, \gamma_{tn}, \gamma_{nt}, \gamma_{nn}$ both points can be formally regarded as one.

We shall now show that analogous results are also valid for three-point unconditional velocity difference PDF $P(\mathbf{v}, V, r, R), V = u(x^{(3)}) u(x^{(1)}), R = x^{(3)} - x^{(1)}$. The relation of this function with the conditional three-point PDFs has the form

$$P(\mathbf{v}, V, r, R) = \gamma_{ttt} P_{ttt} + \gamma_{ttn} P_{ttn} + \ldots + \gamma_{tnt} P_{tnt} + \gamma_{nnn} P_{nnn} \tag{1.9}$$

Here, instead of the remaining four terms, whose form is evident from the structure of the formula, omission dots are inserted.

Probabilities $\gamma_{ttt} \ldots \gamma_{nnn}$ are easily expressed in terms of γ, $D_{\gamma\gamma}$ and $S_{\gamma\gamma\gamma}$ with the aid of relations that are analogous to formulas (1.4). Further, two cases are of interest: 1) $r/L \to 0$, $R \sim L$; 2) $r/L \to 0$, $R/L \to 0$.

We shall assume, as before, that if there is turbulent fluid at one of the points, then the statistical properties of the fluctuations at all the remaining, closely located points are the same as in the turbulent fluid. Then, by analogy with formula (1.6) in the first case we have the relation

$$P_{ntn} = P_{tnn} = P_{ttn}, \qquad P_{ntt} = P_{tnt} = P_{ttt} \tag{1.10}$$

and, hence, formula (1.9) assumes the form

$$P(\mathbf{v}, V, r, R) = (\gamma_{ttt} + \gamma_{ntt} + \gamma_{tnt})P_{ttt} + \gamma_{nnt}P_{nnt}$$

$$+(\gamma_{ttn} + \gamma_{ntn} + \gamma_{tnn})P_{ttn} + \gamma_{nnn}P_{nnn} \tag{1.11}$$

Since $R \sim L$, when finding probabilities $\gamma_{ttt}, \ldots, \gamma_{nnn}$ in a first approximation it can be assumed that the appearance of turbulent fluid at point $x^{(3)}$ is independent of the fluid located at points $x^{(1)}$ and $x^{(2)}$. Taking this into account and also that when $r/L \to 0$ the structure function $D_{\gamma\gamma}$ tends to zero, it is not difficult to show that the following relations are valid

$$\gamma_{ttt} \to \gamma^2, \qquad \gamma_{ntt} = \gamma_{tnt} \to 0, \qquad \gamma_{ntn} = \gamma_{tnn} \to 0$$

$$\gamma_{ttn} = \gamma_{nnt} \to \gamma(1-\gamma), \qquad \gamma_{nnn} \to (1-\gamma)^2, \qquad r/L \to 0, \quad R \sim L \tag{1.12}$$

In the second case we have

$$P_{ttn} = P_{tnn} = P_{ntn} = P_{nnt} = P_{ntt} = P_{tnt} = P_{ttt}$$

$$r/L \to 0, \quad R/L \to 0 \tag{1.13}$$

and formula (1.9) yields

$$P(\mathbf{v}, V, r, R) = (\gamma_{ttt} + \gamma_{ttn} + \gamma_{tnn} + \gamma_{ntn} + \gamma_{nnt} + \gamma_{ntt} + \gamma_{tnt})P_{ttt}$$

$$+\gamma_{nnn}P_{nnn}, \qquad r/L \to 0, \qquad R/L \to 0 \tag{1.14}$$

Probabilities $\gamma_{ttt}, \gamma_{ttn}, \ldots$ are expressed in terms of the moment $S_{\gamma\gamma\gamma}$. In order to find the asymptotic behavior of the latter when $r/L \to 0$, $R/L \to 0$, we consider the identity (when verifying this identity, one must take into account that $[\Gamma(x)]^2 = \Gamma(x)$ follows from the definition of intermittency function)

$$S_{\gamma\gamma\gamma} = \frac{1}{2}\left\{\langle\Gamma(x^{(1)})\Gamma(x^{(3)})\rangle + \langle\Gamma(x^{(2)})\Gamma(x^{(3)})\rangle\right.$$

$$-\langle [\Gamma(x^{(1)}) - \Gamma(x^{(2)})]^2 \Gamma(x^{(3)})\rangle \}$$

Since $1 \geq \Gamma \geq 0$, then

$$\langle [\Gamma(x^{(1)}) - \Gamma(x^{(2)})]^2 \Gamma(x^{(3)})\rangle$$

$$\leq \langle [\Gamma(x^{(1)}) - \Gamma(x^{(2)})]^2\rangle = D_{\gamma\gamma}(r)$$

Since $D_{\gamma\gamma} \to 0$ when $r/L \to 0$, it follows from this inequality and relation (1.4) that

$$S_{\gamma\gamma\gamma} \to \gamma, \qquad r/L \to 0, \qquad R/L \to 0$$

Using conditions $S_{\gamma\gamma\gamma} \to \gamma$, $D_{\gamma\gamma} \to 0$ when $r/L \to 0$, $R/L \to 0$, it can be shown that among $\gamma_{ttt}, \ldots$ nonzero limits exist only for γ_{ttt} and γ_{nnn}, i.e.,

$$\gamma_{ttt} \to \gamma, \qquad \gamma_{nnn} \to 1-\gamma, \qquad r/L \to 0, \qquad R/L \to 0 \tag{1.15}$$

Formula (1.15) shows that in the second case, when finding $\gamma_{ttt}, \ldots, \gamma_{nnn}$, all three points can formally be considered as one.

Substituting the limit dependences (1.12) and (1.15) into relations (1.11) and (1.14) respectively, we obtain the following asymptotic representation for $P(\mathbf{v}, V, r, R)$

$$P(\mathbf{v}, V, r, R) = \gamma^2 P_{ttt} + \gamma(1-\gamma)(P_{ttn} + P_{nnt}) + (1-\gamma)^2 P_{nnn}$$

$$r/L \to 0, \qquad R \sim L \tag{1.16}$$

and

$$P(\mathbf{v}, V, r, R) = \gamma P_{ttt} + (1-\gamma) P_{nnn}, \qquad r/L \to 0, \qquad R/L \to 0 \tag{1.17}$$

§1.3 QUALITATIVE FORM OF PDF OF THE CONCENTRATION OF A PASSIVE CONTAMINANT IN TURBULENT FLOWS

Intermittency exerts a particularly strong influence on the PDF of scalar quantities possessing a limited range of values. In the present section, this influence is analyzed in the case of a dynamically passive contaminant whose concentration will be denoted by the letter z. It is known that in a slightly heated fluid (i.e., when the Archimedean forces are small in comparison with the inertia forces), temperature can be regarded as a passive contaminant. This property is widely used in experimental investigations. Moreover, concentration can be represented by the dimensionless temperature difference $\Delta T/(\Delta T)_{\max}$ (here $\Delta T = T - T_0$, T_0 is the minimum temperature value in the flow).

As evident from §1.1, the structure of energy dissipation ϵ and scalar dissipation N is qualitatively the same. In those flow regions where $\epsilon = 0$, it can be anticipated that $N = 0$ (it should be recalled again that we consider a limit case when Reynolds and Peclet numbers tend to infinity; since the coefficients of molecular transport for gases are close to each other, we shall henceforth talk, for the sake of brevity, only about Reynolds number). Since it is assumed that no contaminant exists in the nonturbulent fluid at the initial moment of time, the existing velocity fluctuations in the nonturbulent fluid cannot generate concentration fluctuations since the diffusion equation does not contain a term equivalent to the pressure gradient in the equation of motion. Therefore, the presence or absence of concentration fluctuations in the nonturbulent fluid is dependent only on the initial conditions. As a result of unidirectional exchange between turbulent and nonturbulent fluids (see the previous section), it can be anticipated that if at $t = 0$, there is no contaminant in the nonturbulent fluid, it will not be present at all subsequent moments of time. This conclusion is confirmed by observations in jets, wakes, and other free flows (see, for example, the works of Uberoi and Singh [1975], Fabris [1979a, b], Breidenthal [1981], Long and Chu [1981], Long, Chu, and Chang [1981]). Hence, the intermittency factors γ and γ_z which are determined by the dynamic and scalar fields, respectively, usually coincide (possible exceptions are analyzed below). The latter state is utilized in many experiments (see, for example, Becker, Hottel, and Williams [1967], Antonia, Prabhu, and Stephenson [1975], La Rue [1974], La Rue and Libby [1974], Bashir and Uberoi [1975], Anderson, La Rue, and Libby [1979]. Nevertheless, it is necessary to clearly realize that intermittency is purely fluid dynamic phenomenon and its description must be based on the Navier-Stokes equations. The properties of the concentration field serve only as indirect signs by which one can judge the appearance of intermittency.

Let us now discuss the cases mentioned above when deviations from equality $\gamma = \gamma_z$ are possible. It is evident from intuitive considerations, based on the physical picture of flow, that the contaminant does not spread, generally speaking, over all the turbulent fluid, since individual "sections" of such a fluid do not necessarily have common boundaries. Observations of a smoke jet spreading in a turbulent atmosphere illustrates this point. It is evident that coefficients γ and γ_z are linked by the inequality

$$\gamma \geq \gamma_z \qquad (1.18)$$

This inequality (1.18) for free turbulent flows is confirmed in the tests of Kuznetsov and Rashchupkin [1977] (Fig. 1.16). The mixing layer was investigated in these tests on the initial section of heated axisymmetric jet discharging into stationary cold air. The width of the dynamic boundary layer on the initial section was an order of magnitude less than the thermal boundary layer owing to the inequality (1.18).

On the bases of this statement, it can be assumed that particularly strong differences between γ and γ_z arise in those cases when the number of the noninterconnected "sections" of the turbulent fluid is large. Such a situation might be encountered, for example, in grid-induced turbulent flows and flows in channels. The number of "sections" in this case is so great that the fluid dynamic intermittency factor is practically indistinguishable from unity, i.e., $\gamma \approx 1$, whereas γ_z can

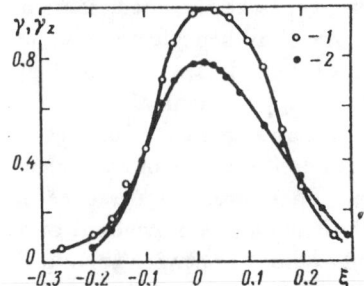

Figure 1.16 Profile of intermittency factor in the mixing layer on the initial section of submerged and heated axisymmetric jet from data by Kuznetsov and Rashchupkin [1977]. Measurements were made at distance $x_1 = 30$ cm from the nozzle edge; thickness of the dynamic boundary layer at the nozzle edge $\delta = 7$ mm; thickness of the temperature boundary layer is one order of magnitude less; $Re_\delta = u_0 \delta / \nu = 3.73 \cdot 10^3$, $u_0 = 8$ m/sec; 1) γ, 2) γ_z, $\xi = x_2/x_1$, origin of the coordinates located at the edge of the nozzle; axis x_2 directed at a right angle to the axis of symmetry toward the submerged space.

be close to zero. The indicated effect is totally caused by the different initial conditions for the scalar and dynamic fields and is not linked with any fundamental characteristics of intermittency. A large amount of work has been devoted to the experimental investigation of the concentration field (Rashchupkin and Sekundov [1978], Gad-el-Hak and Morton [1979], Keffer, Olsen, and Kawall [1977], La Rue and Libby [1981], Meshkov and Shcherbina [1981], La Rue, Libby, and Seshadri [1981], Shcherbina [1982]). In these works the statistically nonhomogenous concentration field in homogeneous turbulence was investigated. Intermittency in these cases strongly affects the characteristics of the scalar field and the process of contaminant transport possesses a number of specific features (Rashchupkin and Sekundov [1978]).

Such cases will not be considered in the present book. The flows of principal interest below are those closer to self-similar flows (the far wake, the main part of a jet) in which quantities γ and γ_z are practically indistinguishable as a result of which subscript z will be henceforward omitted.

If this contaminant is absent from the nonturbulent fluid, the conditional concentration PDF is a delta function. In the simplest case there are two types of regions occupied by the nonturbulent fluid: in some regions $z = 0$, and in the others $z = 1$. For any Re number, however large but finite, such regions do not, in a strict sense, exist. Therefore, to avoid ambiguity, it should be emphasized again that Reynolds number in this reasoning is assumed equal to infinity, and the result of such an analysis will be further used for the approximate description of flows at finite Reynolds number. If the probability of observing the value $z = 1$ in the nonturbulent fluid is small (the far wake, the main part of a jet), the unconditional concentration PDF has the form (Kuznetsov [1972a], Kuznetsov and Frost [1973])

$$P(z) = \gamma P_t(z) + (1-\gamma)\delta(z)$$

$$P_t(z) = \theta(z)F(z) \tag{1.19}$$

Here, P_t is the conditional concentration PDF in the turbulent fluid, F is the smooth function, $\theta(s)$ is the Heaviside function, i.e., $\theta(s) = 0$ when $s < 0$ and $\theta(s) = 1$ when $s > 0$.

The generalization of formula (1.19) is fairly evident (Bray and Libby [1976], Kuznetsov, Lebedev, Sekundov, and Smirnova [1981], Savel'nikov [1979, 1980b], Libby and Bray [1981])

$$P(z) = \gamma_0 \delta(z) + \gamma_1 \delta(z-1) + \gamma_z P_t(z)$$

$$\gamma_0 + \gamma_1 + \gamma_z = 1, \qquad P_t(z) = [\theta(z) - \theta(z-1)]F(z) \tag{1.20}$$

Formula (1.20) is applicable, for example, to the mixing of a large number of closely located jets in a mixing chamber (Kuznetsov, Lebedev, Sekundov, and Smirnova [1981]).

The intermittency factor γ_z and the conditional concentration PDF in the turbulent fluid P_t appearing in formulas (1.19) and (1.20) characterize the degree of mixing at the molecular level. In particular, if the coefficient of molecular diffusion D is equal to zero and the concentration at the initial moment acquires only two values 0 and 1, then we obtain $\gamma_1 = \langle z \rangle$, $\gamma_0 = 1 - \langle z \rangle$, $\gamma_z = P_t = 0$, and $\langle z^n \rangle = \langle z \rangle = \gamma_1$. In this case only turbulent diffusion of the marked fluid particles takes place (see the review in the book by Monin and Yaglom [1965]), a process which is sometimes referred to in the literature as "black-white" mixing (Prudnikov et al. [1971]). The variance of concentration pulsations in "black-white" mixing attains the maximum possible values $\sigma_{\max}^2 = \langle z^2 \rangle - \langle z \rangle^2 = \langle z \rangle (1 - \langle z \rangle)$. It is for this reason that relation $\sigma^2/[\langle z \rangle(1 - \langle z \rangle)]$ is frequently used as a quantitative characteristic of the degree of molecular mixing in turbulent flows (see, for example, Roshko [1976]).

Later, the geometrical interpretation of PDF will be repeatedly used as quantities that are proportional to the volume between two closely located isoscalar surfaces. Let us consider, for more cogency, a statistically homogeneous concentration field. Let V be some sufficiently large volume, and δV_z a volume enclosed between two isoscalar surfaces z and $z+dz$ which are completely located in volume V. Then, because of ergodicity the following relation is true

$$P(z)dz = \lim_{V \to \infty} \frac{\delta V_z}{V} \tag{1.21}$$

Volume δV_z is determined by the obvious formula

$$\delta V_z = dz \int_{S_z} \left|\frac{\partial z}{\partial n}\right|^{-1} dS_z \tag{1.22}$$

Here, S_z is the area of the isoscalar surface $z(x,t) = \text{const}$, n is the unit normal to this surface.

From relations (1.21) and (1.22) we obtain the following formula for the concentration PDF

$$P(z) = \lim_{V \to \infty} \frac{1}{V} \int_{S_z} \left|\frac{\partial z}{\partial n}\right|^{-1} dS_z = \lim_{V \to \infty} \frac{S_z}{V} \left\langle \left|\frac{\partial z}{\partial n}\right|^{-1} \right\rangle_z \tag{1.23}$$

where subscript z of the angular brackets indicates conditional averaging at the specified concentration value.

The generalization of formula (1.23) in the case of a nonhomogeneous field has the form

$$P(z) = \lim_{dV \to 0} \frac{d\langle S_z \rangle}{dV} \left\langle \left|\frac{\partial z}{\partial n}\right|^{-1} \right\rangle_z \tag{1.24}$$

Formulas (1.23) and (1.24) are valid only in those cases when specific concentration values cannot be observed with finite probability. As discussed above, an opposite case is typical for turbulent flows as a result of intermittency when concentration in the regions with finite volume acquires constant values (when Re $\to \infty$). In these regions $\partial z/\partial n = 0$, and the formal acceptance of relation (1.23) yields $P(z) = \infty$. In reality, according to relation (1.20), PDF includes singular additions (note that formula (1.21) remains valid if it is understood in the sense of generalized functions; only the formula for δV_z (1.22) is incorrect).

Formulas (1.21), (1.23), and (1.24) can be modified so that they would be valid for the conditional concentration PDF in the turbulent fluid. For this, one must replace V with $\gamma_z V$ and dV with $\gamma_z dV$ in these formulas. As a result, instead of (1.21), (1.23), and (1.24) we obtain the following relations, respectively,

$$\gamma_z P_t(z) dz = \lim_{V \to \infty} \lim_{Re \to \infty} \frac{\delta V_z}{V} \tag{1.25}$$

$$\gamma_z P_t(z) = \lim_{V \to \infty} \lim_{Re \to \infty} \frac{S_z}{V} \left\langle \left|\frac{\partial z}{\partial n}\right|^{-1} \right\rangle_{t,z} \tag{1.26}$$

$$\gamma_z P_t(z) = \lim_{dV \to 0} \lim_{Re \to \infty} \frac{d\langle S_z \rangle}{dV} \left\langle \left|\frac{\partial z}{\partial n}\right|^{-1} \right\rangle_{t,z} \tag{1.27}$$

Dependences (1.19), (1.20), (1.25) – (1.27) are valid only when Re $\to \infty$. Because of this, the following two problems are of unquestionable interest: 1) what is the qualitative character of the influence of Reynolds number on the concentration probability density, and 2) what is the order of the omitted terms? Let us first analyze the former question. From physical considerations, it is evident that the main change of PDF as a result of the effect of molecular transport will take place in the vicinity of the boundary of the phase space, i.e., near points $z = 0$ and $z = 1$, since the delta function contained in the limit formulas (1.19) and (1.20) will be "smeared" at the limits. The length of this interval with respect to the order of magnitude must coincide with the characteristic value of the amplitude of small-scale fluctuations, determined by the viscous processes z_ν, whose amplitude is stated below. It should be immediately noted that the "smearing" of delta function observed in the experiments considered below can be caused both by the fundamental effect of the processes of molecular transfer and by the inaccuracy of the measurements. The question as to which of the stated factors influences the PDF more requires special consideration in each specific case. Some reasoning of the effect of inaccuracy of measurements on probability density will be given following the discussion of the effect of Reynolds number.

As an illustration of the "smearing" of the delta function, Fig. 1.17 shows the results of tests by Birch, Brown, Dodson, and Thomas [1978] in which the concentration PDF at the edge of a submerged axisymmetric methane jet was measured. A sharp increase of $P(z)$ when $z \to 0$ is seen which can be interpreted as the manifestation of the delta function. Similar results were obtained by Kuznetsov and Rashchupkin [1977], Sreenivasan, Antonia, and Stephenson [1979], Rajagopalan and Antonia [1980], Meshkov and Shcherbina [1981]. As another example, Fig. 1.18 shows the data from measurements of the temperature PDF in the initial section of a heated axisymmetric jet discharging into stationary cold air (Kuznetsov and Rashchupkin [1977]). The measured PDFs at all three points are found to be bimodal, i.e., having two maxima. One of the maxima at each point is caused by the "smearness" of delta function. Indeed, at point $x_2/x_1 = 0$, the maximum at $s = 1.5$ $((z-\langle z\rangle)/\sigma, \sigma^2 = \langle(z-\langle z\rangle)^2\rangle$ is the concentration pulsation variance) corresponds to $z = 0.92$, when $x_2/x_1 = -0.067$, the maximum at $s = 0.85$ corresponds to $z = 0.98$, and when $x_2/x_1 = 0.067$, the maximum at $s = -1.7$ corresponds to $z = 0.15$. Therefore, the maximum when $x_2/x_1 = 0$ and $x_2/x_1 = -0.067$ are manifestations of $\delta(z-1)$, and the maximum when $x_2/x_1 = 0.067$ is of $\delta(z)$. In these estimates the results of measurements of $\langle z\rangle$ and σ given in the work under consideration have been used.

Let us proceed with the discussion of the effects of Reynolds number on the behavior of the concentration PDF in the vicinity of points $z = 0$ and $z = 1$. It is important to emphasize that at a finite Reynolds number, however high it might be, the minimum $Z_{\min}$ and maximum $z_{\max}$ concentration values at any

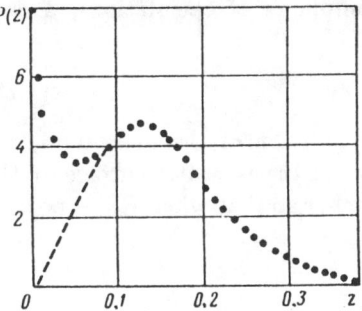

Figure 1.17 Concentration PDF at the edge of a submerged axisymmetric methane jet from data by Birch, Brown, Dodson, and Thomas [1978]. $x_1/d = 10$, $x_2/d = 1.49$, $\text{Re}_d = v_0 d/\nu = 1.6 \cdot 10^4$, $d = 1.3$ cm, $u_0 = 19$ m/sec.

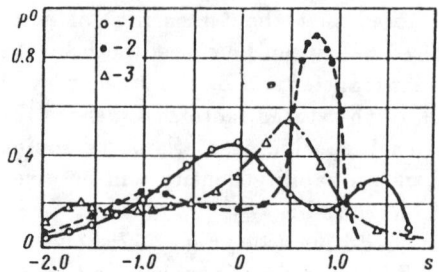

Figure 1.18 Temperature PDF in the mixing layer on the initial part of a submerged heated axisymmetric jet from data by Kuznetsov and Rashchupkin [1977]. *1)* $\xi = 0$, *2)* $\xi = -0.067$, *3)* $\xi = 0.067$, $\xi = x_2/x_1$, $s = (z - \langle z \rangle)/\sigma$, $P'' = \sigma P$, $z = \Delta T/(\Delta T)_{\max}$. The conditions of the tests and the coordinate system are the same as in Fig. 1.16.

finite region differ from zero and unity by virtue of the maximal principle for the diffusion equation‡. It should also be taken into account that at a finite Reynolds number quantities $z_{\min}$ and $z_{\max}$ are functions of the coordinates (and time if the problem is, on the average, nonstationary). The determination of these functions is a considerably complex task and is not considered here.

In order to obtain more concrete conclusions on the behavior of the PDF, information on the concentration field in the vicinity of points $z_{\min}$ and $z_{\max}$ is required. According to formulas (1.23), (1.24) PDF is dependent on the area of the isoscalar surface and quantity $\langle |\partial z/\partial n|^{-1} \rangle_z$. From physical considerations it is clear that the area of the isoscalar surface has finite limit when $z \to z_{\min}$ and $z \to z_{\max}$ (Reynolds number is finite). Thus, the problem is reduced to determining the dependence of $\langle |\partial z/\partial n|^{-1} \rangle_z$ and z. Two simple physical models of the concentration field in the region of extremal points will be considered here. The first model corresponds to the statistically homogeneous concentration field in homogeneous turbulence. The second is intended for the description of free turbulent flows. Let us begin with the discussion of the first model. In it we assume that the concentration field in the vicinity of each extremal point $z = z_m$ is represented in the following form (for the simplicity of calculations, a one-dimensional field is considered)

$$z = z_m + \alpha n^2 \tag{1.28}$$

Coordinate n here is plotted from the point of location of the extremum, the random constant α has the meaning of curvature of the isoscalar surface at the extremal point and is dependent on Reynolds number so that when $\mathrm{Re} \to \infty$ $\alpha \to \infty$.

From formula (1.28) we have

$$\left| \frac{\partial z}{\partial n} \right| = 2 |\alpha(z - z_m)|^{1/2} \tag{1.29}$$

---

‡This principle can be regarded as one of the formulations of the second law of thermodynamics.

Our task is to use relation (1.29) to estimate the conditional mean $\langle |\partial z/\partial n|^{-1}\rangle_z$. For qualitative estimates we ignore the fluctuations of quantity α. As a result, in the vicinity of point $z = z_{\min}$ we obtain the formula

$$\left\langle \left|\frac{\partial z}{\partial n}\right|^{-1}\right\rangle_z \sim |\alpha(z - z_m)|^{-1/2} \tag{1.30}$$

From (1.23) and (1.30) we find

$$P(z) \sim |\alpha(z - z_{\min})|^{-1/2} \tag{1.31}$$

i.e., PDF has integrable singularity at point $z = z_{\min}$. An analogous conclusion is also true for the other boundary point $z = z_{\max}$.

Let us proceed to the discussion of the second model which is based on experimental investigations of the frozen spatial profiles of concentration (or temperature) in free turbulent flows (see Fig. 1.1). It is evident from Fig. 1.1 that a discontinuous change in temperature takes place at the edge of the jet. Analogous results were obtained in the works by Jenkins and Goldschmidt [1976], Chen and Blackwelder [1978], Sreenivasan, Antonia, and Britz [1979].

It can be assumed that the temperature profile in the vicinity of the jump is in some quasi-stationary state as a result of the balance of two oppositely acting factors — diffusion and convection caused by strong fluid dynamic deformation of the medium. An analogous statement for a vortex field in the theory of locally homogeneous and isotropic turbulence was first proposed by Townsend [1951].

The equation of diffusion, written in a system of coordinates linked with the surface $z = z_0 = \text{const}$ on which the discontinuity takes place and the contortion is neglected, and the boundary conditions describing the quasi-stationary concentration profile have the form

$$-kn\frac{dz}{dn} = D\frac{d^2z}{dn^2}, \qquad k > 0$$

$$z(0) = z_0, \qquad z \to 0 \quad \text{when} \quad n \to -\infty \tag{1.32}$$

Here k is a nonaveraged velocity gradient; it is assumed that the turbulent flow is in region $n > 0$. The asymptotic behavior of the solution of equation (1.32) when $n \to -\infty (z \to 0)$ has the form

$$z \to -\left[\frac{dz}{dn}\right]_{n=0} \frac{D}{kn} \exp\left(-\frac{kn^2}{2D}\right) \tag{1.33}$$

Let us differentiate (1.33) with respect to n, express n in terms of z with the aid of (1.33), and substitute the dependence $n(z)$. As a result we find

$$\frac{dz}{dn} \to \left(-\frac{2k}{D}\ln z\right)^{1/2} z, \qquad z \to 0 \tag{1.34}$$

In this estimate, we ignore the fluctuations of dissipation, i.e., we shall assume that $k = \text{const} \sim \sqrt{\langle\epsilon\rangle/\nu}$. Then, from (1.24) and (1.34) we obtain the following relation for PDF

$$P(z) \sim \left[\left(-\frac{2k}{D}\ln z\right)^{1/2} z\right]^{-1} \tag{1.35}$$

From (1.35) we can see that PDF in the second model is limited when $z \to z_{\min}$. Such a feature of $P(z)$ is due to the zero value of concentration not being attained at every finite distance from the boundaries of jet-type flows.

The examples considered above indicate that at a finite Reynolds number, the process of molecular transport influences the form of concentration PDF in a complex manner. In view of this, it is useful to note that in many cases of practical interest the processes under consideration can be significant.

As already indicated earlier, the effect of Reynolds number is particularly significant near the boundaries of the phase space (i.e., points $z = 0$ and $z = 1$), since the PDF in the vicinity of these boundaries tends to delta functions. Therefore, our further discussion will be structured as follows. To start with, we estimate the value z_ν, i.e., that range of concentration values at which the delta function is "smeared", and then find the order of the discarded members in the asymptotic expansion of function $P(z)$ outside the range under study.

It is evident that the value z_ν is determined by the amplitude of the small-scale, viscous-dependent concentration fluctuations. Therefore, to estimate the value z_ν, one can make use of the theory of Kolmogorov [1941] and Obukhov [1941, 1949] according to which z_ν is dependent only on $\langle N \rangle$, $\langle \epsilon \rangle$, ν, and D. We shall assume that $\nu \sim D$, and take into account that replacing z with λz (λ arbitrary) causes the quantity $\langle N \rangle$ to transform by the law $\langle N \rangle \to \lambda^2 \langle N \rangle$. Then, from dimensional considerations we obtain

$$z_\nu \sim \langle N \rangle^{1/2} \left(\frac{\nu}{\langle\epsilon\rangle}\right)^{1/4} \tag{1.36}$$

To estimate $\langle\epsilon\rangle$ and $\langle N \rangle$, we make use of the known relations $\langle\epsilon\rangle \sim q^3/L$, $\langle N \rangle \sim \sigma^2 q/L$. Then

$$\frac{z_\nu}{\sigma} \sim \left(\frac{qL}{\nu}\right)^{-1/4} = \text{Re}^{-1/4} \tag{1.37}$$

Since Reynolds number enters into (1.37) raised to a small power, the effect of the processes of molecular transport on concentration PDF near the boundary points $z = 0$ and $z = 1$ can be very substantial. For example, calculation shows that $z_\nu/\sigma = 0.16$ under the conditions of the tests by Birch, Brown, Dodson, and Thomas [1978], $z_\nu/\sigma = 0.13$ under the conditions of the tests by Kuznetsov and Rashchupkin [1977]. In both cases the values of z_ν are comparable with the range in which the delta function is "smeared" (see Fig. 1.17, 1.18).

Let us now turn to estimating the discarded terms in the asymptotic behavior of the PDF when $\text{Re} \to \infty$ at the internal points of the range $z_{\min} < z < z_{\max}$, i.e., we estimate the addition to function P_t entering into (1.19) and (1.20). In the estimates we make use of the hypothesis concerning the statistical independence of large-scale motions which are self-similar with respect to Reynolds number, and small-scale motions which are determined by molecular viscosity in a turbulent fluid. A detailed discussion of this hypothesis with the appropriate reference to the literature is given in §3.2. We limit ourselves here only in that the introduced hypothesis is an organic part of the Kolmogorov-Obukhov theory of locally homogeneous and isotropic turbulence. In order to apply this hypothesis, we resolve the concentration field into a sum of two fields

$$z = z^{(1)} + z^{(2)}, \quad \langle (z^{(1)})^2 \rangle^{1/2} \sim \sigma, \quad \langle (z^{(2)})^2 \rangle^{1/2} \sim z_\nu \qquad (1.38)$$

one of which, $z^{(1)}$, is large-scale, and the other, $z^{(2)}$, is small-scale locally homogeneous isotropic field. The latter indicates, in particular, that when Re $\to \infty$, the following expression is true

$$\frac{\langle (z^{(2)})^2 \rangle}{\sigma^2} \sim \text{Re}^{-1/2} \qquad (1.39)$$

Division of (1.38) can be effected, for example, as follows. Let us consider cube ω with the center at point x and side ℓ satisfying condition $\ell = \kappa\eta, \kappa \gg 1$, and let us introduce quantities

$$z^{(1)} = \frac{1}{V} \int_\omega z \, dV, \qquad z^{(2)} = z - z^{(1)}$$

Here, V is the volume of region ω. By letting Reynolds number go to infinity when $\kappa = \text{const}$, we obtain representation (1.38).

The use of the hypothesis concerning the statistical independence of fields $z^{(1)}$ and $z^{(2)}$ enables the splitting of the correlation of type $\langle (z^{(1)})^{k_1}(z^{(2)})^{k_2} \rangle$, i.e.,

$$\langle (z^{(1)})^{k_1}(z^{(2)})^{k_2} \rangle = \langle (z^{(1)})^{k_1} \rangle \langle (z^{(2)})^{k_2} \rangle \qquad (1.40)$$

where k_1 and k_2 are any positive numbers. We are now ready to estimate the order of the discarded members in the asymptotic behavior of the concentration probability density.

Indeed, as is generally known, PDF is uniquely determined by its characteristic function, i.e., by the quantity $\langle \varphi \rangle = \langle \exp(i\lambda z) \rangle$. From (1.38), (1.40) we have $\langle \varphi \rangle = \langle \exp(i\lambda z^{(1)}) \rangle \langle \exp(i\lambda z^{(2)}) \rangle$. Since $\langle (z^{(2)})^2 \rangle^{1/2} \ll \sigma$ on account of (1.39), we approximately obtain $\langle \exp(i\lambda z^{(2)}) \rangle = \langle 1 + i\lambda z^{(2)} - 1/2\lambda^2 (z^{(2)})^2 \rangle$. Since field $z^{(2)}$ is locally homogeneous, then $\langle \exp(i\lambda z^{(2)}) \rangle = 1 - 1/2\lambda^2 \langle (z^{(2)})^2 \rangle + \ldots$. Thus, the expansion of the characteristic function has the form

$$\langle \varphi \rangle = \left\langle \exp(i\lambda z^{(1)}) \right\rangle \left(1 - \frac{1}{2}\lambda^2 \left\langle (z^{(2)})^2 \right\rangle + \ldots \right)$$

Since $\langle (z^{(2)})^2 \rangle \to 0$ when Re $\to \infty$, the inverse Fourier transfer of the first term in this expansion by virtue of (1.19) and (1.20) is γP_t. The inverse Fourier transform of the next term in the expansion of the characteristic function by virtue of (1.39) is of the order $Re^{-1/2}$.

Thus, the assumption of self-similarity of function P with respect to Reynolds number is fulfilled with an accuracy of the order of $Re^{-1/2}$.

This accuracy is of the order of one percent even under laboratory conditions in which the Reynolds number $Re = Lq/\nu$ rarely exceeds 10^4. Therefore, the influence of the processes of molecular transfer on the concentration PDF outside of small neighborhoods of points $z = 0$ and $z = 1$ can be neglected. And, conversely, in the neighborhood of these points the effect of Reynolds number is substantial due to two circumstances. First, as established above, at the indicated Reynolds number the range in which delta functions are "smeared" is considerably large $(z_\nu \sim 0.1\sigma)$. Second, since $P \sim \delta(z)$ when $Re \to \infty$, Reynolds number has profound influence on the absolute values of PDF (at finite Reynolds number $P(z)$ has a clearly defined maximum located at $z = 0$; the variation of Reynolds number greatly affects the amplitude of this maximum).

Thus, when interpreting experimental data, one must be very cautious in using the assumption of self-similarity of probability distributions with respect to Reynolds number. This state is frequently ignored. As an example we note the work of Pope [1979a, b] in which a number of variational principles not containing Reynolds number are proposed for the approximation of the measured PDFs in the tests, i.e., functions with viscosity-induced singularities.

The problem of the effect of instrument error on the measured values of $P(z)$ is closely associated with the problem considered above concerning the influence of the processes of molecular transport on the form of probability distribution densities. In a number of cases, the effect of Reynolds number and measurement errors lead to qualitatively identical results. For example, when the width of the window of the comparator[§], used for the measurement of PDF, is different from zero, delta function will be "smeared". The indicated effect can be significant. For example, in the tests by Kuznetsov and Rashchupkin [1977], the width coincides with the amplitude of small-scale pulsations z_ν estimated above. In a number of tests, however, "smearing" of delta functions is caused, first of all, by noise in the measuring apparatus (La Rue and Libby [1974], Rajagopalan and Antonia [1980], Meshkov and Shcherbina [1981]), which must always be taken into consideration when analyzing experimental data. A method to overcome this difficulty is proposed in the works by Bilger, Antonia, and Sreenivasan [1976], Bradshaw [1978], Bilger [1978], Meshkov and Shcherbina [1981], Shcherbina [1982]. The main idea of the method is that the "smeared" probability density by noise in the vicinity of points $z = 0$ and $z = 1$ is approximated by special but adjusted Gaussian curves with areas which yield γ_0 and γ_1, respectively.

This analysis indicates that accounting for the effect of the processes of molecular transport on the form of PDF is coupled with overcoming the considerable

---

[§]Comparator is a device at the output of which the voltage can have one of two possible values: zero or unity. The voltage is equal to unity if the concentration of the contaminant lies in the range $(z, z + \Delta z)$, and zero in the opposite case. *Probability density* is proportional to the mean output voltage of the comparator if Δz — comparator window — is sufficiently small.

difficulties which arise during both theoretical and experimental investigations. The introduction of intermittency enables the bypassing of these difficulties. This represents the main advantage of considering the limit pattern of flow when Re = ∞.

§1.4 QUALITATIVE FORM OF CONCENTRATION PDF UPON COMBUSTION OF A HOMOGENEOUS MIXTURE

With combustion in turbulent flows a number of qualitative features are found in the concentration PDFs of a chemically active contaminant that substantially distinguishes this case from passive contaminant mixing analyzed in the previous section. As will be shown in Chapter 5, the problem can be reduced to the investigation of PDF of a chemically inert contaminant only for diffusion combustion, i.e., when the fuel and air are supplied separately, under some general assumptions. Therefore, the qualitative picture of concentration PDF remains as before.

However, the situation changes significantly upon combustion of premixed gases. Analysis of this case is the purpose of this section. Henceforth, the concentration of the chemically active contaminant will be denoted by letter c. The same notation is also used for the normalized temperature. We shall make the convention that in the latter case $c = 0$ in the fresh mixture and $c = 1$ in combustion products, i.e., $c = (T - T^{(0)})/(T^{(b)} - T^{(0)})$, $T^{(0)}$ is the temperature of the fresh mixture, $T^{(b)}$ is the temperature of the combustion products.

Two limiting combustion mechanisms for homogenous mixtures were formulated in the works by Damköhler [1940], Shchelkin [1943], Shchelkin and Troshin [1965]: 1) combustion takes place in extended zones whose dimensions are comparable with the turbulence scale — a volumetric combustion mechanism and 2) combustion takes place in narrow zones that are strongly contorted and entangled as a result of velocity fluctuations — a flamelet combustion mechanism. It should be specially emphasized that nonaveraged pattern of flow and combustion is considered here. For further information on the models of turbulent combustion, one can refer to the books by Talantov [1975], Shchetinkov [1965], Prudnikov et al. [1971], Il'yashenko and Talantov [1964], Rayshenbakh et al. [1964], Williams [1965], and to the reviews by Andrews, Bradley, and Lwakawamba [1975], Libby and Williams [1981], Abdel-Gayed and Bradley [1981], Bray [1980], Borghi [1980], Bradley [1984].

The "frozen" (instantaneous) temperature distribution in the two cases are considerably different. In volumetric combustion, the temperature change takes place on the scale of order L, and the qualitative form of the concentration PDF with combustion and with mixing without reaction are essentially the same.

Detailed estimates, presented in Chapter 6, indicate that combustion of a homogeneous mixture under conditions realized in technical devices often takes place in accordance with the frontal mechanism. In this case the temperature profile involves a succession of pulses of nearly rectangular form with the same amplitude but different duration. Temperature variation from the lower level to

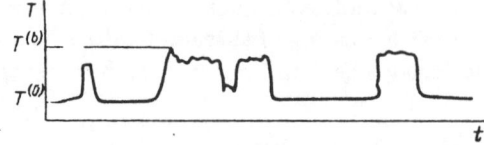

Figure 1.19 Oscillogram of temperature fluctuations in the combustion of a homogeneous mixture from data by Kokushkin [1960]. Experiments were conducted with gasoline-air mixture; excess air coefficient $\alpha = 1.6 - 1.8$; initial temperature 573 K; normal pressure. Mixture burned behind conical stabilizer 6 cm in diameter, located on the exit section of 40 cm diameter tube. Discharge velocity 90 – 110 m/sec. Measurements taken at a distance of 40 cm from the tube exit section. Measurement units along the coordinate axes are arbitrary.

the upper level takes place in a distance δ of the order of thickness of the normal flame front. This is illustrated by the oscillogram of temperature fluctuations which was obtained by Kokushkin [1960] (Fig. 1.19). It is evident from this oscillogram that combustion takes place in accordance with the frontal model.

The structure of the flame surface is illustrated in Fig. 1.20 taken from the same work. In this test, five resistance thermometers located at a distance of 6 mm from each other along a straight line perpendicular to the axis of the plume (axis x_2 in Fig. 1.20) were used. The vertical marks in Fig. 1.20 correspond to the time at which the flame front passes through the sensor. When interpreting Fig. 1.20 on the basis of the known Taylor hypothesis concerning turbulence quenching, time can be linked with the longitudinal coordinate with the aid of relation $x_1 = \langle u_1 \rangle t$. Therefore, the solid lines correspond to the recorded position of the flame front.

Thus, as a first approximation, the flame front can be regarded as a surface on which velocity, density, temperature, and concentration have discontinuities.

It is evident that in the frontal flame mechanism the probability density of the intermediate temperature values ($0 < c < 1$) is proportional to the thermal thickness of the normal flame $\delta \sim a/u_n$, where a is thermal diffusivity, u_n is the velocity of normal flame propagation. These parameters will be discussed in more detail in Chapter 6. The estimates presented there show that parameter δ is small (less than 1 mm). Therefore, the following estimate is true

$$P(c) \sim \delta/L, \qquad \delta = a/u_n, \qquad 0 < c < 1 \tag{1.42}$$

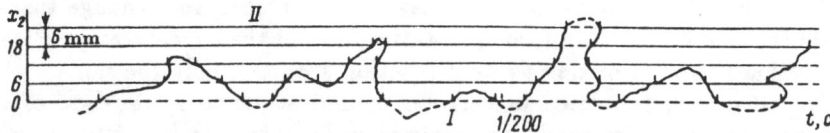

Figure 1.20 "Frozen" position of flame front from data by Kokushkin [1960]. Test conditions are the same as in Fig. 1.19.

Analogous estimate was obtained by Prudnikov in 1960 (see a review of Prudnikov's work in the book by Raushenbakh et al. [1964]). This estimate is not of course true when $c = 0$ or $c = 1$, since it is obvious that when $\delta = 0$, the intermediate values of the temperature are not observed, and PDF has the form

$$P(c) = \gamma_0 \delta(c) + \gamma_1 \delta(c-1), \qquad \gamma_0 + \gamma_1 = 1 \tag{1.43}$$

Relation (1.43) yields the main term of the asymptotic expansion of PDF with respect to the large diameter $L/\delta = u_n L/a$ which can, in the present case, be considered as an analog of Reynolds number.

Formula (1.43) was, apparently, first obtained by Prudnikov [1960] (see the review of Prudnikov's work in the book by Raushenbakh et al. [1964]). Since singular components appear in (1.20), (1.43), it can be assumed that combustion of a homogeneous mixture gives rise to a phenomenon which is similar to intermittency in turbulent flows. This analogy is also justified by the fact that the coefficient of molecular transport enters into the conditions for the applicability of formula (1.43).

It should be emphasized, however, that this analogy is far from complete, since during combustion the fluid before and after the front separating the regions with $c = 0$ and $c = 1$ can be turbulent.

Temperature PDF measurements in homogeneous combustion were conducted by Yoshida and Günther [1980] (natural gas used as the gaseous fuel). One of the probability densities obtained in this work is presented in Fig. 1.21. Two maxima, corresponding to the two "smeared" delta functions in expression (1.43), are clearly seen. The reasons behind the "smearness" are the same as in passive contaminant diffusion (this question was discussed in detail in §1.3). When $0 < c < 1$, the PDF is small. In this work the corollary of formula (1.43) is tested

$$\sigma^2 = \langle c \rangle (1 - \langle c \rangle) \tag{1.44}$$

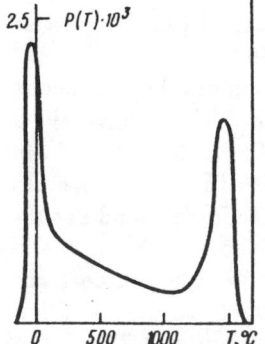

Figure 1.21 Temperature PDF in the combustion of a homogeneous mixture in a Bunsen burner from data by Yoshida and Günther [1980]. $x_1/d = 1.75$, $x_2/d = 0.3$, $\text{Re}_d = u_0 d/\nu = 1.44 \cdot 10^4$, $u_0 = 5.44$ m/sec, $d = 3.97$ cm.

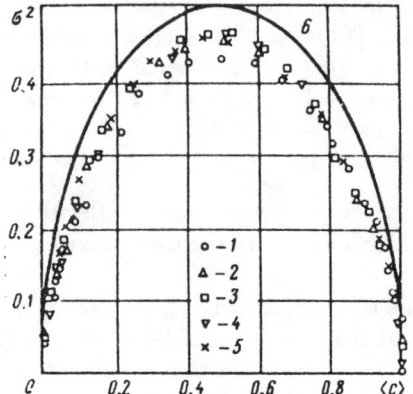

Figure 1.22 Relation of variance of temperature fluctuations with the mean temperature in the combustion of a homogeneous mixture in a Bunsen burner from data by Yoshida and Günther [1980]. *1)* $x_1/d = 3$, $u_1'/u_0 = 6\%$, $\alpha = 1.25$, *2)* $x_1/d = 2.375$, $u_1'/u_0 = 6\%$, $\alpha = 1.18$, *3)* $x_1/d = 1.75$, $u_1'/u_0 = 6\%$, $\alpha = 1.11$, *4)* $x_1/d = 3$, $u_1'/u_0 = 4\%$, $\alpha = 1.25$, *5)* $x_1/d = 1.75$, $u_1'/u_0 = 9\%$, $\alpha = 1.25$, *6)* dependence, described by formula (1.44), α - excess air coefficient, $u_1' = \langle (u_1 - \langle u_1 \rangle)^2 \rangle^{1/2}$, $c = (T - T^{(0)})/(T^{(b)} - T^{(0)})$, $\sigma^2 = \langle (c - \langle c \rangle)^2 \rangle$.

The results of comparison of relation (1.44) with the experimental data are given in Fig. 1.22. It is evident that relation (1.44) is satisfied with an accuracy not less than 10% for every value of $\langle c \rangle$. Similar results are also obtained from the tests by Kalghatgi and Moss [1979]. The experimental results by Bill, Namer, and Talbot [1981] also agree with the flamelet combustion mechanism. In this work, the combustion of ethylene-air mixture in grid-induced turbulence in a channel was investigated. Velocity fluctuations were measured by laser anemometry. Figure 1.23 shows the PDF of the longitudinal velocity which is measured within the combustion zone. The biomodal structure of this PDF is clearly seen. Thereby, at the point with finite probability, two fluids with different statistical properties were observed. Other results of this work will be noted in §6.6.

In contrast to the case discussed in §3.1, in the combustion of a homogeneous mixture, corrections to the PDF must be taken into account. These corrections are called for by the effect of molecular transport. Indeed, the mean value of the rate of chemical reaction cannot, for example, be found with the aid of formula (1.43). Let us analyze this effect making use of the theory of Zel'dovich and Frank-Kamenetskii [1938a, b] which yields, as is generally known, the main term in the asymptotic expansion of the solution of the heat equation with a source (rate of heat release is described by the Arrhenius law) with respect to perturbation $RT^{(b)}/E$, where R is the gas constant, E is the activation energy, and $T^{(b)}$ is the adiabatic combustion temperature. Within the framework of the indicated theory, the thickness of layers δ_c wherein chemical reaction takes place in succession is equal to $\delta_c = \delta RT^{(b)}/E \ll \delta$. For a fixed value of δ and $RT^{(b)}/E \to 0$ it can be assumed that the reactions proceed to surface $c = 1-0$. On one side of this surface — in the combustion products — concentration retains constant value $c = 1$, and on the other side an intermediate value of the concentration is observed. Hence, in the approximation of Zel'dovich and Frank-Kamenetskii, a term proportional to $\delta(c - 1)$ must be included in the expression for the temperature PDF. It is significant that in this reasoning, which is due to Kuznetsov [1976b], the parameter δ is fixed (finite Reynolds number), and, consequently, the term with $\delta(c)$ is absent from the PDF. Note that this feature is overlooked in a number of works, in particular, by Prudnikov (see Raushenbakh et al. [1964], Zimont [1977], Bray [1980], Libby and Bray [1981], Libby and Williams [1981]. Thus, we have

$$P(c) = \gamma_1 \delta(c - 1) + \gamma P_1(c), \quad \gamma_1 + \gamma = 1 \tag{1.45}$$

To avoid confusion it should be emphasized that the subscript t here has a different meaning from the subscript in formula (1.19). Here, it is used only on the basis of the formal analogy between relations (1.19) and (1.45).

The concentration profile in the limit under consideration ($RT^{(b)}/E \to 0$, δ/L is fixed) is such that $\partial c/\partial n \to 0$ when $c \to 0$ and, hence, $P_t(c) \to \infty$ when $c \to 0$, i.e., the conditional PDF has, as seen from (1.24), an integrable singularity at zero.

It is important to emphasize again that the use of the principle of self-similarity with respect to the conditional temperature PDF is, unlike pure mixing, inapplicable in this case.

As in the case of mixing without reactions, when intermittency is induced by large-scale fluctuations of the thin boundary separating turbulent and nonturbulent regions, and again in the case of the combustion of a homogeneous mixture intermittency is caused by large-scale fluctuations of the combustion front. However, despite the formal similarity of formulas (1.45) and (1.19), these types of

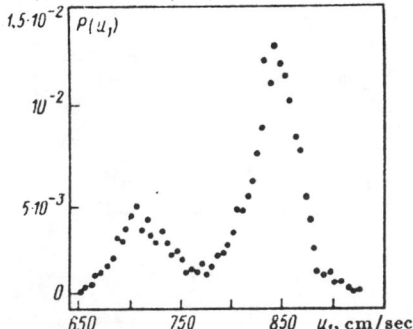

Figure 1.23 PDF of the longitudinal velocity in the combustion of a homogeneous ethylene-air mixture in grid-induced turbulence in a channel from data by Bill, Namer, and Talbot [1981]. Measurements made in a section located at a distance of 7.5 cm from the grid. Size of grid cell $M = 0.5$ cm, channel width 10 cm, $u_0 = 6.84$ m/sec — mean velocity ahead of the grid. Excess air coefficient $\alpha = 1.33$.

intermittency are not interrelated and are induced by totally different causes: fluid dynamic intermittency arises when Re $\to \infty$, and intermittency in case of combustion of a homogeneous mixture arises when $RT^{(b)}/E \to 0$.

CHAPTER
TWO
EQUATIONS FOR PROBABILITY DENSITY FLUCTUATION (PDF)

The first attempt to obtain an equation for PDF in turbulent flow was, apparently, made by Frost [1960] when considering the turbulent combustion of a homogeneous mixture of combustible gases. An equation for the temperature PDF was derived in this work. Upon derivation, in addition to the exact heat equation, additional considerations were enlisted (for example, concerning the Markovian character of the process of turbulent mixing). A similar equation (under the title Langevin model) appeared in the work by Chang [1969]. Later on, this approach was developed in the works by Kuznetsov and Frost [1973] and Frost [1973, 1977].

Exact open equations for n-point probability distribution densities of various fluid dynamic characteristics, obtained from the Navier-Stokes equation, were introduced into the theory of turbulence practically simultaneously in the works of Monin [1967a, b], Lundgren [1967], Novikov [1967], Kuznetsov [1967], Ulinich [1968], Ulinich and Lyubimov [1968]. Subsequently, the equations for PDFs were extended to the case of the Lagrangian description of motion of a medium in the works by Lyubimov [1969], Lyubimov and Ulinich [1970]. The general method of deriving the equations for PDFs in an arbitrary continuum is given in the works by Ievlev [1972, 1975], and Fox [1975] (see also Hill [1976]). A detailed review of the contributions in this field, including closure methods, is included in the articles by Kuznetsov and Sabel'nikov [1981a, b] (see also §2.3 of this book), in *Turbulent Reacting Flows*, edited by Libby and Williams [1980], in the survey by Borghi [1980], and in articles by Sabel'nikov [1985a, 1986].

In the following chapters only two equations will be required: the first — for single-point concentration PDF, and the second — for two-point velocity difference PDF. The method used below in deriving these equations differs from those described in the literature in that intermittency is introduced at the very first stage. This, as was indicated in Chapter 1, avoids the formal difficulties associated with the limit Re $\to \infty$. Such an approach was proposed by Kuznetsov [1972a]. It was used subsequently in the works by Kuznetsov and Frost [1973], Kuznetsov [1977b, 1979a], Sabel'nikov [1979, 1980a, b, 1982b, c], Kuznetsov and Sabel'nikov

[1981a, b], Kuznetsov, Lebedev, Secundov, and Smirnova [1981], O'Brien and Dopazo [1978], O'Brien [1978] and a number of other works.

§2.1 EQUATION FOR CONCENTRATION PDF

Considered in the present section are the processes in three different cases: 1) mixing without chemical reactions; 2) combustion with mixing of fuel and oxidant (diffusion combustion); 3) combustion of premixed combustible components. In the second case, it is assumed that the rates of all chemical reactions are infinitely fast, i.e., the composition and temperature are thermodynamically in equilibrium. In the third case, it is assumed that single-stage combustion takes place at a finite rate. It is supposed that Reynolds number tends to infinity and the Mach number tends to zero. On account of the last assumption, it can be assumed that when computing density ρ, the pressure p is constant, i.e., ρ is dependent only on the temperature and composition (in this approximation the pressure gradient entering into the Navier-Stokes equation is always taken into account).

Let us also assume that all coefficients of molecular transport are equal, and the initial concentration distribution of all substances and the enthalpy and the boundary conditions are similar. The first limitation is not too strong, since the mixing characteristics are weakly dependent on Reynolds number. As is generally known, in the assumptions indicated above for the description of the processes under consideration it is adequate just to specify the fluid dynamic velocity and concentration of one of the substances (Burke and Schumann [1928], Zel'dovich and Frank-Kamenetskii [1938a, b], Schvab [1948], Zel'dovich [1949]).

This conclusion is evident for the case of mixing without reactions.

If chemical reactions take place, the total mass concentrations $c_\alpha (\alpha = 1, 2, \ldots)$ of the atoms of a given type contained in all chemical compounds and the total enthalpy (including the energy of chemical bonds) c_0 are described by identical equations without sources. Since the initial and boundary conditions are usually similar, quantities c_0 and c_α are expressed linearly through the solution of the diffusion equation without source terms. Then, in diffusion combustion the temperature, density, and concentration of all substances can be linked with this solution by means of thermodynamic calculations. Further, it is assumed here that this calculation has been completed.

If the chemical reaction involves a single step, then in the combustion of premixed reactants, the quantity of heat released and the variation of the concentrations of the reacting substances are readily expressed in terms of the concentration of any one of the substances. Consequently, it is sufficient in this case also to specify the distribution of any one characteristic of the process (for example, the temperature).

Thus, in all the cases under consideration, density ρ and the rate of chemical reaction W are expressed in terms of one parameter. We denote this variable by the letter c ($c = z$ is the concentration of the inert contaminant in mixing or in diffusion combustion, $c = T - T^{(0)}/T^{(b)} - T^{(0)}$ upon combustion of homogeneous mixture, T is the temperature, superscripts 0 and b refer to the fresh mixture and the combustion products, respectively). All the results obtained in this chapter remain valid in the case when $W = 0$.

The formulated hypotheses significantly simplify the analysis and, at the same time, are quite adequate for the solution of the problems considered in Chapters 5 and 6.

Let us write the non-averaged equations of diffusion and continuity

$$\rho \frac{\partial c}{\partial t} + \rho u \nabla c = \nabla(\rho D \nabla c) + \rho W \tag{2.1}$$

$$\frac{\partial \rho}{\partial t} + \nabla \rho u = 0 \tag{2.2}$$

Here, D is the coefficient of diffusion, W is the rate of chemical reaction ($W = 0$ upon mixing or diffusion combustion).

Let us introduce quantity $\varphi = \exp(i\lambda c)$, λ is the real number. By definition $\langle \varphi \rangle$ is the characteristic function of concentration probability density. Therefore, if $\langle \varphi \rangle$ is known, the PDF is found with the aid of the inverse Fourier transform

$$P(c) = \frac{1}{2\pi} \int_{-\infty}^{\infty} \langle \varphi \rangle \exp(-i\lambda c) d\lambda$$

We differentiate φ with respect to t. The derivative in this case $\partial c/\partial t$ is expressed from the heat equation (2.1). As a result, we obtain

$$\rho \frac{\partial \varphi}{\partial t} = i\lambda \varphi [-\rho u \nabla c + \nabla(\rho D \nabla c) + \rho W]$$

The first term on the right-hand side of this relation is reduced to the form $\rho u \nabla \varphi$. Therefore, after using the continuity equation (2.2) and averaging we obtain

$$\frac{\partial \langle \rho \varphi \rangle}{\partial t} + \nabla \langle \rho u \varphi \rangle = i\lambda \langle \rho W \varphi \rangle + i\lambda \langle \varphi \nabla(\rho D \nabla c) \rangle \tag{2.3}$$

The second term on the right-hand side of relation (2.3) can be transformed into the following form (Kuznetsov [1967, 1972a], Fox [1971], Pope [1976])

$$i\lambda \langle \varphi \nabla(\rho D \nabla c) \rangle = i\lambda \nabla \langle \rho D \varphi \nabla c \rangle - i\lambda \langle \rho D \nabla c \cdot \nabla \varphi \rangle$$

$$= \nabla \langle \rho D \nabla \varphi \rangle + \lambda^2 \langle \rho N \varphi \rangle, \qquad N = D(\nabla c)^2 \tag{2.4}$$

Transformation (2.4) is significant when deriving the equation for the concentration PDF. It should be noted that this equation is analogous to the transformation which is usually used when deriving the equation for the variance of concentration fluctuations (Corrsin [1951]). The term $\nabla \langle \rho D \nabla \varphi \rangle$ appearing on the right side of (2.4) describes the molecular transport of the moments of the concentration field and is proportional to Re^{-1}. Since the limit $Re \to \infty$ is being considered, this term is henceforth omitted (the remaining terms in the equation are finite, since the PDF and $\langle N \rangle$ when $Re \to \infty$ have finite limits).

Using (2.4), we reduce relation (2.3) to the form

$$\frac{\partial \langle \rho\varphi \rangle}{\partial t} + \nabla \langle \rho u\varphi \rangle = \lambda^2 \langle \rho N\varphi \rangle + i\lambda \langle \rho W\varphi \rangle \tag{2.5}$$

Relations (2.3) and (2.5), as all exact equations for the statistical characteristics in the theory of turbulence, are not closed. In them, in addition to correlation $\langle \rho W\varphi \rangle$ which (under the adopted assumption with respect to the form of the rate of chemical reaction W) is exactly expressed through the probability density $P(c)$ (this circumstance provides the main advantage of using PDFs in the theory of turbulent combustion), correlations $\langle \rho u\varphi \rangle$ and $\langle \rho N\varphi \rangle$ which are not expressed through the required PDF appears.

As regards the correlation $\langle \rho W\varphi \rangle$ we note the following. Most common is the case when in the nonturbulent fluid $W = 0$ and, hence

$$\langle \rho W\varphi \rangle = \gamma \langle \rho W\varphi \rangle_t = \gamma \int \rho W\varphi P_t(c)dc \tag{2.6}$$

Integration in (2.6) is effected over the whole range of c. For the sake of brevity, the integration limits are not shown. Such a simplification will also be used further on.

Let us consider now correlation $\langle \rho u\varphi \rangle$. By definition we have

$$\langle \rho u\varphi \rangle = \int \rho u\varphi P(u,c)d^3u\, dc \tag{2.7}$$

Here, $P(u,c)$ is the joint PDF of velocity and concentration. Reasoning similar to those used in §1.3 when deriving formulas (1.19), (1.20) allows the establishment of the form of this function (Sabel'nikov [1979, 1980b])

$$P(u,c) = \gamma P_t(u,c) + \gamma_0 P_n(u|c=0)\delta(c)$$
$$+ \gamma_1 P_n(u|c=1)\delta(c-1) \tag{2.8}$$

Here $P_t(u,c)$ is the conditional joint PDF of velocity and concentration in the turbulent fluid, $P_n(u|c=0)$ and $P_n(u|c=1)$ are the conditional velocity PDFs in the nonturbulent fluid under condition $c=0$ and $c=1$, respectively.

Substituting (2.8) into (2.7), we obtain

$$\langle \rho u\varphi \rangle = \gamma \int \rho \langle u \rangle_{t,c} \varphi P_t(c)dc + \gamma_0 \rho(0)\langle u \rangle_{n,0}$$
$$+ \gamma_1 \rho(1)\exp(i\lambda)\langle u \rangle_{n,1} \tag{2.9}$$

where

$$\langle u \rangle_{n,0} = \int u P_n(u|c=0)d^3u$$
$$\langle u \rangle_{n,1} = \int u P_n(u|c=1)d^3u$$
$$\langle u \rangle_{t,c} = \int u P_t(u|c)d^3u$$

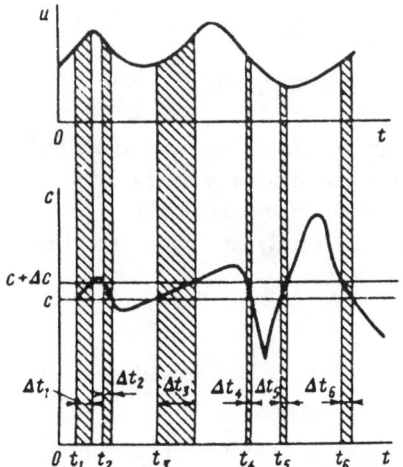

Figure 2.1 Determination of conditionally averaged velocity $\langle u \rangle_{t,c} = \Sigma u(t_k) \Delta t_\lambda / \Sigma \Delta t_\lambda$. Δt_k is the duration of the k-th time interval when the concentration values are in the range c, $c + \Delta c$, $k = 1, 2, 3, \ldots$.

Here, $P_t(u|c)$ is the conditional velocity PDF in the turbulent fluid for a specified value of the concentration, $\langle u \rangle_{n,0}$, $\langle u \rangle_{n,1}$ are the conditionally averaged velocities in the nonturbulent fluid when $c = 0$ and $c = 1$, respectively, $\langle u \rangle_{t,c}$ is the conditionally averaged velocity in the turbulent fluid at the specified value of the concentration. To clarify the meaning of the conditionally averaged characteristics $\langle \ \rangle_{t,c}$, Fig. 2.1 shows schematically the procedure for determining the conditionally averaged velocity $\langle u \rangle_{t,c}$ in the statistically stationary case.

Let us analyze correlation $\langle \rho N \varphi \rangle$. By definition we have

$$\langle \rho N \varphi \rangle = \int \rho N \varphi P(N, c) dN dc \tag{2.10}$$

Here, $P(N, c)$ is the joint PDF of scalar dissipation and concentration. As a result of intermittency, the values $c = 0$, $c = 1$, and $N = 0$ possess a probability which is different from zero. Hence

$$P(N, c) = \gamma P_t(N, c) + \gamma_0 \delta(c) \delta(N) + \gamma_1 \delta(c - 1) \delta(N) \tag{2.11}$$

Here, $P_t(N, c)$ is the joint PDF of scalar dissipation and concentration in the turbulent fluid.

Substitution of relation (2.11) into (2.10) yields

$$\langle \rho N \varphi \rangle = \gamma \langle \rho N \varphi \rangle_t = \gamma \int \rho N \varphi P_t(N, c) dN dc$$

$$= \gamma \int \rho \langle N \rangle_{t,c} \varphi P_t(c) dc \tag{2.12}$$

where

$$\langle N \rangle_{t,c} = \int N P_t(N|c) dN$$

Here, $P_t(N|c)$ is the conditional PDF of scalar dissipation in the turbulent fluid at the specified value of concentration, $\langle N \rangle_{t,c}$ is the conditionally averaged scalar dissipation in turbulent fluid at the specified value of concentration.

Correlations $\langle \varphi \nabla(\rho D \nabla c) \rangle$ are also represented in a fully analogous manner

$$\langle \varphi \nabla(\rho D \nabla c) \rangle = \gamma \langle \varphi \nabla(\rho D \nabla c) \rangle_t$$

$$= \gamma \int \varphi \langle \nabla(\rho D \nabla c) \rangle_{t,c} P_t(c) dc \qquad (2.13)$$

Here, $\langle \nabla(\rho D \nabla c) \rangle_{t,c}$ is the conditional mean value of the divergence of the diffusion flow $\nabla(\rho D \nabla c)$ in the turbulent fluid at the specified value of concentration.

Depending on whether the transformation (2.4) is used or not, two different forms for the concentration PDF are obtained. Consider first relation (2.3). We apply the inverse Fourier transform to this relation and take into account expressions (2.6), (2.9), and (2.13). As a result, we obtain

$$\frac{\partial \rho P}{\partial t} + \nabla \left(\rho \int u P(u,c) d^3 t \right)$$

$$= -\gamma \frac{\partial}{\partial c} \langle \nabla(\rho D \nabla c) \rangle_{t,c} P_t(c) - \gamma \frac{\partial}{\partial c} \rho W P_t(c) \qquad (2.14)$$

Analogously, from relation (2.5) we obtain a second form of the equation

$$\frac{\partial \rho P}{\partial t} + \nabla \left(\rho \int u P(u,c) d^3 u \right) = -\gamma \frac{\partial^2}{\partial c^2} \rho \langle N \rangle_{t,c} P_t - \gamma \frac{\partial}{\partial c} \rho W P_t \qquad (2.15)$$

The integral, appearing in equations (2.14) and (2.15), takes the following form upon the substitution of the expression (2.8) for $P(u,c)$ (Sabel'nikov [1979, 1980b])

$$\int u P(u,c) d^3 u = \gamma_0 \delta(c) \langle u \rangle_{n,0}$$

$$+ \gamma_1 \delta(c-1) \langle u \rangle_{n,1} + \gamma \langle u \rangle_{t,c} P_t(c) \qquad (2.16)$$

The two forms of equation given here for the concentration PDF (2.14) and (2.15) are equivalent. The physical meaning of the different components in (2.14) and (2.15) is sufficiently clear. The first term on the left-hand side of each of equations (2.14), (2.15) results from the non-stationary state (on the average), and the second describes convection and turbulent diffusion. The right-hand side of

(2.14) and (2.15) includes terms characterizing attenuation of concentration non-homogeneity as a result of mixing at the molecular level and of the effect of heat release.

Concluding this section, we present the derivation of the equation for concentration PDF on the basis of its geometrical interpretation given in §1.3. This is useful for understanding the results obtained in Chapters 3, 5, and 6. Furthermore, the intermediate formulas used in this derivation will be repeatedly applied in Chapters 3 and 5. For the sake of simplicity, we consider a statistically uniform concentration field. We compute the rate of volume change δV_c which is enclosed between two closely located isoscalar surfaces c and $c + dc$ which completely lie in some sufficiently large volume V. We have

$$\frac{d\delta V_c}{dt} = \int_{S_c+dc} \mathbf{v} n dS_{c+dc} - \int_{S_c} \mathbf{v} n dS_c = \frac{\partial \int_{S_c} \mathbf{v} n dS_c}{\partial c} dc \tag{2.17}$$

Here, $\mathbf{v}$ is the velocity of motion of the isoscalar surface relative to the gas, $n = \nabla c/|\nabla c|$ is the vector of the unit normal to the isoscalar surface. Velocity $\mathbf{v}$ is found from the diffusion equation written in a coordinate system which is tied to the isoscalar surface $c(x,t) = $ const (Gibson [1968], Klimov [1972])

$$-\rho \mathbf{v} \nabla c = \nabla(\rho D \nabla c) + \rho W$$

$$\mathbf{v} = \mathbf{u}_c - \mathbf{u} \tag{2.18}$$

Here, $\mathbf{u}_c = dx/dt$ is the absolute velocity of points located on the isoscalar surface. From (2.18) we find that

$$\mathbf{v} n = -\rho^{-1} \nabla(\rho D \nabla c)|\nabla c|^{-1} - W|\nabla c|^{-1}$$

Substitution of this expression in (2.17) yields

$$\frac{d\rho \delta V_c}{dt} = \frac{\partial \int_{S_c} [-\nabla(\rho D \nabla c)|\nabla c|^{-1} - \rho W|\nabla c|^{-1}] dS_c}{\partial c} dc$$

$$= \frac{\partial \int_{\delta V_c} [-\nabla(\rho D \nabla c) - \rho W] d\delta V_c}{\partial c} dc \tag{2.19}$$

We divide (2.19) by volume V and let V tend to infinity. Taking into account the definition of PDF (1.21), we obtain

$$\frac{\partial \rho P}{\partial t} = \gamma \frac{\partial}{\partial c} \langle\langle \nabla(\rho D \nabla c) \rangle\rangle_{t,c} P_t - \gamma \frac{\partial}{\partial c} \rho W P_t \tag{2.20}$$

where

$$\langle\langle \nabla(\rho D \nabla c) \rangle\rangle_{t,c} = \frac{1}{\delta V_c} \int_{\delta V_c} \nabla(\rho D \nabla c) d\delta V_c$$

In a statistically homogeneous case, equation (2.20) coincides with equation (2.14) obtained above, since owing to ergodicity the following equality is fulfilled

$$\langle\langle \nabla(\rho D \nabla c) \rangle\rangle_{t,c} = \langle \nabla(\rho D \nabla c) \rangle_{t,c}$$

§2.2 EQUATION FOR TWO-POINT VELOCITY DIFFERENCE PDF IN LOCALLY HOMOGENEOUS TURBULENCE

Let us consider a homogeneous and isotropic turbulence field in incompressible fluid. Let $x^{(1)}$ and $x^{(2)}$ be two arbitrary points separated by a distance within the inertial range, i.e., $\eta \ll r \ll L$, where η is the Kolmogorov scale, L is the integral turbulence scale, $r = x^{(2)} - x^{(1)}$. When deriving the equation for the velocity difference PDF at these points, just as in the previous section, we shall assume that Reynolds number tends to infinity. Thereby, further reasoning pertains only to the main term in the asymptotic expansion of PDF when Re $\to \infty$ which has the form of (1.8).

The derivation of the equation presented here is based on the work of Kuznetsov [1967, 1976a, 1977c] (see also the article by Kuznetsov and Sabel'nikov [1981a]). We introduce the quantity $\varphi = \exp(i\vec{\lambda}\mathbf{v})$, $\mathbf{v} = u(x^{(2)}, t) - u(x^{(1)}, t)$, $\vec{\lambda}$ which is a real vector. The mean value $<\varphi>$ is the characteristic function of the velocity difference PDF at two points.

We differentiate φ with respect to time and express the derivatives $\partial u_k(x^{(1)}, t)/\partial t$ and $\partial u_k(x^{(2)}, t)/\partial t$ resulting from the differentiation of the Navier-Stokes equations

$$\frac{\partial u_k}{\partial t} + u_\ell \frac{\partial u_k}{\partial x_\ell} = -\frac{\partial p}{\partial x_k} + \nu \Delta u_k \qquad (2.21)$$

taken at points $x^{(1)}$ and $x^{(2)}$. In (2.21) p is the pressure divided by the density.

Following the averaging of the expression obtained as a result of the indicated procedure, we obtain

$$\frac{\partial \langle \varphi \rangle}{\partial t} = i\lambda_k \left\langle \varphi \left[u_\ell^{(1)} \frac{\partial u_k^{(1)}}{\partial x_\ell^{(1)}} - u_\ell^{(2)} \frac{\partial u_k^{(2)}}{\partial x_\ell^{(2)}} + \frac{\partial p^{(1)}}{\partial x_k^{(1)}} \right.\right.$$

$$\left.\left. - \frac{\partial p^{(2)}}{\partial x_k^{(2)}} + \nu \Delta_{x^{(2)}} u_k^{(2)} - \nu \Delta_{x^{(1)}} u_k^{(1)} \right] \right\rangle \qquad (2.22)$$

The inertia terms entering the right-hand side of this relationship, following the use of the assumption of turbulence homogeneity and the continuity equation $\nabla u = 0$, are reduced to

$$i\lambda_k \left\langle \varphi \left[u_\ell^{(1)} \frac{\partial u_k^{(1)}}{\partial x_\ell^{(1)}} - u_\ell^{(2)} \frac{\partial u_k^{(2)}}{\partial x_\ell^{(2)}} \right] \right\rangle$$

$$= -\frac{\partial}{\partial x_k^{(1)}} \langle u_k^{(1)} \varphi \rangle - \frac{\partial}{\partial x_k^{(2)}} \langle u_k^{(2)} \varphi \rangle = -\frac{\partial}{\partial r_k} \langle u_k \varphi \rangle = i\frac{\partial^2 \langle \varphi \rangle}{\partial r_k \partial \lambda_k} \qquad (2.23)$$

Let us now manipulate the terms in relation (2.22) characterizing the pressure forces. To this end, we differentiate the Navier-Stokes equations (2.21) twice with respect to x_k and x_i. Making use of the continuity equation, we arrive at Poisson's equation for the pressure gradient

$$\Delta \frac{\partial p}{\partial x_i} = -\frac{\partial^2 u_k u_\ell}{\partial x_i \partial x_k \partial x_\ell} \tag{2.24}$$

From (2.24) we obtain

$$\frac{\partial p^{(1)}}{\partial x_i^{(1)}} = \frac{1}{4\pi} \int \frac{1}{|x^{(3)} - x^{(1)}|} d_{ik\ell} u_k^{(3)} u_\ell^{(3)} d^3 x^{(3)}$$

$$\frac{\partial p^{(2)}}{\partial x_\ell^{(2)}} = \frac{1}{4\pi} \int \frac{1}{|x^{(3)} - x^{(2)}|} d_{ik\ell} u_k^{(3)} u_\ell^{(3)} d^3 x^{(3)} \tag{2.25}$$

where

$$d_{ik\ell} = \frac{\partial^3}{\partial x_i^{(3)} \partial x_k^{(3)} \partial x_\ell^{(3)}}$$

Since the following obvious relation is valid

$$d_{ik\ell} u_k^{(3)} u_\ell^{(3)} = d_{ik\ell} V_k V_\ell, \qquad V = u^{(3)} - u^{(1)}$$

we obtain

$$i\lambda_k \left\langle \varphi \left[\frac{\partial p^{(1)}}{\partial x_\lambda^{(1)}} - \frac{\partial p^{(2)}}{\partial x_k^{(2)}} \right] \right\rangle$$

$$= i\frac{\lambda_k}{4\pi} \int T D_{ik\ell} \varphi P(v, V, r, R) V_k V_\ell d^3 V d^3 R$$

$$T = \frac{1}{R} - \frac{1}{|R - r|}, \qquad R = x^{(3)} - x^{(1)}, \qquad D_{ik\ell} = \frac{\partial^3}{\partial R_i \partial R_k \partial R_\ell} \tag{2.26}$$

Here, $P(v, V, r, R)$ is the three-point velocity difference PDF.

Let us consider the viscous terms in (2.22). Just as in deriving the equation for concentration PDF (see §2.1), they can be written in two different forms. Later on, only one of them will be used, namely, the one which corresponds to the representation of the equation for concentration PDF $P(c)$ in the form of (2.15). In order to obtain this form, we take into account the following identities (it is readily seen that they are analogous to transformations (2.4)).

$$i\lambda_k \varphi \Delta_{x^{(1)}} u_k^{(1)} = -\lambda_i \lambda_j \varphi \frac{\partial u_i^{(1)}}{\partial x_\ell^{(1)}} \frac{\partial u_j^{(1)}}{\partial x_\ell^{(1)}} - \Delta_{x^{(1)}} \varphi$$

$$i\lambda_k \varphi \Delta_{x^{(2)}} u_k^{(2)} = \lambda_i \lambda_j \varphi \frac{\partial u_i^{(2)}}{\partial x_\ell^{(2)}} \frac{\partial u_j^{(2)}}{\partial x_\ell^{(2)}} + \Delta_{x^{(2)}} \varphi$$

which enable the representation of the viscous terms in the required form

$$i\lambda_k \nu \left\langle \varphi \left[\Delta_{x^{(2)}} u_k^{(2)} - \Delta_{x^{(1)}} u_k^{(1)} \right] \right\rangle$$

$$= \lambda_i \lambda_j \left\langle \varphi \left(\epsilon_{ij}^{(1)} + \epsilon_{ij}^{(2)} \right) \right\rangle + 2\nu \frac{\partial^2 \langle \varphi \rangle}{\partial r_\ell \partial r_\ell}$$

$$\epsilon_{ij} = \nu \frac{\partial u_i}{\partial x_\ell} \frac{\partial u_j}{\partial x_\ell} \tag{2.27}$$

The second component on the right-hand side of this relation (it describes the molecular transport of the moments of velocity transfer) when $r \gg \eta$ and $\mathrm{Re} \to \infty$ can be neglected. The first component on the right-hand side of (2.27), as shown by calculations and reasonings similar to those employed in Chapter 1 to obtain the formula for the velocity difference PDF (1.8) from (1.5), has the following form when $\eta \ll r \ll L$

$$\langle \varphi(\epsilon_{ij}^{(1)} + \epsilon_{ij}^{(2)}) \rangle = \gamma \langle \varphi(\epsilon_{ij}^{(1)} + \epsilon_{ij}^{(2)}) \rangle_r \tag{2.28}$$

For the further transformation of relation (2.28) we introduce the conditional PDF $P_{t\epsilon}(\epsilon_{ij}|\mathbf{v})$ of quantity ϵ_{ij} on the condition that points $x^{(1)}$ and $x^{(2)}$ are in the turbulent fluid, and the velocity difference at these points is equal to $\mathbf{v}$. We introduce the notation

$$e_{ij}(\mathbf{v}) = \int \epsilon_{ij}^{(1)} P_{t\epsilon}(\epsilon_{ij}^{(1)}|\mathbf{v}) d\epsilon_{ij}^{(1)} = \int \epsilon_{ij}^{(2)} P_{t\epsilon}(\epsilon_{ij}^{(2)}|\mathbf{v}) d\epsilon_{ij}^{(2)} \tag{2.29}$$

The equality of the integrals in (2.29) follows from considerations of symmetry. Using the Bayes formula for the conditional joint probability density ϵ_{ij} and $\mathbf{v}$ in the turbulent fluid:

$$P_{tt}(\epsilon_{ij}, \mathbf{v}) = P_{t\epsilon}(\epsilon_{ij}|\mathbf{v}) P_{tt}(\mathbf{v}, r)$$

and the introduced notation (2.29), we arrive at the relation

$$\langle \varphi(\epsilon_{ij}^{(1)} + \epsilon_{ij}^{(2)}) \rangle = 2\gamma \int e_{ij}(\mathbf{v}) \varphi P_{tt}(\mathbf{v}, r) d^3 \mathbf{v} \tag{2.30}$$

From formulas (2.23), (2.26), (2.27), and (2.30), we obtain the equation for the characteristic function. Applying the inverse Fourier transform, we find that the velocity difference PDF at two points satisfies the following equation

$$\frac{\partial P}{\partial t} + v_k \frac{\partial P}{\partial r_k} + 2\gamma \frac{\partial^2}{\partial v_i \partial v_j} e_{ij}(\mathbf{v}) P_{tt} + \frac{\partial \pi_k}{\partial v_k} = 0 \tag{2.31}$$

Here

$$\pi_k = \frac{1}{4\pi} \int T D_{kij} V_i V_j P(\mathbf{v}, \mathbf{V}, r, R) d^3 V d^3 R$$

The relation of the unconditional PDF $P(\mathbf{v}, r)$ with the conditional PDF $P_{tt}(\mathbf{v}, r)$ and $P_{nn}(\mathbf{v}, r)$ is given by formula (1.8).

In the inertial range, owing to the equilibrium of small-scale turbulence, the nonstationary components can be omitted.

§2.3 CLOSURE HYPOTHESES OF THE EQUATIONS FOR PDFS

The equations for PDFs obtained in the first two sections of the present chapter, like all averaged equations in the statistical theory of turbulence which follow strictly from the Navier-Stokes and diffusion equations, are not closed (the equation for the characteristic functional is an exception, Monin and Yaglom [1967]). For the unknown functions entering these equations, one can write new equations which will also be open and so on. As a result, an infinite linking chain of equations is obtained similar to the Ivon-Born-Kirkwood-Bogolyubov chain of equations in statistical physics. There have not been many attempts as yet to close this chain of equations. The techniques proposed for approximating the unknown terms are largely purely semiempirical or formal and are, as a rule, based on the analogy with the kinetic theory of gases.

Let us first briefly consider those hypotheses which are used for the closure of the equation for a single-point concentration PDF (a detailed review and analysis of closure hypotheses is included in the work by Sabel'nikov [1985a]). In this case, most difficulties, apparently, arise from the description of mixing at the molecular scale, i.e., the first component which appears on the right-hand side of equation (2.14) or (2.15). Therefore, one very often uses as the initial relation not the exact equations of continuum mechanics, but some model equations based on the qualitative representation of the character of the molecular mixing process in turbulent flows. For example, in the works by Kuznetsov and Frost [1973], Frost [1973], Chang [1969, 1970, 1976], the Langevin equation was used. In this case the first term on the right-hand side of (2.14) acquires the form*

$$-\frac{\partial}{\partial c}\langle \nabla(D\nabla c)\rangle_c P = \beta \frac{\partial}{\partial c}(c - \langle c \rangle)P \qquad (2.32)$$

where β is some function which is independent of concentration.

Analogous expressions have been proposed by Dopazo [1975], Dopazo and O'Brien [1976]. It is shown in these works that (2.32) is an exact expression if the two-point concentration PDF at two close points is normal (for this reason the closure of (2.32) is often referred to in literature as quasi-Gaussian). Then function β has the form $\beta = \langle N \rangle / \sigma^2$, where $\sigma = \sqrt{\langle (c - \langle c \rangle)^2 \rangle}$ is the root-mean-square concentration. It was noted in Chapter 1 that this concentration PDF for a small distance between the points differs fundamentally from the normal distribution (see Fig. 1.11).

Therefore, it is difficult to anticipate that relation (2.32) would yield satisfactory results and, indeed, analysis conducted by Frost [1973], Dopazo [1979], O'Brien [1980a], Sabel'nikov [1982b, 1985a, 1986] confirm this conclusion.

Another approach has been developed in the works of Curl [1963], Pope [1976, 1981b], Kollman and Janicka [1982], Frost [1973, 1977], Janicka, Kolbe, and Kollman [1979], Dopazo [1979], Nedorub, Frost, and Shcherbina [1979] (see also the reviews by O'Brien [1980a, b], Kompaniets, Ovsyannikov, and Polak

---
*Note that Kuznetsov and Frost [1973] used expression (2.32) only in the turbulent fluid.

[1979], Shcherbina [1982]). The term under consideration in these works is practically replaced by an arbitrary expression which is nonlinearly and nonlocally dependent on PDF. Thus, Pope [1976] proposed the expression

$$\frac{\partial}{\partial c}\langle\nabla(D\nabla c)\rangle_c P \sim \frac{q\sigma}{L}\frac{\partial^2}{\partial c^2}\int_{-0}^{c} g\left(\frac{c-c'}{\sigma}\right)P(c')dc'$$

$$\times \int_{c}^{1+0} g\left(\frac{c''-c}{\sigma}\right)P(c'')dc'' \qquad (2.33)$$

where $g(s) = \ln(1+s)$; henceforth symbols -0 and $1+0$ mean that upon integration in agreement with (1.20), the terms proportional to $\delta(c)$ and $\delta(c-1)$ are taken into account, respectively.

The closing relations proposed in the other works, with the exception of the closure from the article by Frost [1973], can be written as follows from the article by Janicka, Kolbe, and Kollman [1979][†]

$$\frac{\partial}{\partial c}\langle\nabla(D\nabla c)\rangle_c P \sim -\frac{q}{L}\left[\int_{-0}^{c}dc'\int_{c}^{1+0}dc''P(c')P(c'')\right.$$

$$\left.\times G(c,c',c'') - P(c)\right], \quad \int_{c'}^{c''}G(c,c',c'')dc = 2, \quad c' \leq c \leq c'' \qquad (2.34)$$

Here, G is some nonnegative function which transforms by the law $G(\lambda c, \lambda c', \lambda c'') = \lambda^{-1}G(c, c', c'')$. The closure of Frost [1973] differs from the general relation (2.34) in that, firstly, function $G(G=1)$ does not satisfy the indicated transformation law and, secondly, $P(c)$ is multiplied by the following factor instead of unity

$$\left[\frac{1}{2}\int_{-0}^{c}(c-c')P(c')dc' + \int_{c}^{1+0}(c''-c)P(c'')dc''\right]$$

Formula (2.34) follows from the model according to which molecular mixing takes place upon contact of two moles with concentrations c' and c''. Under similar contact new moles are formed with concentration c satisfying condition $c' < c < c''$. In particular, if we make the simple assumption that a mole with only one concentration equal to $c = 1/2(c' + c'')$ is formed, i.e., $G = 2\delta[c - 1/2(c' + c'')]$, then we obtain the well known Curl system [1963]

$$\frac{\partial}{\partial c}\langle\nabla(D\nabla c)\rangle_c P \sim -\frac{q}{L}\left\{4\int_{-0}^{c}P(c-c')P(c+c')dc' - P(c)\right\}$$

[†]In their subsequent article, Kollman and Janicka [1982] use (2.34) only in the turbulent fluid. The complete expression for the term describing mixing to the molecular scales also include additional terms accounting for the interaction between turbulent and nonturbulent fluids.

More complex expressions for function G are analyzed by Janicka, Kolbe, and Kollman [1979], Dopazo [1979], Nedorub, Frost, and Shcherbina [1979], Kollman and Janicka [1982], Shcherbina [1982].

The main drawback of a closure of type (2.34) is that they are not based on experimental data. The main goal when constructing these closure relations is to find equations which would permit numerical integration with the help of standard methods (the capabilities of this approach are demonstrated in the works of Pope [1981a], Nedorub, Frost, and Shcherbina [1979], Lockwood and Shah [1982], Kollman and Janicka [1982]).

The description of mixing at the molecular level is not the only problem encountered when analyzing single-point concentration PDF. Another important problem is the description of turbulent diffusion (the second term on the right-hand side of (2.14) or (2.15)). In the equation for $P(c)$ this term is usually written by analogy with the semiempirical theory of turbulent diffusion (Kuznetsov and Frost [1973], Frost [1973, 1977], Pope [1976], Nedorub, Frost, and Shcherbina [1979], Nedorub and Shcherbina [1979], Shcherbina [1982] and a number of other works), i.e.,

$$\int u P(u,c) d^3 u = \langle u \rangle P - D_t \nabla P \tag{2.35}$$

Here D_t is the coefficient of turbulent diffusion.

It is easily seen that formula (2.35) supposes the equality of the transport coefficients for the intermittency, probabilities γ_0, γ_1, and all the moments of the concentration field to the coefficient of diffusion D_t introduced into (2.35). This conclusion, generally speaking, does not agree with the known semiempirical theories. In them, as is generally known, for better correlation of theoretical and experimental data different values of the transport coefficient for $\gamma_0, \gamma_1, \langle c \rangle$, and $\sigma^2 = \langle (c - \langle c \rangle)^2 \rangle$ are used (see, for example, Meshcheryakov [1974], Rodi [1980], Meshcheryakov and Sabel'nikov [1984a, b]).

It should be emphasized that despite the indicated shortcoming, closure (2.35) in a number of cases can apparently be quite satisfactory. Thus, for example, in "black-white" mixing when $\langle c^k \rangle = \langle c \rangle = \gamma_1, k > 0$ (see §1.3), transport factors of all moments coincide exactly. Therefore, it can be assumed that relation (2.35) is applicable when the probability of observing intermediate values of the concentration is small, i.e., at small values of intermittency factor.

The description of turbulent diffusion becomes significantly simpler if the equations for the joint PDF of velocity and concentration are used. A similar approach was developed by Kuznetsov [1976b], Dopazo [1976], Onufriev [1977], Sabel'nikov [1979, 1981, 1983], Pope [1981b], Libby and Bray [1981] (see also the review by Libby and Williams [1981]). In this case, there is no longer any need to introduce some hypotheses concerning the character of turbulent diffusion. However, two new difficulties arise. The first is associated with the multidimensional character of the equation for the joint PDF of velocity and concentration. The second arises when describing pressure fluctuations. As a result of these difficulties, the approach under consideration has not yet led to any specific results.

Let us now consider the closure hypotheses of the equation for velocity PDFs. The equation for single-point velocity PDF was considered by Lundgren [1969], Ievlev [1970, 1975], Onufriev [1977], Sabel'nikov [1982a, d]. In the work by Lundgren, a realization expression was used to describe the pressure forces. This expression agrees with the Krook model in the kinetic theory of gases (see, for example, Cercignani [1975]). Analogous expressions were used by Onufriev [1977]. To approximate viscous forces in the work by Lundgren [1969], it was assumed that the two-point velocity PDF is normal (hypothesis (2.32) considered above is of similar character). The assumption that the n- point velocity PDF differs little from a normal distribution was also used by Ievlev [1975]. In the works by Lundgren and Ievlev the terms describing the processes of molecular transport are proportional to the first derivatives of the PDF with respect to the velocity. In the case of the equation for concentration PDF, such a description corresponds to the representation of the equation in form (2.14).

Unlike the preceding works, Sabel'nikov [1982a, d] used an equation in which the viscous terms are proportional to the second derivatives of the PDF which corresponds to the form adopted in §2.2. The indicated method of representation of the equations for PDF was proposed by Kuznetsov [1967], Ulinich and Lyubimov [1968]. In the work by Sabel'nikov [1982a], the effect of pressure fluctuations was approximated with the aid of a diffusion approximation which is frequently used in problems of the kinetic theory of gases (see, for example, Cercignani [1975]). The resulting equation is similar to the Fokker-Planck equation.

The equations for two-point velocity PDFs were considered by Kuznetsov [1967, 1976a], Lundgren [1975], Ievlev [1970, 1975], Sosinovich [1973, 1974, 1981a, b]. In Lundgren's work [1975] the three-point velocity PDF is expressed in terms of two-point and one-point functions with the aid of a hypothesis to the effect that the three-point correlation function in the group representation of the three-point PDF coincides with the three-point Gaussian correlation function (the indicated closure is readily extended to the case of any n-point probability density, $n > 3$). Such a closure leads to an integrodifferential equation of a considerably complex structure. It is significantly simplified when applied to the inertial range and is reduced to a closed equation for the structure function.

The principal idea of the method proposed by Ievlev [1970] involves a special formulation of the functional form for the conditionally averaged moments which enter into the equation for any n-point PDF ($n \geq 2$). The number of the unknown functions in these approximate expressions coincides with the number of conditions that follow from all of the limit properties of the n-point PDFs (tendency to zero of the semi-invariants for unlimited divergence of the points under consideration is, for example, such a condition). The closure method by Ievlev was used by Alekseev, Ievlev, and Kiselev [1976], Kiselev [1977] in analyzing the decay of homogeneous turbulence in Burgers model, and also in the problem of homogeneous and isotropic turbulence in incompressible fluid. In the works by Kuo and O'Brien [1981] the Ievlev method was used for the description of two-point concentration PDF in chemical turbulence (i.e., stochastic fluctuations of the concentrations in a stationary reacting medium).

In the works of Sosinovich [1973, 1974] the derivation of a closed equation for the two-point velocity PDF is based on nonequilibrium statistical mechanics and quantum field theory (these methods are introduced into the theory by Kraichnan [1959]). The equation thus derived contains quadratic terms which cause insurmountable difficulties when seeking an exact solution. For this reason, in a subsequent work by Sosinovich [1981a, b], simpler closure hypotheses were used leading to a closed equation for the structure function.

There are no empirical constants in the hypotheses proposed in the works under consideration. However, the expressions which are derived from them are considerably cumbersome and, hence, little investigated. The final possibilities of describing turbulence within the frameworks of the derived equations are still unclear.

The considered closure hypotheses have a lot in common. Firstly, the assumption of proximity of the n-point PDFs to the Gaussian functions is often used explicity or implicitly. Experimental data indicate that the assumption is generally untrue (see §1.1). This is mostly pertinent to multipoint PDFs when the distance between the points is small in comparison with the scale of turbulence (see Figs. 1.9 – 1.13). The main reason for deviations from the normal law, as already mentioned in Chapter 1, is the existence of internal and external intermittencies. Secondly, in the majority of the indicated works there is no link with the fundamental theory of locally homogeneous and isotropic turbulence developed by Kolmogorov [1941, 1962a, b], Obukhov [1941, 1962], and, independently, by Onsager [1945, 1949], and Weizsacker [1948]. This particularly pertains to the ideas of universal equilibrium and statistical independence of small- and large-scale motions in a developed turbulent flow which are most fully and clearly expounded in the book by Batchelor [1953]. Furthermore, the end result in some works is to obtain the properties under study by deduction from formally closed equations. Thus, for example, Lundgren [1975] obtained the "two thirds" Kolmogorov-Obukhov law with a numerical factor that is very close to the experimental value. However, bearing in mind that the two-point velocity PDF in the inertial range differs significantly from the normal density, the result by Lundgren does not, admittedly, appear very convincing.

Without denying the value of this direction of investigation, it must be noted that, presently, another approach is apparently more fruitful, whereby the existence of universal equilibrium is assumed primordial. Thereby, the problem is reduced to the investigation and utilization of the properties of universal equilibrium. In view of this, we emphasize that the sequence of equations for the n-point PDFs have, generally speaking, not only one solution but a wide class of solutions, and in order to identify the solutions of physical interest some limitations must be imposed beforehand on the required function.

An analogy can be drawn here with the solution of the problem of equilibrium ensembles in statistical physics. It is generally known (see, for example, Chapter 4 in the book by Balescu [1975]) that it is impossible to solve uniquely this problem within the framework of mechanics. Only the introduction of a statistical assumption — the so-called principle of equal *a priori* probabilities — enables the construction of an equilibrium distribution.

Taking into account these considerations, an attempt is made in the present book to obtain equations for the PDFs of various characteristics of turbulence proceeding from the assumption that there exists a universal statistical equilibrium between large-scale and small-scale pulsations. In other words, the adopted closure relations below are essentially based on the theory of locally homogeneous turbulence which was developed by Kolmogorov [1941, 1962a, b] and Obukhov [1941, 1949, 1962]. This approach is proposed by Kuznetsov [1967, 1972a, 1976a], Ulinich and Lyubimov [1968], Lyubimov and Ulinich [1970].

As is generally known, energy dissipation and scalar dissipation are of fundamental importance in the theory of locally homogeneous turbulence. Therefore, henceforth, great attention will be given to the equations for the PDFs of concentration and velocity difference expressed in the form of (2.15) and (2.31), respectively. From these equations it follows that redistribution of PDF in phase space is of diffusive character, and the scalar dissipation and energy dissipation taken with an inverse sign serve as the coefficients of diffusion. These coefficients are negative, which implies, as it will be shown later, many very unusual properties of the obtained equations.

Thus, the conditionally averaged values of the scalar dissipations $\langle N \rangle_{t,c}$ and the energy dissipation $e_{ij}(v)$ dominate the theory being developed. It is important to emphasize the following: It is not clear from the start that functions $\langle N \rangle_{t,c}$ and $e_{ij}(v)$ possess greater simplicity and universality than the PDFs themselves. The investigation in the following chapters based on the theory of locally homogeneous turbulence, and also some (not many, so far) experimental data indicates that such simplicity and universality do, apparently, exist. Otherwise, the accurate open equations for PDFs would allow the expression of some unknown functions in terms of the others.

In the investigations carried out in Chapters 3 and 4, the condition of non-negativity of the solutions of the equations for PDFs is of no small importance. This condition is not usually specially analyzed. It should be emphasized that the considered condition, from the mathematical point of view, is far from trivial. It substantially narrows the class of probable closures of the equation for probability density (for a known structure of the accurate open equation) by imposing specific limitations on the functional form of the closure relations. Unfortunately, there is no general theory to date which might indicate the form of these limitations.

It should be noted that although equations (2.15) and (2.31) are not closed, they present an accurate connection between the directly measured characteristics of turbulence. This allows, at least in principle, the improvement of the closure as more experimental data are gathered. It appears that the refinement of the theory should take this course in the future.

CHAPTER THREE
PASSIVE CONTAMINANT CONCENTRATION PDF

In this chapter the equation for the concentration PDF of a dynamically passive contaminant is considered. Just as in §1.3, the letter z is used to denote this concentration. Here, the hypotheses used for the closure of this equation are discussed in detail. Analyzed also are the solutions of the closed equation in the case of statistically homogeneous concentration field in free turbulent flows. Three main goals are pursued in the chapter. The first is purely practical and is intended to yield a simple approximate method for determining the concentration PDF and intermittency factor in jets. This problem is solved, as far as possible, without complex mathematical operations. The second goal is to investigate the mathematical properties of the equation for concentration PDF, formulate the boundary value-problem, and show that additional constraints from previously unknown functions, entering via the closing relations, follow from the condition of solvability, since it follows that the developed approach reduces the number of arbitrary functions compared with the usual semiempirical theories for one-point moments. It is not ruled out that the new paths for constructing the closed theory of turbulence will be connected with improving this approach. The third goal is to study the structure of isoscalar surfaces in turbulent flows. Such an investigation enables, firstly, the proposal of an additional method of obtaining boundary conditions for the concentration PDF and identification of their physical meaning and, secondly, the study of the interaction between intermittency and the structure of isoscalar surfaces.

The investigation of concentration PDF assumes that all fluid dynamic characteristics of the flow are specified, i.e., the field of mean velocities and intermittency factor. In the statistically homogeneous case, when the mean concentration is constant, these characteristics are sufficient for the solution of the problem. In turbulent jets, since the mean concentration is unknown, quantity $\langle z \rangle$ must also be added to the parameters which must be specified. The solution of practical problems indicates that it is convenient to change somewhat the statement of the problem. The methods of calculation of the intermittency factor are presently at the initial stage of development, whereas the mean concentration (or, equally, mass flow) can be sufficiently reliably computed from semiempirical models of turbulence. Therefore, it is expedient to assume that the mean concentration is known, and the required function is the intermittency factor. It should be emphasized that the justification for changing the statement of the problem is based only on considerations of convenience, since the fluid dynamic parameter γ is not, generally speaking, connected with the concentration field (see §1.3). The changed statement is fully applicable in jets or wakes, where, as noted in §1.3, the intermittency factors that are determined from the dynamic and scalar fields are practically identical.

The content of the chapter is based on works by Kuznetsov [1972a, 1977b], Sabel'nikov [1980a, 1982b, 1985b], and Kuznetsov and Sabel'nikov [1981b].

§3.1 EQUATION FOR THE CONDITIONAL CONCENTRATION PDF IN TURBULENT FLUIDS. BOUNDARY CONDITIONS

It follows from expression (1.20) that along with smooth functions, generalized functions — the discontinuous Heaviside function, the delta function and its derivative — enter the equation for the unconditional concentration PDF (2.15). This situation suggests that relations of three different types can be obtained from equation (2.15). It is shown below that this is indeed so and one can identify from equation (2.15), first, a separate equation for the conditional concentration PDF, second, equations for probabilities γ_0 and γ_1 and, third, boundary conditions for the conditional PDF when $z = 0$ and $z = 1$. It should be recalled that $z = 0$ and $z = 1$ characterize the range of concentration inside the turbulent fluid, i.e., they are the boundaries of phase space and correspond to the values z in the nonturbulent fluid when $\text{Re} \to \infty$.

The method by which these equations and boundary conditions are obtained was proposed by Kuznetsov [1972a]. In this work it was suggested that the conditionally averaged scalar dissipation $\langle N \rangle_{t,z}$ is independent of z, i.e., $\langle N \rangle_{t,z} = \langle N \rangle_t$ (for more information on this hypothesis see §3.2). However, the applicability of the indicated method is not limited by this case alone.

In accordance with the work of Kuznetsov [1972a], let us multiply equation (2.15), into which expression (1.20) for the unconditional PDF $P(z)$ is substituted, by an arbitrary smooth function of the concentration $\psi(z)$ and integrate for z from $-\infty$ to ∞. Making use of the well-known rules of operations with generalized functions (see, for example, Gel'fand and Shilov [1959]), we obtain

$$\int_0^1 \psi \left[\frac{\partial \gamma F}{\partial t} + \nabla \langle u \rangle_{t,z} \gamma F + \gamma \frac{\partial^2 \langle N \rangle_{t,z} F}{\partial z^2} \right] dz$$

$$+ \left\{ \frac{\partial \gamma_0}{\partial t} + \nabla \langle u \rangle_{n,0} \gamma_0 + \gamma \left[\frac{\partial \langle N \rangle_{t,z} F}{\partial z} \right]_{z=0} \right\} \psi(0)$$

$$+ \left\{ \frac{\partial \gamma_1}{\partial t} + \nabla \langle u \rangle_{n,1} \gamma_1 - \gamma \left[\frac{\partial \langle N \rangle_{t,z} F}{\partial z} \right]_{z=1} \right\} \psi(1)$$

$$-\psi'(0)\gamma[\langle N \rangle_{t,z} F]_{z=0} + \psi'(1)\gamma[\langle N \rangle_{t,z} F]_{z=1} = 0 \qquad (3.1)$$

Here, the prime denotes differentiation with respect to z. We now take into consideration that function $\psi(z)$ is arbitrary. Hence, the expression under the integral sign, and also all the functions ahead of $\psi(0)$, $\psi(1)$, $\psi'(0)$, and $\psi'(1)$ in (3.1) must, each taken separately, be identically equal to zero. As a result, we arrive at the following relations

$$\frac{\partial \gamma F}{\partial t} + \nabla \langle u \rangle_{t,z} \gamma F = -\gamma \frac{\partial^2 \langle N \rangle_{t,z} F}{\partial z^2} \tag{3.2}$$

$$\frac{\partial \gamma_0}{\partial t} + \nabla \langle u \rangle_{n,0} \gamma_0 = -\gamma \left[\frac{\partial \langle N \rangle_{t,z} F}{\partial z} \right]_{z=0} \tag{3.3}$$

$$\frac{\partial \gamma_1}{\partial t} + \nabla \langle u \rangle_{n,1} \gamma_1 = \gamma \left[\frac{\partial \langle N \rangle_{t,z} F}{\partial z} \right]_{z=1} \tag{3.4}$$

$$F = 0, \quad z = 0 \quad \text{and} \quad z = 1 \tag{3.5}$$

It is pertinent to note that when obtaining the boundary conditions (3.5), the assumption that $\langle N \rangle_{t,z} \neq 0$, $z = 0$, $z = 1$ is implicitly used. This assumption is fully justified for the asymptotic theory under consideration when $Re \to \infty$. The results obtained in §1.3 generally assert that $\langle N \rangle_{t,z} \neq 0$, $0 \leq z \leq 1$. Indeed, function $\langle N \rangle_{t,z}$ is the main term of the asymptotic expansion of the conditionally averaged scalar dissipation $\langle N \rangle_z$ when $Re \to \infty$ and, thereby, is independent of Reynolds number. Therefore, the tendency of $\langle N \rangle_{t,z}$ to zero would contradict the conclusion made in §1.3 to the effect that the law according to which $\langle N \rangle_{t,z}$ tends to zero at the boundaries of the phase space is fundamentally dependent on Reynolds number (in formulas (1.29) and (1.34) quantities α and k are dependent on Reynolds number).

Let us elucidate the physical meaning of equations (3.3), (3.4) for probabilities γ_0 and γ_1 and the equations for the intermittency factor which is obtained either by integrating equation (3.2) with respect to all z, or from (3.3) and (3.4) following the use of the relation $\gamma = 1 - \gamma_0 - \gamma_1$

$$\frac{\partial \gamma}{\partial t} + \nabla \langle u \rangle_t \gamma = \gamma \left[\langle N \rangle_{t,z=0} \frac{\partial F(0)}{\partial z} - \langle N \rangle_{t,z=1} \frac{\partial F(1)}{\partial z} \right] \tag{3.6}$$

When transforming the terms on the right side of (3.6), the boundary conditions (3.5) are used. From the condition of non-negativity of the PDF and the boundary conditions (3.5), it follows that $\partial F(0)/\partial z > 0$ and $\partial F(1)/\partial z < 0$. Thereby, the right side of (3.6) is always positive. Hence, equation (3.6) describes the increase of the relative volume of the turbulent fluid as a result of entrainment of the nonturbulent fluid. Analogous reasoning shows that equations (3.3) and (3.4) describe the decrease in the relative volume of the nonturbulent fluid with $z = 0$ and $z = 1$, respectively. With the theory being developed, the rate of change of the volumes of the turbulent and nonturbulent fluids, described by the right sides of equations (3.3), (3.4), and (3.6), are found to be proportional to the flows of concentration PDF at the boundaries of the phase space.

Let us now consider the boundary conditions (3.5). First of all, it must be noted that they are not unexpected, if one takes into account the character of the oscillograms of the instantaneous temperature profiles in the jet obtained by Uberoi and Singh [1975] (see Fig. 1.1). These oscillograms were analyzed earlier in §1.3. Remember that practically discontinuous changes in temperature are clearly seen at the jet boundary. The temperature gradient, therefore, tends to infinity at the boundary. Hence, if the geometrical interpretation of PDF (see formula (1.23)) is used, we arrive at (3.5).

We make one more remark concerning the physical meaning of the boundary layers (3.5). With that end in view, we rewrite formula (1.27) in the form of estimate

$$\gamma P_t(z) \approx \lim_{D \to 0} \frac{d\langle S_z \rangle}{dV} \frac{\sqrt{D}}{\sqrt{\langle N \rangle_t}} \qquad (3.7)$$

When deriving (3.7) it was assumed that $\langle |\partial z/\partial n|^{-1} \rangle_{t,z} \sim \sqrt{D/\langle N \rangle_t}$, i.e., the concentration fluctuations were ignored (this is justified when the moments of the gradient are not excessive: see, for example, Monin and Yaglom [1967]).

Relation (3.7) indicates that $\lim_{\text{Re} \to \infty} d\langle S_z \rangle \text{Re}^{-1/2}$ is finite when $0 < z < 1$. Thereby, the dependence of the area of an element of the isoscalar surface $z = \text{const} \neq 0$ or 1 on Reynolds number has the form $d\langle S_z \rangle \sim \sqrt{\text{Re}}$. If $z = 0$ or $z = 1$, then it follows from (3.5) and (3.7) that for these values $\lim_{\text{Re} \to \infty} d\langle S_z \rangle \text{Re}^{-1/2} = 0$ and that $d\langle S_z \rangle \sim \text{Re}^k$, $0 < k < 1/2$, $z = 0$, $z = 1$. Consequently, when $\text{Re} \to \infty$ the area and, thereby, the degree of distortion of the bounding isoscalar surfaces $z = 0$ and $z = 1$ is substantially weaker than any internal isoscalar surface $0 < z = \text{const} < 1$ located in the turbulent fluid. Thus, the character of an isoscalar surface in turbulent fluids enables the drawing of a fully specific analogy with intermittency, and more precisely, with its qualitative description given in §1.1. Indeed, if the oscillations of bounding isoscalar surfaces are equated to external intermittency, and the oscillations of the internal isoscalar surfaces to the internal intermittency, then one obtains a graphic physical model of the turbulent fluid which is close to the "sponge" pattern proposed in §1.1. This model of turbulent fluid also very clearly illustrates the essence of the averaging procedure which determines the boundary of the turbulent fluid in §1.1, thereby identifying external intermittency. It should be kept in mind that this boundary was selected from the condition of maximum smoothness. The relation of the estimate of isoscalar surface distortion with the indicated condition is fairly obvious.

The discussion of the structure of isoscalar surfaces in turbulent flow will be continued in §3.8. The more rigorous investigation of the problem there confirms the conclusions formulated here.

§3.2 CLOSURE HYPOTHESIS FOR THE COMPONENT DESCRIBING THE MIXING TO THE MOLECULAR LEVEL

Let us turn to the case of statistically homogeneous concentration field (see §3.4). In the equation for PDF, apart from the nonstationary term, only the analyzed component remains. Hence, the qualitative solution in this case is fully determined by the term which describes the process of molecular mixing. The given example clearly indicates that the closure hypothesis with respect to function $\langle N \rangle_{t,z}$ (or $\langle \nabla(D\nabla z) \rangle_{t,z}$) must be thoroughly substantiated, particularly from the physical point of view.

There are no known detailed measurements of $\langle N \rangle_{t,z}$ and $\langle \nabla(D\nabla z) \rangle_{t,z}$. Therefore, general considerations of turbulent flows at high Reynolds numbers become very significant. In this respect, representation of the equation in the form of (2.15), which contains $\langle N \rangle_{t,z}$, is advantageous. This conclusion is based on the fact that in the analysis of the properties of the quantity $\langle N \rangle_{t,z}$, it is natural to use the theory of locally homogeneous and isotropic turbulence by Kolmogorov [1941, 1962a, b] and Obukhov [1941, 1949, 1962]. This theory, as was already discussed in Chapter 1, is based on the cascade mechanism of energy transfer over the turbulent spectrum. As a consequence of this mechanism, small-scale fluctuating motion receives energy from large-scale motion as a result of repeated disintegrations of vortex formations. At very high Reynolds numbers the number of these disintegrations is large. Hence, the natural assumption that the characteristics of energy carrying vortices and the vortices with small dimensions are statistically independent. This statement was formulated by Weizsäcker [1948] and Onsager [1945, 1949] as a hypothesis to the effect that in a homogeneous turbulence the Fourier components with strongly differing wave numbers are statistically independent. A detailed and physically clear statement of this hypothesis is given by Batchelor [1953].

Let us use the indicated hypothesis in the case under consideration. It is evident that the field of concentration z is a macrocharacteristic, and the field of scalar dissipation N is a microcharacteristic of turbulence (henceforth, macrocharacteristic is understood to mean the quantity which is defined by large vortices, and microcharacteristic is a quantity defined by small vortices). According to this hypothesis, these fields are statistically independent at high Reynolds numbers, i.e., function $\langle N \rangle_{t,z}$ is independent of z. Thus, when $Re \to \infty$, we have

$$\langle N \rangle_{t,z} = \langle N \rangle_t \tag{3.8}$$

In view of the significance of hypothesis (3.8), we once again identify the limitations which need to be imposed for its use. First of all, hypothesis (3.8) is used only in the turbulent fluid. In view of this, we note that in the work of Kuznetsov [1967], and Pope [1976] the hypothesis of statistical independence of fields z and N was used without accounting for intermittency, i.e., it was assumed that $\gamma = 1$ and $\langle N \rangle_z = \langle N \rangle$. Pope [1976] and Dopazo [1979] showed that the application of this hypothesis leads to an equation whose solution is characterized by a number of physically improbable properties. Based on this, they came to the conclusion that the hypothesis for statistical independence is inapplicable for approximating the term describing mixing to the molecular level, and also is inexpedient for writing the equation for PDF in the form of (2.15) (a similar viewpoint is widespread in the work of overseas scientists; see, for example, the review by O'Brien [1980a, b]). But it was shown even earlier in the work by Kuznetsov [1972a] that taking intermittency into account, when formulating the hypothesis of statistical independence, automatically eliminates all the difficulties noted by Pope and Dopazo.

In view of its importance for further consideration, the following circumstances must be specially emphasized here. The problems encountered by Pope and Dopazo were caused by the neglect of intermittency and not by the assumption that $\langle N \rangle_z$ is independent of z. This assertion becomes clear if we take into account that the following relationship exists between functions $\langle N \rangle_{t,z}$ and $\langle N \rangle_z$

$$\langle N \rangle_{t,z} = \langle N \rangle_z, \qquad 0 < z < 1 \tag{3.9}$$

and, thereby, hypothesis (3.8) is equally applicable to function $\langle N \rangle_z$. To prove equality (3.9), we consider that the contaminant concentration can serve as an indication of turbulence (see §1.1). Therefore, the conditions that $0 < z = \text{const} < 1$ and that the fluid is turbulent are equivalent. Hence, the conditional means $\langle N \rangle_{t,z}$ and $\langle N \rangle_z$ coincide when $0 < z < 1$. In a nonturbulent fluid $N = 0$, and thus $\langle N \rangle_z = 0$ when $z = 0$ and $z = 1$, i.e., quantity $\langle N \rangle_z$, unlike $\langle N \rangle_{t,z}$ undergoes a discontinuity at the boundaries of the phase space.

It should now be noted that application of hypothesis (3.8) is limited to the cases of a chemically inert contaminant and is untrue for a reacting contaminant. This conclusion is a consequence of the fact that chemical conversions, as noted in the introduction to the book, are concentrated in very narrow zones. Therefore, the statistical properties of small scale concentration fluctuations of a reacting contaminant (or temperature) are determined not by cascade disintegration of vortices of various scales, but by chemical reactions. As a result the conditionally averaged scalar dissipation is found to be explicitly dependent on concentration, i.e., fields c and $N_c = D(\nabla c)^2$ correlate each other. As an example, let us consider diffusion combustion. In this case, as indicated in §2.1 (the proof is presented in Chapter 5), the concentration of all reacting substances c (the subscript is not written for the sake of brevity) are expressed through the concentration of the chemically inert contaminant, i.e., $c = c(z)$. Hence, for the scalar concentration dissipation c we have

$$\langle N_c \rangle_{t,c} = \left(\frac{\partial c}{\partial z} \right)^2 \langle N \rangle_{t,z}, \qquad \frac{\partial c}{\partial z} \neq \text{const}$$

Hence, it is evident that when hypothesis (3.8) is fulfilled, function $\langle N_c \rangle_{t,c}$ is dependent on c.

Let us consider application of the hypothesis concerning the statistical independence of macro- and microcharacteristics for the dynamic problem, i.e., for the field of velocity u and energy dissipation ϵ. The significance of this case is that it enables simpler experimental testing of the hypothesis under consideration. The assumption that $\langle \epsilon \rangle_u = \langle \epsilon \rangle$ was introduced by Kuznetsov [1967], Ulinich and Lyubimov [1968]. Intermittency was taken into account later in the work by Sabel'nikov [1979], in which it was assumed that $\langle \epsilon \rangle_{t,u} = \langle \epsilon \rangle_t$. Since dissipation in nonturbulent fluid is equal to zero and velocity fluctuations are different from zero, the conditional mean $\langle \epsilon \rangle_{t,u}$ and $\langle \epsilon \rangle_u$, in contrast to the conditional mean scalar dissipation $\langle N \rangle_{t,z}$ and $\langle N \rangle_z$ under consideration, are not equal. It is not difficult to see that they are connected with the inequality

$$\langle \epsilon \rangle_{t,u} \geq \langle \epsilon \rangle_u \tag{3.10}$$

From general considerations it is evident that the difference between $\langle \epsilon \rangle_{t,u}$ and $\langle \epsilon \rangle_u$ decreases as the amplitude of velocity fluctuations increases, and in the region of large fluctuations they practically coincide, since such fluctuations take place mainly in the turbulent fluid.

Let us now consider the extent to which the hypothesis on the statistical independence of macro- and microcharacteristics in turbulent fluid is experimentally confirmed. Testing of the hypothesis was carried out in the experiments by Kuznetsov, Rashchupkin [1977], Praskovskii [1982, 1983], Kuznetsov, Praskovskii, and Sabel'nikov [1984a, b]. Measurements were carried out in all these tests in the wake of a circular cylinder. In the first of these works the conditionally averaged squares of the derivatives of concentration and longitudinal velocity, i.e., quantities $\langle (\partial z/\partial x_1)^2 \rangle_z$ and $\langle (\partial u_1/\partial x_1)^2 \rangle_{u_1}$, were measured in the plane of symmetry of the wake, i.e., when $\gamma \approx 1$. In the remaining tests quantities $\langle (\partial u_1/\partial x_1)^2 \rangle_{u_1}$ and $\langle (\partial u_1/\partial x_1)^2 \rangle_{t,u_1}$ were measured at various points of the cross section including the region where intermittency is substantial.

The experimental data of Kuznetsov and Rashchupkin [1977] are presented in Fig. 3.1, and of Kuznetsov, Praskovskii, and Sabel'nikov [1984a, b] in Fig. 3.2. It

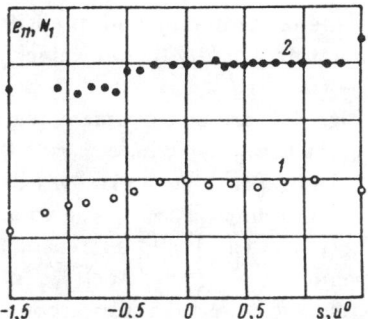

Figure 3.1 Conditionally averaged values of the squared derivatives of concentration and velocity on the plane of symmetry of the wake of a circular cylinder from data by Kuznetsov and Rashchupkin [1977]. $x_1/d = 52.5$, $x_2/d = 0$, $\text{Re}_d = 8 \cdot 10^3$, $d = 8$ mm, $u_0 = 15$ m/sec; 1) $e_{11} = \langle (\partial u_1/\partial x_1)^2 \rangle_{u_1}$, 2) $N_1 = \langle (\partial z/\partial x_1)^2 \rangle_z$, $s = (z - \langle z \rangle)/\sigma$, $u^0 = (u_1 - \langle u_1 \rangle)/\langle (u_1 - \langle u_1 \rangle)^2 \rangle^{1/2}$. Units of measurements along the ordinate axis are arbitrary.

64 TURBULENCE AND COMBUSTION

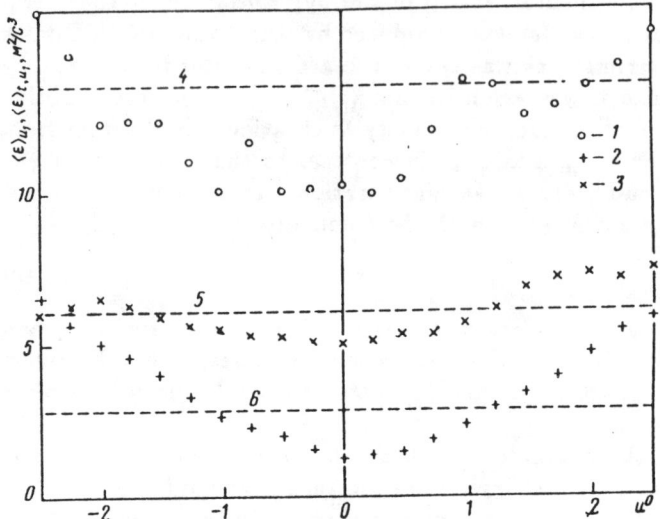

Figure 3.2 Conditionally averaged values of the squared velocity derivative at different distances from the plane of symmetry of the wake of a circular cylinder from data by Kuznetsov, Praskovskii, and Sabel'nikov [1984a, b]. $x_1/d = 38.4$, $Re_d = u_0 d/\nu = 2.5 \cdot 10^4$, $d = 36$ mm, $u_0 = 10.2$ m/sec; 1) $\langle\epsilon\rangle_{u_1}$, $x_2 = 0$, $\gamma \approx 1$, 2) $\langle\epsilon\rangle_{u_1}$, $x_2/d = 3.89$, $\gamma = 0.42$, 3) $\langle\epsilon\rangle_{t,u_1}$, $x_2/d = 3.89$, $\gamma = 0.42$, 4) $\langle\epsilon\rangle$, $x_2 = 0$, 5) $\langle\epsilon\rangle_t$, $x_2/d = 3.89$, 6) $\langle\epsilon\rangle$, $x_2/d = 3.89$, $\epsilon = 15\nu(\partial u_1/\partial x_1)^2$, $u^0 = (u_1 - \langle u_1\rangle)/\langle(u_1 - \langle u_1\rangle)^2\rangle^{1/2}$.

is evident that the conditional mean $\langle(\partial z/\partial x_1)^2\rangle_z$, $\langle(\partial u_1/\partial x_1)^2\rangle_{u_1}$ in the tests of Kuznetsov and Rashchupkin is practically constant with the variation of the level, at which averaging takes place, in the investigated range of concentration and velocity fluctuation amplitudes $|z - \langle z\rangle| \leq 1.5\sigma (\sigma = \langle(z - \langle z\rangle)^2\rangle^{1/2})$, $|u_1 - \langle u_1\rangle| \leq 1.5\langle(u_1 - \langle u_1\rangle)^2\rangle^{1/2}$. In the tests by Kuznetsov, Praskovskii, and Sabel'nikov the variation of quantity $\langle(\partial u_1/\partial x_1)^2\rangle_{u_1}$, in the plane of symmetry and quantity $\langle(\partial u_1/\partial x_1)^2\rangle_{t,u_1}$ at the point where $\gamma = 0.42$ does not exceed 20% in the velocity fluctuations amplitude zone $|u_1 - \langle u_1\rangle| \leq 2.5\langle(u_1 - \langle u_1\rangle)^2\rangle^{1/2}$.

The results of the measurements, as seen in Fig. 3.2, are in agreement with the inequality (3.10). The value of $\langle(\partial u_1/\partial x_1)^2\rangle_{u_1}$ at the point where $\gamma = 0.42$ is always less than the value of $\langle(\partial u_1/\partial x_1)^2\rangle_{t,u_1}$ and rises monotonically with increasing absolute value of the amplitude of the velocity fluctuations. The value of $\langle(\partial u_1/\partial x_1)^2\rangle_{u_1}$ in the range $|u_1 - \langle u_1\rangle| \leq 2.5\langle(u_1 - \langle u_1\rangle)^2\rangle^{1/2}$ increases 5 – 6 times. The monotonic character of the dependence of $\langle(\partial u_1/\partial x_1)^2\rangle_{u_1}$ on the amplitude of velocity fluctuations is easily explained if we take into account that with increasing amplitude of the velocity fluctuations, contribution from the regions with large values of energy dissipation to the conditional mean under consideration increases.

Summing up the results of these experiments, it can be concluded that they, firstly, agree quite satisfactorily with the hypothesis as to the statistical independence of macro- and microcharacteristics in the turbulent fluid and, secondly, confirm the essential importance of intermittency in the formulation of that hypothesis.

The hypothesis of the statistical independence has a wide range of applicability, since it enables "splitting" the large number of correlations between macro- and microcharacteristics of turbulence when Re $\to \infty$ (i.e., to represent the mean value of the product in terms of the product of the means). Let us examine this question following mainly the work of Kuznetsov [1979a].

Let us represent the gradients of velocity and concentration in the form

$$\frac{\partial z}{\partial x_i} = \frac{q}{L}\left[\sqrt{\mathrm{Re}}\,\varphi_i^{(1)} + \varphi_i^{(2)}\right], \quad \left\langle \left[\varphi_i^{(1)}\right]^2 \right\rangle \sim 1$$

$$\frac{\partial u_i}{\partial x_j} = \frac{q}{L}\left[\sqrt{\mathrm{Re}}\,u_{ij}^{(1)} + u_{ij}^{(2)}\right], \quad \left\langle \left[u_{ij}^{(1)}\right]^2 \right\rangle \sim 1$$

$$\lim_{\mathrm{Re}\to\infty} \left\langle \left[\varphi_i^{(2)}\right]^2 \right\rangle_t \mathrm{Re}^{-1} = 0$$

$$\lim_{\mathrm{Re}\to\infty} \left\langle \left[u_{ij}^{(2)}\right]^2 \right\rangle_t \mathrm{Re}^{-1} = 0 \tag{3.11}$$

The values $\langle N \rangle$ and $\langle \epsilon \rangle$ in formulas (3.11) are assumed to tend to finite limits when Re $\to \infty$ (see §1.1) (note that expansions (3.11) are analogous to the representation of the concentration field z in §1.3 in the form of (1.38)). Functions $\varphi_i^{(1)}$ and $u_{ij}^{(1)}$ (multiplied by $\sqrt{\mathrm{Re}}$) characterize the contribution of small eddies to the gradients of concentration and velocity, and functions $\varphi_i^{(2)}$ and $u_{ij}^{(2)}$ characterize the contribution of all intermediate and large eddies. Functions $\varphi_i^{(1)}$ and $u_{ij}^{(1)}$, generally speaking, are dependent on Reynolds number, but the root-mean-square values of these functions are by definition independent of Reynolds number and attain values of the order of unity. The higher order moments of $\varphi_i^{(1)}$ and $u_{ij}^{(1)}$ are dependent on Reynolds numbers owing to internal intermittency. However, this dependence can be neglected providing the order of the moments is not too large (see, for example, Monin and Yaglom [1967]).

Further discussions will consider only the concentration field, since they can be applied to the velocity field practically unchanged. Let us consider the correlation $\langle f\Phi \rangle$ of two arbitrary functions f and Φ. Let one of them (f) be dependent only on z, and the other (Φ) only on ∇z, i.e., $f = f(z)$, and $\Phi = \Phi(\nabla z)$. Let us analyze the limit $\lim_{\mathrm{Re}\to\infty} \langle f\Phi \rangle$. In such a limit intermittency arises and, according to the Bayes formula, the following relation is true

$$\lim_{\mathrm{Re}\to\infty} \langle f\Phi \rangle = \gamma \langle f\Phi \rangle_t + (1-\gamma)\langle f\Phi \rangle_n \tag{3.12}$$

It is evident from (3.11) that the main contribution to the concentration gradient is made by the component containing the function $\varphi_i^{(1)}$. According to the hypothesis, the values $\lim_{\mathrm{Re}\to\infty} \varphi_i^{(1)}$ and $\lim_{\mathrm{Re}\to\infty} z$ are statistically independent in the turbulent fluid. Therefore, correlation $\langle f\Phi \rangle_t$ can be "split", i.e., written as

$$\langle f\Phi \rangle_t = \langle f \rangle_t \langle \Phi \rangle_t \qquad (3.13)$$

Similarly, we obtain

$$\lim_{\text{Re} \to \infty} \langle f\Phi \rangle = \gamma \langle f \rangle_t \langle \Phi \rangle_t + (1-\gamma)\langle f\Phi \rangle_n \qquad (3.14)$$

If $\Phi = 0$ in the nonturbulent fluid, as happens for the quantity N, then the second component on the right side of relation (3.14) does not exist.

It should be emphasized that the hypothesis of statistical independence enables finding the main terms in the asymptotic expansions of the correlations of the micro- and macrocharacteristics in the turbulent fluid when $\text{Re} \to \infty$. When they are equal to zero, to find the first nonzero terms in the expansions of these correlations, one must take into consideration the next terms in the expansions of the concentration and velocity gradients, i.e., functions $\varphi_i^{(2)}$ and $u_{ij}^{(2)}$ in (3.11). These functions, as noted earlier, are determined by the intermediate scales. Therefore, for example, the fields z and $\varphi_i^{(2)}$ cannot be regarded as statistically independent.

There are typical cases of correlation of macro- and microcharacteristics which do not give a result when the hypothesis under consideration is applied to them. These, for example, are the following

1) $\langle f(z)(\nabla z)^{2k+1} \rangle, \qquad k = 0, 1, 2 \ldots$

2) $\left\langle (u_j - \langle u_j \rangle) \dfrac{\partial (u_i - \langle u_i \rangle)}{\partial x_j} \right\rangle$

3) $\left\langle (u_i - \langle u_i \rangle) \dfrac{\partial p}{\partial x_j} \right\rangle$

4) $\langle (u_i - \langle u_i \rangle)\epsilon \rangle$

Upon formal use of the hypothesis we notice that all these correlations are equal to zero. Let us take as an example the second of these correlations. In accordance with expansion (3.11), we have

$$\left\langle (u_j - \langle u_j \rangle) \frac{\partial (u_i - \langle u_i \rangle)}{\partial x_j} \right\rangle \sim \sqrt{\text{Re}} \langle u_{ij}^{(1)} (u_j - \langle u_j \rangle) \rangle$$

where the factor ahead of $\sqrt{\text{Re}}$ is equal to zero by virtue of the hypothesis of statistical independence of macro- and microcharacteristics of turbulence. On the other hand, if use is made of the continuity equation, the analyzed correlation can be reduced to the following form: $\partial \langle (u_i - \langle u_i \rangle)(u_j - \langle u_j \rangle) \rangle / \partial x_j$, i.e., to a three-dimensional derivative of the Reynolds stresses. This derivative in homogeneous flows, as is well known, is different from zero and is of the order q^2/L. Analogous considerations are true also for the other three correlations. Note that, correlations 2 - 4 are of importance in semiempirical theories of turbulence (see, for example, Reynolds [1976], Ginevskii et al. [1978], the book *Prediction Methods for Turbulent Flows* edited by W. Kollman [1980]).

Another type of correlation can also be identified for which the hypothesis of statistical independence of macro- and microcharacteristics does not permit finding the first zero term of the asymptotic expansion when Re $\to \infty$. These correlations contain derivatives of velocity and concentration of an order higher than the first, for example, $\langle u_1 \Delta u_j \rangle$ and $\langle z \Delta z \rangle$; here Δ is Laplace operator. It is this circumstance which makes it difficult to use the theory of locally homogeneous and isotropic turbulence when approximating the conditionally averaged divergence of the diffusive flow $\langle \nabla (D \nabla z) \rangle_{t,z}$ appearing in the equation for the concentration PDF written in the form (2.14).

In conclusion, let us note the following very important aspect. The hypothesis of statistical independence of macro- and microcharacteristics in the turbulent fluid is applicable only for developed and, in a certain sense, "equilibrium" turbulence. In other words, it is necessary that a moving equilibrium should exist between the small and large scales of the motion. This condition, as is generally known, is also necessary for the validity of the theory of locally homogeneous and isotropic turbulence. Hence, if the fields of velocity and concentration at the initial moment of time are in an arbitrary state, then some relaxation time must elapse before the hypothesis can be applied. From dimensional considerations it follows that the relaxation time is of the order L/q. It is appropriate here to consider an analogy with the evolution of the turbulence spectrum from an arbitrary initial state. If use is made of the semiempirical theories of energy transfer over the turbulence spectrum (see, for example, Monin and Yaglom [1967]), it can be shown that the equilibrium range in the energy spectrum also emerges after a time of the order L/q. The remark made is very important in determining the boundaries of applicability of the semiempirical equation for the concentration PDF. Further discussion of this question is included in §3.4.

§3.3 CLOSURE HYPOTHESIS FOR THE CONVECTIVE COMPONENT

The exact representation of integral (2.16) appearing in the convective component in the equation for the concentration PDF as is shown in §2.1, contains a large number of unknown functions. It should be remembered that expression (2.16) includes the conditionally averaged velocity in the nonturbulent fluid $\langle u \rangle_{n,0}$ and $\langle u \rangle_{n,1}$ and the conditionally averaged velocity in the turbulent fluid $\langle u \rangle_{t,z}$. The same reasoning used in the preceding section when deriving equality $\langle N \rangle_{t,z} = \langle N \rangle_z$, $0 < z < 1$, shows that quantities $\langle u \rangle_{t,z}$, $\langle u \rangle_{n,0}$, $\langle u \rangle_{n,1}$, and the conditionally averaged velocity $\langle u \rangle_z$ are connected by the following relations

$$\langle u \rangle_{t,z} = \langle u \rangle_z, \qquad 0 < z < 1$$

$$\langle u \rangle_{n,0} = \langle u \rangle_{z=0}, \qquad \langle u \rangle_{n,1} = \langle u \rangle_{z=1} \qquad (3.15)$$

Note that there is no base whatsoever to assume that

$$\langle u \rangle_{n,0} = \langle u \rangle_{t,z=0}, \qquad \langle u \rangle_{n,1} = \langle u \rangle_{t,z=1}$$

Therefore, function $\langle u \rangle_z$, in the general case, has discontinuities at $z = 0$ and $z = 1$.

It is not yet possible to give a sufficiently sound description of the behavior of function $\langle u \rangle_z$ over the entire interval $0 \leq z \leq 1$. This is mainly due to two reasons. Firstly, extreme shortages in experimental data and, secondly, large errors when measuring $\langle u \rangle_z$ in the region of large amplitudes of concentration fluctuations that are caused by both the effects of molecular transfer (see §1.3) and by the large statistical errors of such rare events as large fluctuation amplitudes. The enumerated problems are particularly serious in flow regions where intermittency is substantial.

Thus, application of the exact expression (2.16) is not as yet possible. Simplifying assumptions must be introduced. In order to gauge the character of the simplifications, we start with the analysis of the simplest case when intermittency is insignificant, i.e., $\gamma \approx 1$, $\gamma_0 \approx 0$, $\gamma_1 \approx 0$. A situation close to this case is observed in the central regions of jet flows. General considerations based on experimental data indicate that PDFs in the case under study differ little from the normal distributions. If it is assumed that the joint PDF of velocity and concentration $P(u, z)$ is described by a normal distribution, then for the conditionally averaged velocity $\langle u \rangle_z$ simple calculations yield the linear dependence

$$\langle u \rangle_z = \langle u \rangle + q\sigma^{-2}(z - \langle z \rangle) \tag{3.16}$$

Here, $q = \langle (u - \langle u \rangle)(z - \langle z \rangle) \rangle$ is the vector of mass flow; $\sigma^2 = \langle (z - \langle z \rangle)^2 \rangle$ is the variance of concentration pulsations.

The assumption that the conditionally averaged velocity $\langle u \rangle_z$ is described by a linear dependence for all values of z, irrespective of whether $P(u, z)$ is a Gaussian curve or not, was made in the work by Kuznetsov [1972a]. The coefficients in this dependence are found as follows. Let the following relation be assumed true

$$\langle u \rangle_z = A + B(z - \langle z \rangle)$$

where vectors A and B are dependent only on time. They can be connected with the fields of the mean velocity and concentration. For this purpose we multiply the relation by $(z - \langle z \rangle)^k P(z)$, $k = 0, 1$ and integrate for all values of z. As a result, we find $A = \langle u \rangle$, $k = 0$; $B = q\sigma^2$, $k = 1$. Thereby, we again arrive at relation (3.16).

Let us compare the linear dependence (3.16) with the results of measurements of the conditionally averaged velocity $\langle u \rangle_z$ near the axis of a submerged axisymmetric jet (Bezuglov [1974], Venkataramani, Tutu, and Chevray [1975], Golovanov [1977], Shcherbina [1982], and in the flow of heated grid-induced turbulence with constant transverse mean temperature gradient (Venkataramani and Chevray [1978], Tavoularis and Corrsin [1981a]. Figure 3.3 depicts the conditionally averaged longitudinal velocity $\langle u_1 \rangle_z$ obtained in the tests by Bezuglov and Golovanov, and Fig. 3.4 and 3.5 depict the conditionally averaged transverse velocity $\langle u_2 \rangle_z$ obtained by Shcherbina and Venkataramani and Chevray, respectively.

From Figs. 3.3 – 3.5, it is evident that the conditionally averaged velocities $\langle u_k \rangle_z$, $k = 1, 2$ are described well by the linear functions $|z - \langle z \rangle| \lesssim 2\sigma$. With large concentration fluctuation amplitudes a deviation from the linear dependence is observed. The character of these deviations is discussed conveniently later.

Summing up, it can be concluded that in a flow region where intermittency is insignificant, the linear dependence is a good approximation for the conditionally averaged velocity $\langle u \rangle_z$. This conclusion, generally speaking, would not have large impact had it not been for the following two conditions. First, it appears that the linear dependence can serve as a fairly good first approximation for the description of $\langle u \rangle_z$ and, hence, in the flow regions where intermittency is substantial the joint PDF of velocity and concentration differs strongly from the normal distribution. Secondly, the discontinuities of function $\langle u \rangle_z$ at the boundaries of the phase space $z = 0$ and $z = 1$ is small, and it can be neglected as a first approximation.

The first assertion is supported by experiments by Sreenivasan and Antonia [1978] and the second is confirmed indirectly*.

Let us consider first the tests by Sreenivasan and Antonia in which $\langle u_1 \rangle_z$ and $\langle u_2 \rangle_z$ were measured at the edge of the flow in axisymmetric coflowing jet at a point where $\gamma = 0.28$. The results of these measurements are given in Fig. 3.6. It is evident that the results of the experiments can indeed be approximated satisfactorily by straight lines. It is clear that as a result of the discontinuities of the conditionally averaged velocity $\langle u \rangle_z$ at points $z = 0$ and $z = 1$, the coefficients in the equations of the straight lines for $\langle u_k \rangle_z (= \langle u_k \rangle_{t,z}, 0 < z < 1)$ are not necessarily equal to $\langle u_k \rangle$ and $q_k \sigma^{-2}$. In particular, if the technique proposed by Kuznetsov [1972a] is used for the determination of the coefficients, we obtain

$$\langle u_k \rangle_{t,z} = \langle u_k \rangle_t + q_{tk} \sigma_t^{-2}(z - \langle z \rangle_t)$$

$$q_{tk} = \langle (u_k - \langle u_k \rangle_t)(z - \langle z \rangle_t)\rangle_t, \quad \sigma_t^2 = \langle (z - \langle z \rangle_t)^2 \rangle_t$$

$$\langle u_k \rangle_z = \langle u_k \rangle_{t,z}, \quad 0 < z < 1$$

The quantitative estimate of the discontinuities of function $\langle u_k \rangle_z$ at the boundaries $z = 0$ and $z = 1$ can be obtained indirectly. With this in mind, we assume that the linear dependence (3.16) is true over the entire range $0 \leq z \leq 1$, and analyze one of the links between the moments which follow from (3.16). This link

---

*An attempt at direct estimate of the value of the discontinuities of function $\langle u \rangle_z$ at the boundary of the phase space was made in recent experiments by Shcherbina and Mogilko [1985]. In these tests, function $\langle u_1 \rangle_z$ was measured near the core of axisymmetric jet at a distance of $x/d = 3 - 4$ from the nozzle exit section by an optical method. It is established that in the vicinity of the boundary of phase space $z = 1$ a fairly sharp change of $\langle u_1 \rangle_z$ takes place which is interpreted by the authors as a discontinuity "smeared" mainly as a result of noise in the measuring device. The value of the discontinuity is computed by means of an approximate technique which is based on a number of assumptions as to the form of the joint PDF of velocity and concentration and PDF of noise in the device. The results indicate that the magnitude of the discontinuities is small. In view of the approximate character of the method of analyzing of the measurement results, it is still early to make a more categoric assertion. It is hoped that further development in the indicated direction will clarify this question.

70 TURBULENCE AND COMBUSTION

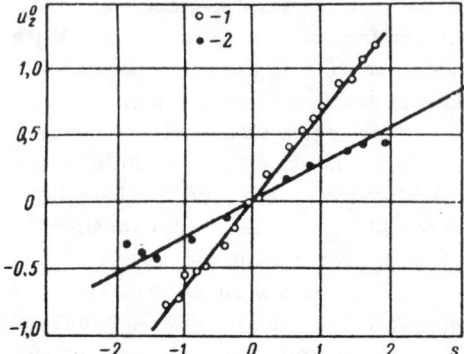

Figure 3.3 Conditionally averaged longitudinal velocity at a specified concentration on the axis and close to the axis of a submerged axisymmetric jet from data by Bezuglov [1974] and Golovanov [1977]. *1)* measurements by Bezuglov, test conditions: $x_1/d = 48$, $x_2/d = 0$, $Re_d = u_0 d/\nu = 3.24 \cdot 10^3$, $d = 3$ mm, *2)* measurements by Golovanov, test conditions: $x_1/d = 30$, $x_2/x_1 = 0.088$, $Re_d = 7.46 \cdot 10^3$, $d = 4$ mm, $u_0 = 28$ m/sec, $u_z^0 = (\langle u_1 \rangle_z - \langle u_1 \rangle)/\sqrt{\langle (u_1 - \langle u_1 \rangle)^2 \rangle}$, $s = (z - \langle z \rangle)/\sigma$.

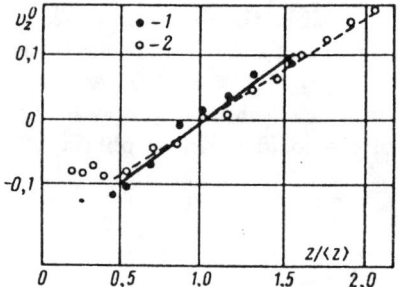

Figure 3.4 Conditionally averaged transverse velocity at a given concentration near the axis of submerged axisymmetric jet from data by Shcherbina [1982]. $x_1/d = 20$, $u_0 = 50$ m/sec, $d = 5$ mm, *1)* $x_2/x_1 = 0.025$, $\sigma/\langle z \rangle = 0.24$, *2)* $x_2/x_1 = 0.1$, $\sigma/\langle z \rangle = 0.545$, $v_z^0 = \langle u_2 \rangle_z / \sqrt{\langle (u_2 - \langle u_2 \rangle)^2 \rangle}$.

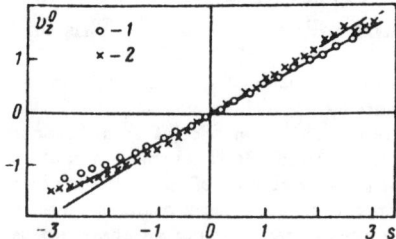

Figure 3.5 Conditionally averaged transverse velocity at a given concentration in grid-induced turbulent flow from data by Venkataramani and Chevray [1976]. $Re_\lambda = 64.4$, $Re_M = u_0 M/\nu = 2 \cdot 10^4$, *1)* $x_1/M = 49.2$, $x_2/M = 0$, *2)* $x_1/M = 55.2$, $x_2/M = 0$, $v_z^0 = \langle u_2 \rangle_z / \sqrt{\langle (u_2 - \langle u_2 \rangle)^2 \rangle}$, $s = (z - \langle z \rangle)/\sigma$, M is the size of grid cell.

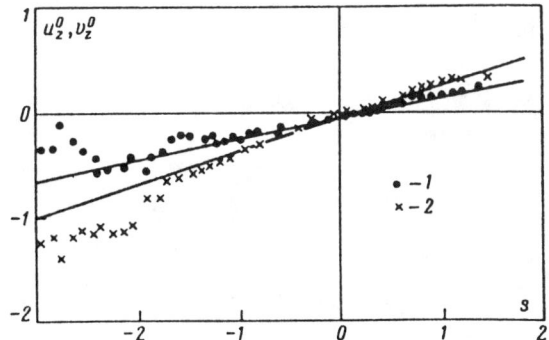

Figure 3.6 Conditionally averaged longitudinal and transverse velocity for a specified concentration at a large distance from the axis of symmetry of coflowing axisymmetric jet from data by Sreenivasan and Antonia [1978]. $x_1/d = 59$, $Re_d = u_0 d/\nu = 4.27 \cdot 10^4$, $d = 2$ cm, $u_0 = 32$ m/sec, $u_\infty = 4.8$ m/sec; measurements conducted at a point where intermittency factor $\gamma = 0.28$; 1) $u_z^0 = (\langle u_1 \rangle_z - \langle u_1 \rangle)/\sqrt{\langle (u_1 - \langle u_1 \rangle)^2 \rangle}$, 2) $v_z^0 = (\langle u_2 \rangle_z - \langle u_2 \rangle)/\sqrt{\langle (u_2 - \langle u_2 \rangle)^2 \rangle}$, $s = (z - \langle z \rangle)/\sigma$.

is found if (3.16) is multiplied by $(z - \langle z \rangle)^2 \times P(z)$ and integrated for all z. As a result, we obtain

$$\langle (u - \langle u \rangle)(z - \langle z \rangle)^2 \rangle = q\sigma^{-2}\langle (z - \langle z \rangle)^3 \rangle$$

This relation connects the flow of concentration fluctuations (the left side of the equality) and the asymmetry of concentration fluctuations $A = \langle (z - \langle z \rangle)^3 \rangle/\sigma^3$. Comparison of the formula with the experimental data in the wake of a circular cylinder (Freymuth and Uberoi [1971], La Rue and Libby [1974]) and in the axisymmetric coflowing jet (Antonia, Prabhu, and Stephenson [1975]) are given in Fig. 3.7 as a dependence of the dimensionless transverse component of the flow vector of the concentration fluctuations $Q_\sigma = \langle (u_2 - \langle u_2 \rangle)(z - \langle z \rangle)^2 \rangle/q_2 \sigma$ on the factor of asymmetry.

The skewness has negative values near the axis or plane of symmetry, and large positive values of A are reached in the regions where intermittency is substantial. Figure 3.7 shows that the experimental data are satisfactorily described by dependence $Q_\sigma = A$, which follows from (3.16), including the region where intermittency is significant.

This result serves as additional indirect support of (3.16) but, furthermore, it indicates (also indirectly) that there is no need to re-evaluate the strength of the discontinuities of the function $\langle u \rangle_z$ and, as a first approximation, it can apparently be simply ignored.

Summing up, we briefly reiterate the two main conclusions of this analysis. The first is that a simple function — a linear dependence (3.16) — is a very good first approximation for $\langle u \rangle_z$ in all regions of the free turbulent flow. The second conclusion is that the discontinuities of function $\langle u \rangle_z$ can be neglected. The significance of the second conclusion is that it permits the representation

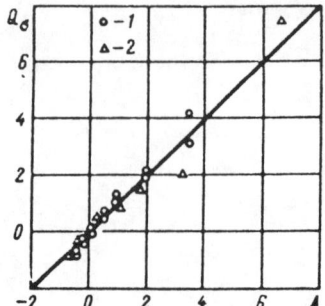

Figure 3.7 The relation of flow of concentration fluctuations in the transverse direction with the factor of asymmetry of concentration fluctuations from data by various authors. *1)* wake of a circular cylinder from data by Freymuth and Uberoi [1971] (test conditions presented in Fig. 3.16) and La Rue and Libby [1974] (test conditions presented in Fig. 1.14); *2)* axisymmetric jet in coflowing stream from data by Antonia, Prabhu, and Stephenson [1975] (test conditions presented in Fig. 1.5); the line corresponds to $Q_\sigma = A$ which follows from (3.16).

of the integral entering the convective component (2.16) of equation (2.15) in a simplified form

$$\int u P(u,z) d^3 u = \langle u \rangle_z P(z) \tag{3.17}$$

This form will be used further. It should be emphasized once more that expression (3.17) is only approximate. It is valid only in the case when $\langle u \rangle_z$ is continuous in the range $0 \leq z \leq 1$.

Relation (3.16) was proposed by Kuznetsov [1972a] (without reference to the conditions when it is valid).

The linear dependence (3.16) for $\langle u \rangle_z$, due to its simplicity, is most appropriate for the solution of the first of the problems indicated at the beginning of the chapter — the development of an approximate description of concentration PDF and intermittency factor in turbulent flows. Investigations conducted in §3.5 indicate that using (3.16) permits on the whole the handling of this problem. However, it is established in §3.5 that the linear dependence also has certain shortcomings. These shortcomings are that PDF fluctuates at large concentration fluctuation amplitudes, i.e., in that region where the results of measurements of $\langle u \rangle_z$ are the least reliable as a result of large errors. By anticipating developments, we note that it follows from the results described in §3.6 that in order to eliminate the tendency for fluctuations, a smaller increase in the absolute value of $\langle u \rangle_z$ is required, for large values of the concentration fluctuation amplitude, than the increase prescribed by the linear function (3.16).

Thus, we arrive at the conclusion that when conducting a rigorous investigation of the concentration PDF, it is necessary to account for the deviation of $\langle u \rangle_z$ from the linear dependence (3.16) in the region of large concentration fluctuations.

The presently existing experimental data do not allow a simple conclusion on the character of the difference under consideration for reasons indicated above (although some tendencies are examined in Figs. 3.3 – 3.5; see below). Therefore, we seek help in general qualitative arguments with the hope that they would enable the identification of some principal features of function $\langle u \rangle_z$. First of all, some information on this function follows naturally from mixing length theory. Let us consider a flow with constant transverse gradient of the mean concentration $\langle z \rangle - \langle z(0) \rangle = K x_2$ (let, for the sake of definiteness, $K > 0$). It is evident that at point $x_2 = 0$, the positive values of concentration fluctuations take place mainly as a result of fluid particles (moles) which arrive from the region $x_2 > 0$. The transverse velocity of such moles is negative. It can also be assumed that the larger the observed value of concentration fluctuation, the farther the region the indicated mole arrives from and, consequently, the larger the absolute value of its transverse velocity. Hence, it can be concluded that under positive concentration fluctuation first, $\langle u_2 \rangle_z < 0$ and, second, $|\langle u_2 \rangle_z|$ increases monotonically with increasing z. The case of negative concentration fluctuations and also the case $K < 0$ may be considered in a similar manner.

The experimental data presented in Figs. 3.3 – 3.6 agree well with the conclusion on the monotonic character of the dependence of the conditionally averaged velocity $\langle u \rangle_z$.

We proceed with the discussion of the flow with uniform mean concentration gradient. Based on the described qualitative picture of such a flow, it is natural to conclude that the moles causing the maximum concentration values at point $x_2 = 0$ are not only the most rapid, but also have the largest spatial scales (of the order of the distance from point $x_2 = 0$ to the point the rapid mole arrives from). Hence, transfer of maximum (and also minimum) concentrations is effected by large vortex formations whose velocities, as is generally known (Townsend [1956]), are limited.

In view of this, we bear in mind the known results obtained in the theory of turbulent diffusion with finite velocity (see a review of this theory in the book by Monin and Yaglom [1965], and also the article by Zimont and Sabel'nikov [1975]). The main inference of this theory is that the limit velocity of concentration transfer is of the order of the root-mean-square of the velocity fluctuation.

In summing up, a hypothesis can be put forward according to which the conditionally averaged velocity $\langle u \rangle_z$ tends to the finite limits $|z - \langle z \rangle|/\sigma \to \infty$. There are presently no experimental data which could have enabled direct verification of this hypothesis. It can only be examined so far by indirect means on the basis that the conclusions reached from an investigation of the concentration PDF are not inconsistent.

It is pertinent to note that although the experimental data in Figs. 3.3 – 3.6 do not yield the means for direct verification of the hypothesis, they do not on the whole contradict the inference to a lower increase of the absolute value of $\langle u \rangle_z$ than by the linear law. This is so, since in the experiments by Golovanov [1977], Venkataramani and Chevray [1976] (see Fig. 3.3 and 3.5) the tendency towards a reduced rate of increase of the conditionally averaged velocity is observed in the region $|z - \langle z \rangle| \gtrsim 1.5\sigma$.

This analysis confirms that the deviations of function $\langle u \rangle_z$ from the linear dependence are significant only at considerably large amplitudes of concentration fluctuations. For this reason, the fine details of the behavior of $\langle u \rangle_z$ in this region will have very little effect on the concentration PDF in the main range of its variation. This state makes the task of qualitative formulation of the deviation of function $\langle u \rangle_z$ from the linear dependence somewhat easier.

We shall later use the following approximation which reflects the principal qualitative feature of function $\langle u \rangle_z$ (Sabel'nikov [1982c])

$$\langle u \rangle_z = \langle u \rangle + q\sigma^{-1}\mu^{-1}[V_0 + s(1 + \omega^2 s^2)^{-1/2}], \quad s = (z - \langle z \rangle)/\sigma \quad (3.18)$$

Here, V_0, μ and ω are coordinate functions. Function $\langle u \rangle_z$, which is described by expression (3.18), is linear at small amplitudes of concentration fluctuations and tends to finite limits when $|s| \to \infty$. Deviations from the linear dependence begin to emerge when $|s| \sim 1/\omega$.

There are two relations between functions V_0, μ, and ω, since the following conditions must be fulfilled identically

$$\int \langle u \rangle_z P dz = \langle u \rangle, \qquad \int (z - \langle z \rangle)(\langle u \rangle_z - \langle u \rangle) P dz = q$$

As indicated in the beginning of the current chapter, the distribution of mean velocity and mass flow when considering the concentration PDF are assumed to be known. Substituting formula (3.18) for $\langle u \rangle_z$ into this condition and taking into account formula (1.20) for the unconditional PDF $P(z)$, we find the following relations

$$V_0 = -\langle s(1 + \omega^2 s^2)^{-1/2} \rangle = -\gamma_0 s_0 (1 + \omega^2 s_0^2)^{-1/2}$$

$$- \gamma_1 s_1 (1 + \omega^2 s_1^2)^{-1/2} - \gamma \int_0^1 s(1 + \omega^2 s^2)^{-1/2} F dz$$

$$\mu = \langle s^2(1 + \omega^2 s^2)^{-1/2} \rangle = \gamma_0 s_0^2 (1 + \omega^2 s_0^2)^{-1/2}$$

$$+ \gamma_1 s_1^2 (1 + \omega^2 s_1^2)^{-1/2} + \gamma \int_0^1 s^2 (1 + \omega^2 s^2)^{-1/2} F dz$$

$$s_0 = -\langle z \rangle/\sigma, \quad s_1 = (1 - \langle z \rangle)/\sigma \quad (3.19)$$

When $\omega = 0$, we obtain $V_0 = 0$, $\mu = 1$ from (3.19), and approximation (3.18) is reduced to the linear dependence (3.16). Hence, function ω reflects the above described effect of the characteristics of large-scale concentration transfer. The following general statement can be made regarding the value of this function. Since the deviation $\langle u \rangle_z$ with the amplitude of concentration fluctuations is significant only at large values of concentration fluctuation amplitudes, it is to be expected that the value of ω will be sufficiently small. This conclusion is subjected to rigorous analysis which is conducted in §§3.6 and 3.7.

One more comment should be made on the verification of the linear dependence (3.16) for $\langle u \rangle_z$ at large amplitudes of concentration fluctuations. The conclusion derived in the present section regarding the need for such verification upon rigorous analysis of the equation for the concentration PDF remains also if the assumption made here of the absence of discontinuities for function $\langle u \rangle_z$ on the boundaries of space phase is rejected.

§3.4 CONCENTRATION PDF IN HOMOGENEOUS TURBULENCE

The case of statistically homogeneous concentration field is, perhaps, one of the simplest problems accessible to complete theoretical analysis within the bounds of equations (3.2) – (3.4). At the same time, in addition to being independently interesting, this example illustrates some of the general properties of the equation for PDF.

First of all, we note that the statistically homogeneous concentration field is an idealization of the process arising in a number of mechanical devices, for example, in the mixing chamber when some substance is added to a channel of constant cross section with the aid of a large quantity of jets at a velocity different from the velocity of the coflowing stream. Usually, the effects of the turbulent mixing in the mixing layers are so great that the flow in the chamber quickly acquires a statistically homogeneous character (i.e., the distributions of the mean parameters are equalized across the channel, and only fluctuations of velocity, concentration, and so on remain). It is useful to consider the limiting situation when the concentration of the contaminant in the initial cross section of the channel randomly attains only two values $z = 0$ and $z = 1$ ($z = 1$ in the jets, $z = 0$ outside the jets), and the initial turbulence is negligibly small. The qualitative picture of the process when $x > x_0$ (x_0 is the coordinate of the initial section) is sufficiently clear. With increasing difference $x - x_0$ at the boundaries of certain volumes in which $z = 0$ or $z = 1$ mixing layers, where velocity turbulent fluctuations arise as a result of velocity shear, form and develop and intermediate concentration values are observed. This process leads to monotonic decrease in concentration fluctuations and monotonic increase in the total volume occupied by the mixing layers. When $x - x_0 \to \infty$, complete mixing occurs to the molecular level, i.e., concentration fluctuations tend to zero and concentration acquires a value equal to the initial mean concentration $\langle z \rangle$ (it is assumed that there is no additional supply of fluid from the walls of the channel into the chamber when $x > x_0$).

This process corresponds to nonstationary statistically homogeneous concentration fields if it is assumed that $t = (x - x_0)/U$, where U is the mean flow velocity in the channel across the section. The concentration PDF in accordance with the above statement at the initial and final moments of time has the form

$$P(z) = \gamma_0 \delta(z) + \gamma_1 \delta(z-1)$$

$$\gamma = 0, \quad \gamma_1 = \langle z \rangle, \quad \gamma_0 = 1 - \langle z \rangle \quad \text{at} \quad t = 0 \tag{3.20}$$

$$P(z) = \delta(z - \langle z \rangle), \quad \gamma = 1, \quad \gamma_1 = \gamma_0 = 0 \quad \text{at} \quad t = \infty \tag{3.21}$$

(strictly speaking, γ must be written here instead of γ_z, but for the sake of brevity subscript z will be henceforth omitted).

Let us explain now what the limit $t \to \infty$ is understood to mean here. When $t \to \infty$, turbulence degenerates (if there is not a source of support) and, thereby, Reynolds number tends to zero. Therefore, the hypothesis of the statistical independence of micro- and macrocharacteristics of turbulence at large values of evolution time will be unacceptable. Therefore, a double limit is assumed here and later when Re $\to \infty$, and then $t \to \infty$.

These tendencies in the behavior of the mean concentration and variance follow immediately from the first two equations for the moments

$$\frac{d\langle z \rangle}{dt} = 0, \quad \frac{d\sigma^2}{dt} = -2\langle N \rangle = -2\gamma \langle N \rangle_t \tag{3.22}$$

The system of equations (3.2) – (3.4) in the statistically homogeneous case acquires the form

$$\frac{\partial \gamma F}{\partial t} = -\gamma \langle N \rangle_t \frac{\partial^2 F}{\partial z^2} \tag{3.23}$$

$$\frac{\partial \gamma_0}{\partial t} = -\gamma \langle N \rangle_t \frac{\partial F(0)}{\partial z} \tag{3.24}$$

$$\frac{\partial \gamma_1}{\partial t} = \gamma \langle N \rangle_t \frac{\partial F(1)}{\partial z} \tag{3.25}$$

Relation (3.23) is the heat equation with a negative coefficient of diffusion. It, thereby, belongs to the inverse parabolic type of equations for which, as generally known (see, for example, Tikhonov and Arsenin [1974]), the global solution of the Cauchy equation (i.e., on the semi-axis $t \geq 0$) does not exist for all initial conditions. Whenever the existence of the solution is established, the Cauchy problem is incorrect. Incorrectness is manifested in the fact that the decrease in the wavelength of the perturbations at the initial conditions leads to an increase in the rate of their rise. As a result, the solution is rapidly distorted.

There is a theorem (Friedman [1964]) on the uniqueness of the solution of the Cauchy problem for the inverse-parabolic equation provided that it exists.

Point $t = \infty$ (since $\langle N \rangle_t \to 0$, $t \to \infty$) is a singular point for equation (3.23). But the indicated theorem is applicable also in the case under consideration, since this singularity can be eliminated with the replacement of variable $\tau = \int_t^\infty \langle N \rangle_t dt$. The integral here converges by virtue of the limitation of the variance of the concentration fluctuations. Indeed, from the equation for σ^2 in (3.22), it follows that $\tau \to 1/2\sigma^2$, $t \to \infty$ (it is assumed here that $\gamma \to 1$, $t \to \infty$).

In order to find the global solution of equations (3.23) – (3.25), which is unique by virtue of the indicated theorem, we consider the inverse Cauchy problem with "initial" conditions (3.21) that are specified at $t = \infty$ instead of the direct Cauchy problem with the initial conditions (3.20) at $t = 0$. When integrating equation (3.23) in the reverse direction, the Cauchy problem becomes proper. The solution corresponding to the "initial" conditions at $t = \infty$ defines the class of initial conditions for which the solution of the direct Cauchy problem exists for all $t \geq 0$. It will be shown below that conditions (3.20) are just such initial conditions under some limitations on function $\langle N \rangle_t$. It is important to emphasize that since $N > 0$, the stated considerations are of a general character and are not linked with the hypothesis of §3.2 on the statistical independence of fields N and z in the turbulent fluid.

The solution of the formulated inverse Cauchy problem is found by the standard Fourier method. The final result has the form (Kuznetsov, Levedev, Sekundov, and Smirnova [1981], Kuznetsov and Sabel'nikov [1981b])

$$\gamma F = 2 \sum_{k=1}^{\infty} \sin \pi k \langle z \rangle \sin \pi k z \exp(-\pi^2 k^2 \tau)$$

$$\gamma_0 = 1 - \langle z \rangle - \frac{2}{\pi} \sum_{k=1}^{\infty} \frac{\sin \pi k \langle z \rangle}{k} \exp(-\pi^2 k^2 \tau)$$

$$\gamma_1 = \langle z \rangle + \frac{2}{\pi} \sum_{k=1}^{\infty} (-1)^k \frac{\sin \pi k \langle z \rangle}{k} \exp(-\pi^2 k^2 \tau)$$

$$\gamma = 1 - \gamma_0 - \gamma_1, \qquad \tau = \int_t^{\infty} \langle N \rangle_t dt \qquad (3.26)$$

Let us analyze two boundary cases: 1) the final stage of degeneration of concentration fluctuations, i.e., $t \to \infty$, and 2) initial stage of degeneration, i.e., $t \to 0$.

In the first case, we have $\gamma \approx 1$, $P \approx P_t$ (it is shown later that these approximate qualities are valid with an accuracy equal to experimental errors, $\tau \to 1/2\sigma^2 \to 0$, and, hence, the series in (3.26) converge extremely slowly. Therefore, the main term of the asymptotic behavior of the solution in the case under study is easier to find directly from equations (3.23) – (3.25). Let us turn first to the equation for PDF (3.23). Since $\lim_{t \to \infty} (\sigma/\langle z \rangle) = 0$, then when determining the term of the asymptotic expansion of PDF valid outside the small neighborhood of the boundary points $z = 0$ and $z = 1$, the boundary conditions (3.5) can be set at $z = -\infty$ and $z = +\infty$. If it is now assumed that the "initial" conditions (3.21) and the constraint $\tau \to 1/2\sigma^2$, $\tau \to 0$, then it is not difficult to see that the first term of the asymptotic behavior is described by a Gaussian curve (Kuznetsov [1967])

$$P(z) \to \frac{1}{\sqrt{2\pi}\sigma} \exp\left[-\frac{(z-\langle z\rangle)^2}{2\sigma^2}\right], \qquad t \to \infty \tag{3.27}$$

This result agrees well with the experimental data by Miyawaki, Tsujikawa, and Uraguchi [1974].

The first terms of the asymptotic behavior of the PDF which are applicable to the vicinity of $z = 0$ and $z = 1$, is found by the known mapping method (see, for example, Mors and Feshbach [1953]). As a result, we obtain

$$P(z) \to \frac{1}{\sqrt{2\pi}\sigma} \left\{ \exp\left[-\frac{(z-\langle z\rangle)^2}{2\sigma^2}\right] - \exp\left[-\frac{(z+\langle z\rangle)^2}{2\sigma^2}\right] \right\}, \; t \to \infty \tag{3.28}$$

in the vicinity of $z = 0$ and

$$P(z) \to \frac{1}{\sqrt{2\pi}\sigma} \left\{ \exp\left[-\frac{(z-\langle z\rangle)^2}{2\sigma^2}\right] - \exp\left[-\frac{(z-2+\langle z\rangle)^2}{2\sigma^2}\right] \right\}$$

$$t \to \infty \tag{3.29}$$

in the vicinity of $z = 1$.

Substituting expressions (3.28), (3.29) in equations (3.24), (3.25), we find the first terms of the asymptotic behavior of probabilities γ_0 and γ_1

$$\gamma_0 \to \sqrt{\frac{2}{\pi}} \frac{\sigma}{\langle z\rangle} \exp\left(-\frac{\langle z\rangle^2}{2\sigma^2}\right)$$

$$\gamma_1 \to \sqrt{\frac{2}{\pi}} \frac{\sigma}{1-\langle z\rangle} \exp\left[-\frac{(1-\langle z\rangle)^2}{2\sigma^2}\right], \qquad t \to \infty \tag{3.30}$$

Hence, when $t \to \infty$, probabilities γ_0 and γ_1 and, thereby, the difference $1 - \gamma$ are exponentially small.

Let us now find the main term of the asymptotic behavior of the solutions of equations (3.23) – (3.25) in another limiting case $t \to 0$. From the general solution of (3.26) it follows that this asymptotic behavior is determined by the character of the dependence of scalar dissipation $\langle N\rangle_t$ on time as $t \to 0$. In order to establish this dependence, let us examine the physical picture of the phenomenon. As indicated at the beginning of this section, when $t > 0$ on the boundaries of the regions with $z = 0$ and $z = 1$, mixing layers arise in which intermediate concentration values appear as a result of molecular diffusion. Scalar dissipation is naturally dependent on the difference of concentrations Δz and on the difference of velocities Δu on the boundaries of the mixing layer, and also on their characteristic width ℓ. From considerations of dimensionality we obtain $\langle N\rangle_t \sim \Delta u (\Delta z)^2/\ell$. We assume that Δu and Δz have finite limits when $t \to 0$ (in particular, $\Delta z \to 1$, $t \to 0$), and the characteristic width ℓ, as dimensional considerations show, is proportional to time: $\ell \sim \Delta u t$, $t \to 0$. Finally, for $\langle N\rangle_t$ we obtain the following formula

$$\langle N\rangle_t = \frac{a}{t}, \qquad t \to 0 \tag{3.31}$$

Here, a is a constant; its value will be found later. Thus, for the variable τ in formulas (3.26) we have

$$\tau \to -a\ln\frac{t}{t_0}, \qquad t \to 0$$

Here, t_0 is the time scale which is determined by large eddies in the mixing layers in the late stage of process development. Therefore, $\tau \to \infty$ when $t \to 0$ and, hence, the first term of the asymptotic solution is determined only by the first terms in formulas (2.36), i.e.,

$$F = \frac{\pi}{2}\sin \pi z$$

$$\gamma = \frac{4}{\pi}\sin \pi\langle z\rangle \exp(-\pi^2 \tau) = \frac{4}{\pi}\sin \pi\langle z\rangle \left(\frac{t}{t_0}\right)^{a\pi^2}$$

$$\gamma_0 = 1 - \langle z\rangle - \frac{2}{\pi}\sin \pi\langle z\rangle \left(\frac{t}{t_0}\right)^{a\pi^2}$$

$$\gamma_1 = \langle z\rangle - \frac{2}{\pi}\sin \pi\langle z\rangle \left(\frac{t}{t_0}\right)^{a\pi^2}, \qquad t \to 0 \tag{3.32}$$

The value of constant a entering into relation (3.31) for scalar dissipation can be found if we introduce the additional supposition that when $t \to 0$, the intermittency factor γ is proportional to the characteristic mixing layer width, i.e., $\gamma \sim t$. This supposition is based on the physical flow pattern and on the definition of quantity γ as the volume fraction occupied by fully turbulent fluid. By equating the exponent in formula (3.32) for γ to unity we obtain $a = \pi^{-2}$.

This theoretical value agrees sufficiently well with an indirect estimate of constant a which can be obtained if we consider the semiempirical relation for scalar dissipation $\langle N\rangle = \chi\langle\epsilon\rangle\sigma^2/q^2$, $\chi \approx 2$ (see, for example, Beguier, Dekeyser, and Launder [1978]) in the self-similar mixing layer. For an estimate we go over this stationary problem to the nonstationary case under consideration with the aid of transformation $t = x/(u_0/2)$. If we now substitute into the expression for $\langle N\rangle$ the experimental values (taken in the region where $\gamma \approx 1$) $\langle\epsilon\rangle = 0.04 u_0^3/x$ and $q^2 = 0.05 u_0^2$ given in the book by Townsend [1956], $\sigma = 0.22$ from the experiments by Rajagopalan and Antonia [1980], we then obtain $\langle N\rangle \approx \chi 0.08/t$. Using for $\langle\epsilon\rangle$ and q^2 the experimental data of Wygnanski and Fiedler [1970]: $\langle\epsilon\rangle = 0.025 u_0^3/x$ and $q^2 = 0.06 u_0^2$, we obtain $\langle N\rangle \approx \chi 0.037/t$. Therefore, in the first case $a \approx 0.16$, in the second case $a \approx 0.074$.

Let us consider one important aspect connected with this solution. The analyzed problem is, in a certain sense, an exceptional case. Indeed, although equation (3.23) includes only the first derivative with respect to time, a solution that satisfies the two conditions (3.20) and (3.21) with respect to the time variable has been attained here owing to the specific initial conditions.

In the case of arbitrary initial conditions, the mixing characters becomes complex. This problem has already been discussed in §3.2 where it was noted that the hypothesis of the statistical independence of N and z in the turbulent fluid is invalid for the turbulent flow existing in the arbitrary initial condition. Before the turbulence reaches an equilibrium structure, some relaxation time, equal in order of magnitude to L/q, must elapse. Only following this does the hypothesis under consideration become valid and, hence, so does equation (3.23).

In the considered problem turbulence is already in the equilibrium state at the initial moment of time (more precisely, the relaxation time is much longer than L/q and tends to zero when Re $\to \infty$) which permits satisfaction of two conditions with respect to the time variable.

It is interesting that the physical considerations of the applicability of the hypothesis of statistical independence of micro- and macrocharacteristics of fluctuating motion in turbulent flows are formally reflected in the properties of the equations (the solution does not exist under arbitrary initial conditions).

In conclusion, on the basis of the problem under consideration, we present a number of qualitative arguments concerning the connection between intermittency and the theory of locally homogeneous turbulence in its unrefined version (Kolmogorov [1941], Obukhov [1941, 1949]). It should be remembered that in this theory it is assumed that $\gamma = 1$, i.e., intermittency is absent. We shall show, in this case, that the equation for the concentration PDF leads to physically absurd results. We first assume that hypothesis (3.8) is true for all z. Then, it is easy to verify that the density of normal distribution (3.27) is the solution of equation (3.23) for all $t > 0$. Since $P \neq 0$ when $-\infty < z < \infty$ for such a solution, it then has no physical meaning.

This difficulty is of a fundamental character, and cannot be bypassed by rejecting hypothesis (3.8). Indeed, let $\langle N \rangle_z$ be a function of concentration. It is essential for function $\langle N \rangle_z$ in the asymptotic theory, in which it is assumed that Re $= \infty$, to be different from zero when $0 \leq z \leq 1$. This assertion follows from the results presented in §1.3 on the behavior of $\langle N \rangle_z$ in the vicinity of the boundaries of the phase space. It is established there that the tendency of $\langle N \rangle_z$ to zero at the boundaries of that space is caused only by the effects of molecular transport and is essentially dependent on Reynolds number (quantities α and k in formulas (1.29) – (1.34) are functions of Re). From the physical standpoint the deviation of $\langle N \rangle_z$ from zero in the entire phase space corresponds to the flow pattern when small-scale turbulent fluctuations are distributed in the space more or less uniformly, i.e., $\epsilon > 0$ and $N > 0$ at all points; when Re $\to \infty$ the velocity and concentration gradients increase without limit in the entire flow (compare with the flow pattern in which intermittency exists — §1.3).

Let us consider the solution of equation (3.23) with the boundary conditions $P = 0$, $z = 0$, $z = 1$. It is evident that for such a solution we will have $\partial \langle N \rangle_z P / \partial z \neq 0$ when $z = 0$ and $z = 1$, i.e., the normalization condition $\int_0^1 P dz = 1$ is not fulfilled, since the "flows" of probability density $\partial \langle N \rangle_z P / \partial z$

beyond the natural boundaries of the phase space $z = 0$ and $z = 1$ are not equal to zero (if we were to seek a smooth solution of (3.23) satisfying the normalization condition and constancy of the mean concentration, then, as can readily be shown, it must additionally be assumed that $\partial P/\partial z = 0$, $z = 0$ and $z = 1$; only the trivial solution $P = 0$, which is physically meaningless, satisfies the boundary conditions $P = \partial P/\partial z = 0$, $z = 0$ and $z = 1$). These flows facilitate the emergence of intermittency or, more strictly, intermittency in the concentration field. The theory of locally homogeneous turbulence in the neighborhood of the phase space boundary cannot be valid anymore, since the interaction between the turbulent and nonturbulent fluids becomes significant here for which, as mentioned in Chapter 1, the direct (and not cascade) effect of large-scale fluctuations on small-scale fluctuations is a characteristic feature (see also Chapter 4).

§3.5 APPROXIMATE DESCRIPTION OF CONCENTRATION PDF IN FREE TURBULENT FLOWS

In the present section, a nonrigorous investigation is carried out of the concentration PDF in turbulent jets on the basis of equation (3.2). In accordance with the conclusions of §3.3 and for the simplification of the analysis it is assumed that the conditionally averaged velocity $\langle u \rangle_z$ is described by the linear dependents (3.16). By means of simple mathematical techniques the qualitative form of the solutions of the equations is established in two characteristic regions of the jet — in the central region where intermittency is of little significance, and at the jet edge where intermittency is substantial.

These results are then used for the solution of a purely practical problem — the development of a simple approximate method for the determination of concentration PDF and intermittency factor in jets.

The content of this section is based on the work by Kuznetsov [1972a] (with some additions from the work by Kuznetsov and Sabel'nikov [1981b] and Sabel'nikov [1982b]).

We shall consider mean quantities in a stationary, slowly expanding, free turbulent flow. A jet or a wake might be such a flow. Let us analyze the concentration PDF in a distant wake or at the main section of a jet. Then, the probability of observing value $z = 1$ can be neglected, i.e., it can be assumed that $\gamma_1 \approx 0$. Suppose that the conditionally averaged velocity $\langle u \rangle_z$ is described by the linear dependence (3.16). Then equations (2.15) and (3.2) for unconditional and conditional concentrations PDFs, respectively, have the form

$$\langle u \rangle \nabla P + \nabla[q\sigma^{-2}(z - \langle z \rangle)P] = -\gamma \langle N \rangle_t \frac{\partial^2 P_t}{\partial z^2}$$

$$\langle u \rangle \nabla \gamma F + \nabla[q\sigma^{-2}(z - \langle z \rangle)\gamma F] = -\gamma \langle N \rangle_t \frac{\partial^2 F}{\partial z^2} \qquad (3.33)$$

Let us analyze a number of special solutions of equations (3.33).

1. Solution on the Axis or in Plane of Symmetry.

Experimental investigations indicate that near the axis or plane of symmetry intermittency is practically nonexistent, i.e., $\gamma \approx 1$ and, hence, $\langle N \rangle_t = \langle N \rangle$, $P(z) \approx P_t(z)$. Furthermore, in this region streamwise mass transfer can be ignored, i.e., it can be assumed that $q_1 \approx 0$. Subscript 1 here refers to the flow direction. It is evident that on the axis or in the plane of symmetry $\langle u_\alpha \rangle = q_\alpha = \partial \langle z \rangle / \partial x_\alpha = \partial \sigma / \partial x_\alpha = 0$, $\alpha = 2.3$. Then, from the first equation in (3.33) we obtain

$$\langle u_1 \rangle \frac{\partial P}{\partial x_1} + P\sigma^{-2}(z - \langle z \rangle)\frac{\partial q_\alpha}{\partial x_\alpha} = -\langle N \rangle \frac{\partial^2 P}{\partial z^2} \tag{3.34}$$

In order to transform the second term on the left side of (3.34), we use the equation for the mean concentration

$$\langle u \rangle \nabla \langle z \rangle + \nabla q = 0 \tag{3.35}$$

On the axis or in the plane of symmetry, from (3.35) we have

$$\frac{\partial q_\alpha}{\partial x_\alpha} = -\langle u_1 \rangle \frac{\partial \langle z \rangle}{\partial x_1}, \qquad \alpha = 2.3 \tag{3.36}$$

As a result, equation (3.34) assumes the form

$$\langle u_1 \rangle \frac{\partial P}{\partial x_1} - \frac{z - \langle z \rangle}{\sigma^2} \langle u_1 \rangle \frac{\partial \langle z \rangle}{\partial x_1} P = -\langle N \rangle \frac{\partial^2 P}{\partial z^2} \tag{3.37}$$

We shall seek a self-similar solution of equation (3.37) of the form $P = g(s)/\sigma$, where g is a function of one variable $s = (z - \langle z \rangle)/\sigma$. Substitution in (3.37) yields the following equation for function g

$$mg'' + (s + h)g' + (hs + 1)g = 0$$

$$m = -\frac{\langle N \rangle}{\frac{1}{2}\langle u_1 \rangle \partial \sigma^2 / \partial x_1}, \qquad h = \frac{\partial \langle z \rangle / \partial x_1}{\partial \sigma / \partial x_1} \tag{3.38}$$

The prime in (3.38) indicates differentiation with respect to s.

It is evident from equation (3.38) that the self-similar solution exists only if parameters m and h are constant. This condition is fulfilled in the main section of a jet and in a distant wake.

Let us elucidate the physical meaning of constants m and h. Parameter m, as can easily be established from the equation for the variance of concentration fluctuations σ^2, is equal to the ratio of twice the scalar dissipation to the absolute value of advection. Experimental data of specific components in the balance equation of concentration fluctuations indicate that $m > 1$ in the flows under consideration. Thus, $m \approx 2.6$ in a distant wake of a circular cylinder (Freymuth and Uberoi [1971]), $m \approx 1.8$ in a plane jet (Bashir and Uberoi [1975]), $m \approx 2$ in axisymmetric jet in coflowing stream (Antonia, Prabhu, and Stephenson [1975]).

Quantities σ and $\langle z \rangle$ at the main section of the jet or in a distant wake change as a function of distance x_1 by an exponential law. Consequently, parameter h

is inversely proportional to the intensity of concentration fluctuations. According to the experimental data (see, for example, La Rue and Libby [1974], Bashir and Uberoi [1975], Becker, Hottel, and Williams [1967], Kusnetsov [1971] and others), it acquires considerably large values: $h = 4 - 5$. This condition suggests an asymptotic expansion of the solution of equation (3.38) with respect to integer exponents of the small quantity h^{-1}. Prior to writing this expansion, we point out some general properties of the solution of equation (3.38).

With this view in mind, we substitute the variables in (3.38)

$$\Pi = g \exp\left[\frac{(s+h)^2}{4m}\right], \quad s_1 = \frac{s+h-2mh}{\sqrt{m}}$$

The equation for function Π has the form

$$\Pi'' + \left(-\frac{1}{4}s_1^2 + \nu + \frac{1}{2}\right)\Pi = 0$$

$$\nu = h^2(m-1) \tag{3.39}$$

The prime here denotes differentiation with respect to s_1. The solution of equation (3.39) is expressed in terms of the parabolic cylinder function (see, for example, Bateman and Erdeylyi [1953a]). Since $m > 1$, then the index of these functions $\nu > 0$. On the basis of the known properties of these functions, we conclude that the condition of non-negativity of the PDF is not fulfilled. Such a shortcoming of the solution is associated with the inaccuracy of approximation of the conditionally averaged velocity $\langle u \rangle_z$ with the linear dependence (3.16) in the region of large values of $|s|$. More thorough investigation shows that the roots of function g lie outside the range of its main variation, namely, when $z > \langle z \rangle + 3\sigma$. The connection between the inaccuracy of the linear approximation of $\langle u \rangle_z$ in the region of large amplitudes of concentration fluctuations and the position of the roots is evident. We should note that as a result of the nonphysical character of the behavior of PDF in the region of concentration fluctuation amplitudes, it cannot, generally speaking, be used for the computation of high-order moments.

Let us now find the asymptotic expansion of the solution of equation (3.38) when $h \to \infty$. We shall seek such a solution in the form of asymptotic series

$$g(s) = g^{(0)}(s) + \frac{1}{h}g^{(1)}(s) + \frac{1}{h^2}g^{(2)}(s) + \ldots \tag{3.40}$$

We substitute series (3.40) into equation (3.38) and equate the coefficients of successive exponents of h to zero. As a result we obtain a recurrent system of equations for functions $g^{(k)}$, $k = 0, 1, 2, \ldots$. The equations for the first three of these functions have the form

$$g^{(0)'} + sg^{(0)} = 0 \tag{3.41}$$

$$g^{(1)'} + sg^{(1)} = -\left(mg^{(0)''} + sg^{(0)'} + g^{(0)}\right) \tag{3.42}$$

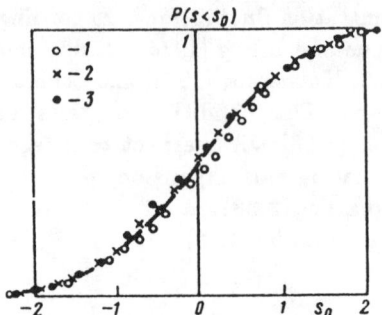

Figure 3.8 Concentration PDF on the axis of submerged axisymmetric jet of carbon monoxide from data by Kuznetsov [1971]. $Re_d = u_0/d\nu = 1.3 \cdot 10^4$, $d = 5$ mm, $u_0 = 20$ m/sec; *1)* $x_1/d = 20$, *2)* $x_1/d = 30$, *3)* $x_1/d = 40$, $s = (z - \langle z \rangle)/\sigma$, $P(s < s_0)$ is the probability of event $s < s_0$. The dashed curve corresponds to a normal distribution.

$$g^{(2)'} + sg^{(2)} = -(mg^{(1)''} + sg^{(1)'} + g^{(1)}) \tag{3.43}$$

Solving equations (3.41) – (3.43) sequentially we obtain

$$g^{(0)} = \frac{1}{\sqrt{2\pi}} \exp\left(-\frac{s^2}{2}\right)$$

$$g^{(1)} = \frac{1-m}{3}(s^3 - 3s)g^{(0)} = -\frac{1}{3}(1-m)\frac{d^3 g^{(0)}}{ds^3}$$

$$g^{(2)} = \frac{1-m}{4}\frac{d^4 g^{(0)}}{ds^4} + \frac{(1-m)^2}{18}\frac{d^6 g^{(0)}}{ds^6} \tag{3.44}$$

It follows from expressions (3.40) and (3.44) that the concentration PDF on the axis or in the plane of symmetry is close to normal distribution. This result agrees well with the experimental data. As an example of the comparison of theory with the experiment, Fig. 3.8 shows the data of measurements of concentration PDF in an axisymmetric jet (Kuznetsov [1971]), and Fig. 3.9 shows the measurement near the plane of symmetry of a distant wake of a circular cylinder (La Rue and Libby [1974]). Analogous experimental data were also obtained by Ibragimov, Petrishcheva, and Taranov [1968], Meshkov and Shcherbina [1979], Golovanov and Shcherbina [1979], Shcherbina [1982].

The deviations from the normal distribution are described by functions $g^{(1)}$ and $g^{(2)}$ in (3.44). The generally accepted qualitative characteristic of these deviations are the skewness A and kurtosis E. From (3.40) and (3.44) we obtain

$$A = \int s^3 g \, ds = h^{-1} \int s^3 g^{(1)} ds = 2(1-m)h^{-1}$$

$$E = \int s^4 g \, ds - 3 = h^{-2} \int s^4 g^{(2)} ds = 6(1-m)h^{-2} \tag{3.45}$$

The value of the skewness calculated with the aid of (3.45) is sufficiently close to the results of experiments. As an example we consider the distant wake of a circular cylinder. Here, $m = 2.6$, $h \approx 4.75$ (Freymuth and Uberoi [1971]) and the formula for A in (3.45) yields $A \approx -0.67$. In experiments by La Rue and Libby [1974] $A \approx -0.4$. For the kurtosis computation by formula (3.45) yields $E = -0.43$ which differs substantially from the experimental value of $E \approx 0.1$ by La Rue and Libby (even the signs disagree). Analogous results are also true for a jet. This disagreement of the theory with the experiment is caused, as noted above, by the rough character of the approximation of the conditionally averaged velocity $\langle u \rangle_z$ by means of the linear dependence (3.16) in the region of large amplitudes of concentration fluctuations.

In conclusion we present, without derivation, a formula which connects the intermittency factor with the intensity of concentration fluctuations on the axis or in the plane of symmetry

$$1 - \gamma = \frac{m^{1/4}}{h\sqrt{\pi}} \left(1 + \frac{1}{\sqrt{m}}\right)^{\nu+1/2} \exp\left[-h^2\left(\sqrt{m} - \frac{1}{2}\right)\right], \quad h \to \infty$$

$$\nu = h^2(m-1) \tag{3.46}$$

It is evident from (3.46) that the difference $1 - \gamma$ is exponentially small when $h \to \infty$. Calculations indicate that $1 - \gamma = 10^{-4} - 10^{-3}$ when $h = 4 - 5$ and $m = 1.8 - 2.6$. Thereby, the assumption made above, namely that $\gamma \approx 1$, is fully justified.

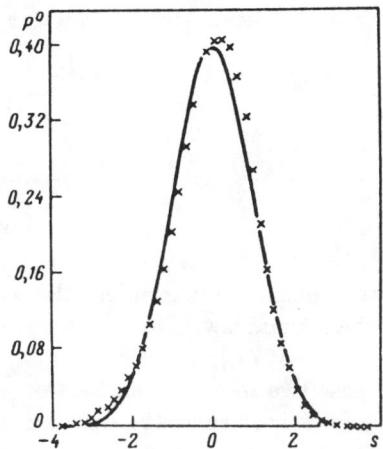

Figure 3.9 Concentration PDF near the plane of symmetry of a wake of circular cylinder from data by La Rue and Libby [1974]. $x_2/\sqrt{(x_1 - x_{10})d} = 0.0275$, $x_{10} = -40d$, $P^0 = \sigma P$, $s = (z - \langle z \rangle)/\sigma$. The curve corresponds to a normal distribution; the crosses denote the experimental data. The conditions of the test are the same as in Fig. 1.14.

2. Solution in the Region of Maximum Transverse Mean Concentration Gradient.

Let us now consider the flow region where the transverse concentration gradient attains its maximum value. It is well known from experiments that the main processes in this region are the creation and dissipation of concentration fluctuations (see, for example, Freymuth and Uberoi [1971]). Therefore, the convective term in the equation for PDF (3.33) can be neglected. Furthermore, in the region under consideration the mass flow in the transverse direction and concentration variance reach maximum values, i.e. $\partial q_\alpha / \partial x_\alpha = \partial \sigma^2 / \partial x_\alpha = 0$, $\alpha = 2.3$. We take into account, as before, that intermittency is also insignificant here, and the transverse mass flow is small, i.e., $\gamma \approx 1$, $q_1 \approx 0$. With the use of these assumptions, the first of the equations in (3.33) takes the form

$$\frac{q_\alpha}{\sigma^2} \frac{\partial}{\partial x_\alpha}(z - \langle z \rangle)P = -\langle N \rangle \frac{\partial^2 P}{\partial z^2} \tag{3.47}$$

Just as earlier in this section, we shall seek the self-similar solution of this equation of the form $P = g(s)/\sigma$. For function g from (3.47), we obtain equation

$$(sg)' - \frac{\langle N \rangle}{q_\alpha \partial \langle z \rangle / \partial x_\alpha} g'' = 0 \tag{3.48}$$

When deriving (3.48), it is assumed that

$$\frac{\partial P}{\partial x_\alpha} = -\frac{1}{\sigma^2} \frac{\partial \langle z \rangle}{\partial x_\alpha} g'$$

Note now that the factor in front of the second derivative in (3.48) is the ratio of the scalar dissipation to the creation of concentration fluctuations. By assumption these values coincide, i.e., the coefficient is equal to unity. Hence, function g satisfies equation

$$g'' + (sg)' = 0 \tag{3.49}$$

The solution of equation (3.49) is

$$g = \frac{1}{\sqrt{2\pi}} \exp\left(-\frac{s^2}{2}\right) \tag{3.50}$$

Thus, in the region of maximum transverse mean concentration gradient the concentration PDF is also distributed according to the normal law. This result agrees with experimental data (see, for example, Kuznetsov [1971]).

In summing up the results obtained in the first two items of this section we conclude that in the region where intermittency is insignificant, the distribution of concentration PDF is close to a normal distribution.

3. Solution on the Edge of the Turbulent Flow.

Let us finally find the conditional concentration probability density $F(P_t = F(z) \cdot \theta(z))$ in the region where intermittency is substantial, i.e., at the edge of the turbulent flow. The principal assumption is based on the results of measurements of the conditionally averaged characteristics in a fully turbulent fluid (see §1.1). Analysis of these results shows that the characteristics under consideration change very little from one point to another. As an example we cite the measurements by Becker, Hottel, and Williams [1967] in a submerged axisymmetric jet which are depicted in Figs. 1.3, 1.4. It is evident that when the ratio x_2/x_1 changes within the range 0.16 – 0.26, the derivative $\partial \gamma/\partial x_2$ exceeds the derivatives $\partial \langle z \rangle_t /\partial x_2$ and $\partial \sigma_t /\partial x_2$ by an order of magnitude. Therefore, derivatives $\partial F/\partial x_k$ in the second of the equations in (3.3) can be neglected in comparison with $\partial \gamma/\partial x_k$. Then we obtain

$$\frac{\partial^2 F}{\partial z^2} + (a_1 - b_1 z)F = 0$$

$$a_1 = \frac{1}{\langle N \rangle}[\langle u \rangle \nabla \gamma - \nabla(q\sigma^{-2}\langle z \rangle \gamma)]$$

$$b_1 = -\frac{1}{\langle N \rangle}\nabla(q\sigma^{-2}\gamma) \qquad (3.51)$$

The solution of equation (3.51) must satisfy the boundary conditions (3.5), i.e., $F = 0$ when $z = 0$ and $z = 1$. As to the boundary conditions when $z = 1$, we note the following: In the flow region under consideration the value of $\langle z \rangle_t$ is small and the ratio $\sigma_t/\langle z \rangle_t$ is of the order of unity (see, for example, the data by Becker, Hottel, and Williams in Figs. 1.3, 1.4). Therefore, the main change of PDF F takes place in the region of small z, namely, $z \sim \langle z \rangle_t \ll 1$. Consequently, the second boundary condition can be approximately set not at the maximum possible value of concentration $z = 1$, but at $z = \infty$.

The solution of (3.51), satisfying the boundary condition $F = 0$ when $z = \infty$, is expressed in terms of Airy function Ai (see, for example, Abramowitz and Stegun [1964])

$$F = R_1 b_1^{1/3} \text{Ai}(\chi)$$

$$\text{Ai}(\chi) = \frac{1}{\pi} \int_0^\infty \cos\left(\frac{t^3}{3} + t\chi\right) dt$$

$$\chi = b_1^{1/3}\left(z - \frac{a_1}{b_1}\right) \qquad (3.52)$$

Here, R_1 is an arbitrary constant. By satisfying the boundary condition $F = 0$, $z = 0$, we obtain the equation Ai $(\nu) = 0$, where $\nu = a_1 b_1^{-2/3}$ is a dimensionless parameter. As is generally known (see Abramowitz and Stegun [1964]), equation Ai $(\nu) = 0$ has a finite number of roots so that $\nu < 0$. Only the largest root of Airy function ν_1 has a physical meaning, since the solution for all other values of ν oscillates in the range $0 < z < \infty$. The approximate value of the largest root $\nu_1 = -2.338$, i.e., $a_1 b_1^{-2/3} = 2.338$. Thereby, it is established that a solution of equation (3.51) does not exist for all a_1 and b_1 or, in other words, for every function $\langle z \rangle$, σ and $\langle N \rangle$.

We now take into account that function F must satisfy the normalization condition and that its first moment is equal to the conditionally averaged concentration $\langle z \rangle_t$, i.e., $\int_0^\infty F\, dz = 1$ and $\int_0^\infty zF\, dz = \langle z \rangle_t$. Hence, we find the value of the constant R_1 and express parameter b_1 through $\langle z \rangle_t$

$$R_1 = \frac{1}{J_0}, \qquad b_1^{1/3} = \frac{b^{1/3}}{\langle z \rangle_t}$$

$$b^{1/3} = \left(\frac{J_1}{J_0} - \nu_1\right), \qquad J_k = \int_{\nu_1}^\infty \chi^k \operatorname{Ai}(\chi)\, d\chi, \qquad k = 0.1$$

Computations yield $J_0 = 1.274$, $J_1 = -0.701$, i.e., $R_1 = 0.785$, $b^{1/3} = 1.788$. The numerical values of the intensity of concentration fluctuations, the skewness and the kurtosis in a fully turbulent fluid are equal to $\sigma_t/\langle z \rangle_t = 0.555$, $A_t = 0.802$, $E_t = 0.694$, respectively. It should be noted that this solution (3.52) has a self-similar form, i.e., $F = 1/\langle z \rangle_t\, f_\infty(\varsigma)$, where

$$f_\infty = R\operatorname{Ai}(\chi), \qquad \chi = b^{1/3}\varsigma + \nu_1, \qquad \varsigma = \frac{z}{\langle z \rangle_t}$$

$$R = R_1 b^{1/3} = 1.403, \qquad b^{1/3} = 1.788, \qquad \nu_1 = -2.338 \tag{3.53}$$

From relation (3.53) it can be concluded that the PDF at the edge of the flow is completely defined by only one parameter — the conditionally averaged concentration $\langle z \rangle_t$.

Let us now consider comparison of these theoretical results and experimental data. The graph of function $f_\infty(\varsigma)$, which is described by formula (3.53), is presented in Fig. 3.10. Plotted here also are the experimental data by Ebrahimi, Günther, and Haberda [1977], Birch, Brown, Dodson, and Thomas [1978]. Let us consider the processing of these data. Since the concentration PDF is "smeared"

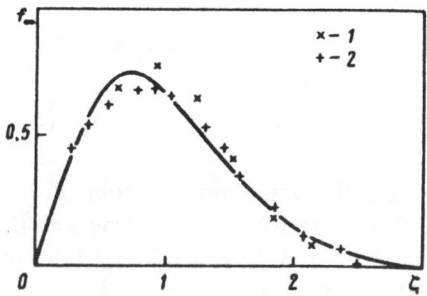

Figure 3.10 Comparison of measured and computed concentration PDFs in a turbulent fluid at a large distance from the axis of a submerged jet. Symbols are experimental data. *1*) Birch, Brown, Dodson, and Thomas [1978], test conditions are presented in Fig. 1.17; *2*) Ebrahimi, Günther, and Haberda [1977]; $x_1/d = 50$, $x_2/d = 6$; $\varsigma = z/\langle z \rangle_t$; the curve is computed.

on the boundary of the phase space $z = 0$ (see §1.3) owing to the processes of molecular mixing and noise in the instruments, the extrapolation of the experimental points lying outside the range of "smearing" is effected at the origin of the coordinates (see Fig. 1.17). The curve thus plotted was normalized and then the conditionally averaged concentration $\langle z \rangle_t$ was determined. From Fig. 3.10 it can be concluded that the agreement between the theoretical probability density and the experimental data is quite satisfactory. The results obtained in §5.1 confirm this conclusion indirectly in a wider range of concentration variation.

The computed intensity of concentration fluctuations in the turbulent fluid $\sigma_t/\langle z \rangle_t = 0.555$ is a good agreement with the experimental data by Becker, Hottel, and Williams [1967], La Rue and Libby [1974], Antonia, Prabhu, and Stephenson [1975], (see Figs. 1.4 – 1.6, and also the graph of the dependence of $\sigma_t/\langle z \rangle_t$ on the intermittency factor γ, plotted with the use of data from the indicated three works, in the article by Kuznetsov [1977b]. The calculated value of the skewness $A_t = 0.802$ also agrees well with the measurements by La Rue and Libby [1974] ($A_t = 0.8 - 0.9$), Antonia, Prabhu, and Stephenson [1975] ($A_t = 0.5$). Noticeable deviation arises when comparing the calculated and experimental values of the kurtosis: $E_t = 0.694$ is the theoretical value, $E_t = -0.1$ (La Rue and Libby [1975], $E_t = (-0.4) - (-0.5)$ (Antonia, Prabhu, and Stephenson [1975]). This deviation can be explained by the inaccuracy of the assumptions leading to the equation for the concentration PDF and by the serious difficulties in the accurate measurement of the high order amount in the turbulent fluid at small values of intermittency factor. In these circumstances the processes of molecular transport have particularly strong influence on the results of measurements of the intermittency factor and the concentration probability factor. As shown in §1.3, these processes lead to delta functions in (1.19) for the unconditional PDF $P(z)$ "smeared" in the range on the order $\sigma \text{Re}^{-1/4}$. Therefore, for small values of the intermittency factor γ, by virtue of the limitation of the resolving power of the device, only "smearing" of delta function is recorded. To illustrate the effect under study the result of measurement by La Rue and Libby [1974] of the concentration PDF in the wake of a cylinder is presented in Fig. 3.11. The measurement was made at a point where $\gamma = 0.175$, i.e., at the edge of the turbulent flow. It is seen from Fig. 3.11 that only the "smeared" delta function could be recorded in the indicated work.

4. Approximate Method of Determination of Concentration PDF and Intermittency Factor. In practical applications associated mainly with the calculation of flows of reacting gas, it is important to have an approximate method for the determination of concentration PDF and intermittency factor. Several such methods are known in the literature (see, for example, Vilyunov and Dik [1976], Borghi [1980], and others). In these studies, the equation for the PDF is not used at all. Instead, the functional form of PDF is given *a priori* and is usually assumed to be universal in all regions of the turbulent flow. Such an assumption enables the determination of PDF from the first two moments which can be calculated by means of the traditional semiempirical theory of turbulence.

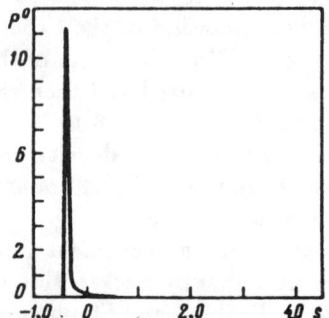

Figure 3.11 Concentration PDF at a large distance from the plane of symmetry in the wake of circular cylinder from data by La Rue and Libby [1974]. Test conditions are presented in Fig. 1.14, $P^0 = \sigma P$, $\xi = 0.431$, $\gamma = 0.175$, $s = (z - \langle z \rangle)/\sigma$.

The main conclusion of the investigation conducted above is that the form of concentration PDF in free turbulent flows is significantly different in the regions of strong and weak intermittency. This conclusion indicates that the region of applicability of the indicated approximate methods is very limited.

In the present item another approach to the approximate determination of the concentration PDF and intermittency in turbulent flows is proposed (see the work by Kuznetsov et al. [1977]. It is based on the simplified analysis of the equation for PDF presented in the first three items of the section. It takes into account the qualitative influence of intermittency on the form of PDF. Just as in the approximate methods mentioned earlier, the method described below requires knowledge of two moments of concentration $\langle z \rangle$ and σ^2. Let us first consider the methods of determining the intermittency factor in the main section of the jet or in a distant wake where it can be assumed that $\gamma_1 = 0$. The method is based on the exact formula which connects the intensity of concentration fluctuations $\sigma/\langle z \rangle$, intermittency factor γ, and the intensity of concentration fluctuations in the fully turbulent fluid $\sigma_t/\langle z \rangle_t$

$$\gamma = \frac{(1 + \sigma_t^2/\langle z \rangle_t^2)\langle z \rangle^2}{\langle z \rangle^2 + \sigma^2} \tag{3.54}$$

To elucidate, we note that formula (3.54) follows from equality $\langle z^2 \rangle = \gamma \langle z^2 \rangle_t$ if $\langle z^2 \rangle = \langle z \rangle^2 + \sigma^2$, $\langle z^2 \rangle_t = \langle z \rangle_t^2 + \sigma_t^2$ are substituted into its right and left sides, respectively, and equality $\langle z \rangle_t = \langle z \rangle / \gamma$ is taken into account.

In accordance with the results obtained in the first three items of the section, the value of the sum $1 + \sigma_t^2/\langle z \rangle_t^2$ varies very little across a jet (wake): $1.04 \lesssim 1 + \sigma_t^2/\langle z \rangle_t^2 \lesssim 1.31$. Therefore, we shall consider that the value of the sum $1 + \sigma_t^2/\langle z \rangle_t^2$ is constant and equal to the limit value 1.31. Then, formula (3.54) enables the calculation of intermittency factor if the conditionally averaged moments $\langle z \rangle$ and σ^2 are known. When the intensity of the fluctuations is small, namely when $\sigma/\langle z \rangle < 0.555$, calculation from formula (3.54) under the assumption that $\sigma_t/\langle z \rangle_t = 0.555$ yields a value $\gamma > 1$ which is physically meaningless. Such a situation arises in the central region of the flow near the axis or plane of symmetry where intermittency is insignificant. In these cases, it is simply assumed that $\gamma = 1$. The adopted assumption is quite justified, since by virtue of formula (3.46), the inaccuracy in specifying the ratio $\sigma_t/\langle z \rangle_t$ in this region relates only at small deviations of intermittency factor from unity.

Finally, we obtain the following algebraic approximation for the intermittency factor in terms of $\langle z \rangle$ and σ^2

$$\gamma = 1.31 \frac{\langle z \rangle^2}{\langle z \rangle^2 + \sigma^2}, \qquad \frac{\sigma}{\langle z \rangle} \geq 0.555$$

$$\gamma = 1, \qquad \frac{\sigma}{\langle z \rangle} < 0.555 \tag{3.55}$$

Formulas differing from (3.55) only in the value of the coefficient were proposed by Zimont, Meshcheryakov, and Sabel'nikov [1978, 1981], Kent and Bilger [1977].

As an illustration of the effectiveness of the described approximate method of determining the intermittency factor, Fig. 3.12, taken from the work by Kent and Bilger [1977], shows a comparison of formula (3.55) with the dependence between γ and $\sigma/\langle z \rangle$ from the results of measurements of these parameters in a submerged axisymmetric jet (Becker, Hottel, and Williams [1967]), in coflowing axisymmetric jets (Antonia, Prabhu, and Stephenson [1975]), and in a far wake of a circular cylinder (La Rue and Libby [1974]). An analogous correlation dependence between γ and $\sigma/\langle z \rangle$, as shown in tests by Drake, Pitz, and Lapp [1984], is also true in a diffusion flame.

Let us now discuss the approximate method of describing concentration PDF. The method is based on the form of PDF being qualitatively different in the regions with high and low intermittency. Hence, jets and wakes can be divided into two regions. In the central region wherein intermittency is insignificant, PDF is described by a Gaussian curve. Near the boundaries of a jet or a wake, where intermittency is important, the conditional concentration PDF in the turbulent fluid is expressed in terms of Airy formula (3.53). In summing up, we obtain an approximate description of concentration PDF and intermittency factor in turbulent jets and wakes

$$P(z) = \frac{1}{\sqrt{2\pi}\sigma} \exp\left[-\frac{(z-\langle z \rangle)^2}{2\sigma^2}\right]; \qquad \gamma = 1, \qquad \frac{\sigma}{\langle z \rangle} \leq 0.555 \tag{3.56}$$

and

$$P = \gamma P_t(z) + (1-\gamma)\delta(z), \qquad \gamma < 1$$

where

$$P_t(z) = R \frac{1}{\langle z \rangle_t} \text{Ai}(\chi), \qquad \chi = b^{1/3}\varsigma + \nu_1, \qquad \varsigma = \frac{z}{\langle z \rangle_t}$$

$$R = 1.4, \qquad b^{1/3} = 1.79, \qquad \nu_1 = -2.338$$

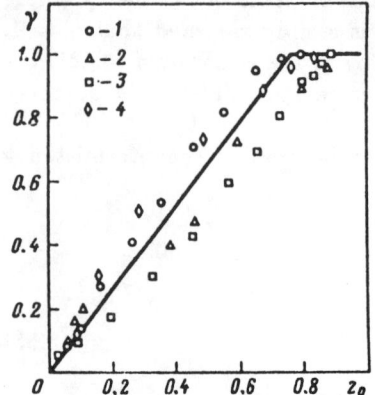

Figure 3.12 Dependence of intermittency factor on the mean concentration and variance in free turbulent flows from data by different authors. *1*) submerged axisymmetric jet, Becker, Hottel, and Williams [1967] (test conditions shown in Fig. 1.3); *2*) wake of a circular cylinder, La Rue and Libby [1974] (test conditions the same as in Fig. 1.14); *3, 4*) coflowing axisymmetric jet, Antonia, Prabhu, and Stephenson [1975] (*3*) $u_0/u_\infty = 6.6$, *4*) $u_0/u_\infty = 2.9$, u_∞ is the velocity of the coflowing stream (test conditions shown in Fig. 1.5)); the solid straight lines correspond to dependence (3.55), $z_0 = (1+\sigma^2/\langle z\rangle^2)^{-1}$.

$$\gamma = 1.31 \frac{\langle z\rangle^2}{\langle z\rangle^2 + \sigma^2}, \qquad \langle z\rangle_t = \frac{\langle z\rangle}{\gamma}, \qquad \frac{\sigma}{\langle z\rangle} > 0.555 \tag{3.57}$$

It should be recalled that formulas (3.56) and (3.57) apply in those regions of the flow where $\gamma_1 = 0$, i.e., in the main portions of the jet. They are also useful in the initial section where $\langle z\rangle < 0.5$. In the region where $\langle z\rangle > 0.5$, substitution $z \to 1 - z$ must be made.

Thus, relations (3.56) and (3.57) solve the problem of a simple approximate description of concentration PDF and intermittency factor. The calculation of $\langle z\rangle$ and σ^2 appearing in formulas (3.56) and (3.57), as already noted, can be carried out by means of the semiempirical theory of turbulence.

§3.6 MATHEMATICAL PROPERTIES OF THE EQUATION FOR CONCENTRATION PDF IN FREE TURBULENT FLOWS. STATEMENT OF THE BOUNDARY-VALUE PROBLEM

The present section is devoted to a more rigorous (than in §3.5) mathematical investigation of the equation for concentration PDF in free turbulent flows. A refined approximation of the conditionally averaged velocity $\langle u\rangle_z$ in the region of large amplitude concentration fluctuations is used in the analysis (3.18). Discussed are such general qualitative properties of the equation as singular points, existence of self-similar solutions, and formulation of the boundary-value problem. Noted are the available analogies with the case of a statistically homogeneous concentration field which was considered in §3.4. A significantly important part of the analysis is played by the nonlocal properties of the equation. It is shown that the condition of solvability of the boundary-value problem allows the finding of two unknown functions entering into the closure relations. In the present, as well as in the next section (wherein is presented a numerical solution of the formulated boundary-value problem), two principal objectives are pursued. The first is to substantiate the approximate method of analysis of the equation described in §3.5. The second objective is to show, by considering the equation for the concentration PDF, that according to the development proposed in this book the construction of a closed theory of turbulence can be connected. Presently, at least a decrease of the number of arbitrary functions compared with the semiempirical theories is attained for one-point moments. Note that this investigation is associated with a large amount of cumbersome calculations, and also with the use of a number of nonformal qualitative considerations. The material of this section is intended, first of all, for a reader who is interested in the very nonstandard mathematical structure of the equations for PDFs which are obtained with the aid of the theory of locally homogeneous and isotropic Kolmogorov-Obukhov turbulence, and in the opportunities which such equations (or equations with similar properties) present in the solution of the problem of closure in the theory of turbulence. Other readers can skip this section and go directly to §3.7, wherein the numerical solution of the self-similar problem is presented and the main results of the study are enumerated briefly.

1. Main Equations. Let us consider, just as in the previous section, stationary, slowly expanding, free turbulent flows which have an axis or plane of symmetry (jet, wake). The distinctive feature of such flows is the nonuniformity of the statistical characteristics of turbulence which makes the analysis of the equation for concentration PDF considerably more difficult. Let us ignore the mass flow in the transverse direction, i.e., we shall suppose that $q_1 = 0$. Assume that the conditionally averaged velocity $\langle u \rangle_z$ is described by the refined dependence (3.18), (3.19). Then, from equations (3.2) – (3.4) we obtain

$$\langle u \rangle \frac{\partial \gamma F}{\partial x} + \langle v \rangle \frac{\partial \gamma F}{\partial y} + \frac{1}{y^i} \frac{\partial y^i \langle v' \rangle_z \gamma F}{\partial y} = -\gamma \langle N \rangle_t \frac{\partial^2 F}{\partial z^2} \qquad (3.58)$$

$$\langle u \rangle \frac{\partial \gamma_0}{\partial x} + \langle v \rangle \frac{\partial \gamma_0}{\partial y} + \frac{1}{y^i} \frac{\partial y^i \langle v' \rangle_{z=0} > \gamma_0}{\partial y} = -\gamma \langle N \rangle_t \frac{\partial F(0)}{\partial z} \qquad (3.59)$$

$$\langle u \rangle \frac{\partial \gamma_1}{\partial x} + \langle v \rangle \frac{\partial \gamma_1}{\partial y} + \frac{1}{y^i} \frac{\partial y^i \langle v' \rangle_{z=1} \gamma_1}{\partial y} = \gamma \langle N \rangle_t \frac{\partial F(1)}{\partial z} \qquad (3.60)$$

Here

$$x = x_1, \quad y = x_2, \quad u = u_1, \quad v = u_2, \quad v' = v - \langle v \rangle$$

$$\langle v' \rangle_z = \langle v \rangle_z - \langle v \rangle = q_y \sigma^{-1} \mu^{-1} V, \quad q_y = q_2, \quad V = V_0 + s(1 + \omega^2 s^2)^{-1/2}$$

$$s = (z - \langle z \rangle)/\sigma$$

$$V_0 = -\langle s(1+\omega^2 s^2)^{-1/2}\rangle = -\gamma_0 s_0 (1+\omega^2 s_0^2)^{-1/2}$$

$$-\gamma_1 s_1 (1+\omega^2 s_1^2)^{-1/2} - \gamma \int_0^1 s(1+\omega^2 s^2)^{-1/2} F\, dz$$

$$\mu = \langle s^2(1+\omega^2 s)^{-1/2}\rangle = \gamma_0 s_0^2 (1+\omega^2 s_0^2)^{-1/2}$$

$$+\gamma_1 s_1^2 (1+\omega^2 s_1^2)^{-1/2} + \gamma \int_0^1 s^2(1+\omega^2 s^2)^{-1/2} F\, dz$$

$$s_0 = -\langle z\rangle/\sigma, \qquad s_1 = (1-\langle z\rangle)/\sigma \tag{3.61}$$

Parameter i is equal to zero for the flows having a plane of symmetry, and to unity for axisymmetric flows.

When solving equations (3.58) – (3.60), as noted at the start of this chapter, the mean velocity field, mass flow, and scalar dissipation are assumed to be specified. Thereby, functions $\langle u\rangle$, $\langle v\rangle$, q_y, and $\langle N\rangle_t$ are further assumed to be known. Note that the mass flow q_y and mean concentration $\langle z\rangle$ are linked by the equation of turbulent diffusion

$$\langle u\rangle \frac{\partial \langle z\rangle}{\partial x} + \langle v\rangle \frac{\partial \langle z\rangle}{\partial y} + \frac{1}{y^i}\frac{\partial y^i q_y}{\partial y} = 0 \tag{3.62}$$

Therefore, it can be assumed that the distribution of mean concentration is specified instead of the mass flow.

The required functions are the conditional concentration PDF in the turbulent fluid F, intermittency factor γ and one of the probabilities γ_0 and γ_1 (the other can be found with the aid of relation $\gamma_0 + \gamma_1 + \gamma = 1$). These functions are determined from equation (3.58), from one of equations (3.59) or (3.60) and the normalization condition of PDF F. The integrals of the required function F appear in the coefficients of equations (3.58) – (3.60), namely, quantities σ, μ, and V_0 (see (3.61)). Hence, (3.58) – (3.60) is an integrodifferential system of equations.

In these equations the function ω, appearing in the expression for the conditionally averaged velocity $\langle u\rangle_z$ (3.18), is also unknown. This function, as will be shown below in the example of the self-similar case (distant wake, main section of the jet), is found from the condition of solvability of the boundary-value problem. Let us present the general characteristics of equation (3.58). Just like the equation analyzed in §3.4, it is an inverse parabolic (the role of time is played by the coordinate x, the longitudinal mean velocity $\langle u\rangle$ is positive, and the coefficient of the second derivative with respect to z is negative). Let us consider the singular points of equation (3.58). Such points, just as in the statistically homogeneous case are, generally speaking, two — a point at an infinite distance $x = \infty$ and the origin of the coordinates $x = 0$ (it should be remembered that the origin of the coordinate, by definition, is at the exit section of the jet). It is essential that the singular point $x = \infty$ can be eliminated by the following substitution of the variables: $X = \int_x^\infty \langle N \rangle_t / \langle u \rangle d\xi$, $Y = y$. Indeed, if the known laws of decay of quantities $\langle N \rangle_t$ and $\langle u \rangle$ are to be used when $x \to \infty$ (see, for example, Townsend [1956], Hinze [1959]), we then find that the ratio $\langle N \rangle_t / \langle u \rangle$ is proportional to x^{-2} in a plane jet, x^{-3} in the axisymmetric jet, and x^{-2} in the wake of a circular cylinder. Hence, in all cases integral $\int_x^\infty \langle N \rangle_t / \langle u \rangle d\xi$ is convergent; thereby, one can use the theorem indicated in §3.4 when analysing statistically homogeneous cases concerning the uniqueness of the solution of the Cauchy problem for the inverse parabolic equation.

The second singular point $x = 0$ arises when the mixing layer in the initial section of the jet is close to the self-similar state. Then, as it is easily established with the help of dimensional considerations $\langle N \rangle_t \sim u_0/x$, $x \to 0$ (see formula (3.31) in §3.4). The asymptotic behavior of the solution of equations (3.58) – (3.60) for a jet when $x \to 0$ is found in §3.7 by means of a numerical method.

Principal attention in the current section will be given to the analysis of only the main term of the asymptotic expansion of PDF when $x \to \infty$. From general considerations, it can be concluded that the first term of the asymptotic behavior is the self-similar solution of equation (3.58). The following experimental data support such a conclusion.

It is well known that in free turbulent flows the moments of the distributions of velocity and concentration when $x/d \gg 1$ (d is the width or diameter of the jet for random flows, diameter of the cylinder for a wake) are described with high accuracy by self-similar dependencies (see, for example, Townsend [1956], Hinze [1959]). Hence, concentration PDF must have a similar behavior. Experiments conducted by Kuznetsov [1971], Golovanov and Shcherbina [1979], La Rue and Libby [1981], Sreenivasan [1981], Shcherbina [1982] confirm this conclusion.

The asymptotic behavior under consideration plays the same role as does the asymptotic behavior (3.27) for the statistically homogeneous case (it is not difficult to see that (3.27) is also a self-similar solution of equation (3.23)).

The global solution obtained by solving the inverse Cauchy problem determines the initial conditions for which a solution of the direct Cauchy problem exists on the semi-axis $x > 0$ (see analogous reasonings for the statistically homogeneous case). For a jet, when concentration in the exit section of the nozzle possesses only two values $z = 0$ and $z = 1$ ($z = 0$ outside the jet, $z = 1$ inside the jet), the asymptotic behavior of the global solution when $x \to 0$ is the self-similar solution for the mixing layer.

Let us derive an equation which is satisfied by the self-similar asymptotic behavior of equation (3.58) in the vicinity of an infinitely distant point $x = \infty$. Similarly, equations for the self-similar problem in the mixing layer can also be obtained (see Sabel'nikov [1982b]).

2. The Self-Similar Problem. The first terms in the asymptotic expansions of the mean velocities $\langle u \rangle$ and $\langle v \rangle$ in the vicinity of point $x = \infty$ in free turbulent flows have the self-similar form

96 TURBULENCE AND COMBUSTION

$$\langle u \rangle = u_s u(\xi), \qquad \langle v \rangle = u_s v(\xi), \qquad \xi = \frac{y}{\ell(x)} \qquad (3.63)$$

for a jet and

$$\langle u \rangle = u_0 - u_s u(\xi), \qquad \langle v \rangle = 0 \qquad (3.64)$$

for a distant wake of a circular cylinder. Here $u_s = u_0(d/x)^\beta$, $\beta = (1+i)/2$, u_0 is the outlet velocity from the nozzle in the case of jet flows and velocity of free flow for a wake, $\ell(x)$ is the characteristic scale: $\ell(x) = x$ for a jet, $\ell(x) = (xd)^{1/2}$ for a wake.

The first terms in the asymptotic expansion of functions $\langle z \rangle$, σ^2, and $\langle N \rangle_t$ is written in the form

$$\langle z \rangle = z_s Z(\xi), \qquad \sigma^2 = z_s^2 \Sigma^2(\xi)$$

$$\langle N \rangle_t = \frac{u_s z_s^2}{\ell} n_t(\xi), \qquad z_s = \frac{u_s}{u_0} = \left(\frac{d}{x}\right)^\beta \qquad (3.65)$$

According to the assumption of self-similarity, the main terms in the expansion of functions $F(z, x, y)$, γ, γ_0, and γ_1 have the form

$$F(z, x, y) = \langle z \rangle_t^{-1} f(\varsigma, \xi), \qquad \varsigma = \frac{z}{\langle z \rangle_t}$$

$$\gamma = \gamma(\xi), \qquad \gamma_0 = 1 - \gamma, \qquad \gamma_1 = 0, \qquad \frac{x}{d} \to \infty \qquad (3.66)$$

It should be emphasized that the first term of the asymptotic expansion under consideration is useless in the vicinity of the boundary of phase space $z = 1$ (it does not satisfy the boundary condition $F = 0$, $z = 1$). The main term of the asymptotic behavior of the solution in this region can be found, just as in the statistically homogeneous case, by the mapping method (this method can be used since equation (3.58) is invariant with respect to the transformation $z \to -z + \text{const}$). As a result, we obtain in the vicinity of $z = 1$

$$F \to \frac{1}{\langle z \rangle_t} \left[f\left(\frac{z}{\langle z \rangle_t}, \xi\right) - f\left(\frac{2-z}{\langle z \rangle_t}, \xi\right) \right]$$

The first term of the asymptotic expansion γ_1 is found from equation (3.60) by using this relation.

Thus, to determine the first terms of the asymptotic expansions of the solutions of equations (3.58) – (3.60) it suffices to find the self-similar solution of equation (3.58) which is a function of two variables $f(\varsigma, \xi)$. Substituting relations (3.63) – (3.66) into (3.58) and executing cumbersome but simple transformations, we obtain an equation for function $f(\varsigma, \xi)$

$$A = \frac{\partial \gamma f}{\partial \xi} + B\gamma \frac{\partial f}{\partial \varsigma} = \gamma \frac{\partial^2 f}{\partial \varsigma^2} + C\gamma f \qquad (3.67)$$

The coefficients in equation (3.67) are described by the following expressions

$$A = \beta \xi^{-i} \Psi Z_t^2 n_t^{-1}[1 - h\mu^{-1}V], \qquad Z_t = \frac{Z}{\gamma}$$

$$B = \beta \xi^{-i} \left\{ -\frac{\partial \Psi}{\partial \xi} Z_t - \Psi \frac{\partial Z_t}{\partial \xi} [1 - h\mu^{-1}V] \right\} Z_t n_t^{-1} \varsigma$$

$$C = -B\varsigma^{-1} + Z_t^2 n_t^{-1} \left[\xi^{-i} \frac{\partial \xi^i W}{\partial \xi} - \varsigma \frac{\partial Z_t}{\partial \xi} Z_t^{-1} \frac{\partial W}{\partial \varsigma} \right] \qquad (3.68)$$

Here

$$W = \beta \xi^{-i} \Psi h \mu^{-1} V$$

$$h = \frac{Z}{\Sigma} = \sqrt{\frac{\gamma}{I_2 - \gamma}}, \qquad I_k = \int_0^\infty \varsigma^k f \, d\varsigma$$

$$V = V_0 + s(1 + \omega^2 s^2)^{-1/2}, \qquad s = h(\varsigma - \gamma)\gamma^{-1}$$

$$V_0 = (1 - \gamma) h (1 + \omega^2 h^2)^{-1/2} - \gamma \int_0^\infty s(1 + \omega^2 s^2)^{-1/2} f \, d\varsigma$$

$$\mu = (1 - \gamma) h^2 (1 + \omega^2 h^2)^{-1/2} + \gamma \int_0^\infty s^2 (1 + \omega^2 s^2)^{-1/2} f \, d\varsigma \qquad (3.69)$$

$\Psi = \xi^i \beta^{-1}(u\xi - v)$ is the stream function for random flows, $u = \xi^{-i} \partial \Psi / \partial \xi$; $\Psi = \xi$ — for a wake.

When deriving (3.69) the algebraic relation between the mass flow and mean concentration $q_y = \beta \xi^{-i} Z \Psi z_s u_s$ obtained following a single integration of the self-similar equation for the mean concentration (3.62), is used.

The self-similar function $f(\varsigma, \xi)$, as follows from (3.5) and from the natural assumption of the existence of moments I_k when $k > 0$, must satisfy the following boundary conditions

$$f(0, \xi) = 0, \qquad \lim_{\varsigma \to \infty} \varsigma^\ell f = 0 \qquad (3.70)$$

where ℓ is an arbitrary positive number.

3. General Properties of the Equation for the Self-Similar Problem. If we disregard the fact that the required function f enters into coefficients A, B, C, then (3.67) is a parabolic equation. Variable ξ in this case plays the role of time-like coordinate. It is known from the theory of standard parabolic equations that the Cauchy problem is applicable to them, i.e., the initial conditions are given for one value of the time-like coordinate.

The problem under consideration is essentially different from the classical case as a consequence of the special properties of the coefficient for the time-like coordinate. Part of these properties is caused by the symmetry of the problem. First of all, it follows from (3.68) that $A = 0$ when $\xi = 0$. Furthermore, by virtue of the symmetry, condition $\partial f/\partial \xi = 0$ must be fulfilled when $\xi = 0$. Thereby, equation (3.67) on this line in the phase space degenerates into an ordinary integrodifferential equation. Further, as seen from (3.68), $A \sim \xi$ when ξ is small. Therefore, equation (3.67) is satisfied by solutions of type $\xi^k g(\varsigma)$, where k is the arbitrary number, g is some function. With the exception of case $k = 0$, such solutions are physically meaningless. Thus, line $\xi = 0$ is singular. Since integral $\int_0^\xi A^{-1} d\xi$ is divergent, this singularity cannot be eliminated by transforming the coordinate ξ. Thus, PDF on line $\xi = 0$ cannot be specified arbitrarily (the situation here is similar to the one occurring in determining in §3.4 the asymptotic behavior of the solution when $t \to 0$ in a statistically homogeneous case and when $x \to 0$ in jets; see §3.7).

An analogous situation also arises on line $\xi = \infty$. For an explanation we revert to the repeatedly cited measurements of the conditionally averaged moments of the concentration field in turbulent fluid (Becker, Hottel, and Williams [1967]). The results of these measurements are shown in Figs. 1.3 and 1.4 from which it is evident that the deviation from statistical uniformity in the turbulent fluid in steady equilibrium is extremely insignificant. For example, measurements carried out by Becker, Hottel, and Williams [1967] in a submerged axisymmetric jet indicate that when y/x changes in the range $0.16 - 0.26$, the intermittency factor changes by 30 times, the mean concentration changes by 60 times, whereas the conditionally averaged concentration in the turbulent fluid $\langle z \rangle_t$ changes only by two times (see Fig. 1.4). The dimensionless quantity $\sigma_t/\langle z \rangle_t$ — intensity of concentration fluctuations in the turbulent fluid — in these tests within the indicated range of y/x is practically constant. Analogous results were obtained in other experiments (La Rue and Libby [1974], Antonia, Prabhu, and Stephenson [1974], Fabris [1979a, b]). The oscillograms of the "frozen" temperature profiles, obtained in a submerged plane jet by Uberoi and Singh [1975] and given in Fig. 1.1, also agree with the pattern described above.

Thus, it can be concluded that equilibrium is sufficiently rapidly established between two competing processes — entrainment of the nonturbulent fluid and mixing at the molecular level of the fluid entrained from the surrounding medium with the gas fluid.

It is evident from physical considerations that each entrainment act of nonturbulent fluid can only increase nonhomogeneity in the distribution of the fluid dynamic quantities in the turbulent fluid. Molecular mixing, on the contrary, facilitates the establishment of homogeneity. As measurements show, the total entrainment rate of the turbulent fluid is independent of Reynolds number when $Re \gg 1$ (Townsend [1956]). Hence, from the two processes under consideration, entrainment of the nonturbulent fluid is the limiting process. Mixing at the molecular level, if it can be thus expressed, is tuned to the change in rate of entrainment by the corresponding reorganization of the small-scale structure of turbulence (see analogous considerations in the work by Broadwell and Briedenthal [1982]).

Based on the aforementioned, it is natural to assume that when $\xi \to \infty$ all dimensionless combinations of the conditionally averaged moments in the turbulent fluid tend to finite values. It is evident that this is possible only when function $f(\varsigma, \xi)$ has a finite limit when $\xi \to \infty$, i.e., the following relation is true

$$\lim_{\xi \to \infty} f(\varsigma, \xi) = f_\infty(\varsigma) \tag{3.71}$$

It should be emphasized that assumption (3.71) is weaker than the assumption of total statistical nonhomogeneity in the turbulence fluid, since it follows from relation (3.71) that moments $\langle z^k \rangle_t$ ($k > 0$) can be dependent on the transverse coordinate y ($\langle z \rangle_t$ is dependent on y).

The coefficients in (3.67) must have such a form so that it would degenerate into an ordinary integrodifferential equation when $\xi = \infty$. Function f_∞ is determined from this equation. Hence, just as on line $\xi = 0$, function f on line $\xi = \infty$ cannot be specified arbitrarily.

Thus, line $\xi = \infty$ is also singular, and this singularity is nonapparent. This conclusion formally results from the integral $\int_\xi^\infty A^{-1} d\xi$ being divergent. Indeed, it follows from expression (3.96), which is given below, that $A \sim Z(dZ/d\xi)^{-1}$, i.e., $\int A^{-1} d\xi \sim \ln Z$. Therefore, equation (3.67) is satisfied by a solution of the form $Z^k \theta(\varsigma)$, where θ is some function. When $k < 0$ these solutions are physically meaningless.

In addition to the above mentioned properties, coefficient A in (3.67) possesses one more very nontrivial feature. It is linked with the geometry of the regions of sign constancy. To simplify the discussion of this question we shall assume here that the conditionally determined transverse velocity $\langle v \rangle_z$ is described by the linear dependence (3.16), i.e., in (3.69) we must substitute $v_0 = \omega = 0$, $\mu = 1$, $V = s$. Such an assumption is fully justified when identifying the qualitative features of the problem, since, as was established in §3.3, the linear dependence very satisfactorily describes the experimental data in the range of fluctuation amplitudes $|z - \langle z \rangle| < 1.5\sigma$.

By adopting this assumption, it is not difficult to find the regions of constant sign of A from the relation for coefficient A in (3.68). These regions are described by the inequalities

$$\begin{aligned} A &> 0 \quad \text{at} \quad \leq \varsigma \leq I_2(\xi), \quad 0 < \xi < \infty \\ A &< 0 \quad \text{at} \quad I_2(\xi) < \varsigma < \infty, \quad 0 < \xi < \infty \\ A &= 0 \quad \text{at} \quad \xi = 0, \quad \xi = \infty, \quad \varsigma = I_2(\xi) \end{aligned} \tag{3.72}$$

Note now that the second moment $I_2(\xi)$ is a limited function of ξ when $0 \leq \xi \leq \infty$ (the limit when $\xi = \infty$ follows from assumption (3.71)), i.e., $I_2 < \infty$. Furthermore, from the definition of I_2, we have

$$I_2 = \frac{\langle z^2 \rangle_t}{\langle z \rangle_t^2} = 1 + \frac{\sigma_t^2}{\langle z \rangle_t^2}, \qquad \sigma_t^2 = \langle (z - \langle z \rangle_t)^2 \rangle_t$$

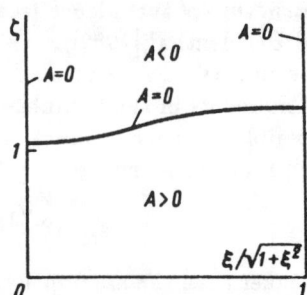

Figure 3.13 Schematic diagram of the regions of constant sign of the coefficient of time-like coordinate in equation (3.67). For convenience of representation of the asymptotic form of the region when $\xi \to \infty$ a change of the variable is effected to transfer the infinitely distant straight line $\xi = \infty$ to a finite distance from the origin of the coordinates.

i.e., inequality $I_2 > 1$ is true. Thus, we conclude that the region of constant sign of A is unlimited in the direction of ξ. The qualitative form of these regions is illustrated in Fig. 3.13. The effect of the deviation of function $\langle v \rangle_z$ from the linear dependence (3.16) in the region of large amplitudes of concentration fluctuations on the position in phase space of the line on which coefficient A tends to zero is analyzed in items 5 and 6 of the present section. It is shown there that this effect does not change the qualitative form of the regions of constant sign of A.

On the basis of this analysis it can be concluded that both directions of the time-like coordinate ξ for the parabolic equation (3.67) are equally valid. Thus, apparently, for this equation the boundary-value problem must be set not only with respect to variable ς, but also with respect to coordinate ξ.

The exposed property of coefficient A and the presence of the singular lines $\xi = 0$ and $\xi = \infty$ indicate that equation (3.67) possesses essentially nonlocal properties. Let us consider this aspect in more detail. Consider for the sake of definiteness a small neighborhood of the singular line $\xi = 0$. Note that on this line (3.67) is expressed as an ordinary integrodifferential equation whose solution is function $f(\varsigma, 0)$. In the vicinity of line $\xi = 0$ a Taylor series can be constructed with respect to variable ξ for the required solution $f(\varsigma, \xi)$. The coefficients of this series are dependent on variable ς and also satisfy the ordinary integrodifferential equations. However, such a solution will not generally satisfy condition (3.71) and will be singular in the vicinity of line $\xi = \infty$. Thus, the case when $\lim_{\xi \to \infty} \gamma = \infty$, $\lim_{\xi \to \infty} f = \infty$ is not excluded. Hence, it follows that condition (3.71) is satisfied only for a specific connection between the coefficients in equation (3.67). Such a connection is obviously obtained from the solution of the boundary-value problem with respect to variable ξ and, hence, is of nonlocal character. It should be noted that this result is, in a sense, analogous to the condition of solvability of a two-point boundary-value problem for an ordinary linear differential equation of the second order with a free parameter appearing in the coefficients of the equation.

Such a problem is well known and is solvable only for specific values of the free parameter. The difference is that (3.67) is a partial differential equation for which the boundary conditions are given on lines $\xi = 0$ and $\xi = \infty$, and that the condition of solvability of the boundary-value problem here enables finding an integral function.

Besides this connection, a second one exists between the coefficients in (3.67). It follows from the condition of normalization of function $f(\varsigma, \xi)$, and has the form of integral equation $I_0 = 1$ and is, hence, also of nonlocal character. Note that condition $I_1 = 1$, following from the definition of the conditionally averaged concentration in the turbulent fluid $\langle z \rangle_t$, is fulfilled automatically when $I_0 = 1$. In order to prove this, equation (3.67) ought to be multiplied by ς and integrated with respect to ς from zero to infinity. From the obtained result of the differential equation for I_1, it is not difficult to show that $I_1 = 1$ if only $I_0 = 1$.

In conclusion it should be indicated that the parabolic equation with analogous properties of the coefficient in the time-like coordinate are encountered in the asymptotic theory of boundary layer separation (Neiland [1971], Sychev [1972], Stewartson [1974] and in the theory of nonstationary boundary layers (Stewartson [1951], Holl [1965], Wang [1983]).

4. Statement of the Boundary-Value Problem.

The notions presented in the previous item are of qualitative character and cannot obviously be regarded as a strict proof of the existence of a solution of equation (3.67). The main purpose of these arguments is to show that for the parabolic equation (3.67) only the boundary-value problem with respect to time-like coordinate (i.e., additional conditions must be set for values of the time-like coordinate at $\xi = 0$ and $\xi = \infty$) can be correct (in a mathematical sense). The conditions of symmetry and the results of the analysis conducted above lead to the following statement of the boundary-value problem. In the region $0 \leq \xi < \infty$, $0 < \varsigma < \infty$ a non-negative function $f(\varsigma, \xi)$ is sought which satisfies equation (3.67) and the conditions

$$f(0, \xi) = 0, \qquad \lim_{\varsigma \to \infty} \varsigma^k f = 0$$

$$\frac{\partial^{2\ell+1} f(\varsigma, 0)}{\partial \xi^{2\ell+1}} = 0, \qquad \ell = 0, 1, 2, \ldots$$

$$\lim_{\xi \to \infty} f(\varsigma, \xi) = f_\infty(\varsigma), \qquad I_0 = \int_0^\infty f \, d\varsigma = 1 \qquad (3.73)$$

Note that the distributions of mean velocities (i.e., function Ψ), mean concentration Z and scalar dissipation n_t are assumed to be specified. The coefficients of equation (3.67) include, in addition to the indicated quantities, two previously unknown functions of one variable — intermittency factor $\gamma(\xi)$ and function $\omega(\xi)$ which characterizes the process of turbulent diffusion (it appears in the expression (3.18) for the conditionally averaged velocity $\langle v \rangle_z$ refined in the region of large amplitudes of concentration fluctuations). Featured also in formulas (3.68) and

(3.69) for the coefficients of equation (3.67) are quantities μ, V_0, Σ, which are expressed in terms of various integrals of the PDFs. The two indicated functions are found from two additional connections based on the following conditions.

The first connection is the condition of solvability of the boundary-value problem with respect to the time-like coordinate ξ. The second connection between γ and ω is specified by the normalization condition (last condition in (3.73)).

We should particularly note the essential role played by the requirement of non-negativity of f in the condition of solvability of the boundary problem. The character of the limitations following from this condition will be clearly seen in the analysis of equation (3.67), in items 5 and 6 of the present section, on the singular lines $\xi = 0$ and $\xi = \infty$. For the sake of definiteness let us consider line $\xi = \infty$. The analysis conducted in item 6 shows that there is a finite set of solutions on this line with only one (non-negative) meaningful (see also §3.5, item 3). It is to be expected that an analogous situation also arises from solving the general boundary-value problem (3.73). Numerical calculation (see §3.7) supports this conclusion. Therefore, the condition of non-negativity of the solution plays an important part in the formulation of the boundary-value problem.

Thus, function ω characterizing the specific feature of a large-scale process of turbulent concentration transfer is found from the solvability condition of the boundary-value problem, i.e., the process under consideration is of essentially nonlocal character. The developed approach hence provides the opportunity to take into account the nonlocal effect of large-scale pulsations on turbulent mixing.

In conclusion to this item, we note that the formulated boundary-value problem differs considerably from the boundary-value problems for parabolic equations of mixed type that are described in the literature (see, for example, Ventsel [1975], Kislov [1980]). The theoretical results here pertain to the equations which are considered in bounded rectangular regions and have no singularities and the line on which the type of the equation changes is known beforehand.

The numerical solution of the formulated boundary-value problem is obtained in §3.7. This solution is the only argument so far supporting the existence of a solution for formula (3.67). In order to understand the qualitative structure of the solution of the equation, we analyze the solution on the singular lines $\xi = 0$ and $\xi = \infty$ in the remaining items of the section.

5. Solution of the Singular Line $\xi = 0$. Since $A = 0$ when $\xi = 0$, we obtain from (3.67)

$$f'' + a_1 \varsigma f' + a_1(1 + h\mu^{-1}V)f = 0, \qquad V = V_0 + s(1 + \omega^2 s^2)^{-1/2} \qquad (3.74)$$

where $a_1 = h^2/\gamma m$, $h^2 = Z^2/\Sigma^2 = \gamma/I_2 - \gamma$, $m = \gamma n_t/\beta u \Sigma^2$ for a jet, $m = 2\gamma n_t/\Sigma^2$ for a wake; the prime in (3.74) denotes differentiation with respect to ς.

The physical meaning of parameters m and h in equation to (3.74) was discussed in §3.5. It is important to bear in mind that m is equal to the ratio of twice the scalar dissipation to the absolute value of advection, and h is inversely proportional to the intensity of concentration fluctuations. In the analyzed flows here, as indicated in §3.5, $m > 1$: $m \approx 2.6$ in a distant wake of a circular cylinder,

$m \approx 1.8$ in a plane jet, $m \approx 2$ in an axisymmetric jet in coflowing stream. The values of parameter h lie within the range $4-5$, i.e., h is a sufficiently large value.

Let us first point out the general properties of the solution of equation (3.74). Among these properties, the most significant is that the nontrivial solution of equation (3.74) exists only upon fulfilling the strict inequality $\gamma < 1$. This important conclusion follows from the relation which is obtained if (3.74) is multiplied by ς, integrated for ς from zero to infinity, and use is made of the normalization condition

$$f'(0) = a_1 h \mu^{-1} \left[\int_0^\infty s(1+\omega^2 s^2)^{-1/2} f \, d\xi + h(1+\omega^2 h^2)^{-1/2} \right] (1-\gamma) \qquad (3.75)$$

Hence, it is obvious that $f'(0) = 0$ when $\gamma = 1$. Since $f(0) = 0$, only one trivial solution exists when $\gamma = 1$. Thus, from this theory it follows that $\gamma < 1$ at all points of the flow, although this inequality does not exclude the existence of extended areas in which the difference $1-\gamma$ is negligibly small. Later in this item, it will be shown that $1 - \gamma \to 0$ when $h \to \infty$.

Let us proceed with the discussion of the general properties of equation (3.74). We shall find the number of parameters on which its solution is dependent. We use the following reasonings for this purpose. We shall assume that the values of quantities ω, h, γ, μ, and V_0 appearing in the coefficients of (3.74) are totally arbitrary so far and, hence, (3.74) is a linear differential equation. By using standard methods it is easy to establish that one of the linearly independent solutions of this equation diminishes exponentially when $\varsigma \to \infty$ (proportionally to $\exp(-1/2a_1\varsigma^2)$, and the second diminishes in an algebraic manner (proportionally to $\varsigma^{-\ell}$, $\ell = 1 + h\mu^{-1}|\omega|^{-1}$). Consequently, the boundary condition $\lim_{\varsigma \to \infty} \varsigma^k f = 0$ in (3.73) is nontrivial, and the boundary-value problem is solvable for specific values of ω. The condition of non-negativity of function $f(\varsigma, 0)$ will be fulfilled only for one of these values. The value of ω thus found and function $f(\varsigma, 0)$ contain the unknowns h, γ, μ, and V_0 as parameters. Since for specified h, γ, μ, and V_0 equation (3.74) is linear and the boundary conditions (3.73) are homogeneous, function $f(\varsigma, 0)$ is determined with an accuracy up to an arbitrary multiplier. We now take into consideration that function $f(\varsigma, o)$ must be normalized, and the integrals of $f(\varsigma, 0)$ enter into the expression for quantities μ, V_0, h, γ, and the aforementioned arbitrary multiplier. Thus, the solution of equation (3.74) is found with one arbitrary parameter. The intermittency factor is convenient to select as that parameter. The connection between γ and h in this case is found from relation (3.75).

This analysis enables, first, understanding the role played by the refinement of the linear dependence proposed in §3.3, of the conditionally averaged velocity $\langle u \rangle_z$ in the region of large amplitudes of concentration fluctuations when determining the non-negativity of the solution and, second, identifies the cause of the oscillation of the solutions of the equation when using the linear dependence (3.16) for $\langle u \rangle_z$. The equation for this case is obtained from (3.74) if we substitute $V_0 = \omega = 0$, $\mu = 1$, $V = s$. Then, both linearly independent solutions of the equation satisfy the boundary condition $\lim_{\varsigma \to \infty} \varsigma^k f = 0$ and, hence, the boundary-value problem is always solvable. It is this circumstance which is responsible for the requirement of non-negativity of function f not being satisfiable (note that case $m > 1$ is being considered).

Let us analyze the asymptotic expansion of equation (3.74) with respect to the integer exponents of the small quantity h^{-1} when $h \to \infty$.

We first consider the asymptotic solution which applies outside the small neighborhood of points $\varsigma = 0$ and $\varsigma = \infty$. Before this we note that the analysis of the uniform asymptotic behavior, whose results are presented below, indicates that the difference $1 - \gamma$ is exponentially small when $h \to \infty$. Hence, when determining the asymptotic behavior of the coefficients of equation (3.74), it can be assumed that $\gamma = 1$.

Let us substitute the variables $g(s) = h^{-1}f(\varsigma)$, $s = h(\varsigma - 1)$ in (3.74). The equation for function g has the form

$$mg'' + (s+h)g' + (h\mu^{-1}V + 1)g = 0$$

$$V = V_0 + s(1 + \omega^2 s^2)^{-1/2}$$

$$V_0 = -\int_{-h}^{\infty} s(1 + \omega^2 s^2)^{-1/2} g \, ds$$

$$\mu = \int_{-h}^{\infty} s^2 (1 + \omega^2 s^2)^{-1/2} g \, ds$$

$$g = 0, \qquad s = -h, \qquad s = \infty \tag{3.76}$$

The prime here denotes differentiation with respect to s.

Equation (3.76) differs from the equation analyzed in §3.5 only by the coefficient in front of function g. Therefore, in order to repeat the analysis of the asymptotic behavior here, we consider only the differences which arise as a result of refining function $\langle u \rangle_z$ in the region of large concentration fluctuations. Just as in §3.5, the asymptotic behavior of the solution and the value proper of ω are sought in the form of series

$$g = g^{(0)} + h^{-1}g^{(1)} + h^{-2}g^{(2)} + \ldots$$

$$\omega = h^{-1}\Omega_1 + h^{-3}\Omega_3 + \ldots \tag{3.77}$$

Then, as it follows from (3.76), the asymptotic expansions of quantities V_0 and μ have the form

$$V_0 = \frac{1}{2}\Omega_1^2 A h^{-2} + \ldots$$

$$\mu = 1 - \frac{1}{2}\Omega_1^2(3 + E)h^{-2} + \ldots$$

$$A = A_0 + A_1 h^{-1} + \ldots$$

$$E = E_0 + E_1 h^{-1} + E_2 h^{-2} + \ldots \qquad (3.78)$$

Here, A and E are the factors of symmetry and kurtosis, i.e.,

$$A = \int s^3 g \, ds, \qquad E = \int s^4 g \, ds - 3$$

When deriving relation (3.78), the function g was required to satisfy the following conditions

$$\int g \, ds = \int s^2 g \, ds = 1, \qquad \int s g \, ds = 0$$

Substituting series (3.77) and (3.78) into equation (3.76) and equating the coefficients for the successive exponents of h to zero yield a recurrent system of equations for function $g^{(k)}$, $k = 0, 1, 2, \ldots$. The equations for the first two of these functions agree with equations (3.41) and (3.42) written in §3.5 i.e., $g^{(0)}$ and $g^{(1)}$ are described by expressions (3.44). In order to obtain the equation for function $g^{(2)}$, one must add the term $1/2\Omega_1^2(s^3 - 3s)g^{(0)}$ to the right side of (3.43).

Thus, refinement of the dependence of $\langle u \rangle_z$ in the region of large concentration fluctuations influences the solution only from the third term of the asymptotic expansion. Function $g^{(2)}$ compared with (3.44) has an additional component which is equal to $1/4\Omega_1^2 s^2 (1/2s^2 - 3)g^{(0)}$. It is now obvious that the main term of the asymptotic behavior for the skewness is described as before by formula (3.45) from §3.5, i.e., $A_0 = 0$, $A_1 = 2(1 - m)$ in (3.78). The expression for the first term in the asymptotic behavior of the kurtosis differs from (3.45) owing to the additional contributions in the expression for function $g^{(2)}$, as calculations show, by the value $15/8 \, \Omega_1^2 h^{-2}$, i.e.

$$E_0 = E_1 = 0, \qquad E_2 = 6(1 - m) + \frac{15}{8}\Omega_1^2 \qquad (3.79)$$

In connection with the expression for the kurtosis, the following must be noted. As was found in §3.5, when using the linear dependence (3.16) for $\langle u \rangle_z$, the theoretical values of the kurtosis differ strongly from the experimental values. Even the signs do not agree — theory yields $E < 0$, and the experiment $E > 0$. Therefore, the encouraging fact is that the additional contribution to the kurtosis resulting from refinement of the linear dependence for $\langle u \rangle_z$ in the region of large concentration fluctuations has a positive sign. In order to determine the magnitude of this contribution quantitatively Ω_1 appearing in (3.79) must be known.

The value of Ω_1 is found from consideration of the equilibrium asymptotic behavior of the solution of equation (3.74) on the entire semi-axis $\varsigma > 0$. By somewhat anticipating developments, we note that one of the results of the study of this asymptotic behavior is the conclusion that when $m \gtrsim 1.6$, the coefficients Ω_1 and Ω_3 in the asymptotic series (3.77) for function ω acquire very large values.

Thus, for example, $\Omega_1 = 5.33$, $\Omega_3 = 300$ when $m = 2.6$. For this reason, the uniform asymptotic behavior, generally speaking, is of little use for quantitative estimates when $m \geq 1.6$ and $h = 4 - 5$. This circumstance does not detract from the significance of the uniform asymptotic behavior as a means of studying the qualitative features of the solution. To obtain quantitative results in these cases, one must resort to numerical solution of the boundary-value problem.

Let us proceed to the analysis of the uniform asymptotic behavior. Moreover, taking into account the earlier remark, we limit ourselves to the main stages of the analysis. Here, just as when considering a nonuniform asymptotic behavior, we can substitute $\gamma = 1$ in the coefficients of equation (3.74). The difference $1 - \gamma$ is found by using relation (3.75) following the determination of the asymptotic behavior of the solution.

Let us change the variables in (3.74)

$$\Pi = f \exp\left(\frac{1}{4m} h^2 \varsigma^2\right)$$

The equation and the boundary conditions for function Π have the form

$$\Pi'' - h^4 Q \Pi = 0$$

$$Q = \frac{1}{m}\left(\frac{\varsigma^2}{4m} - h^{-1}\mu^{-1}V - \frac{1}{2h^2}\right)$$

$$V = V_0 + s(1 + \omega^2 s^2)^{-1/2}, \qquad s = h(\varsigma - 1)$$

$$\Pi = 0, \qquad \varsigma = 0, \qquad \varsigma = \infty \tag{3.80}$$

Equation (3.80) contains a large parameter and its solution can be represented as an asymptotic series (see, for example, Fedoryuk [1983]). Further, only the main term of this series is written. For this purpose it suffices to account for two terms in the expansion of function Q

$$Q = Q_0 + h^{-2} Q_2$$

$$Q_0 = \frac{1}{m}\left[\frac{\varsigma^2}{4m} - (\varsigma - 1)d\right]$$

$$Q_2 = -\frac{1}{2m} - \frac{3}{2m}\Omega_1^2(\varsigma - 1)d + \frac{1}{m}(\varsigma - 1)^3 d^3 \Omega_1 \Omega_3$$

$$d = [1 + \Omega_1^2(\varsigma - 1)^2]^{-1/2} \tag{3.81}$$

In the derivation of (3.81), expansions (3.77) and (3.78) are taken into account.

It is not difficult to see that the non-negative solution of the boundary-value problem (3.80) when $h \to \infty$ exists only when function Q_0 is non-negative and tangent to the semi-axis $\varsigma > 0$ or, in other words, equation (3.80) has a turning point of the second order. Indeed, if the strict inequality $Q > 0$, $\varsigma > 0$ is true, then the homogeneous boundary conditions satisfy only the trivial solution $\Pi \equiv 0$. On the other hand, if there is a region where $Q_0 < 0$, the solution is found to be a strongly oscillating function.

The coordinates of the turning point ς_n and parameter Ω_1 are found from the equations corresponding to the condition of tangency of Q_0 with semi-axis $\varsigma > 0$: $Q_0 = Q_0' = 0$, $\varsigma = \varsigma_n$. The solution of this system at various values of parameter $m > 1$ is shown in Fig. 3.14.

The subsequent procedure for determining the first term of the asymptotic behavior of the solution in the presence of a turning point is well known (see, for example, Morse and Feshbach [1953]). To start with, one must find the external expansion of the solution. This expansion is a linear combination of two linearly independent, quasiclassical solutions of the equation. For the sake of brevity, this expansion is not presented here. This expansion is useful outside the small vicinity of the turning point. Next, the internal (local) expansion, which is valid in the small vicinity of the turning point, is found. The internal expansion for the turning point of the second order is expressed in terms of the parabolic cylinder function (Morse and Feshbach [1953]). Analysis indicates that the joining of two expansions is possible if the value of Ω_3 (with an accuracy to exponential small terms) is equal to the root of equation $\nu = 0$, where $\nu = -1/2 - Q_2(\varsigma_n)[2Q_0''(\varsigma_n)]^{-1/2}$ is the index of the parabolic cylinder function. Dependence of Ω_3 on parameter m is shown in Fig. 3.14.

Note that the internal expansion is actually necessary only for the determination of the value of Ω_3, since the function g is insignificantly different from the Gaussian function, the main variation of the solution when $h \to \infty$ takes place in the region $|s| \lesssim 3$, i.e., in the small vicinity of $|\varsigma - 1| \lesssim 3/h$ of point $\varsigma = 1$, and the distance from the transition point ς_n to point $\varsigma = 1$ is independent of h. Hence, we conclude that all singularities of the solution can be described with

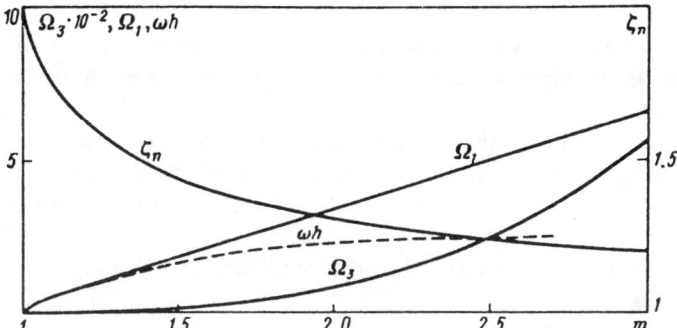

Figure 3.14 Computation of function ω appearing in expression (3.18) for the conditionally averaged velocity at a specified concentration. Curves Ω_1, Ω_3, ς_n are the result of the asymptotic solution of (3.74). Curve ωh is found with the aid of numerical solution of (3.74) when $h = 4.75$.

the aid of the external expansion in the region $0 \leq \varsigma < \varsigma_n$. Two constants in this expansion can be found by using the boundary condition $f = 0$, $\varsigma = 0$ and the normalization condition (moreover, the computation of the integral is effected by means of the saddle-point method). The final expression for the main terms of the asymptotic behavior of the solution has the form

$$f = CQ_0^{-1/4}[\exp(h^2 J_1 + J_2) - \exp(-h^2 J_1 - J_2)] \exp\left(-\frac{h^2}{4m}\varsigma^2\right)$$

$$J_1(\varsigma) = \int_0^\varsigma Q_0^{1/2} d\varsigma$$

$$J_2(\varsigma) = \frac{1}{2}\int_0^\varsigma Q_2 Q_0^{-1/2} d\varsigma, \qquad 0 \leq \varsigma < \varsigma_n$$

$$C = Q_0^{1/4}(1) \exp\left[-h^2\left(S_1 - \frac{1}{4m}\right) - S_2\right](2\pi)^{-1/2} h$$

$$S_1 = J_1(1), \qquad S_2 = J_2(1) \tag{3.82}$$

The second component in square brackets in the expression for f must be taken into account only in the vicinity of point $\varsigma = 0$.

Let us now obtain the relationship between intermittency factor γ and parameter h by using (3.82). For this purpose, we compute the derivative f' when $\varsigma = 0$ and substitute the result into the left side of relation (3.75). If we take into account that the limit of the right side of (3.75) when $h \to \infty$ is equal to $h^4 m^{-1}(1+\Omega_1^2)^{-1/2}(1-\gamma)$, then as a result we find

$$1 - \gamma = \frac{m^{1/4}(1+\Omega_1^2)^{3/8}}{h\sqrt{\pi}} \exp\left[-h^2\left(S_1 - \frac{1}{4m}\right) - S_2\right] \tag{3.83}$$

From formula (3.83) it follows that the difference $1 - \gamma$ is exponentially small. Thereby, the assumption made when obtaining the first term of the asymptotic expansion has been proved.

In accordance with relation (3.83), the small values of the intensity of concentration fluctuations (i.e., large values of h) on the axis or in the plane of symmetry is caused by the condition that the significance of intermittency in this region is small.

In the case when $\langle u \rangle_z$ is described by the linear dependence (3.16), i.e., $\Omega_1 = \Omega_3 = 0$, computation yields

$$S_1 = \frac{1}{4m} - \frac{1}{2} + \sqrt{m} - (m-1)\ln\left(1 + \frac{1}{\sqrt{m}}\right)$$

$$S_2 = -\frac{1}{2}\ln\left(1 + \frac{1}{\sqrt{m}}\right)$$

As a result, we obtain formula (3.46) which is presented in §3.5.

Let us now address some of the results of numerical integration of equation (3.74). They are presented in Figs. 3.14 and 3.15. Depicted in Fig. 3.14 is the dependence of function ω on parameter m for the case $h = 4.75$ (i.e., when $\sigma/\langle z \rangle = 0.21$ which is characteristic for jet flows). When $m = 2.6$ (wake of a circular cylinder), we have $\omega \approx 0.52$. This value can be regarded as sufficiently small. Indeed, from expression (3.18) we conclude that the main deviation of the conditionally averaged velocity from the linear dependence (3.16) occurs outside the range $|s| \gtrsim 1/\omega \approx 2$, i.e., practically outside the region of the main variation of PDF. With decreasing parameter m, as seen from Fig. 3.14, the value of ω drops and, hence, the range of concentration values with the linear dependence for $\langle u \rangle_z$, increases. The discussion of the behavior of ω will be continued in §3.7 following the numerical solution of the boundary-value problem.

Figure 3.15 depicts the calculated dependencies of the intensity of concentration fluctuations $\sigma_t/\langle z \rangle_t$, skewness A_t and kurtosis E_t, and also function ω of the intermittency factor for the case $m = 2.6$ which corresponds to the flow in the wake of a circular cylinder.

Let us compare the theoretical and experimental values of the skewness and kurtosis. Measurements of A_t and E_t in a wake were conducted by La Rue and Libby [1974]. Obtained in these tests were the values $A_t = -0.4$ and $E_t = 0.1$. The intensity of concentration pulsations $\sigma/\langle z \rangle$ amounted to 0.21, i.e., $h = 4.75$. Since the difference of unconditional and conditional means can be neglected here, then $\sigma_t/\langle z \rangle_t \approx 0.21$. According to the results of calculations shown in Fig. 3.15, for such an intensity of concentration pulsations $1 - \gamma \approx 10^{-3}$, $A_t = -0.5$, $E_t = 0.13$. It can be concluded that the calculated values of A_t and E_t agree well with the measurements of La Rue and Libby. It should be remembered here that, as shown in §3.5, the use of a linear dependence for $\langle u \rangle_z$ yields the value $E = -0.43$ for the kurtosis. Thus, refinement of the linear dependence of $\langle u \rangle_z$ in the region

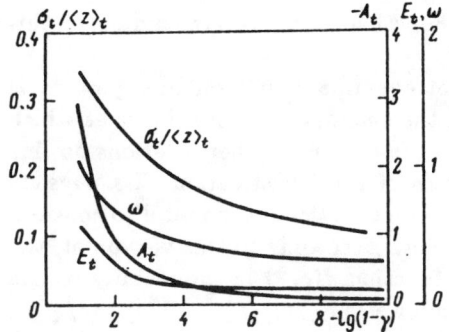

Figure 3.15 Calculated values of the intensity of concentration fluctuations, skewness, and kurtosis in the turbulent fluid and function ω on the plane of symmetry in the wake of a circular cylinder.

of large amplitudes of concentration fluctuations enables elimination of the large difference between the theoretical and experimental values of the kurtosis.

Direct comparison of the calculated and measured values of intermittency factor is not possible, since this quantity has not yet been measured with the required accuracy.

In conclusion, let us consider the effect of deviation of function $\langle u \rangle_z$ from the linear dependence (3.16) on the position of line $\varsigma_0(\xi)$ on which factor A in a time-like coordinate ξ in equation (3.67) tends to zero. The form of this line, as established in item 3, determines to a great extent the qualitative properties of equation (3.67). It should be remembered that for the linear dependence (3.16) $\varsigma_0(\xi) = I_2(\xi)$ (see item 3). In the case under consideration the equation of line $\varsigma_0(\xi)$, as can be seen from the expression for A in (3.68), can be written in the implicit form

$$h[V_0 + s(1+\omega^2 s^2)^{-1/2}] = \mu, \qquad s = h\frac{\varsigma_0 - \gamma}{\gamma} \qquad (3.84)$$

Let us find the asymptotic behavior of the solution of this equation on the line $\xi = 0$ when $h \to \infty$. We substitute expansions (3.77), (3.78) into (3.84). As a result we obtain

$$\varsigma_0 = 1 + h^{-2} - \frac{1}{2}\Omega_1^2(3 + A_1)h^{-4} + \ldots, \qquad \xi = 0$$

$$A_1 = 2(1-m) \qquad (3.85)$$

Since $I_2 = 1 + h^{-2}$, we conclude from (3.85) that when $h \approx 5$ the value $\varsigma_0(0)$ practically coincides with the second moment.

6. Solution on the Singular Line $\xi = \infty$. A special solution of equation (3.67) on the singular line $\xi = \infty$ satisfying assumption (3.71) was obtained in §3.5 by means of a nonrigorous method. Function $f_\infty(\varsigma)$ corresponding to this solution has the form of (3.53).

In this item a rigorous investigation of equation (3.67) on line $\xi = \infty$ is conducted. Such investigation allows, first, the elucidation of the limits at which the solution obtained in §3.5 is valid and, second, to find other solutions on line $\xi = \infty$. Assumption (3.71) lies at the basis of the investigation. Its physical substantiation is included in item 3 of the present section. It is useful to consider the assumption from another formal viewpoint. According to this viewpoint, the main term of the asymptotic expansion of function $f(\varsigma, \xi)$ in the vicinity of line $\xi = \infty$ is described by a self-similar dependence, i.e., $f(\varsigma, \xi) \to \xi^{\beta_1} f_\infty(\varsigma/\xi^{\beta_2})$, $\xi \to \infty$, β_1 and β_2 are constants. Function f must satisfy the normalization condition and the consequence from the definition of $\langle z \rangle_t$

$$I_0 = 1, \qquad I_1 = 1, \qquad I_k = \int_0^\infty f \varsigma^k d\varsigma, \qquad k = 0, 1$$

It is not difficult to realize that this is possible only if $\beta_1 = \beta_2 = 0$. As a result we arrive at assumption (3.71).

Let us proceed with the analysis of equation (3.67) on line $\xi = \infty$. It is evident that assumption (3.71) is fulfilled only when $\lim_{\xi \to \infty} A\partial f/\partial \xi = 0$ (indeed, if this condition is violated, then function f is dependent on ξ when $\xi \to \infty$). In this case equation (3.67) on line $\xi = \infty$ acquires the form

$$f''_\infty + \alpha_1(\varsigma) f'_\infty + \alpha_0(\varsigma) f_\infty = 0$$

$$\alpha_1 = -\lim_{\xi \to \infty} B, \qquad \alpha_0 = \lim_{\xi \to \infty} \left(C - \frac{1}{\gamma} \frac{d\gamma}{d\xi} A \right) \qquad (3.86)$$

the prime in (3.68) denotes differentiation with respect to ς.

It is easy to establish that the boundary-value problem $f_\infty = 0$, $\varsigma = 0$, $\varsigma = \infty$ is solvable only when $\alpha_0 \neq 0$. Indeed, otherwise, from (3.86) it follows that

$$f'_\infty(\varsigma) = f'_\infty(0) \exp\left[\int_0^\varsigma \alpha_1(t) dt \right]$$

i.e., derivative f'_∞ does not change the sign on the semi-axis $\varsigma > 0$ and, thereby, $f_\infty \neq 0$ when $\varsigma = \infty$.

Hence, it can be concluded that at least one of the components appearing in the expression for coefficient α_0 must be different from zero. Coefficients α_0 and α_1 in (3.86) are determined by the first terms of the asymptotic expansions of the mean concentration Z, scalar dissipation in the turbulent fluid n_t, function Ψ, function ω and intermittency factor γ in the vicinity of an infinitely distant point $\xi = \infty$.

The question concerning the asymptotic behavior of function Ψ is the simplest to solve. We have $\Psi = \xi$ in a wake of a circular cylinder, $\Psi \to $ const when $\xi \to \infty$ in jets (see, for example, Schlichting [1960]). The first term of the mean concentration in accordance to the experimental data is presented in the following general form

$$Z \to Z_0 \exp[-p_1(\xi)], \qquad \xi \to \infty \qquad (3.87)$$

The pre-exponent Z_0 in this expression tends to zero or to infinity more slowly than some exponent ξ. The measurements in the axisymmetric jet, the results of which are given in Fig. 1.3, enable the conclusion to be drawn that here $p_1(\xi) \sim \xi^2$. It is to be expected that function p_1 is also close to the quadratic in other flows.

The determining role in further investigation is attributed to the asymptotic behavior of the conditionally averaged scalar dissipation in the turbulent fluid. Let us specify this asymptotic behavior on the basis of the experimentally established property of weak change of the conditionally averaged quantities (see §§1.1, 3.5, and item 3 of the present section). In accordance with this property, it is natural to assume that the main term of the asymptotic behavior of function n_t is of the same character as the pre-exponential multiplier Z_0 in the asymptotic behavior (3.87) for the mean concentration, i.e.

$$\text{const } \xi^{k_1} < n_t < \text{const } \xi^{k_2}, \qquad \xi \to \infty \tag{3.88}$$

Here, k_1 and k_2 are some constants which can be either positive or negative. This assumption does not contradict the known experimental data to date which is depicted in Figs. 1.7 and 3.16. Figure 1.7 shows the results of direct measurements of the conditionally averaged squared temperature derivative in the wake of a circular cylinder (Fabris [1979a, b], and Fig. 3.16 shows the profile of the scalar dissipation in a wake which is found by indirect means by calculation using formula $\langle N \rangle_t = \langle N \rangle / \gamma$ from the data of two experimental works (the measurements of $\langle N \rangle$ were conducted by Freymuth and Uberoi [1971], the measurements of γ were conducted by La Rue and Libby [1974]).

It is, however, pertinent to emphasize that assumption (3.88) is based so far on a limited amount of experimental data. Therefore, generally speaking, one cannot totally exclude the probability that in a number of cases the asymptotic behavior of the scalar dissipation will be described by a relation differing from (3.88). An analysis corresponding with this case was conducted in the works by Kuznetsov and Sabel'nikov [1981b] and Sabel'nikov [1982b].

Condition $\alpha_0 \neq 0$, as is seen from the relation for coefficient A in (3.68), is fulfilled if

$$\frac{1}{\gamma} \frac{d\gamma}{d\xi} \beta \frac{\Psi}{\xi^i} \frac{Z_t^2}{n_t} \to -a, \qquad \xi \to \infty$$

$$\beta = \frac{1+i}{2}, \qquad Z_t = \frac{Z}{\gamma} \tag{3.89}$$

Here a is some positive constant.

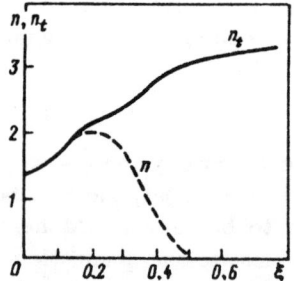

Figure 3.16 Profile of the conditionally averaged scalar dissipation over a fully turbulent fluid in the wake of a circular cylinder, calculated from measurement results of the unconditionally averaged scalar dissipation by Freymuth and Uberoi [1971], and intermittency factor by La Rue and Libby [1974]. The conditions of the tests by La Rue and Libby are the same as in Fig. 1.14. The conditions of the tests by Freymuth and Uberoi: $x_1/d = 1140$, $Re_d = u_0 d/\nu = 960$, $d = 0.28$ cm, $u_0 = 6.1$ m/sec, $x_{10} = -50d$, $n_t = \langle N \rangle_t (x_1 - x_{10})^2 / u_0 d$, $n = \langle N \rangle (x_1 - x_{10})^2 / u_0 d$, $\langle N \rangle = \gamma \langle N \rangle_t$, $\xi = x_2 / \sqrt{(x_1 - x_{10})d}$.

The first term of the asymptotic behavior of the intermittency factor is found from (3.89). If (3.87) and (3.88) are taken into consideration, then following integration of (3.89) we obtain

$$\gamma \to \left[\frac{dp_1}{d\xi}\frac{\Psi\beta}{\xi^i n_t a}\right]^{1/2} Z, \qquad \frac{dp_1}{d\xi} > 0 \qquad (3.90)$$

It follows from this formula that the main term of the expansion of the conditionally averaged concentration over the turbulent fluid is equal to

$$Z_t = \frac{Z}{\gamma} \to \left[\frac{dp_1}{d\xi}\frac{\beta\Psi}{\xi^i n_t a}\right]^{-1/2} \qquad (3.91)$$

Featured on the right-hand side of (3.91) are slowly changing (presumably) functions. Hence, the decrease of the conditionally averaged concentration Z_t with increasing ξ takes place very slowly. This result agrees with the property indicated above of weak deviation from the statistical homogeneity in the fully turbulent fluid. According to this property, the rapid change of the unconditionally averaged quantities are completely determined by the intermittency factor.

It should be noted that formula (3.91) and the analysis conducted herewith are meaningful with some upper limitation on function n_t. This limitation can be obtained if we take into account that inequality $Z_t < 1$ must be satisfied. The latter is satisfied if

$$n_t < \frac{\beta\Psi}{\xi^i a}\frac{dp_1}{d\xi} \qquad (3.92)$$

In particular, if $p_1 \sim \xi^2$, $\xi \to \infty$, then $n_t <$ const ξ^3 for a wake of a circular cylinder, $n_t <$ const ξ for a plane jet, and $n_t <$ const for an axisymmetric jet. The limitation (3.92) is not significant if one accounts for the initial assumption (3.88).

We must now specify the first term of the asymptotic behavior of function ω entering into formula (3.18) for the conditionally averaged velocity $\langle u \rangle_z$. We shall assume that this is described by the following dependence

$$\omega^2 \to \omega_\infty^2 I_2 \gamma \qquad (3.93)$$

Here ω_∞ is a constant. To elucidate (3.93) we note that the form of the main term of expansion ω is selected from the condition that when $\xi \to \infty$ there exists a finite limit of the product $\omega^2 s^2$, $s = z - \langle z \rangle/\sigma = h(\varsigma - \gamma)/\sigma$, $h^2 = \langle z \rangle^2/\sigma^2 = \gamma/(I_2 - \gamma)$, appearing in (3.18). If (3.93) is true, then $\omega^2 s^2 \to \omega_\infty^2 \varsigma^2$, $\xi \to \infty$.

Substitution of the first terms of the expansions of functions Ψ, Z, ω, γ, and n_t (3.87), (3.88), (3.90), (3.93) into the expressions for coefficients A, B, and C in (3.68) yields the following result

$$\lim_{\xi \to \infty} B = \lim_{\xi \to \infty} C = 0$$

$$\lim_{\xi \to \infty} \frac{1}{\gamma} \frac{d\gamma}{d\xi} A = b\varsigma(1+\omega_\infty^2 \varsigma^2)^{-1/2} - a$$

$$b = \frac{a}{I_2 \mu}, \qquad \mu I_2 = \int_0^\infty \varsigma^2 (1+\omega_\infty^2 \varsigma^2)^{-1/2} f d\varsigma \tag{3.94}$$

Hence, for coefficients α_0 and α_1 in equation (3.86) we obtain

$$\alpha_0 = -b\varsigma(1+\omega_\infty^2 \varsigma^2)^{-1/2} + a, \qquad \alpha_1 = 0 \tag{3.95}$$

Prior to writing the final equation for function f_∞, let us consider the asymptotic behavior of coefficient A in front of the derivative with respect to the time-like coordinate ξ. This asymptotic behavior enables establishing the particular character of line $\xi = \infty$, and also finding the limit value to which the equation of the line, on which coefficient A goes to zero, tends. From (3.94) it follows that the main term of the asymptotic behavior A has the form

$$A \to -\left(\frac{dp_1}{d\xi}\right)^{-1} [b\varsigma(1+\omega_\infty^2 \varsigma^2)^{-1/2} - a]$$

$$= Z \left(\frac{dZ}{d\xi}\right)^{-1} [b\varsigma(1+\omega_\infty^2 \varsigma^2)^{-1/2} - a], \qquad \xi \to \infty \tag{3.96}$$

Using expression (3.96), we obtain $\int A^{-1} d\xi \sim \ln Z$. Since $\ln Z \to -\infty$, $\xi \to \infty$, the indicated integral is divergent. Hence, the singularity on line $\xi = \infty$ cannot be eliminated with the aid of variable transformation. This property of coefficient A has already been discussed in item 3 of the present section.

The asymptotic behavior of line $\varsigma_0(\xi)$, $\xi \to \infty$ on which $A = 0$, as evident from (3.96), is equal to

$$\varsigma_0^2 \to \frac{(\mu I_2)^2}{1 - \omega_\infty^2 \mu^2 I_2^2}, \qquad \xi \to \infty \tag{3.97}$$

Note that this expression is greater than zero, since from the determination of μI_2 (3.94) the following inequality is obtained

$$\mu I_2 = \int_0^\infty \varsigma^2 (1+\omega_\infty^2 \varsigma^2)^{-1/2} f_\infty d\varsigma < \frac{1}{\omega_\infty} \int_0^\infty \varsigma f_\infty d\varsigma = \frac{1}{\omega_\infty}$$

On the basis of (3.85) and (3.97), we conclude that the refinement of dependence $\langle u \rangle_z$ in the region of large amplitudes of concentration fluctuations does not qualitatively change the form of line $A = 0$ which is established in item 3 when using a linear dependence for $\langle u \rangle_z$.

Let us return now to equation (3.86) for f_∞. Substitution of the coefficients α_0 and α_1 into this equation yields

$$f_\infty'' + [a - b\varsigma(1+\omega_\infty^2 \varsigma^2)^{-1/2}] f_\infty = 0, \qquad b = \frac{a}{I_2 \mu}$$

$$f_\infty = 0, \quad \varsigma = 0, \quad \varsigma = \infty \tag{3.98}$$

Simple analysis shows that one of the linearly independent solutions of equation (3.98) increases exponentially when $\varsigma \to \infty$ and, hence, does not satisfy the boundary condition when $\varsigma = \infty$. Therefore, if the value of parameter ω_∞ is specified, then the solution of the boundary-value problem exists only for specific values of constant a. The condition of non-negativity of f_∞ can be satisfied at one of these values.

Function f_∞ decays most rapidly when $\omega_\infty = 0$. The equation for f_∞ in this case assumes the form

$$f''_\infty + (a - b\varsigma)f_\infty = 0 \tag{3.99}$$

It is not difficult to verify that equation (3.51), analyzed in §3.5, reduces to (3.99) following transformation of the variables $b_1 = b/\langle z\rangle_t^3$, $a_1 = a/\langle z\rangle_t^2$, $F = f_\infty/\langle z\rangle_t$. Hence, the solution of equation (3.99) is described by formula (3.53). The value of constant a in the first term of the expansion of intermittency factor (3.90) when $\omega_\infty = 0$ is equal to $a = 7.847$.

The solution of the boundary-value problem (3.98) for case $\omega_\infty \neq 0$ is obtained numerically by Sabel'nikov [1985b]. The results of calculations of PDF and a number of its integral characteristics for several values of parameter ω_∞ are presented in Figs. 3.17 and 3.18.

It is evident from Fig. 3.17 that with increasing ω_∞ the rate of reduction of PDF decreases for large amplitudes of fluctuations. As a result, the intensity of concentration fluctuations, skewness, and kurtosis increase (see Fig. 3.18). Thus taking into account the effect of large-scale concentration transport in the dependence for $\langle u\rangle_z$ leads to an increase in the probability of large amplitudes of concentration fluctuations.

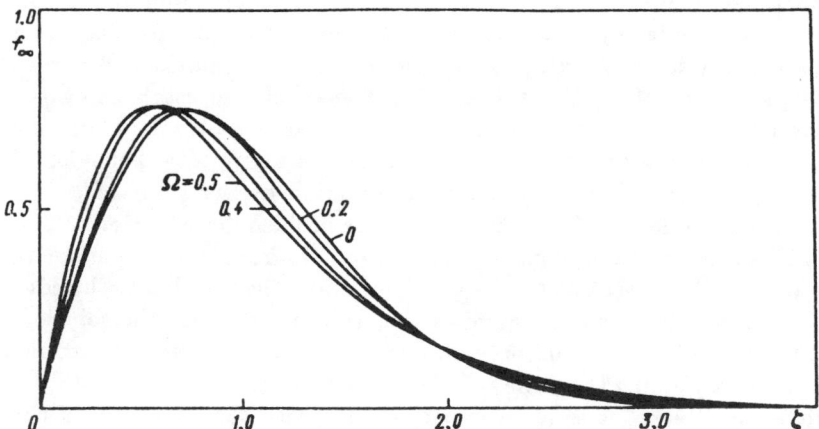

Figure 3.17 Calculated concentration PDF in the turbulent fluid at a large distance from the axis or plane of symmetry. $\Omega = \omega_\infty b^{-1/3}$.

116 TURBULENCE AND COMBUSTION

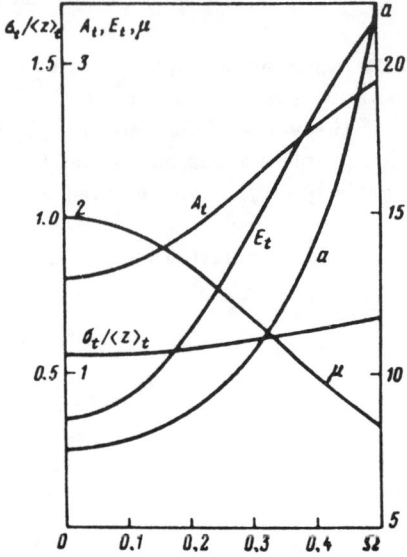

Figure 3.18 Calculated values of the first characteristic value, intensity of fluctuations skewness and kurtosis in the turbulent fluid at a large distance from the axis or plane of symmetry. $\Omega = \omega_\infty b^{-1/3}$.

At the same time, it should be noted that for not-very-large amplitudes of concentration fluctuations the form of PDF changes weakly depending on the value of parameter ω_∞ in the investigated range; the intensity of concentration fluctuations also changes very weakly. This fact and comparison with the experimental data given in §3.5 enable concluding that it is apparently expedient, heretofore, to assume that $\omega_\infty = 0$.

Condition (3.88) played an important role when obtaining equation (3.98). Hence, it can be concluded that the approximate method in §3.5 for obtaining equation (3.51) is applicable only upon satisfaction of this condition. Rigorous analysis of the solutions of (3.67) on line $\xi = \infty$ enabled also establishing the physical meaning of constant a in equation (3.98). As seen from (3.90), the value of this constant determines the first term of the asymptotic behavior of the intermittency factor. Therefore, the determination of the value of a is a prerequisite in the formulation of the boundary-value problem for equation (3.67).

Let us dwell also on an interesting aspect that follows from the comparison of approximate (see §3.5) and rigorous methods of obtaining the solution at the edge of the turbulent flow. From the viewpoint of the rigorous approach the solution, described by formula (3.53), is valid only on line $\xi = \infty$. On the other hand, the approximate method yielding the same solution does not include such a limitation in an explicit form. Hence, it is to be expected that with the exception of the region near the axis or plane of symmetry of the flow, where the intermittency factor is sufficiently close to unity, the solution of (3.67) differs very insignificantly from function (3.53). The numerical solution of the boundary-value problem confirms this conclusion (see §3.7).

Finally, we note that the conditions of applicability of the approximate method do not feature the assumption of self-similarity. Therefore, function (3.53) can be used as a first approximation for PDF in a fully turbulent fluid and at the edge of non-self-similar turbulent flows. The formulated conclusions justify the proposed approximate method in §3.5 for the determination of concentration PDF and intermittency factor in turbulent jets.

§3.7 NUMERICAL SOLUTION OF THE BOUNDARY-VALUE PROBLEM

The main property of the parabolic equation (3.67), as shown in §3.6, is that both directions along the time-like coordinate ξ are equally valid. Therefore, the main requirement of the numerical method which is intended for the solution of this equation is for the difference approximation to correctly reflect this circumstance. For this purpose, a sufficiently simple finite difference method was employed here which is widely used for the calculation of the boundary layer with local zones of reverse flows (Carter [1974], Kleinberg and Steger [1974]).

Let us briefly explain the idea of the method. It is to the effect that the difference approximation of the derivative with respect to the time-like coordinate ξ varies depending on the sign of coefficient A: when $A > 0$ backward differences are used, and when $A < 0$ forward differences are used (in computational fluid dynamics such difference approximation is usually referred to as undirectional or counterflow differences; see Roache [1976]). Similarly, depending on the sign of B the first derivative with respect to ξ is also approximated. The second derivative with respect to ς is approximated in the usual manner. It is evident from qualitative considerations that the described finite difference approximation with respect to the time-like coordinate ξ enables achieving sufficiently exact correspondence between the regions of influence and dependence of the differential equation (3.67) and the difference scheme. The difference scheme approximates the differential equation (3.67) with a first order of accuracy.

In the calculation, instead of the infinite region $0 \leq \xi < \infty$, $0 \leq \varsigma < \infty$ the rectangle $0 \leq \xi \leq \xi_{\max}$, $0 \leq \varsigma \leq \varsigma_{\max}$ is considered, i.e., the boundary conditions when $\xi = \infty$ and $\varsigma = \infty$ in (3.73) are transferred to lines $\xi = \xi_{\max}$ and $\varsigma = \varsigma_{\max}$, respectively.

The nonlinear system of finite difference equations are solved by the relaxation method (see, for example, Kalitkin [1978]). At initial (zero) iteration PDF at all the internal nodes of the difference mesh (ξ_i, ς_j) are specified by formula

$$f_{ij} = \gamma_1 Z(0) \Sigma^{-1}(0)(2\pi)^{-1/2} \exp\left(-\frac{1}{2}s_j^2\right) + (1-\gamma_i)f_\infty(\varsigma_j)$$

$$s_j = Z(0)\Sigma^{-1}(0)(\varsigma_j - 1), \qquad i = 1,\ldots, I-1; \qquad j = 2,\ldots, J-1$$

$$\xi_I = \xi_{\max}, \qquad \varsigma_J = \varsigma_{\max}$$

Here, $f_\infty(\varsigma)$ is described by relation (3.53), the values of γ, Z, and Σ on the left boundary of the integration domain $\xi_1 = 0$, i.e., $\gamma(0)$, $Z(0)$, and $\Sigma(0)$ at the start of iteration are taken from experiments. The value of the intermittency factor on the right boundary of the integration domain ξ_I, i.e., γ_I is specified by the asymptotic formula (3.90). Function ω at the start of iteration is considered equal to zero. The variance and integrals of function f appearing in quantities V_0 and μ (3.69) are obtained by the trapezoidal method.

Let distribution $f_{ij}^{(k)}$, $\gamma_i^{(k)}$, and $\omega_i^{(k)}$ be known at the k-th iteration. Now, on lines $\xi_i = $ const starting with $i = 1$ and ending $i = I - 1$ the boundary-value problem is solved at the characteristic values for ω_i by the linearization method (see, for example, Kalitkin [1978]). Upon transition to the neighboring line $\xi_{i+1} > \xi_i$ the values $f^{(k+1)}$, $\omega^{(k+1)}$ which are obtained anew at $\xi = \xi_i$ are used. The value of the intermittency factor at the $(k + 1)$-th iteration is then found from the normalization condition $I_0 = 1$. In all the calculations the lower relaxation is used for convergence with a parameter equal to 0.01. Iteration ends when the value of the sum of the nodes of the difference mesh at the current and preceding iteration divided by the number of internal nodes of the difference mesh (i.e., by $(I - 1)(J - 2)$) becomes less than the specified small number (equal to 10^{-3}).

We also note the following aspects. When the finite difference scheme is formulated, the condition of symmetry (3.73) is set in a simplified form. It is assumed that on the axis or in the plane of symmetry of the flow only the first derivative of f with respect to ξ tends to zero, i.e., instead of $\partial^{2\ell+1} f / \partial \xi^{2\ell+1} = 0$, $\ell = 0, 1, 2, \ldots$, when $\xi = 0$ it is assumed that $\partial f / \partial \xi = 0$. To increase the accuracy of the numerical solution in the vicinity of line $\xi = 0$, where the intermittency factor is close to unity, function $f(\varsigma, \xi)$ is represented as a product of two functions one of which is equated to $\exp(-\varphi s^2)$, $s = (z - \langle z \rangle)/\sigma$. The value of constant φ is selected during the process of calculation from the condition of weak change of the second cofactor in the range $-h \leq s \leq 0$.

With the aid of the described method the numerical solution is obtained for a distant wake of a circular cylinder, for axisymmetric and plane submerged jet, and also for the mixing layer of plane-parallel flows. It is important to bear in mind that the solution of the last problem is the asymptotic behavior of the solutions of equations (3.58) – (3.60) for a jet when $x \to 0$. The equations for the self-similar mixing layer are presented by Sabel'nikov [1982c]. The self-similar coordinate ξ here is the ratio $\xi = \nu/x$ and the self-similar PDF is described by expression

$$F(z, x, y) = \sigma_t^{-1} g(s_t, \xi), \qquad s_t = \frac{z - \langle z \rangle_t}{\sigma_t}$$

where $\sigma_t^2 = \langle z^2 \rangle_t - \langle z \rangle_t^2$ is the variance of concentration fluctuations in the turbulent fluid. The cited work shows that the equation for function g degenerates into an ordinary differential equation on lines $\xi = -\infty$ and $\xi = \infty$, which are singular, analogously to lines $\xi = 0$ (axis or plane of symmetry) and $\xi = \infty$ in jets and wakes. Functions $g(s_t, -\infty)$ and $g(s_t, \infty)$ just as $f_\infty(\varsigma)$ in §3.5 (see (3.53)) are expressed in terms of Airy function. If $\lim_{\xi \to \infty} \langle z \rangle = 1$, $\lim_{\xi \to \infty} \langle z \rangle = 0$, then $g(s_t, \infty)$ is determined in the range $-[I_2(\infty) - 1]^{-1/2} \approx -1.802 \leq s_t < \infty$ and is connected with $f_\infty(\varsigma)$ by the following relation: $g(s_t, \infty) = [I_2(\infty) - 1]^{-1/2} f_\infty(\varsigma)$, $\varsigma = 1 + [I_2(\infty) - 1]^{-1/2} s_t$. Function $g(s_t, -\infty)$ is linked with $g(s_t, \infty)$ by the obvious equality $g(s_t, -\infty) = g(-s_t, \infty)$. The statement of the boundary-value problem is effected the same way as in flows with an axis or plane of symmetry.

Figures 3.19 and 3.20 show some results of the calculations for a wake of a circular cylinder and the mixing layer of two plane-parallel flows obtained in the articles by Sabel'nikov [1982b, c]. Plotted on the same figures are the results of measurements in a wake carried out by Freymuth and Uberoi [1971], La Rue and Libby [1974], and in the mixing layer on the initial section of a submerged heated plane jet — Rajagopalan and Antonia [1980]. It is evident that they agree very satisfactorily with the computed results. The profiles of the mean velocities, concentration and scalar dissipation, averaged over the turbulent fluid, were specified in the calculations. For this purpose the experimental data used were those of La Rue and Libby [1974], Freymuth and Uberoi [1971] (wake of a cylinder) and Rajagopalan and Antonia [1980] (mixing layer).

Measurements of scalar dissipation in the mixing layer are unknown in the literature; therefore, function $\langle N \rangle_t$ was found with the aid of formula $\langle N \rangle_t = \langle N \rangle / \gamma$, where $\langle N \rangle$ is described by the empirical relation $\langle N \rangle = \chi \sigma^2 \langle \epsilon \rangle / q^2$, $\chi = 2$ (see, for example, Beguier, Dekeyser, Launder [1978]). To clarify we note that this relation is none other than the assumption of proportionality of the time scale of concentration fluctuations $\sigma^2 / \langle N \rangle$ and ordinary time scale of turbulence $q^2 / \langle \epsilon \rangle$.

The values of concentration variance σ^2 and intermittency factor γ entering the presented formula for $\langle N \rangle_t$ were taken from the experiments by Rajagopalan and Antonia [1980]. Turbulence energy dissipation $\langle \epsilon \rangle$ and velocity variance are taken from the experiments by Wygnanski and Fiedler [1970][†].

The profile of the scalar dissipation in the turbulent fluid $\langle N \rangle_t$ in a wake of a circular cylinder is presented in Fig. 3.16. This was obtained on the basis of the measurements of the unconditionally averaged scalar dissipation $\langle N \rangle$ by Freymuth and Uberoi [1971] and the intermittency factor γ by La Rue and Libby [1974] from formula $\langle N \rangle_t = \langle N \rangle / \gamma$. Note that profile $\langle N \rangle_t$ which is determined by such an indirect method is analogous to the profile of the conditionally averaged squared temperature derivative $\langle (\partial T / \partial t)^2 \rangle_t$. Direct measurements of the latter were conducted by Fabris [1979a, b]. These data are presented in Fig. 1.7.

It follows from the calculations that the concentration PDF differs little from the normal distribution in the entire region where the intermittency factor is close to unity. This conclusion agrees with the data of measurements of the skewness and kurtosis which were obtained in the region under study by La Rue

---

[†]Note that the method described here of the determination of scalar dissipation in the mixing layer with σ taken from the experimental data is selected only to simplify the calculations. Under a more rigorous approach the scalar dissipation must also be found in the process of numerical solution from the presented formula (moreover, $\langle N \rangle_t$ is recalculated anew at each new iteration). The good agreement of the calculation by the simplified method with the experimental values of σ is an indication that the said refinement does not alter the solution noticeably.

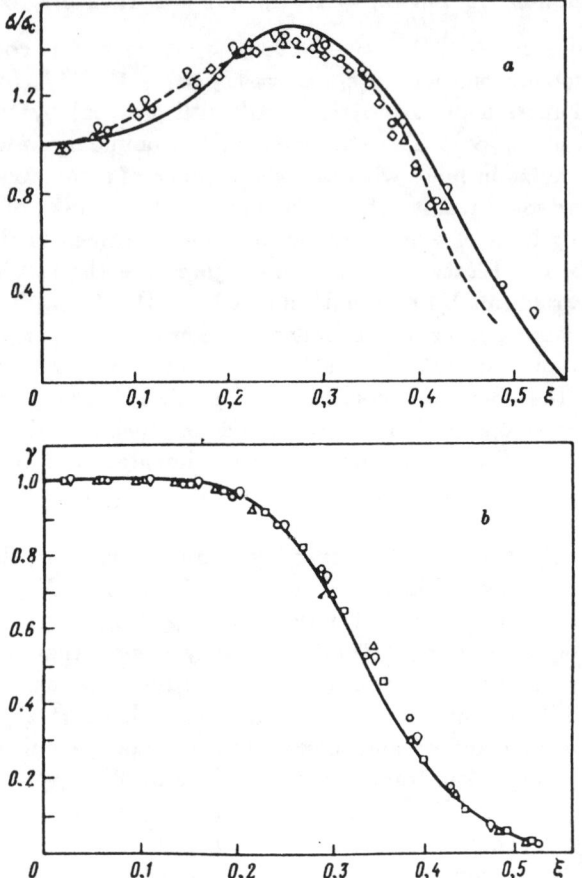

Figure 3.19 Comparison of the results of calculation of root-mean-square concentration fluctuations and intermittency factor in a wake of a circular cylinder with the experimental data of La Rue and Libby [1974, 1976], and Freymuth and Uberoi [1971]; a) solid line — calculated σ/σ_c; symbols — measurement results of σ/σ_c by La Rue and Libby [1974] (test conditions are shown in Fig. 1.14); dashed line — measurement results of σ/σ_c by Freymuth and Uberoi (test conditions are shown in Fig. 3.16) $\xi = x_2/\sqrt{(x_1 - x_{10})d}$, σ_c — root-mean-square value of concentration fluctuations in the plane of symmetry of a wake; b) solid line — calculated γ; symbols — measurement results of γ by La Rue and Libby [1976] (test conditions are shown in Fig. 1.14 $\xi = x_2/\sqrt{(x_1 - x_{10})d}$. Calculation conditions: $\xi_I = 1$, $\varsigma_J = 3$, $I = 51$, $J = 121$.

and Libby [1974], Antonia, Prabhu, and Stephenson [1975], Venkataramani and Chevray [1978], Birch, Brown, Dodson, and Thomson [1978]. This conclusion is also supported by direct measurements of the PDF whose results are presented in the indicated works and also in the articles by Kuznetsov [1971], Shcherbina

[1982], Golovanov and Shcherbina [1979], Meshkov and Shcherbina [1979] and a number of other works.

The calculated results also enable the conclusion that the transformation of the concentration PDF into the limit dependence which is given by formula (3.53) occurs in a fairly narrow region. Thus, for a wake, this occurs in the range $0.3 \leq \xi \leq 0.45$ (Fig. 3.21). Hence, as a first approximation, two regions can be identified wherein the qualitative form of the PDF is substantially different. In the first region, where $\gamma \approx 1$, the PDF is close to the normal, and in the second, where intermittency is substantial, the PDF is described by expression (3.53).

This conclusion has great practical significance, since it justifies the principal assumptions used in §3.5 when developing a simple approximate method of describing the concentration PDF in turbulent jets.

Concluding the discussion of the numerical solution, we briefly dwell on the results of calculation of function ω which is featured in expression (3.18) for the conditionally averaged velocity $\langle v \rangle_z$. The character of variation of this function if found to be identical in all the flows under study. Let us describe this on the example of flow in a wake. The maximum value of ω attained when $\xi = 0$ is $\omega(0) \approx 0.5$ (in jets and mixing layer the maximum value of ω is somewhat less; see also Fig 3.14). Near the plane of symmetry, where the intermittency factor

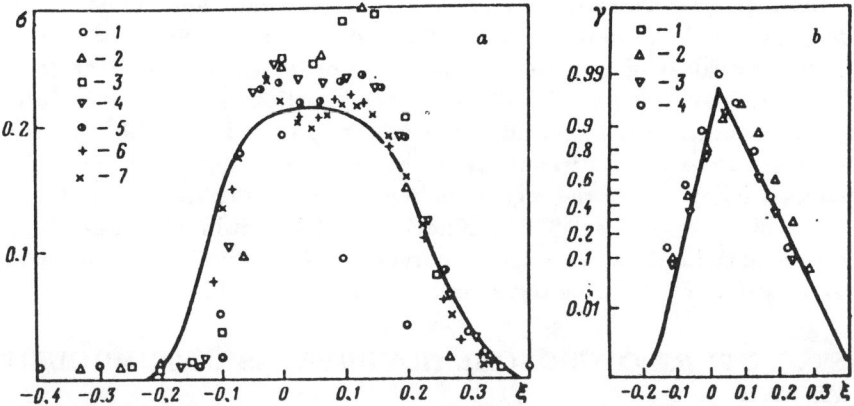

Figure 3.20 Comparison of the results of calculations of root-mean-square of concentration fluctuations and intermittency factor in the mixing layer with the experimental data by Rajagopalan and Antonia [1980] obtained in the mixing layer in the initial section of submerged heated plane jet. Symbols – experimental data. Test conditions: $Re_d = u_0 d/\nu = 2.65 \cdot 10^4$, $d = 2.54$ cm, $u_0 = 16.1$ m/sec; a) the solid curve-calculated σ; symbols — measurement results of σ; 1) $x/d = 1$, 2) $x/d = 1.5$, 3) $x/d = 2$, 4) $x/d = 2.5$, 5) $x/d = 3$, 6) $x/d = 3.5$, 7) $x/d = 4$, $\xi = y/x$; b) the solid curve — calculated γ; symbols — measurement results of γ; 1) $x/d = 1$, 2) $x/d = 2$, 3) $x/d = 3$, 4) $x/d = 4$, $\xi = y/x$. Calculation conditions $\xi_1 = -0.3$, $\xi_I = 0.4$, $I = 101$, $J = 121$; coordinate system as in Fig. 1.16; the scale along the ordinate axis is selected so that the normal distribution (error function) would be represented by a straight line.

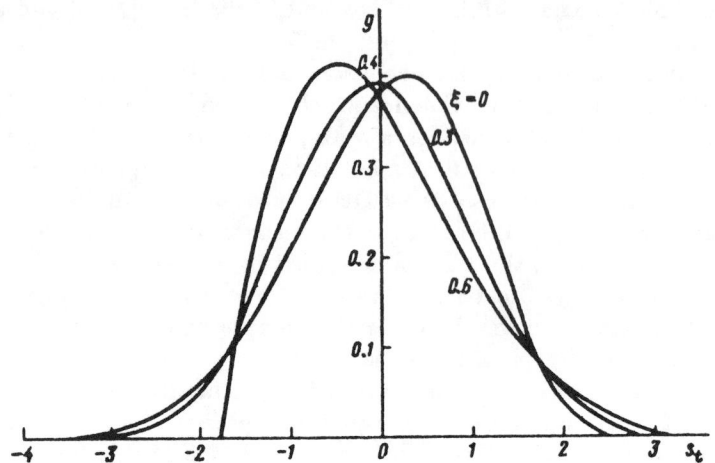

Figure 3.21 Calculated concentration PDFs in a turbulent fluid at various distances from the plane of symmetry of a self-similar wake of a circular cylinder. $g = \sigma_t P_t = \sigma_t f/\langle z \rangle_t$, $s_t = (z - \langle z \rangle_t)/\sigma_t$.

is close to unity, function ω drops slowly with increasing ξ. A sharp drop takes place in the region where restructuring of PDF to the limit form (3.53) occurs. For large ξ, i.e., at the edge of the wake, function ω is practically equal to zero.

The nonzero values of ω causes the deviation of $\langle v \rangle_z$ from the linear dependence (3.16). It is evident from (3.18) that these deviations start to take effect in the range of concentration amplitudes $|z - \langle z \rangle|/\sigma \gtrsim 1/\omega$. If $\omega \approx 0.5$, the linear dependence describes $\langle v \rangle_z$ satisfactorily in the range $|z - \langle z \rangle|/\sigma \lesssim 2$. This range encompasses a large part of the region wherein the main variation of PDF takes place. Thus, as expected (see §3.3), the deviation of $\langle v \rangle_z$ from the linear dependence is substantial only at very large concentration amplitudes. In this sense, the obtained values of ω can be regarded as sufficiently small.

§3.8 STRUCTURE OF ISOSCALAR SURFACES IN TURBULENT FLOWS

The geometrical interpretation of the concentration PDF given in §1.3 and the hypothesis considered in §3.2 concerning the statistical independence of the distributions of scalar dissipation N and concentration z in the turbulent fluid enable the elucidation of the structure of isoscalar surfaces in turbulent flows. As was indicated in the introduction to this book, the investigation of the indicated problem is of interest for two reasons. First, it allows the derivation of the boundary conditions (3.5) in a new fashion and ascribing to them a clear physical interpretation. Second, as a result of such an investigation an additional view on intermittency appears. A preliminary discussion of these questions was included in §3.1.

For the sake of simplicity we first limit ourselves to the case of statistically homogeneous concentration fields, and then present a generalization of the obtained results for a nonhomogeneous case.

1. Isoscalar Surfaces Located Inside the Turbulent Fluid. To start, let us consider the isoscalar surfaces located inside the turbulent fluid, i.e., $z \neq 0$ and $z \neq 1$. For the sake of brevity these isoscalar surfaces are henceforth called internal.

The following two characteristics will be of main interest — the mean area $\langle S_z \rangle$ of the isoscalar surface $z(x,t) = \text{const}$ and the mean distance $d\langle h_z \rangle$ between two close isoscalar surfaces $z = \text{const}$ and $z + dz = \text{const}$. Assuming statistically independent concentrations and absolute value of the concentration gradient, we obtain from formula (1.26)

$$\gamma P_t(z) = \lim_{V \to \infty} \langle S_z \rangle V^{-1} \left\langle \left| \frac{\partial z}{\partial n} \right|^{-1} \right\rangle_t, \qquad 0 < z < 1 \tag{3.100}$$

It is readily seen that the expression for $d\langle h_z \rangle$ in the same assumption has the following form

$$d\langle h_z \rangle = dz \left\langle \left| \frac{\partial z}{\partial n} \right|^{-1} \right\rangle \tag{3.101}$$

In relations (3.100) and (3.101), just as in §1.3, n is the normal to the isoscalar surface $z = \text{const}$, and V is the volume inside which the isoscalar surface under consideration is located.

Let us find the dependence of the area $\langle S_z \rangle$ and the distance $d\langle h_z \rangle$ on Reynolds number. For this purpose, we neglect the fluctuations of scalar dissipation in (3.100) and (3.101). This is fully justified, since it is generally known that accounting for these fluctuations when determining the lower order moments leads to only small corrections (see, for example, Monin and Yaglom [1967]). Then we can write $\langle |\partial z/\partial n|^{-1} \rangle_t \sim \sqrt{D/\langle N \rangle_t}$ (D is the coefficient of molecular diffusion), and, hence, from expression (3.100), (3.101) we find

$$\frac{\langle S_z \rangle L}{V} \sim \gamma P_t(z) L \sqrt{\frac{\langle N \rangle_t}{D}} = \gamma P_t \sqrt{\frac{\langle N \rangle_t L}{q}} \, \text{Re}^{1/2}$$

$$\frac{d\langle h_z \rangle}{L} \sim \frac{1}{L} \sqrt{\frac{D}{\langle N \rangle_t}} dz = dz \sqrt{\frac{q}{\langle N \rangle_t L}} \, \text{Re}^{-1/2} \tag{3.102}$$

Here, as before, L is the integral turbulence scale, q is the root-mean-square value of velocity pulsations. Since the functions $\langle N \rangle_t$, γ, and $P_t(z)$ are independent of Reynolds number (see Chapter 1), then from (3.102) we obtain the following estimate for the internal isoscalar surfaces

$$\frac{\langle S_z \rangle L}{V} \sim \mathrm{Re}^{1/2}, \qquad \frac{d\langle h_z \rangle}{L} \sim \mathrm{Re}^{-1/2}, \qquad 0 < z < 1 \tag{3.103}$$

It is assumed here that $P_t(z) \neq 0$, $0 < z < 1$.

Relations (3.103) show that with increasing Reynolds number the mean area of an arbitrary internal isoscalar surface $\langle S_z \rangle$ ($0 < z < 1$) increases without restriction, and the mean distance $d\langle h_z \rangle$ between two close internal isoscalar surfaces decreases without restriction. It should be emphasized that the estimate of the area in (3.103) is inapplicable for the limit isoscalar surface $z = 0$ and $z = 1$, since from (3.102), by virtue of the boundary conditions (3.5), it formally follows that $\langle S_z \rangle = 0$. Evidently, the equality to zero here must be understood in the sense that the increase of the area of limit isoscalar surfaces when $\mathrm{Re} \to \infty$ takes place more slowly than $\sqrt{\mathrm{Re}}$ (see also §3.3). In order to find a qualitative estimate of the area of these surfaces, one must revert to more subtle considerations. These are discussed in the next item of this section.

It is interesting to compare formulas (3.103) with the well known results of Batchelor [1952] and Batchelor and Townsend [1956] obtained on the assumption that $D = 0$

$$\frac{\langle S_z \rangle L}{V} \sim \exp\left(\frac{tq}{L}\sqrt{\mathrm{Re}}\right)$$

$$\frac{d\langle h_z \rangle}{L} \sim \exp\left(-\frac{tq}{L}\sqrt{\mathrm{Re}}\right) \tag{3.104}$$

From formulas (3.103), (3.104) it is apparent that in both the effect of Reynolds number is qualitatively the same. However, when $D = 0$, this effect is expressed significantly more strongly than when $D \neq 0$. The explanation of this difference is provided by the following reasonings. When $D = 0$, the isoscalar surface is composed of the same fluid particles. If $D \neq 0$, then the particles on the isoscalar surface are lost, i.e., molecular diffusion smoothes the wrinkles forming on such a surface.

Relations (3.100) and (3.101) along with the assumption $\langle |\partial z/\partial n|^{-1}\rangle_t \sim \sqrt{D/\langle N \rangle_t}$ also establish the variation of $\langle S_z \rangle$ and $d\langle h_z \rangle$ with time if the concentration PDF $P_t(z,t)$ is known. Let us consider the corresponding formulas for two limiting cases analyzed in §3.4. At the initial stage of decay, from relations (3.32), (3.102), and relation (3.31) for the scalar dissipation $\langle N \rangle_t = a/t$, $a = 1/\pi^2$, we obtain

$$\frac{\langle S_z \rangle}{V} \sim \frac{1}{t_0}\sqrt{\frac{t}{D}}, \qquad d\langle h_z \rangle \sim \sqrt{Dt}\, dz \tag{3.105}$$

It should be recalled that t_0 here is the time scale which is determined by large vortices in the mixing layer at a late stage of development of the flow.

At the final stage of decay of concentration fluctuations (not to be confused with the final stage of turbulent decay see remark in §3.4), by virtue of (3.27), the following relations are true

$$\frac{\langle S_z \rangle}{V} \sim \frac{1}{\sigma} \exp\left[-\frac{(z-\langle z \rangle)^2}{2\sigma^2}\right] \left(\frac{\langle N \rangle_t}{D}\right)^{1/2}$$

$$d\langle h_z \rangle \sim \sqrt{\frac{D}{\langle N \rangle_t}} \qquad (3.106)$$

From relations (3.105) and (3.106) it is evident that the mean distance between isoscalar surfaces increases monotonically. At the same time when $D = 0$, according to expression (3.104), this distance decreases monotonically. When $D \neq 0$ at the initial stage of decay, the mean area of any isoscalar surface increases. At the concluding stage the mean areas of all isoscalar surfaces, with the exception of surface $z = \langle z \rangle$, decrease. The mean area of surface $z = \langle z \rangle$ always decreases with time. When $D = 0$, as follows from (3.104), $\langle S_z \rangle$ always increases. Note that formula (3.104) is approximately true also when $D \neq 0$ until the effect of molecular diffusion starts to play a noticeable role, and then it will be impossible to ignore the process of destruction of some fluid particles on the isoscalar surface.

2. Limit Isoscalar Surfaces. Let us dwell now on the estimation of the mean areas $\langle S_0 \rangle$ and $\langle S_1 \rangle$ of the limit isoscalar surfaces $z = 0$ and $z = 1$. It should be noted that the formal determination of the indicated isoscalar surfaces present some known difficulties. Here, surfaces S_0 and S_1 are understood to mean the surfaces to which the isoscalar surfaces $z_0 = \text{const } z_\nu = \text{const } \sigma \, \text{Re}^{-1/4}$ and $z_1 = 1 - \text{const } \sigma \, \text{Re}^{-1/4}$ tend when $\text{Re} \to \infty$, respectively. Such a definition is quite natural from the viewpoint of the analysis in §1.3 where it is shown that the effects in the phase space caused by molecular diffusion are substantial in a narrow region located near the boundaries of the phase space $z = 0$ and $z = 1$ and the characteristic dimension of these regions is of the order $\sigma \, \text{Re}^{-1/4} = z_\nu$ (see formula (1.36)).

Since the exchange between turbulent and nonturbulent fluids is unidirectional (remember that fluid particles cannot leave the region of vortex flow, i.e., the ambient always flows into the turbulent fluid), then the normal velocity component of the medium relative to the isoscalar surfaces $z = z_0$ and $z = z_1$ has a determined sign. This circumstance sharply differentiates the isoscalar surfaces $z = z_0$ and $z = z_1$ from the internal isoscalar surfaces located in the turbulent fluid.

The dependence of $\langle S_0 \rangle$ and $\langle S_1 \rangle$ on Reynolds number can be established by means of relation (2.17), which describes the rate of volume change between isoscalar surfaces. For this purpose, we integrate (2.17) for z from some level z, which is varied later, to ∞. We have $dV_z/dt = \int v n dS_z$, where V_z is the volume wherein concentration is larger than or equal to z, $n = \nabla z/|\nabla z|$. This relation can be represented in the form

$$\frac{dV_z}{dt} = v_\Sigma S_z, \qquad v_\Sigma = \frac{1}{S_z}\int \mathbf{v}\, n dS_z \qquad (3.107)$$

where v_Σ is the averaged volume of the normal velocity component of the motion of surface $z = $ const relative to the medium (averaging is effected over surface S_z).

Let us first estimate the amplitudes of velocity fluctuations of the motion of surface $z = $ const relative to the medium. For this purpose, we turn to equation (2.18). As evident from relation (1.38), the quantities ∇z and Δz entering (2.18) are determined by large-scale fluctuations. Therefore, the amplitudes of the fluctuations of the quantities under consideration can be found from the theory of locally homogeneous turbulence. Then, considering that $\nu \sim D$ and neglecting the fluctuations of scalar dissipation and energy dissipation and energy dissipation, we obtain $|\nabla z| \sim z_\nu/\eta$, $|\Delta z| \sim z_\nu/\eta^2$ where η is the Kolmogorov scale. Using this estimate we conclude from (2.18) that

$$|v_n| \sim [\langle \epsilon \rangle \nu]^{1/4} \sim q \, \mathrm{Re}^{-1/4}, \qquad v_n = \mathbf{v} \, \mathbf{n} \tag{3.108}$$

This relation yields the estimate of the amplitude of velocity fluctuations of any isoscalar surface relative to the medium and says nothing about the sign of this velocity.

It can be assumed from general considerations that the sign of this velocity for internal isoscalar surfaces vary randomly. Therefore, quantity v_Σ entering (3.107) is less than $q \, \mathrm{Re}^{-1/4}$. The results obtained above indicate that velocity v_Σ is even much less than $q \, \mathrm{Re}^{-1/4}$. Indeed, since PDF is independent of Reynolds number and $\langle S_z \rangle \sim \mathrm{Re}^{1/2}$ when $z \neq 0$ or $z \neq 1$, then it follows from (3.107) that $v_\Sigma \sim q \, \mathrm{Re}^{-1/2}$, i.e., the large positive values of v_n are compensated for by large negative values.

The limit isoscalar surfaces behave differently from this pattern. Indeed, since the trajectory of particles cannot leave the boundaries of the turbulent fluid, the sign of the velocity of motion of similar surfaces relative to the medium is always the same. For this reason, formula (3.108) is useful for estimating the value of v_Σ. Then, from (3.107) and (3.108) we obtain

$$\frac{\langle S_z \rangle L}{V} \sim \mathrm{Re}^{1/4}, \qquad z=0, \qquad z=1 \tag{3.109}$$

The generalization of the above results for the case of homogeneous turbulence is obvious. For this it is required to make the substitutions $\langle S_z \rangle \to d\langle S_z \rangle$ and $V \to dV$ only in the main relations. Thereby, the character of variation of quantities $d\langle S_z \rangle$ and $d\langle h_z \rangle$ when $\mathrm{Re} \to \infty$ remains as before.

3. Relation Between Structures of Isoscalar Surfaces and the Boundary Conditions and Intermittency. From formulas (3.100) and (3.109) it follows that

$$P_t = 0, \qquad z=0, \qquad z=1, \qquad \mathrm{Re} \to \infty$$

i.e., we have the boundary conditions (3.5) obtained in §3.1 from the equation for the concentration PDF (2.15) by means of a formal mathematical technique. The conclusion presented here shows that these boundary conditions are fulfilled for finite Reynolds numbers with an accuracy of $Re^{-1/4}$.

The significance of formula (3.109) is not only (more accurately, not so much) that it gives another method of derivation of the boundary conditions. Its principal value is in imparting clear physical meaning to these boundary conditions.

Furthermore, formulas (3.103) and (3.109) enable conducting a certain analogy between the limit and internal isoscalar surfaces, on the one hand, and external and internal intermittency on the other (see the end §3.1). Indeed, according to (3.103) and (3.109) the area and, thereby, the degree of distortion of the limit isoscalar surfaces is significantly less than the intermediate surfaces. Hence, if the external intermittency is set by the fluctuations of the limit isoscalar surfaces and the internal intermittency by the fluctuations of the intermediate isoscalar surfaces, then an intermittency model which is qualitatively close to the sponge pattern, proposed in §1.1, is obtained.

The described model of the turbulent fluid also explains the meaning of the averaging procedure (considered in §1.1) as a means of determining the boundaries of the turbulent fluid and identification of external intermittency. The maximum smoothness of the indicated boundary and the degree of contortion of the limit isoscalar surfaces are closely interrelated.

An interesting mathematical example, aiding the physical picture of the structure of isoscalar surfaces and illustrated in turbulent flows at high Reynolds

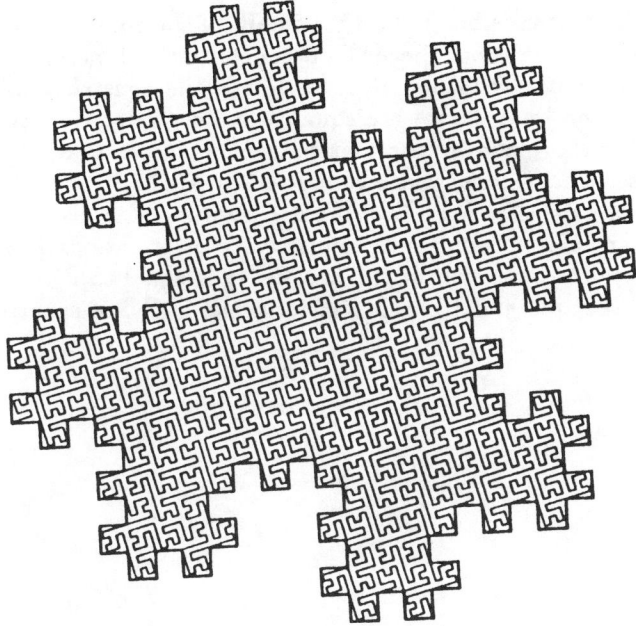

Figure 3.22 Combination of two polygons with a very large number of sides (the figure is adapted from an article by Mandelbrot [1977]).

Figure 3.23 Initial combination (on the left) and first step in the procedure (on the right) leading to the pattern in Fig. 3.22 (the figure is adapted from an article by Mandelbrot [1977]).

numbers, is included in the work by Mandelbrot [1977]. Let us dwell briefly on the discussion of this example. Figure 3.22, adapted from the work by Mandelbrot, depicts a combination of two polygons with very large number of sides. This combination represents some intermediate step in an infinite procedure when one of the polygons is intensively contorted around itself (similar to the procedure upon construction of a curve filling a plane), and the second borders with the first. Similar polygons can be regarded as some form of isoscalar surfaces. One of the polygons (the external) corresponds to the boundary of the turbulent fluid, and the internal corresponds to the isoscalar surface located in the turbulent fluid. Polygon contortion corresponds to fluid dynamic deformations. The initial combination and first step of this construction are depicted in Fig. 3.23 (also adapted from the work by Mandelbrot). Further details of the construction are given in the work under consideration. Since the perimeter of both polygons are multiplied by a number greater than one at each step of the construction, the lengths of the corresponding limit curves are equal to infinity. But it turns out that the extent of filling the plane by the first polygon is larger than the degree of curvature of the second. Computations conducted by Mandelbrot yield the quantitative character of this assertion. Both limit curves are fractals that are obtained by means of an infinite number of similarity transformations (see §1.1). The Hausdorff-Bezikovich dimension of the first curve is equal to $d = 2$, the second $d \approx 1.37$. The example under consideration is, naturally, of special character.

CHAPTER
FOUR
STATISTICAL CHARACTERISTICS OF SMALL-SCALE TURBULENCE

The study of the characteristics of small-scale turbulent motions is of considerable interest for three reasons. First, the motion of small vortices is the simplest type of turbulent flow and, hence, its description meets with the least difficulty. Furthermore, the behavior established when investigating small-scale turbulence is important to understand the structure of an arbitrary turbulent flow. Second, the study of the characteristics of small-scale fluctuations can be useful when creating the so-called estimate turbulence models. In this approach, large-scale motions are described by equations that are similar to the Navier-Stokes equations, and the small-scale motions are characterized by turbulent viscosity which is dependent on the energy and scale of the small vortices. Third, as mentioned in the introduction, understanding the structure of small-scale motions is important when studying chemical reactions in turbulent flows. For example, in combustion, chemical conversion occurs in very thin zones and, hence, the structure of such zones are determined, first of all, by small-scale fluctuations.

Let us recall the main principles of the theory of small-scale turbulence. The first step in constructing this theory belongs to Richardson [1922, 1926], who formulated the qualitative concepts of the cascade character of energy transfer over the turbulence spectrum, i.e., small vortices attain energy as a result of successive breaking of large vortices. The quantitative description of this process was given by Kolmogorov [1941] and Obukhov [1941]. It is based on the use of the structure functions and their spectra, i.e., quantities of the form

$$D_{ij}(r) = \langle [u_i(x) - u_i(x+r)][u_j(x) - u_j(x+r)] \rangle$$

$$E_{ij}(\vec{k}) = \int D_{ij} \exp(i\vec{k}r) d^3r$$

($\vec{k}$ is the wave number). It is evident that the characteristics of large-scale motions vary slowly at distances of the order r if r is sufficiently small compared with the integral scale of turbulence. Therefore, the velocity difference at two closely located points provides the most suitable description of small-scale fluctuations. Hence, the study of the structure functions and their spectra is of great interest.

Kolmogorov and Obukhov assumed that when Reynolds number is very high, r can be chosen as small as required and the chain of successive disintegrations of large vortices, which leads to the creation of vortices with dimension r, will turn out to be very long. Therefore, the parameters describing small-scale fluctuations depend very weakly on the structure of large-scale turbulence and thus must be statistically homogeneous and isotropic. They also assumed that one of the most important quantities characterizing the progressive disintegration of vortices is the mean rate of energy transfer over the spectrum of turbulence. In equilibrium conditions this rate is equal to the energy dissipation $\langle \epsilon \rangle$. Another important quantity determining the structure of small-scale turbulence is the coefficient of kinematic viscosity ν. Therefore, on the basis of dimensional considerations and statistical isotropy the following formula is obtained

$$D_{ij} = (\langle\epsilon\rangle r)^{2/3} \left[f\left(\frac{r}{\eta}\right) \frac{r_i r_j}{r^2} + g\left(\frac{r}{\eta}\right) \delta_{ij} \right], \qquad \eta = \frac{\nu^{3/4}}{\langle\epsilon\rangle^{1/4}}$$

where η is the so-called Kolmogorov length scale which characterizes the influence of viscous effects on the structure of small-scale turbulence, f and g are arbitrary functions of a single argument r/η. On the basis of the known principle of self-similarity of turbulent flows with respect to Reynolds number, it is natural to assume that when $L \gg r \gg \eta \sim L\,\text{Re}^{-3/4}$, i.e., in the so-called inertial range, the structure function is independent of viscosity. Consequently, $f = \text{const}$, $g = \text{const}$ when $r/\eta \gg 1$. Hence, the following formula was obtained

$$D_{11} = C(\langle\epsilon\rangle r)^{2/3}$$

where C is a constant which is referred to frequently as the Kolmogorov constant.

Extension of the stated concepts to the case of structure functions of arbitrary order does not present difficulties. For example, we can derive the relation

$$\langle v^n \rangle \sim (\langle\epsilon\rangle r)^{n/3}, \qquad \mathbf{v} = u(x) - u(x+r)$$

The next step in the development of the theory of small-scale turbulence was made by Landau (see Landau and Lifshitz [1954]) who noticed that energy dissipation is distributed in the turbulent flow in a random fashion. Therefore, their estimate $v^n \sim \langle \epsilon r \rangle^{n/3}$ can be given which, upon averaging, yields $\langle v^n \rangle \sim \langle \epsilon^{n/3} \rangle r^{n/3}$.

The analysis conducted in the book by Monin and Yaglom [1967] indicated that refinement is inessential if the dissipation PDF does not possess some anomalous features. Experiments directed towards the explanation of this question showed that such features do indeed exist (Batchelor and Townsend [1949], Gurvich and Zubkovskii [1963], Kuo and Corrsin [1971], Wingaard and Pao [1975], Gibson and Masiello [1975], Kholmyanskii [1970, 1972], Champagne, Pao, and Wygnanski [1976] and others) as was indicated in §1.1. These features are caused by the internal intermittency of turbulent flows which has been discussed in Chapter 1.

Therefore, these concepts were refined in the works of Kolmogorov [1962a, b], Obukhov [1962]. The refinements were based on the consideration of energy dissipation ϵ_r averaged over some region (for example, a sphere) with the characteristic dimension r. In these works a number of propositions were put forward to the effect that the random quantity ϵ_r is distributed according to a lognormal law. This conclusion allows computing quantity $\langle \epsilon^{n/3} \rangle$ and, thereby finding a structure function of the type $\langle v^n \rangle$. The lognormal character of energy dissipation PDF follows also from the similarity hypothesis proposed by Yaglom [1966], Gurvich and Yaglom [1967]. In the case under consideration we have $\langle \epsilon_r^n \rangle \approx \langle \epsilon \rangle^n (r/L)^{-\mu n(n-1)/2}$, where μ is a constant. Hence it follows that the dependence of the structure functions of the type $\langle v^n \rangle$ on the distance between the points, particularly high-order structure functions ($n \gg 1$), differs from the dependence predicted by the unrefined theory of locally homogeneous turbulence.

The limited scope of this book does not allow conducting a detailed analysis of all refinements of the theory of small-scale turbulence. Therefore, we shall single out only the main results. 1) Within the framework of all the theories, refinements lead to little correction to the "two-thirds" law, i.e., to $D_{11} = C(\langle \epsilon \rangle r)^{2/3}$. This law is presently well confirmed experimentally in widely varying turbulent flows (see, for example, the literature review by Monin and Yaglom [1967]). 2) Refinements are found to be significant only when considering high-order structure functions. In particular, in all the proposed schemes the kurtosis of dissipation fluctuations tend to infinity with unlimited increases of Reynolds number (Yaglom [1966], Novikov and Stewart [1964], Novikov [1971] and others). This conclusion is confirmed experimentally (Gibson and Masiello [1975], Champagne, Pao, and Wygnanski [1976] and others) and is usually explained by the internal intermittency of turbulent flows. 3) Although experimental data indicate that energy dissipation PDF is close to the lognormal (Gurvich and Zubkovskii [1963], Chen [1971], Gibson and Masiello [1975] and others), a number of works casts doubt on the validity of this law (Novikov [1971], Kraichnan [1974], Mandelbrot [1974, 1975], Gibson and Masiello [1975]).

The analysis presented below is based on the principles of the theory of locally homogeneous turbulence. Principal attention is given to the analysis of those corrections to this theory that are dictated by intermittency. The phenomenon under consideration is understood within the definition adopted in Chapter 1, i.e., external intermittency is analyzed. The introduced assumption is based on the results of the investigation of Chapter 3 in which it is established that the factor of intermittency γ is less than unity in all regions of turbulent flows. Numerical calculations conducted in Chapter 3 showed the existence in jets and wakes of extended domains in which the intermittency factor differs so little from unity as to make it impossible to detect these differences with the current level of measuring devices. Hence, this approach does not contradict the known experimental data which indicate that there are regions in jets, wakes, and boundary layers, etc., in which, one would think, $\gamma = 1$. Therefore, it seems that the phenomenon under study ought to be accounted for even upon analysis of the local turbulence structure.

As was shown in Chapter 1, cascade transfer of energy over the turbulence spectrum is absent in the nonturbulent fluid. Hence, it is evident that the Kolmogorov theory [1941] and all its refinements, considered above, are incorrect when describing the phenomena in the nonturbulent fluid. Therefore, the assumptions made below are formulated only for the turbulent fluid.

One of the main assumptions results from the hypothesis formulated by Kolmogorov [1962a, b] stating that in the inertial range the probability distribution of quantity

$$b_k = \frac{u_k(x) - u_k(x+r)}{u_k(x) - u_k(x+R)}$$

(not summed with respect to k) when $R \gg r$ depends only on r/R. This hypothesis requires only slight refinement, since b_k is so defined that it is not a tensor quantity.

For the solution of the problem, one must use the equation derived in Chapter 2 for the velocity difference PDF at two points in the inertial range. This equation is analyzed on the basis of the adopted hypothesis. The main information for its solutions is obtained when analyzing the component describing pressure fluctuations. It is established that this component results in the appearance of singularities in the PDF. A number of arguments have been put forward indicating that these singularities are caused by intermittency.

The main results, expounded below, have been published in works by Kuznetsov [1976a, 1977c] and Kuznetsov and Sabel'nikov [1981a].

§4.1 SIMILARITY HYPOTHESIS

In order to formulate the main hypothesis and obtain its principal consequences, let us consider points 1 – 5 with the coordinates $x^{(1)}$, $x^{(2)}$, $x^{(3)}$, $x^{(4)}$, $x^{(5)}$ in the turbulent fluid. Let $r = x^{(2)} - x^{(1)}$, $R = x^{(3)} - x^{(1)}$, $S = x^{(4)} - x^{(1)}$, $L = x^{(5)} - x^{(1)}$, $\mathbf{v} = u(x^{(2)}) - u(x^{(1)})$, $V = u(x^{(3)}) - u(x^{(1)})$, $U = u(x^{(4)}) - u(x^{(1)})$, $w = u(x^{(5)}) - u(x^{(1)})$. Here, all velocities are considered at the same moment of time. Points 1 and 5 are situated at a distance of the order of the turbulence integral scale, i.e., $L \sim \langle (u - \langle u \rangle)^2 \rangle^{2/3} / \langle \epsilon \rangle$, and the velocity difference in them w is of order of the fluctuating velocity of the energy carrying vortices. Points 1 – 4 are located at a distance belonging to the inertial range of the spectrum, i.e., $\eta \ll r \ll L$, $\eta \ll R \ll L$, $\eta \ll S \ll L$, and $\eta = \nu^{3/4} \langle \epsilon \rangle^{-1/4}$. Let also $P_{tt}(\mathbf{v}, r|V, R)$ be the conditional velocity difference PDF at points 1 and 2 provided that these points are in the turbulent fluid and the velocity difference at points 1 and 3, also in the turbulent fluid, is equal to V.

Let us formulate the main similarity hypothesis, namely, let us assume that when $r \ll R$ quantity P_{tt} is dependent only on $\mathbf{v}, r, V, R$. Then, from considerations of dimensionality we deduce

$$P_{tt} = v^{-3} P_0 \left(\frac{v}{V}, \frac{r}{R}, \varphi\right) \qquad \varphi = \frac{vr}{vr} \qquad (4.1)$$

where P_0 is a dimensionless function.

Prior to deriving a formal consequence of this hypothesis, let us elucidate its physical meaning.

From (4.1) it is evident that this assumption is close to the hypothesis formulated by Kolmogorov [1962a, b], since it considers that the statistical characteristics of quantity v/V are universally dependent on r/R. Unlike the quantity v_i/V_i (not summed with respect to i) used by Kolmogorov, considered here is the quantity v/V which is a vector. Therefore, the formulated similarity hypothesis is true in any system of coordinates. Its physical meaning is obvious. Parameters V and R characterize large scale motion; v, r characterize small-scale motion. The "life" time of the first motion is of the order $\tau_1 = R/V$, and the second is of the order $\tau_2 = r/v$. Since $r \ll R$, then $\tau_1 \gg \tau_2$. Therefore, it is assumed that the small-scale motion is in a statistical equilibrium with the large-scale motion. Since $VR/\nu \gg 1$ then we can select $M(M \gg 1)$ points so that $R \gg R^{(1)} \gg \ldots \gg r$. Then, all that is said about the interrelation of the motions with scales R and r is also pertinent to the interaction of motions with scales $R^{(i)}$ and $R^{(i+1)}$, i.e., the process of energy transport over the turbulence spectrum is of cascade character. Since at high Reynolds number the number of cascade "links" is also large, then the characteristics of small scale motions are statistically isotropic, i.e., independent of the orientation of vectors V and R and, therefore, are determined only by quantities V and R.

Let us now find the formal consequences of this hypothesis which follow from the fact that P_{tt} is the conditional PDF. For this, we consider the velocity differences at three pairs of points 1 and 2, 1 and 3, 1 and 4 and make use of the Bayes theorem

$$P_{tt}(v, r|U, S) = \int P_{tt}(v, r|V, R) P_{tt}(V, R|U, S) d^3V \qquad (4.2)$$

where $r \ll R \ll S \ll L$.

Relation (4.2) is a nonlinear integral equation for whose solution we use the following technique. We multiply (4.2) by $2\pi v^{n+2}$ and integrate for v from 0 to ∞ when $\varphi = \text{const}$, i.e., we introduce the quantity

$$F = 2\pi \int_0^\infty v^{n+2} P_{tt} dv = V^n G\left(n, \frac{r}{R}, \varphi\right)$$

where G is a dimensionless function. The last equality in this formula follows from (4.1). Further, the following quantity will also be required

$$P_{tt}^0 = v^2 \int_{-1}^1 P_{tt} d\varphi$$

$$F^0 = \int_{-1}^1 F d\varphi = V^n G^0\left(n, \frac{r}{R}\right)$$

where $P_{tt}^{(0)}$ is the PDF of the absolute value of the velocity difference. Note that function F is the Mellin transform from P_{tt}. By utilizing the properties of such a transform (Bateman and Erdeylyi [1953b], we obtain

$$P_{tt} = \frac{1}{i}\int_{-i\infty}^{i\infty} v^{-n-3} F \, dn, \qquad P_{tt}^0 = \frac{1}{i}\int_{-i\infty}^{i\infty} v^{-n-1} F^0 \, dn \qquad (4.3)$$

By using the introduced notation we reduce (4.2) to the form

$$G\left(n, \frac{r}{S}, \varphi\right) = G\left(n, \frac{r}{R}, \varphi\right) G^0\left(n, \frac{R}{S}\right)$$

$$G^0 = \int_{-1}^{1} G\left(n, \frac{R}{S}, \varphi\right) d\varphi \qquad (4.4)$$

The relation (4.4) is a functional equation which is more easily solved than the initial equation (4.2). Indeed, by taking the logarithm of (4.4) and differentiating the outcome with respect to r and R, we obtain

$$\frac{\partial^2}{\partial r \partial R} \ln G\left(n, \frac{r}{R}, \varphi\right) = 0$$

Integrating this equation yields

$$G = H(n, \varphi) \left(\frac{r}{R}\right)^{Q(n,\varphi)} \qquad (4.5)$$

where H and Q are the integration "constants". Let us take the logarithm of (4.4) and differentiate the obtained result with respect to R and S. We similarly obtain

$$G^0 = H_1(n) \left(\frac{R}{S}\right)^{q(n)} \qquad (4.6)$$

where H_1 and q are the new integration "constants".

Substituting (4.5) and (4.6) into (4.4) we obtain

$$H_1(n) \left(\frac{R}{S}\right)^{Q(n,\varphi)-q(n)} = 1$$

By varying (R/S) when $n = $ const, we conclude that this equality can be implemented only when $Q = q$. Taking into account that $Q = q$, we also obtain $H_1 = 1$. Thus, we arrive at the final result

$$F = V^n H(n, \varphi) \left(\frac{r}{R}\right)^{q(n)}, \qquad H_1 = \int_{-1}^{1} H(n, \varphi) d\varphi = 1 \qquad (4.7)$$

From (4.3) and (4.7) we also find that

$$P_{tt}^0 = \frac{1}{i}\int_{-i\infty}^{i\infty} v^{-n-1} V^n \left(\frac{r}{R}\right)^{q(n)} dn \tag{4.8}$$

Let us compare (4.1) and (4.7). From the similarity hypothesis and considerations of dimensionality it follows that P_{tt} is dependent on only three variables. An additional limitation based on P_{tt} being the conditional PDF allows reducing the number of variables to two.

Let us explain the meaning of functions H and q. From (4.7) and the definition of function F it follows that

$$\langle v_N^{n-m} v_L^m \rangle_{V,t} = V^n \left(\frac{r}{R}\right)^{q(n)} \int_{-1}^{1} H(n,\varphi)\varphi^m (1-\varphi^2)^{(n-m)/2} d\varphi \tag{4.9}$$

where $v_L = v\varphi$ is the projection of vector v on vector r, $v_n = v\sqrt{1-\varphi^2}$ is the component of vector v perpendicular to vector r. The symbol $\langle\ \rangle_{V,t}$ corresponds to averaging under the condition that points 1 and 2 are in the turbulent fluid, and the velocity difference at points 1 and 3, also located in the turbulent fluid, is equal to V. Thus, $q(n)$ characterizes the dependence of the structure function of the order n on the distance between the points, and $H(n,\varphi)$ describes the relation between the various structure functions of the same order. It is also evident that since the scale characteristics of the process are determined by the PDF of the absolute value of the velocity difference P_{tt}^0, then the main information is described by function $q(n)$ (it is evident from (4.8) that P_{tt}^0 is completely determined by this function). Therefore, the main purpose of further investigation is to elucidate the dependence of q on n.

The first limitation on the form of this dependence is described by equalities

$$q(0) = 0, \quad q(3) = 1 \tag{4.10}$$

Condition $q(0) = 0$ follows from the fact that $\int P_{tt} d^3 v = 1$. Condition $q(3) = 1$ was obtained by Kolmogorov [1941] from the Karman-Hovars equation (the third order structure function is linearly dependent on the distance). This condition is discussed in detail in Monin and Yaglom [1967].

The following limitation can be obtained from the Cauchy-Bunyakovskii inequality

$$\langle v^{m+n} \rangle_{V,t}^2 \leq \langle v^{2m} \rangle_{V,t} \langle v^{2n} \rangle_{V,t}$$

Using (4.7) we reduce this inequality to the form $2q(m+n) \geq q(2m) + q(2n)$. Assuming $m = n + \Delta n$, $\Delta n \to 0$ we, hence, obtain

$$\frac{d^2 q}{dn^2} \leq 0 \tag{4.11}$$

The last and, as will be seen later, the most important limitation follows from the definition of function F. In order to formulate this limitation, parameter n will be regarded as a complex number, i.e., we consider how a quantity of the following form is dependent on the distance r

$$\langle v^n \rangle_{V,t} = \langle v^\alpha \cos(\beta \ln v) \rangle_{V,t} + i \langle v^\alpha \sin(\beta \ln v) \rangle_{V,t}$$

where $n = \alpha + i\beta$.

Let us introduce the natural assumptions: 1) P_{tt} is a bounded function; 2) when $v \to \infty$ the probability density P_{tt} tends to zero more rapidly than any exponent of v, i.e., there exist structure functions of all orders. Consider the expression

$$\frac{\partial F}{\partial n} = 2\pi \int_0^\infty v^{n+2} \ln v \, P_{tt} dv$$

It follows from the assumptions made that this integral converges uniformly when $\text{Re } n > -3$, i.e., F and, hence, H and q are analytical functions in the complex half-plane $\text{Re } n > -3$.

§4.2 RELATION OF THE CHARACTERISTICS OF TURBULENT AND NONTURBULENT FLUIDS

Investigations conducted in §4.1 show that only the similarity hypothesis and the formal limitations, following from the physical meaning of the various quantities, define a wide class of PDFs. The lognormal law is a special case in which q is a quadratic polynomial (Monin and Yaglom [1967]). Then, from (4.10) and (4.11) we obtain

$$q(n) = \left(\frac{1}{3} + 3q_2\right) n - q_2 n^2, \qquad q_2 > 0 \tag{4.12}$$

where q_2 is a positive number.

Physical considerations reveal that the most important limitations are described by the equation of motion which has not yet been considered. Therefore, it is natural to use relation (2.31). In such an approach a number of difficulties arise in view of the fact that unconditional PDF P appears in (2.31) and the similarity hypothesis is true only for the conditional PDF P_{tt}. Consequently, a separate equation must be obtained describing the properties of the turbulent fluid.

The mathematical procedure for deriving the equation for P_{tt} directly from the equations of motion involves the rearrangement of the operations of differentiation and conditional averaging. It can be shown that it leads to the appearance of additional terms in the equation for P_{tt} characterizing the exchange of mass, momentum, and energy between the turbulent and nonturbulent fluids. The approximation of these terms causes considerable difficulties. A similar problem also arises upon analysis of the equations for the conditional one-point moments (Libby [1975], Dopazo [1977], Sabel'nikov [1979, 1980b, 1985c], Dopazo and O'Brien [1979], Byggstoyl and Kollman [1981], Duhamel [1981]). Without enlisting additional hypotheses it is impossible to obtain the relations which are essentially different from equation (2.31) in this approach.

A more fruitful approach is the one used in Chapter 3 when deriving the equation for the concentration PDF in the turbulent fluid. However, another difficulty arises in this case. The fact is that the technique developed in Chapter 3 is based on the identification of the regular and singular (proportional to δ-function) components in the equation for the unconditional concentration PDF. To solve the problem under consideration, this technique is inappropriate, since singular components are absent in the expression for the velocity difference PDF which is caused by pressure fluctuations generating velocity fluctuations in the turbulent fluid. Thus, one must find a rule which allows obtaining a relation for the conditional PDF P_{tt} from the equation for the unconditional PDF P. The present section is devoted to the solution of this task.

Let us elucidate the main idea in the example of the continuity equation $\partial \langle v_i v_j \rangle / \partial r_j = 0$, namely let us consider how to obtain a relation for $\langle v_i v_j \rangle_{w,t}$ from this equation. Note that the operations of differentiation and averaging when $w = $ const cannot commute (w is the velocity difference at two other points located at a distance of the order of the integral scale L). This difficulty is insignificant, since when analyzing small-scale fluctuations (i.e., the values of v), it can be assumed that the characteristics of large-scale perturbations (i.e., the values of w) play the role of the averaged motion (otherwise the similarity hypothesis is invalid). Hence, it follows that if $r \ll L$, then the operations of differentiation and conditional averaging when $w = $ const are rearrangeable, i.e., $\partial \langle v_i v_j \rangle_w / \partial r_j = 0$. Hence, from (1.8) we obtain

$$\frac{\partial}{\partial r_j} [\gamma \langle v_i v_j \rangle_{w,t} + (1-\gamma)\langle v_i v_j \rangle_{w,n}] = 0$$

It follows from (4.9) that the first component in this relation is proportional to $r^{q(2)} \approx r^{2/3}$, and it can be concluded from (1.7) that the second component is proportional to r^2. Since both components are dependent on r in different ways, they are separately equal to zero, i.e.,

$$\frac{\partial}{\partial r_j} \gamma \langle v_i v_j \rangle_{w,t} = 0$$

Thus, in this case the operating rule is: the relation which is valid for the unconditional structure function is valid for the description of the conditional structure function if symbol $\langle\ \rangle$ is replaced by $\gamma \langle\ \rangle_{w,t}$.

This rule also applies in the general case with some limitations. Indeed, let us multiply (2.31) by $2\pi v^{n+2}$ and integrate over all v when $\varphi = $ const. The resulting relation again includes the structure functions which are conditionally averaged over the turbulent and nonturbulent fluids and, as seen from (1.7) and (4.9), two groups of components are featured, one of which is proportional to $r^{q(n)}$ and the other to r^n. Since $q(n) \not\equiv n$, both groups of components are separately equal to zero. Hence, we obtain two relations: one includes the structure functions averaged over the turbulent fluid, and the other includes only the structure functions averaged over the nonturbulent fluid. It is evident that replacement of symbol $\langle\ \rangle$ with symbol $\gamma \langle\ \rangle_{w,t}$ corresponds to the substitution of the unconditional two-point PDF $P(v,r)$ with quantity $\gamma_{tt} P_{tt} = \gamma P_{tt}$ (here we take into account relation (1.4) from which it follows that $\gamma_{tt} \to \gamma$ when $r/L \to 0$). Similarly, the unconditional three-point PDF $P(v,r,w,L)$ ought to be replaced by quantity $\gamma_{ttt} P_{ttt}$.

Note that, just as in this case and that considered in Chapter 3, the procedure for the separation of the characteristics of turbulent and nonturbulent fluids is essentially the same, since equation (3.2) for the conditional concentration PDF $P_t(z)$ is also obtained from equation (2.15) for the unconditional PDF $P(z)$ by means of replacing P with $\gamma P_t (F - P_t$ when $0 < z < 1)$ in (2.15)).

Let us now consider the limitations which must be remembered upon utilization of the procedure of separating the characteristics of both fluids. The most important limitation is linked to the fact that function $q(n)$, just as any analytical function, which differs from a constant has singularities in the complex plane n. One of these points is located at $n \to \infty$, since it follows from the definition of function F that $F \to \infty$ when $n \to \infty$.

Numerous examples, both in physics generally and in fluid dynamics in particular, indicate that the appearance of singularities is always caused by the violation of the principal assumption of the theory. For example, the appearance of singularity $n = \infty$ is caused by the fact that when $n \to \infty$ quantity $\langle v^n \rangle$ is determined by the fluctuations whose amplitude tends to infinity. The description of such fluctuations within the framework of incompressible fluid is impossible*.

Another reason behind the appearance of singularities can be the inaccuracy of the similarity hypothesis. To elucidate let us consider a hypothetical case when there is only one singularity, that at $n = \infty$. As an example we cite formula (4.12) which is true for the lognormal law. The procedure used above to separate the characteristics of both fluids is true for all finite n. Hence, for all structure functions averaged over the turbulent fluid we can obtain relations which do not feature structure functions averaged over the nonturbulent fluid (and vice versa). Physically, this means that either the nonturbulent fluid does not exist at all or both fluids exist separately without influencing each other. Such a picture of the phenomenon contradicts experimental data.

Hence, it must be assumed that there are singularities in function $q(n)$ located at finite points n. The conditionally averaged structure functions of both types at these points are essentially linked, i.e., such points characterize the interaction of turbulent and nonturbulent fluids. These arguments will be verified in §4.3 where it is shown that: 1) the existence of such singularities must follow from the equations of motion and the similarity hypothesis; 2) the assumptions made upon formulation of the similarity hypothesis are invalid in the vicinity of the singularities.

Let us dwell now on less significant limitations of the applicability of the procedure of separating the characteristics of turbulent and nonturbulent fluids.

---

*The theory of laminar boundary layers can serve as another example. Even when there is no discontinuity (in particular, on a thin plate in a flow at zero angle of attack), a singularity is located at the initial point of the boundary layer, since its thickness δ changes according to the law $\delta \sim \sqrt{x}$, i.e., there is a singularity $x = 0$ in the equations of the boundary layer.

On the face of it, this procedure is inapplicable if $q(n) = n$ at some $n = n_0$ (then the structure functions of both types are equally dependent on r). However, if n_0 is not a singularity of function $q(n)$, then the relations obtained when $n \neq n_0$ can be analytically extended to point $n = n_0$, i.e., no limitations whatsoever arise.

It should also be noted that formulas (4.7) are true only when $r \to 0$, i.e., they yield only the first terms of the asymptotic expansion of various structure functions. If the subsequent terms are to be accounted for, then the series for quantities $\langle v^n \rangle_t$ and $\langle v^n \rangle_n$ can contain components that are equally dependent on r. This means that the operations of differentiation and conditional averaging are rearrangeable only when $r \to 0$.

§4.3 INTERACTION BETWEEN TURBULENT AND NONTURBULENT FLUIDS

Relation (2.31) contains three terms describing inertia forces, energy transport over the turbulence spectrum, and pressure fluctuations. Only the first term is exactly expressed in terms of the required two-point probability density. The similarity hypothesis enables drawing a number of important conclusions concerning the remaining terms and, thereby analyzing the features of the interaction between the turbulent and nonturbulent fluids.

Let us first consider energy transport over the turbulence spectrum, i.e., the components in (2.31) which are proportional to quantity

$$e_{ij} = \int \epsilon_{ij}^{(1)} P_{t\epsilon}(\epsilon_{ij}^{(1)} | \mathbf{v}, r) d\epsilon_{ij}$$

where $P_{t\epsilon}$ is the conditional dissipation PDF at point 1 provided that points 1 and 2 are in the turbulent fluid, and the velocity difference at them is equal to $\mathbf{v}$. Thus, e_{ij} is the conditionally averaged energy dissipation. Quantity e_{ij} is determined by the small-scale fluctuations with a dimension of the order of the Kolmogorov scale. These fluctuations are in equilibrium with large-scale fluctuations whose characteristics $\mathbf{v}$ and r are assumed specified. It follows from the similarity hypothesis that only two characteristics of large-scale motions, namely v and r, can affect small-scale fluctuations. Because of the principle of self-similarity of turbulent flows with respect to Reynolds number, viscosity ν is not a determining parameter. Hence, from dimensional considerations we obtain

$$e_{ij} = k \frac{v^2}{r} \delta_{ij} \tag{4.13}$$

where k is some constant. This relation is analogous to the formula used in the semiempirical theory of turbulence upon closure of the equation for fluctuation energy ($\langle \epsilon_{ij} \rangle = \text{const } q^3/L$, where q is the fluctuating velocity, L is the integral turbulence scale).

Let us estimate constant k by linking it with constant C in the "two thirds" law. By definition $C = r^{-2/3} \langle v_L^2 \rangle \langle \epsilon_{ij} \rangle^{-2/3}$. From formula (1.8) we obtain $\langle v_L^2 \rangle = \gamma \langle v_L^2 \rangle_t$, since $\langle v^2 \rangle_t \gg \langle v^2 \rangle_n$ when $r \to 0$. Using the continuity equation and formula (4.9), we obtain $\langle v_L^2 \rangle_t = \langle w^2 \rangle_t \times (r/L)^{q(2)}/[3+q(2)]$. From (4.10) and (4.13) we obtain

$$\langle \epsilon_{ij} \rangle = \gamma \langle \epsilon_{ij} \rangle_t = 3k\gamma r^{-1} \langle v^3 \rangle_t = \frac{3k\gamma \langle w^3 \rangle_t}{L}$$

i.e.,

$$C = (3k)^{-2/3}[3+q(2)]^{-1}\gamma^{1/3}\left(\frac{r}{L}\right)^{q(2)-2/3} \langle w^2 \rangle_t \langle w^3 \rangle_t^{-2/3} \qquad (4.14)$$

It should be remembered that w is the velocity of the energy carrying vortices and, therefore, it can be interpreted as the fluctuating velocity at one point. This circumstance allows using one-point velocity probability distribution in the turbulent fluid in the estimates of the moments of quantity w.

It is evident from the obtained formula that quantity C can only tentatively be called constant, since $q(2) \neq 2/3$.

Experimental data indicate that the difference $q(2)-2/3$, however, constitutes a value of the order of several hundredths, i.e., very small (Monin and Yaglom [1967]). Therefore the dependence of C and r is weak and will be henceforth, disregarded. Using the value $C = 1.9^\dagger$, recommended by Monin and Yaglom [1967], and assuming the one-point velocity PDF to be normal, from (4.14) we have $k \sim 1.5 \times 10^{-2} \gamma^{1/2}$.

It can be seen from this estimate that the constant k is small. This conclusion is of importance as is evident from the following considerations. The component $v_j \partial P/\partial r_i$ entering into (2.31) in accordance with the inertia component in the equation of motion is dependent only on the first derivatives with respect to PDF, and the numerical coefficients which appear in it are equal to unity. Component $\partial^2 e_{ij} P/\partial v_i \partial v_j$ describes energy transport over the turbulent spectrum. The second derivatives appear in this component and, hence, it largely determines the structure of formula (2.31). Since the component under consideration is proportional to a small constant k, we arrive at the classical problem involving the perturbation of a highest derivative. The physical meaning of this result is clear: energy dissipation is a relatively slow process. Note that a similar phenomenon is also observed in nonhomogeneous turbulent flows. For example, the angle of expansion of a submerged jet is small and, hence, its energy dissipation is small. This example is an indication that there is a deep-rooted link between constant C and the constants characterizing the rate of expansion of nonhomogeneous jet-type flows, the intensity of turbulence in these flows and so on.

Let us now analyze dissipation fluctuations which play, as is generally known, a very important role in the theory of locally homogeneous turbulence. Using dimensional considerations and formula (4.9) we have

$$R_{\epsilon\epsilon} = \langle \epsilon_{ij}^{(1)} \epsilon_{ij}^{(2)} \rangle \sim \langle v^6 \rangle r^{-2} \sim r^{-\mu}, \qquad \mu = 2 - q(6) \qquad (4.15)$$

---

†It is thereby assumed that constant C is independent of γ. The validity of such an assumption is not quite evident at the present time (see §4.5).

This result is known from the work of Kuznetsov [1976a], Frisch, Sulem, and Nelkin [1978].

Let us now proceed to deducing the equation for P_{tt}. We make use of the procedure of separating the characteristics of turbulent and nonturbulent fluids formulated in §4.2. In accordance with this procedure we must replace $P(\mathbf{v}, r)$ in (2.31) by $\gamma_{tt} P_{tt}(\mathbf{v}, r|w, L)$, $P(\mathbf{v}, r, V, R)$ by $\gamma_{ttt} P_{ttt}(\mathbf{v}, r, V, R|w, L)$. It should be recalled that $r = x^{(2)} - x^{(1)}$, $R = x^{(3)} - x^{(1)}$, $L = x^{(5)} - x^{(1)}$, and $\mathbf{v}, V, w$ are the differences of the velocities at points 1 and 2, 1 and 3, 1 and 5, respectively; moreover $r \ll L$, $R \ll L$, L is the order of the integral turbulent scale. We take into account that $\gamma_{tt} \to \gamma$ when $r/L \to 0$. Using (2.31) and (4.13), we obtain

$$\gamma v_k \frac{\partial P_{tt}}{\partial r_k} + \frac{2k\gamma}{r} \frac{\partial^2}{\partial v_\ell^2} v^3 P_{tt} + \frac{\partial \pi_k}{\partial v_k} = 0$$

$$\pi_k = \frac{1}{4\pi} \int T D_{kij} \gamma_{ttt} V_i V_j P_{ttt} d^3 V d^3 R$$

$$T = \frac{1}{R} - \frac{1}{|R-r|}, \qquad D_{kij} = \frac{\partial^3}{\partial R_k \partial R_i \partial R_j} \tag{4.16}$$

The first term in this relation characterizes the inertia forces, the second the energy dissipation, and the third the pressure fluctuations. Since π_k is proportional to the integral taken over the whole of R, the pressure fluctuations created by vortices of all dimensions affect the motion with dimension r.

Hence, it is evident that the validity of the similarity hypothesis, just as in the entire theory of small-scale turbulence, requires careful analysis. Indeed, the cascade character of energy transport over the turbulence spectrum can be realized only if the regions $R \ll r$ and $R \gg r$ (i.e., very large or very small vortices) do not contribute to the integral included in (4.16). In other words, the convergence of this integral must be analyzed. The problem under consideration is solved when investigating the behavior of the integrand in the regions $r \ll R$ and $r \gg R$.

In order to solve this task, let us identify three regions: 1. $(R > ar)$, 2. $(r/a < R < ar)$, 3. $(R < r/a)$, where $a \gg 1$. Quantity π_k is represented in the form

$$\pi_k = \pi_k^{(1)} + \pi_k^{(2)} + \pi_k^{(3)}$$

$$\pi_k^{(1)} = \frac{1}{4\pi} \int_{R>ar} T D_{kij} \gamma_{ttt} V_i V_j P_{ttt} d^3 V d^3 R$$

$$\pi_k^{(2)} = \frac{1}{4\pi} \int_{r/a<R<ar} T D_{kij} \gamma_{ttt} V_i V_j P_{ttt} d^3 V d^3 R$$

$$\pi_k^{(3)} = \frac{1}{4\pi} \int_{R<r/a} T D_{kij} \gamma_{ttt} V_i V_j P_{ttt} d^3 V d^3 R$$

By definition $\pi_k^{(1)}$ characterizes pressure fluctuations that are created by vortices with a dimension far exceeding r, $\pi_k^{(2)}$ is the pressure fluctuations that are generated by vortices with a dimension of the order of r, $\pi_k^{(3)}$ is the pressure fluctuations that are induced by vortices with a dimension far less than r.

Quantities $\pi_k^{(1)}$ and $\pi_k^{(3)}$ determining the convergence of the integral under consideration can be expressed in terms of the two-point PDF with the aid of the similarity hypothesis. Note that by virtue of the Bayes theorem, P_{ttt} can always be represented in two equivalent forms

$$P_{ttt}(\mathbf{v}, r, V, R|w, L) = P_{tt}(\mathbf{v}, r|V, R)P_{tt}(V, R|w, L)$$

or

$$P_{ttt}(\mathbf{v}, r, V, R|w, L) = P_{tt}(V, R|\mathbf{v}, r)P_{tt}(\mathbf{v}, r|w, L)$$

In so doing, one must, however, bear in mind that if r/R is arbitrary then the similarity hypothesis and the formulas derived in §4.1 are invalid for the description of P_{tt} and such a representation does not make the solution of the problem easier. And if $r \gg R$ or $r \ll R$, then the similarity hypothesis is valid. Thus, we obtain

$$P_{ttt} = P_{tt}(\mathbf{v}, r|V, R)P_{tt}(V, R|w, L), \qquad L > R \gg r$$

$$P_{ttt} = P_{tt}(V, R|\mathbf{v}, r)P_{tt}(\mathbf{v}, r|w, L), \qquad R \ll r < L \tag{4.17}$$

where Mellin transform F of P_{tt} ($F = 2\pi \int_0^\infty v^{n+2} P_{tt} dv$) is given by formula (4.7).

It is now readily seen that the pressure fluctuations which are created by the smallest vortices ($R \ll r$) do not influence the motion of large vortices, i.e.,

$$\pi_k^{(3)} = \frac{\gamma}{4\pi} \int_{R<r/a} V_i V_j T D_{kij} P_{tt}(V, R|\mathbf{v}, r) P_{tt}(\mathbf{v}, r|w, L) d^3V d^3R$$

$$= \frac{\gamma}{4\pi} P_{tt}(\mathbf{v}, r|w, L) \int_{R<r/a} T D_{kij} \langle V_i V_j \rangle_{v,t} d^3 R = 0$$

Here, the second equality follows from the second relation in (4.17), and the third equality from the continuity equation which, as was shown, is also true for the conditional structure functions. Also taken into account is the equality $\gamma_{ttt} = \gamma$ when $r \gg R$ which follows from (1.15).

Analysis of component $\pi_k^{(1)}$ is conveniently conducted by using Mellin transform from (4.16). We multiply (4.16) by $2\pi v^{n+1}$ and integrate for v from 0 to ∞ when $\varphi = \text{const}$. Taking into account (4.7) and equality $\pi_k^{(3)} = 0$, which has just been derived, we obtain

$$\varphi q(n) H(n, \varphi) + (1 - \varphi^2) \frac{dH(n, \varphi)}{d\varphi} + 2kn(n-1)H(n, \varphi)$$

$$+ 2k \frac{d}{d\varphi}(1-\varphi^2)\frac{dH(n,\varphi)}{d\varphi} + J_1 + J_2 = 0$$

$$J_s = \frac{2\pi r}{\gamma w^n}\left(\frac{L}{r}\right)^{q(n)} \int_0^\infty v^{n+1} \frac{\partial \pi_k^{(s)}}{\partial v_k} dv, \qquad s = 1.2 \qquad (4.18)$$

Here the first two components describe the inertia forces and the next two components, which are proportional to k, describe energy transport over the spectrum. Component J_2 characterizes pressure fluctuations that are created by the vortices with a dimension of the order of r.

The most interesting features of equation (4.18) result from component J_1 which describes the pressure fluctuations created by vortices with a dimension far exceeding r. Its computation is based on formula (4.7), on the first relation in (4.17), and on equation (1.12) from which it follows that $\gamma_{ttt} = \gamma^2$ when $r \ll R \sim L$. The result is presented by Kuznetsov [1976a], and the detailed calculations are included in the work by Kuznetsov and Sabel'nikov [1981a]. Since all manipulations are cumbersome, we shall present only the final formula

$$J_1 = -\frac{\gamma}{3}(1+\delta q)\left[(\delta q + 3)\int_{-1}^1 H(n,\varphi)\varphi^2 d\varphi - 1\right]$$

$$\times \left\{(1-\varphi^2)\frac{dH(n-2,\varphi)}{d\varphi} - (n+1)\varphi H(n-2,\varphi)\right\}$$

$$\times r^{2-\delta q} \int_{ar}^\infty (\delta q - 2) R^{\delta q - 3} dR$$

$$\delta q = q(n) - q(n-2) \qquad (4.19)$$

It is evident that the integral in (4.19) converges only when Re $(\delta q) \leq 2$. Then, J_1 and the other components in (4.18) are independent of r.

Consequently, two variables are possible. In the first, function δq is an entire function (it has no singularities when $|n| < \infty$). Then, from the principle of the maximum, if follows that condition Re $(\delta q) \leq 2$ is fulfilled only when $\delta q = $ const. Condition $\delta q = $ const is a difference equation. Function $q = $ const $\cdot n + $ const is a special solution of this equation. An arbitrary periodic function with a period of 2 can be added to this solution. Similar solutions when lm $n = 0$ do not satisfy condition (4.11) and, hence, must not be considered. Therefore, using (4.10), we obtain $q = 1/3n$, i.e., we arrive at the unrefined Kolmogorov theory [1941]. This variant, as already noted in Chapter 1, does not agree with the experimental data and, furthermore, possesses the following feature which, apparently, cannot be characteristic for real turbulent flows. Indeed, from the second formula in (4.3) we find that

$$P_{tt}^0 = \delta\left[v - V\left(\frac{r}{R}\right)^{n/3}\right]$$

i.e., it follows from the adopted similarity hypothesis and the equations of motion that the absolute value of the velocity difference is not a random value.

Hence, functions $\delta q(n)$ and $q(n)$ have singularities at finite n. Thus, from the equation of motion it follows that the velocity difference PDF is not described by the lognormal law (function (4.12) is an entire function and condition Re $(\delta q) \leq 2$ is violated when Re $n < 5/2 - 1/3q_2$).

As mentioned in §4.2, the appearance of singularities in function $q(n)$ is caused by the interaction between the turbulent and nonturbulent fluids. We shall show that the structure of the solutions of equation (4.18) in the vicinity of both singularities truly reflects the principal features of this interaction which directly follow from the Navier-Stokes equations. It was shown in Chapter 1 that this interaction possesses three characteristics. First, it is caused by pressure fluctuations. Second, it follows from (1.7) that $\langle v^2 \rangle_n / \langle v^2 \rangle_t \sim r^{2-q(2)} \approx r^{4/3} \to 0$ when $r \to 0$. This means that intensive fluctuations in the turbulent fluid interact with the fluctuations in the nonturbulent fluid which have small amplitude. Third, the small-scale fluctuations in the nonturbulent fluid receive energy directly from large-scale energy carrying vortices (coefficient A_2 in (1.7) is proportional to the velocity gradient of these vortices, i.e., $\langle v^2 \rangle_n \sim A_2$).

All these characteristics are also observed in the case under study. Indeed, the appearance of singularities in function $q(n)$ is caused by the divergence of integral (4.16) which describes pressure fluctuations. It is also evident that fluctuations with small amplitude correspond to these points. Indeed, let n_1 be the first singularity of function $\delta q(n)$. The appearance of such a point can be connected only with the fact that integral $2\pi \int_0^\infty v^{n+2} P_{tt} dv$ diverges when $v \sim 0$ and $n = n_1$, i.e., the structure of the fluctuations with small amplitude is of decisive significance. Finally, when $n \leq n_1$, condition Re $(\delta q) \geq 2$ is fulfilled, since, otherwise, there is no singularity. Then, from (4.19) it follows that when $n \approx n_1$ the main contribution to the integral is made by region $R \gg r$, i.e., the large-scale energy carrying motions exert direct influence on the slow small-scale vortices. Hence, it follows that the structure of the solutions of equation (4.18) in the vicinity of the singularities of function $q(n)$ reflects the interaction between the turbulent and nonturbulent fluids.

This conclusion is exceptionally important, since the interaction between vortices of different scales in the vicinity of the singularities are of direct, and not cascade, character and, hence, not only the adopted similarity hypothesis is invalid, but also is the entire theory of locally homogeneous turbulence. The question regarding the validity of this theory outside the singular points remains open, which is evident from the following considerations. Since $q(n)$ is an analytical function, then its real (q_r) and imaginary (q_i) parts satisfy the Laplace equation $\Delta q_r = \Delta q_i = 0$, where $\Delta = \partial^2/\partial n_r^2 + \partial^2/\partial n_i^2$, $n_r =$ Re n, $n_i =$ Im n. The solutions of this equation possess its distinctive nonlocal properties, i.e., the structure of the singularity determines the character of the solution at all distances from it. Therefore, the influence of large-scale pressure fluctuations leading to the appearance of singularities is also reflected in the structure of small-scale fluctuations with all amplitudes[‡]. It is significant that this influence is not weakened under an arbitrarily small ratio of the scales of small and large vortices.

Weak traces of such an influence remain only at large distances from the singularities, i.e., strictly speaking, the theory of locally homogeneous turbulence can be true only when describing the structure functions of sufficiently high order.

Let us analyze the question in more detail. From (4.18) and (4.19) it is evident that the first singularities of functions $H(n,\varphi)$ and $\delta q(n) = q(n) - q(n-2)$ coincide. Since $H(n,\varphi)$ is an analytical function of n in the region Re $n > -3$, then Re $n_1 < -3$, where n_1 is the first singularity of function $\delta q(n)$. So far as $\delta q = q(n) - q(n-2)$, the first singularity $(n = n_2)$ of function $q(n)$ is located in the region Re $n < n_1 - 2 = -5$ (Re $n_1 < -3$). This estimate has an important practical significance, since the investigation of the behavior of function $q(n)$ at large distances $\delta n = n - n_2$ from the singularity is of greatest interest. For example, for the structure function describing the energy of fluctuations we have $n = 2$, $\delta n = 7$, and for the structure function describing dissipation energy we obtain $n = 6$, $\delta n = 11$, i.e., $|\delta n| \gg 1$. Hence it is evident: the principal features of small-scale fluctuations are determined by the structure of function $q(n)$ at large distances from the first singularity. In other words, the details of the interaction between turbulent and nonturbulent fluids are of no significance. Thus, it is possible to construct a simple theory describing the velocity difference PDF. On the other hand, in many practical investigations, one has to estimate the dependence on distance of the structure functions whose order varies slightly in comparison with $|\delta n|$. For example, the structure functions for which n is in the range 2 – 6 are mainly measured in experiments. For the description of such structure functions, one can assume, with sufficient accuracy, that q is a quadratic polynomial of n, i.e., formula (4.12) which corresponds to the lognormal law is approximately true.

It is readily seen that the only constant q_2 describing this law is of small magnitude. Indeed, experimental data presented below in §4.5 indicate that constant μ in formula (4.15) does not exceed 0.5. Then from (4.12), (4.15) we find $q_2 = \mu/18 \leq 0.03$. Hence, we arrive at the important conclusion: there are at least two small constants in the theory of locally homogeneous turbulence. One of them (q_2), as follows from (4.15) and equality $q_2 = \mu/18$, characterizes dissipation fluctuations. The second constant k appears in formula (4.13) describing the mean value of dissipation; this constant is connected with "constant" C in the "two thirds" law. As the estimates presented at the beginning of this section show, the values of k and q_2 are of the same order. This coincidence is not, apparently, accidental and, is rather associated with the fact that the first singularity of function $q(n)$ is located at a large distance. Physically this means that the turbulent and nonturbulent fluids interact weakly.

[‡]Considered here is function $q(n)$ describing the dependence of integral $F = 2\pi \int_0^\infty v^{n+2} P_{tt} dv$ on the distance. It is obvious that the values of F for large $|n|$ are determined by the structure of function P_{tt} for large v (and vice versa). Thus, quantity $|n|$ characterizes the amplitude of the fluctuations.

Note now that both constants under consideration appear in equation (4.18). It is to be expected that this equation defines some connection between k and μ (or q_2). This analysis indicates some important features of this connection. Indeed, from (4.18) and (4.19) it can be seen that the inertia components are independent of the intermittency factor γ, and one of the components describing the pressure fluctuations (J_1), is proportional to γ. Hence, the coefficients in equation (4.18) are essentially dependent on γ. Generally speaking, this means that all the constants considered above (k, C, μ, q_2) are dependent on γ. From the physical standpoint the universality of these constants is fairly natural, since the character of the interaction between the turbulent and nonturbulent fluids must depend on how the fluid fills the volume under study.

§4.4 EFFECT OF VISCOSITY ON THE STRUCTURE OF SMALL-SCALE PULSATIONS

Let us first analyze the effect of viscosity on small-scale fluctuations in the inertial range $(L \gg r \gg \eta)$. It is evident from physical considerations that even at large distances between the points under consideration $(r \gg \eta)$, owing to the random quality of the process, such situations are possible when v and, hence, the local Reynolds number vr/ν are small. Thus, the effect of viscosity appears in the vicinity of the singularities of function $q(n)$.

This remark creates a whole set of problems associated with the need to refine equation (4.16). First, we must take into account that quantity k characterizing the conditionally averaged energy dissipation is constant only when $vr/\nu \gg 1$, i.e., $k = k(vr/\nu)$, where $k \to \text{const}$ when $vr/\nu \to \infty$. Second, we must refine the hypothesis adopted when computing the value of $\pi_k^{(1)}$ which characterizes the pressure fluctuations generated by large-scale motion. As a first approximation, similar refinements can be based on the following considerations. Let us turn to the second relation in (4.16) which determines quantity π_k. When $R \gg L$ (L is the integral turbulence scale), the three-point PDF P_{ttt} is the product of one-point and two-point PDFs (fluctuations at point 3 are independent of the velocity difference at points 1 and 2). Then, from the continuity equation it can be shown that region $R \gg L$ does not contribute to the integral appearing in the definition of quantity π_k. Hence, as a first approximation, it can be assumed that during calculations of π_k the integration is over the region $R \leq L$. If it is assumed, as before, that the first relation in (4.17) is true in this region, the upper limit in (4.19) must be replaced by a value of the order of L. This means that quantity J_1 in the vicinity of the singularities is dependent on r/L.

Thus, two parameters enter equation (4.16) vr/ν and r/L, and when analyzing the statistical characteristics of turbulence in the inertial range of its spectrum the problem of computation of the double limit arises when $vr/\nu \to \infty$ and $r/L \to 0$. In the phase space region under consideration, such a computation cannot be carried out uniquely, since function $q(n)$ has singularities. Thus, the same problems as in Chapter 1 are encountered.

Let us now consider the effect of viscosity on the structure of the more small-scale fluctuations ($r < \eta$). The statistical characteristics of the velocity gradient are of greatest practical interest. The remaining part of this section will be devoted to the solution of this problem.

Let points 1, 2, 3 be located in the turbulent fluid, the distances between points 1 and 2, 1 and 3 are equal to r and R, respectively, and, furthermore, condition $\eta \ll r \ll R \ll L$ is implemented, i.e., the scales of r and R belong as before to the inertial range. Let us analyze the parameters on which the conditionally averaged moment is dependent

$$K_n = \left\langle \left[\sqrt{\left(\frac{\partial u_i}{\partial x_j}\right)^2}\right]^n \right\rangle_{v,t}$$

Quantity K_n is determined by the more small-scale fluctuations with a dimension that is comparable with quantity η. Therefore, it is natural to assume that K_n is dependent only on v, r, ν. Then from dimensional considerations we obtain

$$K_n = v^n r^{-n} M_n \left(\frac{vr}{\nu}\right)$$

where M_n is a dimensionless function. Using the Bayes theorem we obtain

$$K_n(V, R) = V^n R^{-n} M_n \left(\frac{VR}{\nu}\right)$$

$$= \int K_n(v, r) P_{tt}(v, r|V, R) d^3 v \qquad (4.20)$$

where P_{tt} is determined by formulas (4.3), (4.7) which are valid in the inertial range. We seek a solution in the form

$$M_n = \left(\frac{vr}{\nu}\right)^{x(n)}$$

Substituting this formula in (4.20) and accounting for (4.7), we obtain

$$q[n + x(n)] + x(n) - n = 0 \qquad (4.21)$$

Hence it is evident that the dependence of the moments of the velocity gradient on Reynolds number is determined by function $q(n)$ characterizing the structure function in the inertial range.

As was shown in §4.3, when n is not too large, formula (4.12) can be used as a sufficiently good approximation. Let us take into account that $q_2 = \mu/18 \ll 1$. Then, by ignoring the terms of the order q_2^2, we obtain from (4.21)

$$x(n) = \frac{n}{2} + \frac{3}{32} \mu(n^2 - 2n) \qquad (4.22)$$

This formula corresponds to the lognormal law. In particular, for the dissipation moments from (4.22) it follows that

$$\langle \epsilon^n \rangle_{v,t} = v^{3n} r^{-n} \left(\frac{vr}{\nu} \right)^{3\mu n(n-1)/8}$$

By extrapolating this relation in the region $r \sim L$ (L is the integral scale), we obtain

$$\langle \epsilon^n \rangle \sim \langle \epsilon \rangle^n \left(\frac{L}{\eta} \right)^{\mu n(n-1)/2}$$

$$P(\epsilon) = \frac{1}{\sqrt{2\pi}\epsilon\sigma_\epsilon} \exp\left\{ -\frac{(\ln(\epsilon/\langle\epsilon\rangle) + \sigma_\epsilon^2/2)^2}{2\sigma_\epsilon^2} \right\}$$

$$\sigma_\epsilon^2 = \mu \ln \frac{L}{\eta} \qquad (4.23)$$

Formula (4.23) is often used when analyzing the measurements of the velocity gradient (Frenkiel and Klebanoff [1975] and others). It is based on the approximate relation (4.22) and is, hence, true only when $q_2 = \mu/18 \ll 1$. The formula does not describe the fluctuation of quantity ϵ with very large amplitude even when $q_2 \ll 1$, since (4.23) is a consequence of the approximate relation (4.12) which is unsuitable for large n, i.e., for large fluctuation amplitudes.

Indeed, let us investigate the character of behavior of function $q(n)$ at large positive n (Im $n = 0$). By introducing the notation $y = n + x$, we obtain from (4.21)

$$2n = y + q(y) \qquad (4.24)$$

The solution of this equation, generally speaking, is ambiguous as may be seen from the analysis of the signs of the derivative dn/dy when $y \sim 1$ and when $y \to \infty$. Indeed, since q_2 is small, then $q'(2) \approx 1/3$. Therefore from (4.24) we obtain $dn/dy > 0$ when $y \sim 1$. Let us now consider the case when $y \to \infty$. Let $\alpha = q(y)/y$. Two variants are obviously possible. In the first, we have $\alpha > -1$ when $y \to \infty$. Then, $dn/dy > 0$ both when $y \sim 1$ and when $y \to \infty$. In the second, we obtain $\alpha < -1$ when $y \to \infty$. Then $dn/dy = 1 + \alpha < 0$ when $y \to \infty$, i.e., dn/dy tends to zero for finite y and, hence, functions $y(n)$ and $x(n)$ are ambiguous.

Let us illustrate this conclusion by the example of the lognormal law ($\alpha = -\infty$ when $y = \infty$). From (4.12), (4.24) it follows that $dn/dy = 0$, $n = n_3 = (4/3 + 3q_2)^2/(4q_2)$ when $y = (4/3 + 3q_2)/(2q_2)$. In the small vicinity of the point under consideration we obtain $x = y - n = \text{const} + \text{const}\sqrt{n_3 - n}$, i.e., function $x(n)$ has a singularity and, hence, the moments of the velocity gradient, whose order exceeds n_3, do not exist.

Hence it follows that all the moments of the velocity gradient exist only when the following asymptotic dependence ($n \to \infty$) is true

$$q = \text{const} + \alpha n, \qquad \alpha > -1 \tag{4.25}$$

One interesting feature of the PDFs for which formula (4.25) is true can be singled out. From (4.3), (4.25) it can be shown that the maximum possible velocity difference ($P_{tt}^0 = 0$ when $v > v_m$) exists at two points in the inertial range of the spectrum. Dependence $v_m(r)$ is described by expression $v_m \sim r^{-\alpha}$.

It should also be noted that in the case under consideration ($\alpha > -1$), the differentiability of the nonaveraged velocity field follows from the existence of all the moments of the velocity gradient (Monin and Yaglom [1967]). In the other case ($\alpha < -1$) it is not possible to come to any conclusions concerning the differentiability of the nonaveraged velocity field, since the smooth random function might not have some moments (Monin and Yaglom [1967]). It should be remembered that the differentiability of the solutions of the Navier-Stokes equations has not been proved. A number of arguments are put forward to the effect that such solutions might be nondifferentiable (Ladyzhenskaya [1970]).

§4.5 EXPERIMENTAL INVESTIGATION OF INTERMITTENCY AND THE STRUCTURE OF THE FINE-SCALE PART OF THE TURBULENCE SPECTRA

Theoretical considerations were presented in the previous sections indicating significant deviations from the universal dependences of the Kolmogorov-Obukhov theory. This conclusion is of fundamental significance since fine-scale velocity fluctuations determine the dissipation and, hence, considerably affect the dynamics of the turbulent flow as a whole. The present section is devoted to the experimental verification of the Kolmogorov-Obukhov hypothesis on the universality of the fine-scale structure of turbulence. Preliminary results are included in this work by Kuznetsov, Praskovskii, and Sabel'nikov [1984a, b] and the final data are partially published in the works by Kuznetsov, Praskovskii, and Sabel'nikov [1987, 1988][§].

1. Analysis of the known experimental data. The structure of the fine-scale part of the turbulence spectra has been investigated experimentally in many works (see, for example, the reviews by Champagne [1978], Yaglom [1981], Antonia, Chambers, and Anselmet [1984]). The experimental data indicate that the Kolmogorov-Obukhov theory is qualitatively correct and with the variation of the scale r (the distance between the points) within a wide range of values it correctly describes the order of change of the main characteristics of turbulence. As noted by Yaglom [1981], there is now no basis for any doubts as to the universality of the applicability of the "two thirds" law to all turbulent flows with sufficiently large values of Reynolds number. At the same time, analysis of the literature reveals noticeable variations between the results obtained in different studies.

[§]It is particularly pertinent to mention the role of A. A. Praskovskii, who carried out the most laborious part of these tests and without whose participation the material of this section would have never been completed.

To confirm the above, we examine tables 1 and 2. The first table, borrowed from the review by Yaglom [1981] and supplemented by data from later work of Antonia, Satyaprakash, and Chambers [1982], encompasses the results of a large number of measurements of the coefficient C in the "two thirds" law. The second table encompasses the exponent μ in the correlation function of energy dissipation $R_{\epsilon\epsilon}$ (4.15). The data in tables 1 and 2 show the scatter of the experimental values of C and μ. If one, according to Yaglom [1981], excludes from the discussion the values of C given in rows 1, 2, 10, 14, 16, 17, and 23 of table 1 as questionable and unreliable, the remaining values of C would be within the range 1.6 – 2.5. The constant μ, as evident from table 2, varies within the range 0.18 – 0.5. Of course, this scatter can be attributed to the usual experimental errors but this variation of C and μ exceeds the accuracy of measurements. The theoretical analysis presented in §4.2 and §4.3 shows that the scatter could have deeper physical origins. It should be noted that even Yaglom [1981] does not totally rule out such a possibility.

Let us now consider the results of the experimental verification of the hypothesis on local isotropy of turbulence which occupies a central position in the Kolmogorov-Obukhov theory. It should be emphasized that sufficiently satisfactory experimental results have not yet been obtained in support of this hypothesis which is evident, for example, from the studies by Mestayer [1982], Antonia, Anselmet, and Chambers [1986], Browne, Antonia, and Shah [1987], Gibson, Friehe, and McConnells [1977]. It follows from these studies that in laboratory sheared flows some of the corollaries of local isotropy are not fulfilled and, in particular, the relations are violated between the mean values of the squares of the spatial derivatives of the various velocity components. It is significant that according to the available data the degree of deviation from local isotropy with increasing Reynolds number does not decrease and remains approximately constant. To illustrate this conclusion, tables 3 and 4, reproduced from the work by Browne, Antonia, and Shah [1987], show the data of various authors for the ratios of the squares of the derivatives (denoted here $x_3 = z$, $u_3 = w$)

$$K_1 = 2\left\langle \left(\frac{\partial u}{\partial x}\right)^2 \right\rangle \bigg/ \left\langle \left(\frac{\partial v}{\partial x}\right)^2 \right\rangle$$

$$K_2 = 2\left\langle \left(\frac{\partial u}{\partial x}\right)^2 \right\rangle \bigg/ \left\langle \left(\frac{\partial w}{\partial x}\right)^2 \right\rangle$$

$$K_3 = 2\left\langle \left(\frac{\partial u}{\partial x}\right)^2 \right\rangle \bigg/ \left\langle \left(\frac{\partial u}{\partial y}\right)^2 \right\rangle$$

$$K_4 = 2\left\langle \left(\frac{\partial u}{\partial x}\right)^2 \right\rangle \bigg/ \left\langle \left(\frac{\partial u}{\partial z}\right)^2 \right\rangle$$

SMALL-SCALE TURBULENCE 151

Table 4.1   Experimental values of the constant C

N	C	Authors	Type of flow
1	2	3	4
1.		Gurvich [1960a][c]	atmospheric layer over land
2.		Gurvich [1960b, 1962][a]	atmospheric layer over land
3.		Zubkovskii [1962][a]	atmospheric layer over land
4.		Grant, Stewart, and Moilliet [1962][a]	ocean flow
5.		Pond, Stewart, and Burling [1963][a]	atmospheric layer over sea
6.		Stewart [1963][c]	atmospheric layer over land
7.		Gibson [1963][a]	axisymmetric jet
8.		Bradbury [1965][a]	plane jet
9.		Pond, Smith, Hamblin, and Burling [1966][a]	atmospheric layer over sea
10.		Kistler and Vrebalovich [1966][a]	behind a grid in a tube
11.		Record and Cramer [1966][b]	atmospheric layer over land
12.		Uberoi and Freymuth [1969][a]	wake behind a cylinder
13.		Uberoi and Freymuth [1970][a]	wake behind a sphere
14.		Gibson, Stegan, and Williams [1970][a]	atmospheric layer over sea
15.		Van Atta and Chen [1970][b]	atmospheric layer over sea
16.		Van Atta and Chen [1970][a]	atmospheric layer over sea
17.		Van Atta and Chen [1970][c]	atmospheric layer over sea
18.		Stewart, Wilson, and Burling [1970][c]	atmospheric layer over sea
19.		Nasmyth [1970][a]	ocean flow
20.		McBean, Stewart, and Miyake [1971][a]	atmospheric layer over land
21.		Wyngaard and Cote [1971][a]	atmospheric layer over land
22.		Pekvin and Pond [1971][c]	atmospheric layer over sea
23.		Sheikh, Tennekes, and Lumly [1971][a]	lower atmosphere ($z = 100$ m) from a plane
24.		Sheikh, Tennekes, and Lumly [1971][a]	lower atmosphere ($z = 100$ m) from a tower
25.		Boston and Burling [1972][a]	atmospheric layer over land
26.		Kaimal, Wyngaard, Izumi, and Cote [1972][a]	atmospheric layer over land
27.		Kholmyanskii [1972][a]	atmospheric layer over land
28.		Schedwin, Stegan, and Gibson [1974][a]	behind a grid in a tube
29.		Bashir and Uberoi [1975][a]	plane jet
30.		Helland, Van Atta, and Stegen [1977][a]	behind a grid in a tube
31.		Helland, Van Atta, and Stegen [1977][a]	axisymmetric jet
32.		Williams and Paulson [1977][a]	atmospheric layer over land
33.		Champagne, Freihe, La Rue, and Wyngaard [1977][a]	atmospheric layer over land
34.		Champagne [1978][a]	atmospheric layer over land
35.		Champagne [1978][a]	axisymmetric jet
36.		Caulliez [1980][a]	atmospheric layer over land

152 TURBULENCE IN COMBUSTION

Table 4.1 Continued

N	C	Authors	Type of flow
1	2	3	4
37.		Result of analysis of the Tsimlyansk expedition data of the Institute of Atmospheric Physics at the USSR Academy of Sciences	atmospheric layer over land
38.		Antonia, Satyaprakash, and Chambers [1982][a]	axisymmetric jet

[a] Constant C determined from the longitudinal spectrum.
[b] Constant C determined from the longitudinal structure function.
[c] Constant C determined from the asymmetry of velocity difference.

Table 4.2 Experimental values of the constant μ

N		Authors	Type of flow
1	2	3	4
1.		Gurvich and Zubkovskii [1963, 1965]	atmospheric layer over land
2.		Pond and Stewart [1965]	atmospheric layer over sea
3.		Freihe, Van Atta, and Gibson [1971]	wake behind a cylinder
4.		Kholmyanskii [1972]	atmospheric layer over land
5.		Gibson and Maseillo [1972]	atmospheric layer over land
6.		Wyngaard and Pao [1975]	atmospheric layer over land
7.		Champagne, Pao, and Wignanski [1976]	mixing layer of a plane jet
8.		Gagne and Hopfinger [1979]	axisymmetric jet, plane channel
9.		Antonia, Phan-Thien, and Satyaprakash [1981]	axisymmetric jet
10.		Antonia, Satyaprakash, and Hussain [1982]	plane and axisymmetric jets
11.		Antonia, Rajagopalan, Browne, and Chambers [1982]	plane jet
12.		Antonia, Satyaprakash, and Chambers [1982]	plane and axisymmetric jet, atmospheric layer over land
13.		Anselmet, Gagne, Hopfinger, and Antonia [1984]	axisymmetric jet, flow in plane channel
14.		Chambers and Antonia [1984]	atmospheric layer over land
15.		Kuznetsov, Praskovskii, and Sabel'nikov [1984a, b]	wake behind a cylinder

Table 4.3 Values of the ratios K_1 and K_2

Type of flow	Experimental conditions	K_1	K_2	Authors
1	2	3	4	5
Quasihomogeneous sheared flow	$Re_\lambda = 160$			Tavalaris and Corsin [1981a]
Boundary layer	$Re_\delta = 50000$ $y/\delta = 0.08$ $y/\delta = 0.32$			Verollet [1972]
Pipe	$Re_d = 50000$, near the wall at half radius distance from the wall			Laufer [1954]
	$Re_d = 500000$, near the wall at half radius distance from the wall			
	$Re_d = 90000$, near the wall at half radius distance from the wall			Lawn [1971]
Axisymmetric jet	$Re_d = 100000$ $y/x = 0$ $y/x = 0.05$ $y/x = 0.1$			Wygnanski and Fiedler [1969]
Plane jet (without external flow)	$Re_\lambda = 990$ $y/x = 0$ $y/x = 0.05$ $y/x = 0.1$			Gutmark and Wygnanski [1976]
	$Re_\lambda = 204$ jet axis			Antonia, Rajagopalan, Browne, and Chambers [1982]
	$Re_\lambda = 300$ jet axis $Re_\lambda = 200$ half width			Everitt and Robins [1978]
Plane jet (with external flow)	Strong jet, jet axis half width Weak jet, jet axis half width			Everitt and Robins [1978]
Mixing layer	$Re_\lambda = 330$			Champagne, Pao, and Wygnanski [1976]
Two-dimensional wake behind a circular cylinder	$Re_d = 2700$, $y/d = 0$ $y/d = 6$ $Re_d = 19050$, $y/d = 0$ $Re_d = 8400$, across wake section			Fabris [1974] Champagne [1978] Townsend [1948]

Table 4.4 Values of ratios K_3 and K_4

Type of flow	Experimental conditions	K_3	K_4	Authors
Boundary layer	$Re_\delta = 50000$ $y/\delta = 0.08$ $y/\delta = 0.32$			Verollet [1972]
Pipe	$Re_d = 50000$, near the wall at half radius distance from the wall $Re_d = 500000$, near the wall at half radius distance from the wall $Re_d = 90000$, across the entire section			Laufer [1954] Lawn [1971]
Axisymmetric jet	$Re_d = 100000$ $y/x = 0$ $y/x = 0.05$ $y/x = 0.1$			Wygnanski and Fiedler [1969]
Plane jet (without external flow)	$Re_\lambda = 204$, jet axis			Antonia, Browne, and Chambers [1984]

For locally isotropic turbulence $K_1 = K_2 = K_3 = K_4 = 1$. It is apparent that the quantitative deviations from local isotropy turn out to be independent of the Reynolds number Re_λ in the range $40 < Re_\lambda < 10^3$.

Deviation from local isotropy can lead to noticeable errors when determining energy dissipation with the aid of the relationship which is true for isotropic turbulence

$$\langle \epsilon \rangle = 15\nu \left\langle \left(\frac{\partial u_1}{\partial x_1} \right)^2 \right\rangle$$

As an example, Fig. 4.1 shows the results of measurements of the ratio $\epsilon^0 = \langle \epsilon \rangle / 15\nu \langle (\partial u_1 / \partial x_1)^2 \rangle$ in a wake behind a circular cylinder conducted by Browne, Antonia, and Shah [1987]. Measured in this work were the nine components in the expression for $\langle \epsilon \rangle$ yielding the main contribution to energy dissipation. The other three terms in this expression were estimated with the aid of isotropic relations.

It is evident that the assumption of local isotropy in the central region of the wake, where intermittency is insignificant, lowers dissipation by 45%, and at the edge of the wake, where intermittency is significant, by 80%.

At the present time, one can only make some assumptions for the reasons behind the anisotropy in the fine-scale turbulence structure and its maintenance at very high Reynolds numbers that are characteristic for atmospheric and ocean flows. The deviation from local isotropy is most often explained by the direct influence on fine-scale motion of large-scale anisotropic structures, which are often referred to as coherent and are always present in shear flows. Such an effect particularly strongly influences the fine-scale structure of the temperature field (see, for example, Shrinivasan, Antonia, and Brits [1979], Antonia, Anselmet, and Chambers [1986], Asizyan and Koprov [1985]).

One of the possible explanations of the indicated effect can be obtained from the following model. Let the main part of dissipation be concentrated near some system of distorted surfaces with the derivatives normal to these surfaces of the velocities being much greater than the derivatives in the tangential direction (such surfaces can be interpreted as tangential discontinuities). It is obvious that the velocity derivatives will be statistically isotropic only if the orientation of the vector normal to surface n, which is a random quantity, is equiprobable in all directions. Let us now imagine that there are very weak deviations from the equiprobable orientation of vector n. Then, the deviations from isotropy are found, generally speaking, to be significant since the velocity derivatives in a direction normal to the surface are much greater than in the tangential directions. It is also evident that within the framework of the given model the ratio of the velocity derivatives in the normal and tangential directions increases without restriction with increasing Reynolds number. Under such conditions, a constant anisotropy of the velocity derivatives indicates that with increasing Reynolds number the orientation of vector n approaches isotropic orientation. Therefore, the anisotropy of fine-scale turbulence structure which is observed in experiments is not, generally speaking, an indication of fundamental flaws in the Kolmogorov-Obukhov theory although it indicates the need for considerable refinement of the theory.

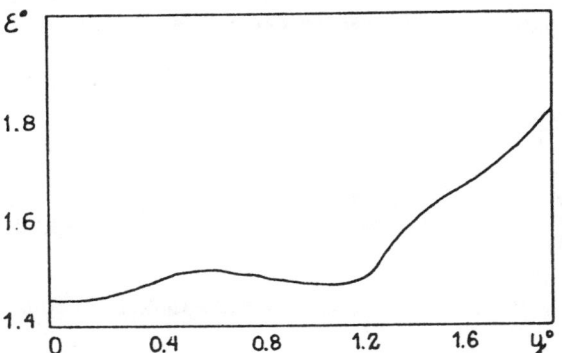

Figure 4.1 Distribution of the ratio of dissipation of energy to the dissipation calculated from the isotropic relation $\langle \epsilon \rangle = 15\nu \langle (\partial u_1/\partial x_1)^2 \rangle$ in a wake behind a circular cylinder from data by Browne, Antonia, and Shah [1987]. $x/d = 420$, $\text{Re}_d = u_0 d/\nu = 1170$, $d = 2.67$ mm, $u_0 = 6.7$ m/sec, $\text{Re}_\lambda = 40$ on the plane of symmetry of the wake, $\text{Re}_\lambda = 80$ on the wake boundary, L_0 is the linear scale equal to the distance from the plane of symmetry of the wake for which velocity defect is equal to half the value on the plane of symmetry, $y^0 = y/d$.

This analysis of the experimental data and the theoretical results presented in §§4.1 – 4.3 show that there are apparently some flaws in the Kolmogorov-Obukhov theory and, hence, further experiments to verify the theory are needed.

2. Conditions of the experiment and processing technique. Let us consider the difficulties arising when measuring the characteristics of turbulence in the inertial and viscous ranges of the spectrum. In such experiments one must satisfy a number of very rigid and contradictory requirements for recording and processing the signals. Reference to and discussion of the possible sources of errors when measuring fine-scale characteristics of turbulence can be found, for example, in the studies by Champagne [1978], Antonia, Chambers, and Anselmet [1984]. The main problems which must be solved here are: 1) to ensure high locality (spatial resolution) of the measurements; 2) to ensure sufficient frequency and 3) dynamic ranges, i.e., high ratio of the useful signal and noise; 4) to select the correct averaging time.

These difficulties are caused by the very wide range of values of fluctuations of the energy dissipation. Therefore, errors which are associated with the spatial averaging of the velocity field arise. With increasing amplitude of energy dissipation fluctuations, the amplitude of oscillations of the minimum scale of hydrodynamic nonhomogeneity increases and the higher the number of the measured moment, the greater this effect. Let us estimate the role of the indicated factors (Kuznetsov, Praskovskii, and Sabel'nikov [1984a, b]). The analysis presented in §4.4 indicates that when determining the moments of energy dissipation whose order are not too high, formula (4.23) can be used.

Let it be required to measure quantity $\langle \epsilon^n \rangle$ with accuracy Δ. Then, from (4.23) it follows that the measuring device must record the values of dissipation in the range $(0, \epsilon_1)$, where

$$\epsilon_1 = \langle \epsilon \rangle \exp\left[\sigma^2\left(n - \frac{1}{2}\right) + \sigma h(\Delta)\right]$$

$$\sigma^2 = \mu \ln \frac{L}{\eta}, \qquad \frac{1}{\sqrt{2\pi}} \int_{k(\Delta)}^{\infty} \exp\left(-\frac{s^2}{2}\right) ds = \Delta \qquad (4.26)$$

This estimate is obtained if it is assumed that the measured value of ϵ is equal to zero for $\epsilon > \epsilon_1$.

Hence it follows that the dimension of region ℓ over which instrument averaging is carried out (for example, the length of the anemometer wire) must not exceed the value $\ell = \text{const } \nu^{3/4}\epsilon_1^{-1/4}$. This value can be substantially less than the Kolmogorov scale η, since dissipation in the region under consideration can substantially exceed $\langle \epsilon \rangle$. In order to conduct finite calculations some level of accuracy is assumed, say $\Delta = 0.2$. Let us consider again the measurement of the mean energy dissipation. By using the data for energy dissipation spectrum (see Fig. 4.2), it can be concluded that 80% of energy dissipation is concentrated in that region of the spectrum which satisfies condition $k\eta < 0.4$ (k is the wave number).

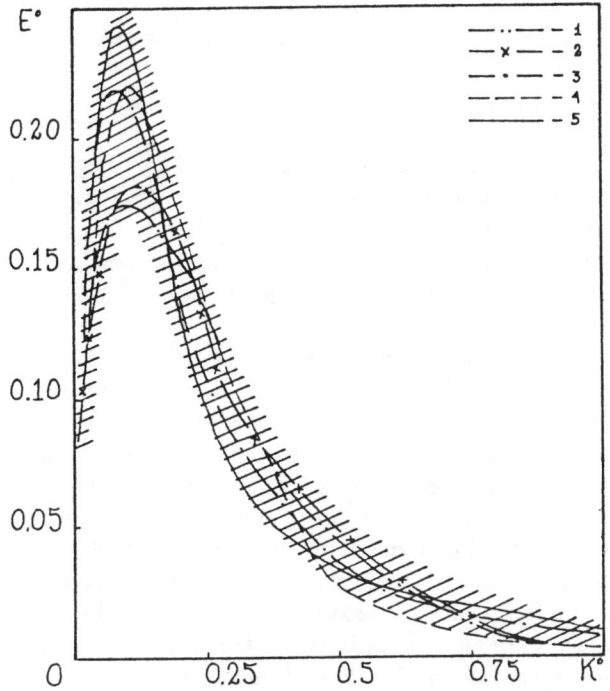

Figure 4.2 Energy dissipation spectra from data by Kuznetsov, Praskovskii, and Sabel'nikov. *1)* axisymmetric wake ($y/d = 0.41$, $\langle(u_1 - \langle u_1\rangle)^2\rangle^{1/2}/\langle u_1\rangle = 0.07$, $\text{Re}_\lambda = 888$, $L = 0.48$ m, $\eta = 0.34$ mm, $\gamma = 0.46$); *2)* three-dimensional wake ($y/d = 0$, $\langle(u_1 - \langle u_1\rangle)^2\rangle^{1/2}/\langle u_1\rangle = 0.68$, $\text{Re}_\lambda = 524$, $L = 0.56$ m, $\eta = 0.3$ mm, $\gamma = 0.63$); *3)* mixing layer ($y/x = 0$, $\langle(u_1 - \langle u_1\rangle)^2\rangle^{1/2}/\langle u_1\rangle = 0.21$, $\text{Re}_\lambda = 1660$, $L = 1.21$ m, $\eta = 0.21$ mm, $\gamma = 0.89$); *4)* plane wake ($y/d = 3.2$, $\langle(u_1 - \langle u_1\rangle)^2\rangle^{1/2}/\langle u_1\rangle = 0.051$, $\text{Re}_\lambda = 162$, $L = 0.1$ m, $\eta = 0.24$ mm, $\gamma = 0.7$); *5)* boundary layer ($y/\delta = 0.21$, $\langle(u_1 - \langle u_1\rangle)^2\rangle^{1/2}/\langle u_1\rangle = 0.11$, $\text{Re}_\lambda = 140$, $L = 0.1$ m, $\eta = 0.13$ mm, $\gamma \approx 1$). The shaded area shows the range of the experimental data by other authors and generalized by Monin and Yaglom [1967]. $E^0 = (\eta k)^2 E_1(k)/\langle\epsilon\rangle\nu^5)^{1/4}$, $k^0 = k\eta$.

Thus, for measuring the quantity $\langle\epsilon\rangle$ with an accuracy of 20%, the length of the wire must not be more than $\ell = 2.5\eta$. It can be assumed that a similar condition must be observed when measuring moments of higher order, the only difference being that the mean value of the Kolmogorov scale η must be replaced by the value $\nu^{3/4}\epsilon_1^{-1/4}$, i.e., $\ell = 2.5\nu^{3/4}\epsilon_1^{-1/4}$.

When $\Delta = 0.2$, we have $h = 0.84$. From the estimate $\langle\epsilon\rangle = [1/3(\langle u_k^2\rangle - \langle u_k\rangle^2)]^{3/2}/L = u^3/L$ and the known equality $\langle\epsilon\rangle = 15\nu\langle(\partial u_1/\partial x_1)^2\rangle$ we obtain $L/\eta = 15^{-3/4}\text{Re}_\lambda^{3/2}$, where $\text{Re}_\lambda = u\lambda/\nu$, λ is the turbulence microscale. It is now easy to find the maximum value of the length of the anemometer wire. We have $\ell/\eta = 2.5(\langle\epsilon\rangle/\epsilon_1)^{1/4}$. Since $n = 2$ when measuring the energy dissipation spectrum, then from (4.26) we have $\ell/\eta = 0.5$ when $\text{Re}_\lambda = 300$ and $\mu = 0.5$,

$\ell/\eta = 0.2$ when $Re_\lambda = 5000$ and $\mu = 0.5$, $\ell/\eta = 1.2$ when $Re_\lambda = 300$ and $\mu = 0.2$, $\ell/\eta = 0.8$ when $Re_\lambda = 5000$ and $\mu = 0.2$.

In the results presented below an attempt was made to satisfy these requirements more fully as far as possible.

Investigated were various sheer flows in large size wind tunnels. The first tunnel had an elliptical nozzle with 24 m and 14 m long axes. The working section was open. The intensity of turbulence of the oncoming flow was equal to 0.6 – 0.8%. The second tunnel had a closed working section of 1× 1 m square cross section and its length was 4 m. Intensity of turbulence of the oncoming flow was equal to 0.02 – 0.04%.

Three flows were investigated in the first tunnel. Flow 1 is an axisymmetric wake behind a cylinder of diameter $d = 0.975$ m and length 10 m. The axes of the cylinder and the tunnel were coincident. The velocity of the oncoming flow $u_0 = 10.3$ m/sec, $Re = u_0 d/\nu = 6.7 \times 10^5$. Measurements were conducted at distance $x_1/d = 16.8$ from the end face of the cylinder. Flow 2 was obtained by the fluid flow around the same cylinder perpendicular to its generatrix. The cylinder was mounted in the plane of nozzle exit section symmetrically along the major axis of the nozzle section. Measurements were carried out transverse to this axis in a section at a distance of $x_1/d = 19.7$ from the axis of the cylinder. The wake was three-dimensional, since there were clearances between the walls of the tube and the cylinder ends. The effects caused by the three-dimensional character of the averaged flow were not controlled. This flow is referred to as a spatial wake. Measurements here were conducted for $u_0 = 9.89$ m/sec, $Re = 6.4 \times 10^5$. Flow 3 is a jet mixing layer flowing out of the first wind tunnel at a velocity of u_0 with stationary air. For this flow $u_0 = 10.9$ m/sec, $Re = u_0 x_1/\nu = 1.5 \times 10^7$, $x_1 = 20$ m is the distance from the nozzle exit section to the section in which measurements were conducted.

Two other flows were investigated in the second wind tunnel. Flow 4 is a plane behind a circular cylinder of diameter $d = 5$ cm in a stream of velocity $u_0 = 8.12$ m/sec. Measurements were conducted in section $x_1/d = 58.6$, $Re = u_0 d/\nu = 2.7 \times 10^4$. Flow 5 is a boundary layer on the walls of the working section of the tunnel. Measurements were conducted at a distance of 2.99 m from the nozzle exit section at an external flow velocity of $u_0 = 8.13$ m/sec, $Re = u_0 \delta/\nu = 2.1 \times 10^4$, where $\delta = 38.2$ mm is the thickness of the boundary layer.

A DISA constant temperature anemometer 55AOI was used. The sensitive element was made from platinum coated tungsten of 2.2 μm diameter welded to a DISA sensor 55A22. The main measurements were conducted with a wire $\ell = 0.33$ mm long at relative heating of 1.8. Wires of length $\ell = 1$ and 3 mm were also used.

The velocity was measured by a single-wire sensor on the assumption that Taylor's hypothesis on "frozen turbulence" is true, i.e., the dependence of the longitudinal component of velocity u_1, recorded in the course of the experiments, on time t and the longitudinal coordinate x_1 has the form $u_1 = u_1(x_1 - \langle u_1 \rangle t)$. The selection of such a method is due to the necessity of determining the small difference of large values when measuring structure functions. Therefore, there are two sources of errors. The first arises upon recording of the signal from two or more wires placed at different points. This is associated with the inaccuracies of measurements of the calibration curves of different wires. Since this source of errors is difficult to eliminate, a single-wire probe was used. Such a method, generally speaking, does not permit elimination of the second source of errors which are caused by the nonlinear character of the calibration curve. Correction of nonlinearity does not lead to an increase in accuracy, since the signal from the anemometer is a function of the values $u_{\|}^2 + u_{\perp}^2$, where $u_{\perp}$ and $u_{\|}$ are the perpendicular and parallel components of the velocity to the wire. Therefore, correction of the nonlinearity does not exclude the quadratic velocity components and under high turbulence intensity the value of $u_{\|}$ must be measured, requiring the use of a multiple-wire sensor and again leading to the need to measure a small difference of large values. Hence, the only way to increase accuracy is to conduct tests in flows with low turbulence intensity. This also leads to the simultaneous elimination of the errors that are caused by the inaccuracies of the Taylor hypothesis since these errors are known (Monin and Yaglom [1967]) to be proportional to the square of turbulence intensity $u'/\langle u_1 \rangle$ where $u' = \sqrt{\langle (u_1 - \langle u_1 \rangle)^2 \rangle}$.

In view of the above, attention was directed during the processing of the results of the measurements toward the analysis of the data which were obtained at the points where the following conditions were fulfilled $|u_0 - \langle u_1 \rangle|/u_0 < 0.13$; $u'/\langle u_1 \rangle < 0.11$. These conditions were fulfilled at all the points of the axisymmetric, spatial, and plane wakes (flows 1, 2, 4). These conditions were violated in the mixing layer (flow 3) of the stationary medium and in the boundary layer (flow 5) at the points located in the region near the wall.

Turbulence intensity, if the indicated points are excluded from consideration, was varied within the range 3.5 – 11.5%, i.e., considerably less than in the majority of tests (on the axis of a plane and axisymmetric jet the intensity reaches 26 – 30%, see, for example, Antonia, Satyaprakash, and Hussain [1982], in the atmosphere up to 18 – 30% — Antonia, Satyaprakash, and Hussain [1982], Champagne [1978]). The results of measurements at the points of high turbulence intensity were used for comparison purposes. At these points the effect of nonlinearity on various statistical characteristics of the fine-scale turbulence structure was analyzed. The maximum turbulence intensity (21%) was observed in the mixing layer. It was established that increasing the intensity of fluctuations does not practically affect the various scale characteristics in the inertial range. In particular, the variation of the exponents in the structure functions does not exceed 5%, although the absolute values of the structure functions are somewhat dependent on turbulence intensity. Therefore, in the following sections, in addition to the results obtained at the points with low intensity (such points are in the majority), some data are also presented relating to the points at which turbulence intensity exceeded 11%.

The upper limit of the frequency range of the anemometer is 10 kHz. In this case the inaccuracy of the measurements was determined by a long wire which is evident from the following estimate. Due to insufficient resolution of the wire, oscillations with a frequency $f_\ell = \langle u_1 \rangle / 2\pi \ell$ are incorrectly recorded. It is evident that $f_\ell < 5$ kHz when $\langle u_1 \rangle < 10$ m/sec and $\ell = 0.33$ mm, i.e., the frequency range of the thermoanemometer has a margin.

The output signal of the anemometer was recorded on a measuring magnetograph by frequency modulation which provided a frequency range of up to 12 kHz. The recorded signals were passed through a filter of low frequency with an upper frequency limit $f_c = 6.4$ kHz. This frequency exceeded the Kolmogorov frequency $f_k = \langle u_1 \rangle / 2\pi\eta$. Taking into account that the maximum of the dissipation spectrum corresponds to frequency $0.1\,f_k$ (Fig. 4.2), it can be assumed that the requirement for frequency range is completely fulfilled.

The filter characteristic discrimination amounted to 48 dB/octave. The recorded signal was transmitted to the computer via an analog-digital converter (12 digits + sign) with a scanning frequency $f_s = 32$ kHz in order to reduce signal noise. The ratio of the mean square level of the signal to the noise level at the input into the analog-digital converter was equal to 80 – 250 corresponding to the ratio of 5 – 15 for the derivatives (these values are at the limit of capabilities of modern measuring devices and are comparable with the best of the published studies). The statistical characteristics were computed with the aid of a computer from a sample of 1,704,000 readings which corresponded to a record of 53.25 sec duration (specially conducted computations, in which the length of record was decreased artificially, showed that the averaging results were statistically stable for the moments and for the derivative up to 8 orders inclusive).

The three-dimensional statistical characteristics in the direction of the mean velocity of flow were reconstructed with respect to time with the aid of the Taylor hypothesis according to which it was assumed that

$$r = x_1^{(2)} - x_1^{(1)} = \langle u_1 \rangle (t^{(2)} - t^{(1)})$$

$$\frac{\partial u_1}{\partial x_1} = -\frac{1}{\langle u_1 \rangle} \frac{\partial u_1}{\partial t}$$

The integral scale L was found from the known formula

$$L = \langle u_1 \rangle (u')^{-2} \int_0^\infty \langle [u_1(t+\tau) - \langle u_1 \rangle][u_1(t) - \langle u_1 \rangle] \rangle d\tau$$

and the Taylor microscale λ from the relation

$$\lambda^2 = (u')^2 / \langle (\partial u_1 / \partial x_1)^2 \rangle$$

The last scale enters into the number $\mathrm{Re}_\lambda = u'\lambda/\nu$ traditionally used in the investigations of the fine-scale structure of turbulence.

The derivative $\partial u_1 / \partial t$ was computed with the aid of cubic splines. For comparison purposes this derivative was also computed at some points by a difference scheme of second order

$$\frac{\partial u_1}{\partial t} = \frac{u_1(t + \Delta t) - u_1(t - \Delta t)}{2\Delta t}$$

where Δt is the analog-to-digital conversion interval which corresponded to a scanning frequency of 32 kHz (this frequency is five times larger than the upper limit of the analyzed frequencies (6.4 kHz)). The values of all the statistical characteristics of the derivative $\partial u_1/\partial t$, determined by two differentiation methods, coincided with an accuracy of up to three significant numbers.

Energy dissipation was calculated on the assumption of local turbulence isotropy, i.e.,

$$\langle \epsilon \rangle = 15\nu \left\langle \left(\frac{\partial u_1}{\partial x_1}\right)^2 \right\rangle$$

This assumption, as noted above, is not generally fulfilled in shear flows (see table 2 and Fig. 4.1). Its use, however, does not change the conclusion drawn below on the nonuniversality of constant C in the "two third" law.

Indeed, let us turn to Figs. 4.1 and 4.13. From the first it is evident that as we move away from the plane of symmetry of the wake, i.e., with decreasing intermittency factor, the use of the isotropic relation lowers the true value of $\langle \epsilon \rangle$ more and more. Since $C \sim \langle [u_1(x_1) - u(x_1 + r)]^2 \rangle / \langle \epsilon \rangle^{2/3}$ we can come to the conclusion that high values of C are obtained in Fig. 4.13 and these values are the higher the less is γ, i.e., the variation of C is in reality even larger than the variation following from Fig. 4.13.

When processing the results of measurements it was assumed in the correlation function of energy dissipation that this same formula is valid for the unaveraged values of ϵ and $(\partial u_1/\partial x_1)^2$. Such an assumption did not alter the character of the dependence on various parameters.

The main experimental conditions for flows 1 – 5 are presented in table 5. It is evident that the ratio ℓ/η varies in flows 1 – 4 within the range 0.8 – 1.6 and in flow 5 within the range 0.9 – 2.5 (the value $\ell/\eta = 2.5$ is reached only at one point, at the other points $\ell/\eta < 1.6$), i.e., ℓ/η is either at a level which is characteristic for the more accurate of the known laboratory tests or somewhat in excess of this level. Since $\ell/\eta \sim 1$, it might appear that the accuracy of the measurements are not sufficiently high. This, however, is not so, since the experimental data generalized by Monin and Yaglom [1967] and the data obtained in the present work show that the maximum value of the dissipation spectrum $k^2 E_1(k)$ is reached when $k\eta \approx 0.1$ (see Fig. 4.2), i.e., the maximum contribution to dissipation is due to the fluctuations whose scale is approximately one order higher than the length of the wire. The effect of the spatial resolution of the measurements on some characteristics of turbulence will be considered below in section 3.

The main information was obtained when investigating flows 1 – 3 for which, as shown in table 5, the Reynolds numbers are high and the integral scales of turbulence are approximately one order higher than in the laboratory tests of other authors. Reynolds numbers Re_λ in flows 4 and 5 are lower. Therefore, comparison of the results obtained in both groups of flows enables its influence to be studied.

162 TURBULENCE IN COMBUSTION

Table 4.5 Experimental conditions

Flow	L, m	$Re_L = \frac{u'L}{\nu}$	λ, mm	Re_λ	η, mm	ℓ/η	γ	$\frac{u'}{\langle u_1 \rangle}$, %	$\frac{\|u_0 - \langle u_1 \rangle\|_{max}}{u_0}$
1 Axisymmetric wake	0.44-0.48	$1.0\text{-}2.5\ 10^4$	16-21	380-1080	0.32-0.42	0.8-1.0	0.21-0.54	3.5-8.8	0.11
2 Spatial wake	0.56-0.68	$1.6\text{-}2.2\ 10^4$	13-15	340-520	0.30-0.40	0.8-1.1	0.21-0.63	3.6-6.9	0.13
3 Mixing layer	1.0-1.2	$0.5\text{-}1.2\ 10^5$	16-21	1040-1660	0.21-0.33	1.0-1.6	0.33-0.89	7.0-21	0.33
4 Plane wake	0.10-0.13	$2.8\text{-}4.5\ 10^3$	6.0-6.1	160-200	0.22-0.24	1.4-1.6	0.7-1.0	5.1-6.9	0.12
5 Boundary layer	0.10-0.28	$3.3\text{-}6.8\ 10^3$	3.1-6.1	70-150	0.13-0.36	0.9-2.5	0.1-1.0	2.2-11.4	0.26

3. Longitudinal structure function of the second order and correlation function of energy dissipation. The initial version of the Kolmogorov-Obukhov theory was based on the hypothesis that the fine-scale structure of turbulence in the inertial range is determined by only one characteristic of the large-scale fluctuations, namely, by the mean energy flow from large-scale to small-scale fluctuations. In equilibrium conditions this flow is equal to the mean energy dissipation $\langle \epsilon \rangle$. Therefore, from considerations of dimensionality in the inertial interval $\eta \ll r \ll L$, the following formulas are obtained

$$\langle v_1^n \rangle = \text{const}(\langle \epsilon \rangle r)^{n/3}, \qquad v_1 = u_1(x_1 + r) - u_1(x_1) \tag{4.27}$$

$$R_{\epsilon\epsilon} = \langle \epsilon(x_1)\epsilon(x_1 + r) \rangle = \text{const}\langle \epsilon \rangle^2 \tag{4.28}$$

A special case of relation (4.27) is the well-known "two thirds" law

$$D_{11} = \langle v_1^2 \rangle = C(\langle \epsilon \rangle r)^{2/3} \tag{4.29}$$

where C is a universal constant (Kolmogorov constant).

As already mentioned, tests have confirmed the validity of the "two thirds" law (although, as evident from table 1, the scatter of the experimental values is very large) but do not confirm formula (4.27) when $n > 3$ and formula (4.28). Therefore, a number of refinements which are discussed by Monin and Yaglom [1967] and in §4.1 – §4.3 were formulated. In the refined Kolmogorov-Obukhov theory the effect of large scale fluctuations on fine-scale fluctuations is described not by one but by two parameters, for example, $\langle \epsilon \rangle$ and L.

The character of these refinements follows from the general relation (4.9). During the derivation of the latter it was assumed that the fine-scale structure of turbulence is also determined by two parameters. Indeed, let L be the order of the integral scale of turbulence. Then V is the characteristic speed of the energy-carrying fluctuations. As is generally known, the statistics of this quantity do not possess anomalous features as is the case with the lognormal law which is characteristic of fine-scale fluctuations. It can hence be assumed that $\langle V^n \rangle \sim \langle V^2 \rangle^{n/2}$. Thus, assuming that $\langle \epsilon \rangle \sim \langle V^2 \rangle^{3/2}/L$ when averaging in (4.9), (4.15) we obtain

$$\langle v_1^n \rangle = \text{const}(\langle \epsilon \rangle L)^{n/3}(r/L)^{q(n)} \tag{4.30}$$

$$R_{\epsilon\epsilon} = C_\epsilon \langle \epsilon^2 \rangle (r/L)^{-\mu} \tag{4.31}$$

where C_ϵ, μ are universal constants and $q(n)$ is a universal function of n. It is significant that the difference $q(2) - 2/3$ is very small; therefore, the corrections to the "two thirds" law can be neglected.

In the spectral formulation the "five thirds" law (Obukhov [1941]) is equivalent to the "two thirds" law (4.29)

$$E_1 = \frac{C}{4.02}\langle \epsilon \rangle^{2/3} k^{-5/3}, \qquad k = 2\pi f/\langle u_1 \rangle \tag{4.32}$$

where f is the frequency and E_1 is the longitudinal energy spectrum.

Constant C was determined from both the results of measurements of the longitudinal structure function of the second order D_{11} and the longitudinal spectra E_1. Both measurement techniques yielded practically identical results. It is established that formulas (4.29) and (4.32) are valid for variations of r and k by two orders in flows 1 – 3 and by one order in flows 4 and 5. As an illustration Fig. 4.3 shows the longitudinal spectrum at one point on the mixing layer (flow 3).

Constants C_ϵ and μ were determined from the results of measurements of the correlation function of dissipation $R_{\epsilon\epsilon}$. It is revealed that formula (4.31) is valid when r varies by two orders in flows 1 – 3 and by one order in flows 4 and 5. As an illustration, Fig. 4.4 shows function $R_{\epsilon\epsilon}$ at one point in the axisymmetric wake (flow 1). It has been shown by measurements that the inertial interval for the quantities D_{11} and $R_{\epsilon\epsilon}$ is worked out from the same values of r.

The results of the measurements were approximated by formulas

$$\ln D_{11} = a + \frac{2}{3}\ln r, \qquad \ln R_{\epsilon\epsilon} = b - \mu \ln r$$

Constants a, b, μ were found by the least squares method which enabled to compute C, C_ϵ, and μ.

Table 5 gives an idea about the resolution of the measurements in various sections. The effect of such a resolution was tested in all flows. In each flow tests were carried out at several points including those with the least and the largest intermittency factor. For this purpose, sensors with different wire length were used $\ell = 0.33$ mm, 1 mm, and 3 mm. The measurement errors in all cases were found to be approximately identical. A typical example is illustrated in Figs. 4.5 and 4.6 in which the results of methodical investigations in the axisymmetric wake are shown (flow 1). Furthermore, in the case under consideration, the input of the signal into the computer was averaged by integration over the variable time period τ with $\ell = \langle u_1 \rangle \tau$.

From Figs. 4.5 and 4.6 it can be concluded that, firstly, the "natural" averaging obtained by increasing the length of the wire is equivalent, as a first approximation, to the "artificial" averaging with the aid of a computer and, secondly, the use of wires of lengths less than 2.5 η yields sufficient accuracy of measurement of the constants C, C_ϵ, and μ. Note also that since $\mu \approx 0.2$, then, as the above conducted estimates show, it suffices for resolution to have $\ell/\eta \approx 1$ for measurements with an accuracy of 20%.

The method of representation of the experimental data is based on the concepts expounded in §4.3 from which it followed that the nonuniversality of the fine-scale structure of turbulence is caused by intermittency. Therefore, the results of the measurements were not associated with the location of the points but with the intermittency factor γ. Let us consider further the technique and results of measurements of this characteristic.

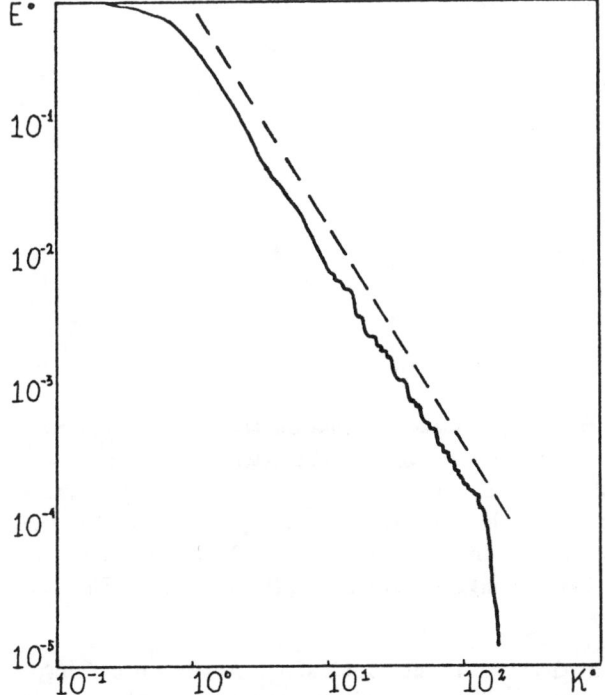

Figure 4.3 Longitudinal energy spectrum in the mixing layer from data by Kuznetsov, Praskovskii, and Sabel'nikov, $y/x = -0.05$, $\text{Re}_\lambda = 1420$, $L = 1.15$ m, $\eta = 0.25$ mm, $\gamma = 0.52$, $\langle(u_1 - \langle u_1 \rangle)^2\rangle^{1/2}/\langle u_1 \rangle = 0.11$. The solid curve is the result of measurements, the dashed straight line corresponds to the dependence $E_1 \sim k^{-5/3}$, $E^0 = E_1(k)/E_1(0)$, $k^0 = kL$.

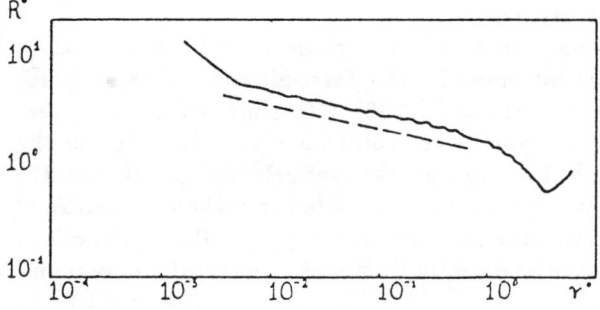

Figure 4.4 Correlation function of energy dissipation in axisymmetric wake from data by Kuznetsov, Praskovskii, and Sabel'nikov $y/d = 1.03$, $\text{Re}_\lambda = 376$, $L = 0.45$ m, $\eta = 0.41$ mm, $\gamma = 0.21$, $\langle(u_1 - \langle u_1 \rangle)^2\rangle^{1/2}/\langle u_1 \rangle = 0.035$. The solid curve is the measurement results, the dashed straight line corresponds to dependence $R_{\epsilon\epsilon} \sim r^{-0.22}$, $R^0 = R_{\epsilon\epsilon}(r)/\langle\epsilon^2\rangle$, $r^0 = r/L$.

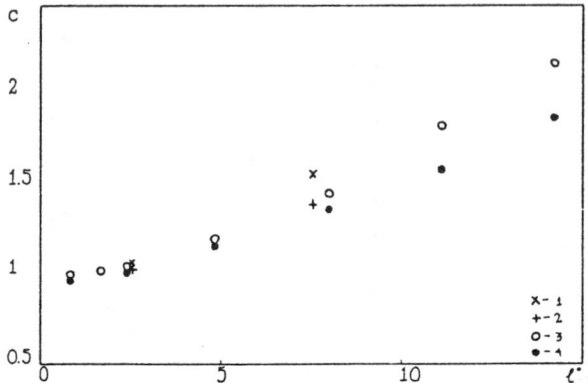

Figure 4.5 The effect of resolution of the measurements on the value of the Kolmogorov constant C in the "two thirds" law from data by Kuznetsov, Praskovskii, and Sabel'nikov. Points 1 and 2 are obtained from spectra measurement and structure functions, respectively. Points 3 and 4 are obtained following averaging of the signal in the computer from the spectra and structure functions, respectively. Measurements were conducted in axisymmetric wake. Experimental conditions are the same as in Fig. 4.4. $\ell^0 = \ell/\eta$.

4. Intermittency function and its characteristics. As shown in Chapter 1, the measurement of intermittency characteristics is associated with a whole series of fundamental difficulties. These difficulties turned out to be far greater than was noted in Chapter 1. These have not been totally overcome at the present time and, therefore, the intermittency factor measured with the aid of the method described at the end of the present section must be understood as just a parameter ordering the results of measurements of various fine-scale characteristics of turbulence. Prior to discussing this technique, it is useful to describe three difficulties that are encountered in the measurements.

We note first that when the turbulent and nonturbulent fluids are separated, at least three quantities must be introduced. The determination of these quantities *a priori* is quite arbitrary (Praskovskii [1982]). To start with, there must be introduced a detector function whose values permit one to judge whether the fluid is turbulent. It was established in the tests that the selection of the detector function is not fundamental. Investigation was conducted in a three-dimensional wake (flow 2, $y/d = 1.23$) and in the mixing layer (flow 3, $y/x = -0.05$). Functions $|\partial u_1/\partial t|$ and $(\partial u_1/\partial t)^2$ were chosen as detector functions. The results were found to be identical.

Then, as indicated in Chapter 1, intermittency can be determined only by excluding from consideration the viscous effects for which the detector function must be averaged with respect to the varied time interval. From the practical standpoint it was found convenient to replace averaging by high-frequency filtering. For this purpose, a digital filter was used (eighth order Butterworth sinusoidal filter) cutting off all frequencies above a certain limit frequency f_F

whose value can be varied. The results obtained upon changing f_F are presented in Fig. 4.7, where the measured spectra of the intermittency function $E_{\gamma\gamma}(k)$ are shown. Here, as everywhere further on, the quantity $|\partial u_1/\partial t|$ has been used as a detector function (see Fig. 4.9). It is convenient to dwell later on the method of selection of the threshold value j_{cr}, which when exceeded, the fluid is considered to be turbulent. We shall only note that this value was selected to be suitably low. It has been established that when the limit frequency f_F is changed by two orders, the intermittency factor γ varies by just 10%, and in order to maintain a constant intermittency factor with the variation of f_F it is sufficient to slightly change the threshold value j_{cr} (also within 10%). The data presented in Fig. 4.7 were obtained by just such a small variation of j_{cr}.

Figure 4.7 shows that there are two exponential sections $E_{\gamma\gamma} \sim k^{-\varsigma}$ ($\varsigma = -0.25$ in the present case) and $E_{\gamma\gamma} \sim k^{-2}$ in the spectrum of the intermittency function in the region of large wave numbers with practically any smoothing. We note that in the tests by La Rue and Libby [1976] conducted at significantly lower Reynolds numbers only the last section is recorded (Fig. 1.14).

The measured values of various characteristics of intermittency are most strongly affected by the selection of the third quantity, namely, the threshold value j_{cr} (remember that the fluid is considered turbulent if $|\partial u_1/\partial t| > j_{\mathrm{cr}}$). This conclusion is evident from Fig. 4.8, from which it follows that although the change of j_{cr} does not change the results qualitatively, the quantitative value of γ can vary noticeably.

The reason for the indicated difficulty is clear from Fig. 4.9 which shows the oscillograms of fluctuations of the velocity and the modules of its derivatives in

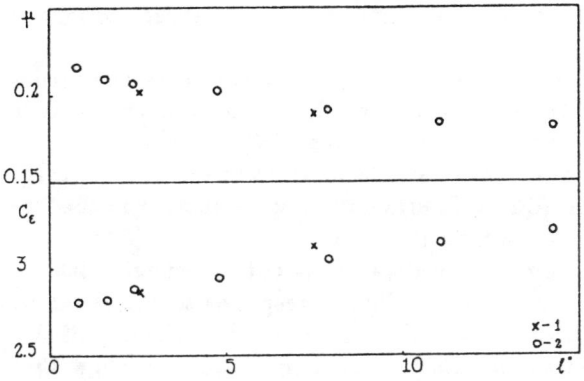

Figure 4.6 The effect of resolution of measurements on the values of constants μ and C_ϵ in the correlation function of energy dissipation from data by Kuznetsov, Praskovskii, and Sabel'nikov. Measurements were conducted in axisymmetric wake. Experimental conditions are the same as in Fig. 4.4. Points 1 are obtained from the measurement of the correlation function of energy dissipation. Points 2 are obtained following the averaging of the signal in the computer from the correlation function of energy dissipation. $\ell^0 = \ell/\eta$.

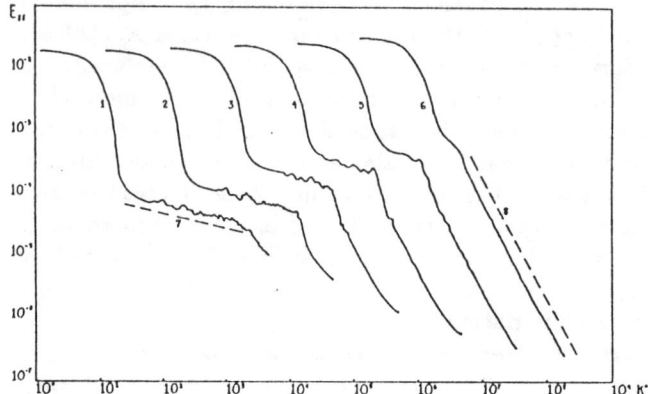

Figure 4.7 Spectra of the intermittency function for various values of the upper frequency limit of the smoothing filter from data by Kuznetsov, Praskovskii, and Sabel'nikov. Measurements were conducted in the mixing layer. Experimental conditions are the same as in Fig. 4.3. *1)* $f_F = 10/8f_k$, *2)* $f_F = 3/8f_k$, *3)* $f_F = 1/8f_k$, *4)* $f_F = 1/16f_k$, *5)* $f_F = 1/32f_k$, *6)* $f_F = 1/80f_k$. The dashed straight lines 7 and 8 correspond to the dependences $K^{-0.25}$ and K^{-2}, respectively. Curves 2, ... 6 are shifted along the abscissa relative to curve 1 by 10, ... 10^5, respectively. $k^0 = kL$.

a three-dimensional wake. It is evident from the figure that there are indeed extended regions in which the fluctuations of the velocity derivative have different character. However, an exact determination is difficult due to the fact that significant fluctuations of the velocity derivative are observed in some sections that are naturally considered as nonturbulent (in particular, one of these sections is located in $1 \leq t^0 \leq 1.7$ in Fig. 4.9).

The fluctuations in the indicated sections are not associated with the turbulence of the external flow, since the fluctuation velocity far outside the wake is one order less than inside the wake. Since the scale of turbulence in the free flow is not less than inside the wake, the dissipation inside the wake is at least three orders less than in the free stream (direct determination of dissipation in the free stream was not possible because of instrument noise).

Apparently, the phenomenon under consideration is not associated with instrument noise. Indeed, in Fig. 4.9 the dotted line corresponds to threshold j_{cr} which is selected in accordance with the technique expounded at the end of this section. Evidently, this threshold has been selected wisely. Estimates show that in the present case j_{cr} exceeds by roughly 4 times the noise level in the signal which is proportional to $|\partial u_1/\partial t|$.

It is not excluded that the indicated difficulty is a fundamental one, i.e., associated with some specific features of turbulence. For example, it might turn out that a small vortex in the external flow increases significantly as fluid particles move toward the turbulent region.

Such a viewpoint is confirmed quite indirectly by an experiment in which all the front and rear boundaries of the "turbulent" intervals were fixed, i.e., those

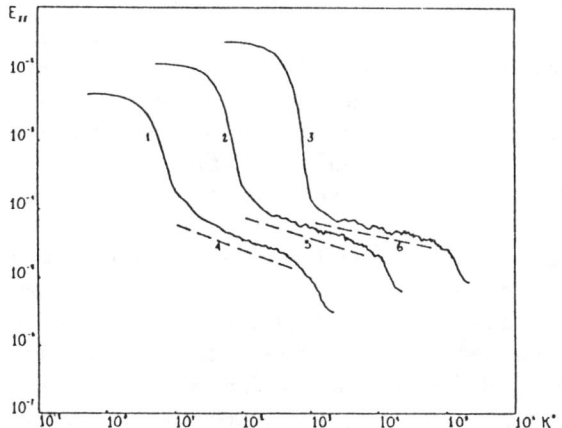

Figure 4.8 The effect of threshold level on the spectrum of the intermittency function from data by Kuznetsov, Praskovskii, and Sabel'nikov. Measurements were conducted in axisymmetric wake, the detector function is $|\partial u_1/\partial t|$, weak smoothing $f_F = \frac{10}{8}f_k$, $y/d = 0.615$, $\langle u_1 - \langle u_1 \rangle)^2\rangle^{1/2}/\langle u_1 \rangle = 0.059$, $Re_\lambda = 745$, $L = 0.48$ m, $\eta = 0.366$ mm, $\gamma = 0.372$. *1)* $j_{cr}^0 = 0.920$ — elevated threshold level, the corresponding value of $\gamma = 0.178$; *2)* $j_{cr}^0 = 0.473$ — threshold calculated by the iteration method described in the text (see the end of the section), the corresponding value of $\gamma = 0.372$; *3)* $j_{cr}^0 = 0.247$ low threshold level and the corresponding value of $\gamma = 0.619$ $j_{cr}^0 = j_{cr}/\langle(\partial u_1/\partial t)^2\rangle^{1/2}$. The dashed straight lines 4, 5, and 6 correspond, respectively, to dependences $k^{-0.37}$, $k^{-0.30}$, $k^{-0.21}$, $k^0 = kL$. Curves 2 and 3 are shifted along the abscissa relative to curve 1 by 10 and 10^2 times, respectively.

values of t at which discontinuity of the intermittency function $\Gamma(t)$ takes place from zero to unity and from unity to zero, respectively (see Fig. 4.9). The location of the boundaries (separately for the front and for the rear boundaries) were assumed as the origin of the coordinate relative to which averaging of energy dissipation was carried out over all the boundaries. The results of the described procedure are depicted in Fig. 4.10. It is evident that the boundary between turbulent and nonturbulent fluid is not sharp. The thickness of the transition region for standard smoothing ($f_F = 1/8 f_k$) is of the order $0.1\,L$. It is also evident that the character of change of energy dissipation, averaged under such condition, was found to be independent of the filtering level if one selects the mean extent of the "turbulent" region $L_\gamma = \langle u_1 \rangle \langle \tau \rangle$ ($\langle \tau \rangle$ is the mean time duration of this region) as the normalizing scale. It is important that depending on the filter level f_F the dimension L_γ strongly varies (in the given case by approximately 7 times).

Thus, the dissipation field in the vicinity of the boundary of the turbulent fluid possesses scale similarity, i.e., the spatial distribution of dissipation exhibits a specific behavior. This scale similarity does not allow reproduction of a unique boundary for the turbulent fluid.

170 TURBULENCE IN COMBUSTION

We finally note that the reason behind these difficulties in measuring the characteristics of intermittency can be due to the use of the derivative $\partial u_1/\partial t = 1/\langle u_1 \rangle \partial u_1/\partial x_1$ as a detector function instead of dissipation of turbulence energy ϵ (when $\text{Re} \gg 1\epsilon$ is proportional to the square of the vorticity see the beginning of §1.1). The fact is that the damping rate ϵ and $\partial u_1/\partial x_1$ become very different with increasing distance from the boundary of the turbulent fluid.

The above can be illustrated by the following simple flow model for the outside turbulent fluid. Let us consider the inviscid (i.e., $\text{Re} = \infty$) vortex-free flow in a half space behind the plane boundary at which a random field of velocity potential is specified (Phillips [1955]). It is assumed that this field is statistically stationary and homogeneous in the plane boundary. The mean square of the derivative $\langle (\partial u_1/\partial x_1)^2 \rangle$ in the region under consideration is not, unlike the vorticity equal to zero and is not expressed in terms of the spectral function at the boundary. The main contribution to $\langle (\partial u_1/\partial x_1)^2 \rangle$ is made by the high-frequency part of the spectrum which is described by the "five thirds" law (4.32). Computations show that

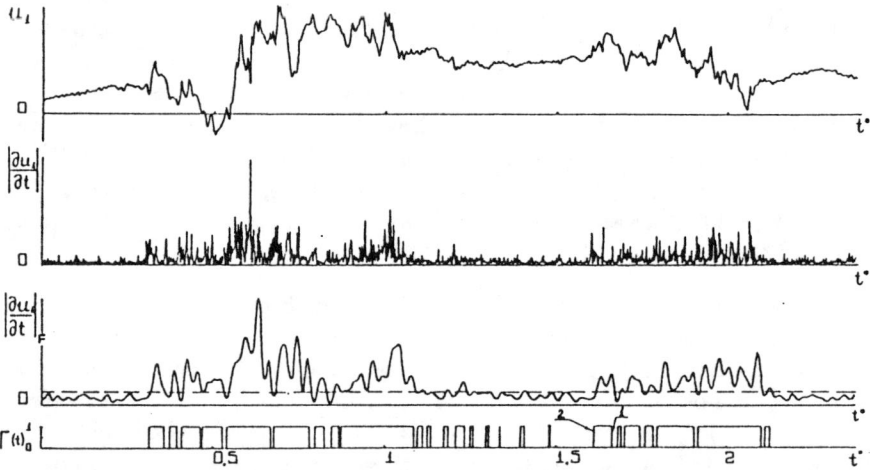

Figure 4.9 On the determination of the intermittency function. Presented in the figure consecutively from top to bottom are the oscillograms of fluctuations of the components of the velocity, the module of its derivative with respect to time, the module of the derivative with respect to time for standard smoothing $f_F = 1/8 f_k$ and the functions of intermittency from data by Kuznetsov, Praskovskii, and Sabel'nikov. Measurements were conducted in a three-dimensional wake. $y/d = 0$, $\text{Re}_\lambda = 524$, $\langle u_1 - \langle u_1 \rangle)^2 \rangle^{1/2}/\langle u_1 \rangle = 0.068$, $L = 0.563$ m, $\eta = 0.30$ mm, $\gamma = 0.63$, $t^0 = t\langle u_1 \rangle/L$. The dotted line, shown on the oscillogram of the module of the velocity derivative with respect to time for standard smoothing $|\partial u_1/\partial t|_F$, corresponds to the threshold calculated by the iteration method described in the text (see the end of the section); 1 and 2 are the front and rear boundaries of the "turbulent" interval.

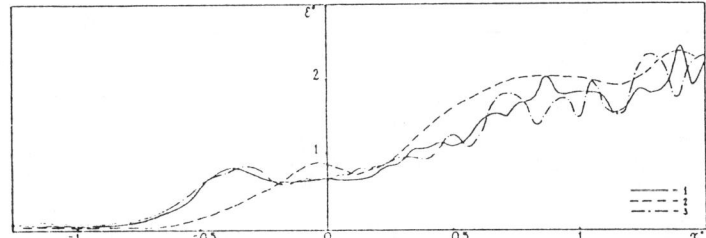

Figure 4.10 Profiles of conditionally averaged energy dissipation in the vicinity of the front boundary of the "turbulent" region from data by Kuznetsov, Praskovskii, and Sabel'nikov. Measurements conducted in a three-dimensional wake, $y/d = 1.23$, $\langle(u_1 - \langle u_1 \rangle)^2\rangle^{1/2}/\langle u_1 \rangle = 0.042$, $Re_\lambda = 403$, $L = 0.66$ m, $\eta = 0.38$ mm, $\gamma = 0.24$, *1)* $f_F = 1/8 f_k$, $L_\gamma = 0.058 L$, *2)* $f_F = 1/2 f_k$, $L_\gamma = 0.016 L$, *3)* $f_F = 1/16 f_k$, $L_\gamma = 0.12 L$, $x^0 = x/L_\gamma$, $\epsilon^0 = \epsilon/\langle \epsilon \rangle$.

$$\left\langle \left(\frac{\partial u_1}{\partial x_1}\right)^2 \right\rangle \sim \left(\frac{x_2}{L}\right)^{-4/3}, \quad x_2 \ll L$$

where x_2 is the distance from the boundary, L is the turbulence scale.

Let us make the rough assumption that this formula remains valid even for finite Reynolds numbers up to a distance $x_2 \approx 10\eta$.

From this formula two conclusions can be made. Firstly, additional confirmation of the validity of threshold selection follows. Indeed, when x_2 is varied within the range $10\eta \leq x_2 \leq 100\eta$, which is equal to the width of the "transition" zone under standard signal smoothing, the quantity $\langle(\partial u_1/\partial x_1)^2\rangle$ according to the formula must vary by 20 fold which is in agreement with the data presented in Fig. 4.10. Secondly, when the derivative $\partial u_1/\partial x_1$ is used as an indicator of turbulence, the reliable measurement of the characteristics of intermittency requires Reynolds numbers significantly larger than those in the analyzed tests.

In view of the above, the method described below yields only a very imprecise and incorrect determination of the intermittency factor. Let us describe this method. The quantity $|\partial u_1/\partial t|$ was used as a detector function (see Fig. 4.9). A signal proportional to $|\partial u_1/\partial t|$ is smoothed which, as noted in Chapter 1, is necessary for the correct identification of intermittency. During smoothing frequencies above $f_F = 1/8 f_k$ were cut off in the signal $|\partial u_1/\partial t|$, where f_k is the Kolmogorov frequency. It was assumed that the intermittency function has the form $\Gamma = \theta(j - j_{cr})$, where $\theta = 1$ for $j > j_{cr}$, $\theta = 0$ for $j < j_{cr}$, and j is a function which is obtained upon smoothing of $|\partial u_1/\partial t|$.

It was assumed that $j_{cr} = \varphi\sqrt{\langle j^2 \rangle_n}$, where subscript n refers to averaging in the nonturbulent fluid and φ is a coefficient of proportionality. Since $\langle j^2 \rangle_n$ is known beforehand, this method involves iteration. It was assumed that the coefficient φ is a function of γ selected by comparing the results of measurements by the given method and Townsend's method [1956]. This coefficient had the form $\varphi = 1.2\sqrt{1 - \gamma^4}$. The method under consideration is described in the works of Kuznetsov, Praskovskii, and Sabel'nikov [1984a, 1988].

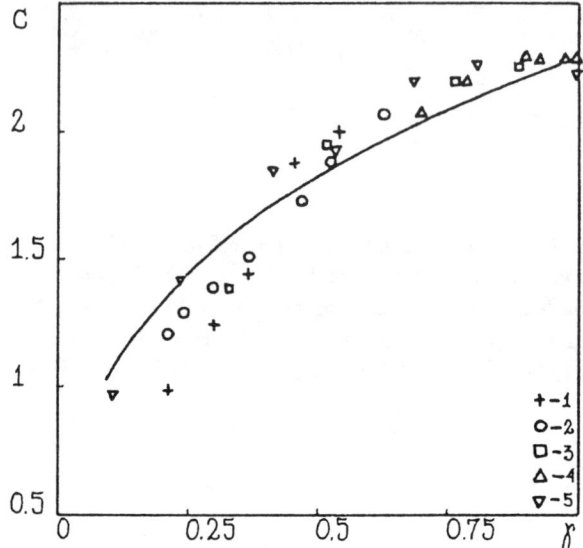

Figure 4.11 The dependence of the Kolmogorov constant C on the intermittency factor from data by Kuznetsov, Praskovskii, and Sabel'nikov. Points 1 – 5 correspond to the flow numbers in table 5. The curve corresponds to the dependence $C = \text{const } \gamma^{1/3}$.

5. Nonuniversality of the constants characterizing the inertial range of the turbulence spectrum. The results of measurements of the constants C, C_ϵ, and μ in flows 1 – 5 are presented in Figs. 4.11 – 4.13. It is evident that the values of C, C_ϵ, μ vary approximately by twofold, i.e., are not universal constants. The effect of the Reynolds number Re_λ which varies in the experiments by more than one order is not detected since all the data are generalized by introducing only one parameter γ. The results of measurements of C and μ are in the same range of values as presently available data. The results of measurements of C_ϵ are not known. These results lead to the conclusion that the scatter of the data obtained earlier is not associated with the experimental errors but is rather of objective character. Thereby, one of the conclusions made in §4.3 is confirmed.

We note that the nonuniversality of constants C, C_ϵ, μ can be linked with the inapplicability of the Kolmogorov-Obukhov theory to nonturbulent fluids. Therefore, it is useful to quantitatively analyze the results of measurements of the unconditionally averaged characteristics assuming that the fine-scale structure of the fluctuations inside the turbulent fluid is universal. Then the symbol of unconditional averaging $\langle \ \rangle$ in (4.29) and (4.31) ought to be replaced by a symbol of conditional averaging in the turbulent fluid, i.e.,

$$\langle v_1^2 \rangle_t = C^0 (\langle \epsilon \rangle_t r)^{2/3} \tag{4.33}$$

$$\langle \epsilon(x_1 + r)\epsilon(x_1) \rangle_t = C_\epsilon^0 \langle \epsilon \rangle_t^2 (r/L)^{-\mu_0} \tag{4.34}$$

where C^0, C_ϵ^0, μ_0 are universal constants. Taking into account that $\epsilon = 0$ in the nonturbulent fluid and $\langle \epsilon \rangle_t = \langle \epsilon \rangle / \gamma$ and also the relation which is a consequence of formula (1.8)

$$\langle v_1^2 \rangle = \gamma \langle v_1^2 \rangle_t, \qquad r/L \to 0$$

we obtain

$$C = C^0 \gamma^{1/3}, \qquad C_\epsilon = C_\epsilon^0 / \gamma, \qquad \mu_0 = \mu$$

The dependences $C \sim \gamma^{1/3}$ (see also (4.14)), $C_\epsilon \sim 1/\gamma$ are depicted in Figs. 4.11 and 4.13 by the solid curves. Figure 4.11 shows that constant C^0, as a first approximation, is universal since the deviations of the experimental points from the curve obtained on the assumption that $C^0 = \text{const}$ are small (recall that this result is obtained on the assumption of local isotropy; taking into account the deviations from local isotropy which were mentioned at the beginning of the section — see Fig. 4.1 — can change this). The structure of the dissipation field in the turbulent fluid is not universal, since μ is dependent on γ and noticeable deviation of the points in Fig. 4.13 from the curve is observed, i.e., $C_\epsilon^0 \neq \text{const}$.

Since C, C_ϵ, μ are similarly dependent on the intermittency factor in different flows, it might appear that a definitive verification of the Kolmogorov-Obukhov theory is obtained. Such a conclusion can be premature for two reasons. Firstly, the results of measurements may be affected by flaws in the determination of the intermittency factor mentioned in the previous section. The difficulties of measuring this quantity are of fundamental character. These difficulties, as shown in

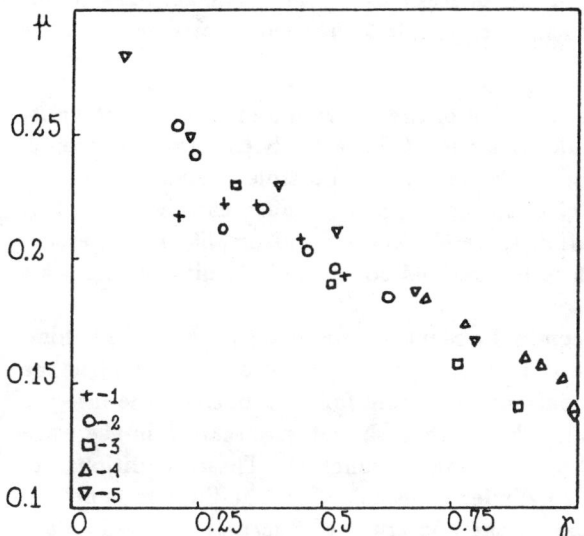

Figure 4.12 The dependence of constant μ in the correlation function of energy dissipation on the intermittency factor from data by Kuznetsov, Praskovskii, and Sabel'nikov. Points 1 – 5 correspond to the flow numbers in table 5.

174 TURBULENCE IN COMBUSTION

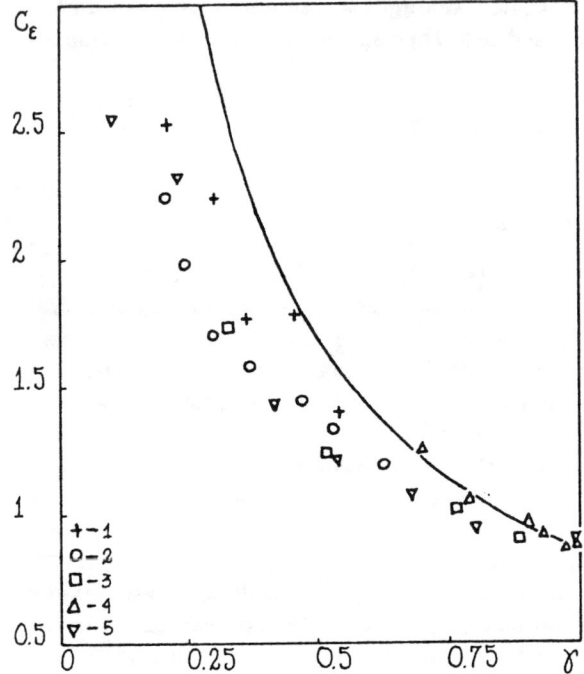

Figure 4.13 The dependence of constant C_ϵ in the correlation function of energy dissipation on the intermittency factor from data by Kuznetsov, Praskovskii, and Sabel'nikov. Points 1 – 5 correspond to the flow numbers in table 5. The curve corresponds to the dependence $C_\epsilon = \text{const}/\gamma$.

§1.1, are caused by the need for separation of the internal and external intermittences and, as shown from the results of section 4, have not been totally overcome. Therefore, these results are tentative. Secondly, the possible explanation of the correlation obtained in the tests of constants C, μ, C_ϵ, and γ can be in the fact that in the flows under consideration there is a certain universality in the structure of large-scale motions which is not related to the universality of fine-scale motions.

To elucidate this point let us conduct a mental experiment in the wake behind a plane cylinder which is displaced normal to its plane of symmetry. Let the motion of the cylinder $y_0(t)$ be a stationary random function of time; the integral time scale T_0 of this function is much larger than the integral scale T in the wake behind a stationary cylinder and y_0 is a limiting function. These conditions are selected so that the motion of the cylinder does not affect the flow dynamics in the wake, and the large-scale part of the spectrum of fluctuations would have substantially nonequilibrium character.

Averaging is carried out in two stages. In the first stage all the results are averaged with respect to time t so that $T_0 \gg t \gg T$. Then, we obtain the same dependences as in a wake behind a stationary cylinder. For example, dependence $\mu(\gamma)$ will be given by the graph depicted in Fig. 4.12, where γ will have the form $\gamma = \gamma_0(y + y_0)$, and γ_0 is the intermittency factor distribution in the wake behind a stationary cylinder. In the second stage, averaging is with respect to time far in excess of T_0, which is equivalent to averaging with respect to the probability density of cylinder displacement $P(y_0)$. For example

$$R_{\epsilon\epsilon} \sim \int r^{-\mu} P(y_0) dy_0, \qquad \mu = \mu[\gamma_0(y + y_0)]$$

$$\gamma = \int \gamma_0(y + y_0) P(y_0) dy_0 \qquad (4.35)$$

Since y_0 is limited, then $\gamma_m > 0$, where γ_m is the minimum possible value of function $\gamma_0(y + y_0)$ for $y = $ const. Since μ is a diminishing function of γ, then when $r \to 0$, the main contribution to the first integral in (4.35) is made by such displacements of the cylinder which correspond to the minimally possible γ_0. Therefore, when $r \to 0$, from (4.35) we obtain $R_{\epsilon\epsilon} \sim r^{-\mu_m}$, where $\mu_m = \mu(\gamma_m)$. Furthermore, the link between γ_m and the mean value of γ, determined by the second formula in (4.35), can be arbitrary. Consequently, the data presented in Figs. 4.11 - 4.13 can be an indication of some universality of the large-scale structure of turbulence in these flows. Such universality might be absent in other flows and under different conditions.

6. Structure functions of velocity and correlation functions of energy dissipation levels. The n-th order longitudinal structure functions

$$D_n(r) = \langle |u_1(x_1 + r) - u_1 x_1|^n \rangle = \langle |v_1|^n \rangle$$

were measured at two-three points of each of the five flows. The degree n changed in the range $0.125 \leq n \leq 8$. In contrast to the generally accepted definition of the structure function (see, for example, Monin and Yaglom [1967]), the written expression includes the absolute value of the velocity difference at two points. Such a change enabled statistically stable moments of odd order to be obtained and, thereby, the function $q(n)$ which defines the dependence of the structural functions on the distance between the points (see formula (4.30)) to be determined for all values of n in the indicated range (in the inertial range the exponents n_1 and n_2 in the expressions $\langle v_1^{n_1} \rangle \sim r^{n_1}$ and $\langle |v_1|^{n_2} \rangle \sim r^{n_2}$ are obviously equal).

The experimental values of function $q(n)$ are shown in Fig. 4.14 in the form of the dependence $q^0 = (q - n/3)/n$ on n. Let us compare these data with known theories. For the initial version of the Kolmogorov-Obukhov theory $q = n/3$ and, hence, $q^0 = 0$. If the lognormal law (4.12) is true, we have the linear dependence $q^0 = q_2(3 - n)$, $q_2 = \mu/18$. For the model by Novikov and Stewart [1964], which later became known as the β-model in the work by Frisch, Sulem, and Nelkin [1978], we have $q = n/3 - \mu(n - 3)/3$, $q^0 = -\mu(n - 3)/3n$. The theoretical dependences corresponding to the lognormal law and β-model are also depicted

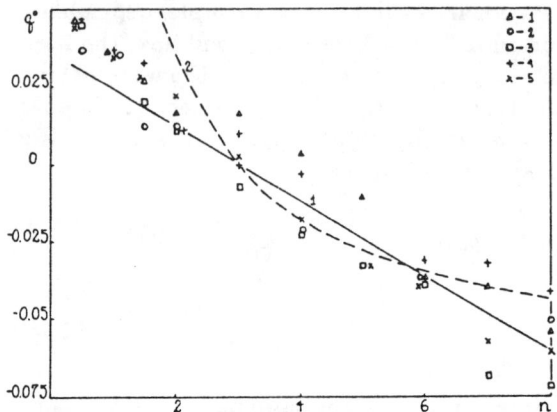

Figure 4.14 Function $q(n)$ characterizing the dependence of the n-th-order structure function of the velocity on the distance between the points from data by Kuznetsov, Praskovskii, and Sabel'nikov. Points *1, 2*) axisymmetric wake, experimental conditions the same as in Figs. 4.2 and 4.4, respectively; *3*) three-dimensional wake, experimental conditions the same as in Fig. 4.2; *4, 5*) mixing layer, experimental conditions the same as in Figs. 4.3 and 4.2, respectively. The straight line corresponds to the log-normal law with $\mu = 0.2$, the curve corresponds to β-models with the same value of μ, $q^0 = [q(n) - \frac{1}{3}n]/n$.

in Fig. 4.14 for the "mean" value of $\mu = 0.2$. The latter dependence is shown only in the region $n \geq 2$, since it does not satisfy the condition $q(0) = 0$ (see the first equality in (4.10)) and, therefore, is inapplicable in the vicinity of point $n = 0$.

The experimental data presented in Fig. 4.14 exhibit considerable scatter which, as established in section 5, is not caused by measurement errors but by the nonuniversality of the fine-scale structure of turbulence. Despite this scatter of the data, it is evident that within the range of values $0.125 \leq n \leq 8$ for the function $q(n)$, the quadratic dependence (4.12) can be used as a first approximation with some mean value of the constant μ or, which is the same, assuming q_2 to be a constant. It is apparent that the β-model correlates somewhat poorer with the experiment.

These conclusions are in agreement with the results of other authors, for example, Anselmet, Gagne, Hopfinger, and Antonia [1984].

Let us now consider the results of experimental verification of formula (4.15). Figure 4.15 shows the results of measurements of the ratio of constant μ to the difference $2 - q(6)$ in flows $1 - 5$. It is seen that the difference $2 - q(6)$ in all cases exceeds the constant μ and thus that these experiments do not confirm formula (4.15).

It should be noted that this formula was experimentally checked in the works by Antonia, Satyaprakash, and Chambers [1982], Anselmet, Gagne, Hopfinger, and Antonia [1984] who established that formula (4.15) is fulfilled with an accuracy of up to the error of measurement.

The reason for such contradictory conclusions is not yet clear and for the solution of this vital problem additional experiments are required. The need for additional verification is apparent from the experiments in which the following quantities were measured

$$R^{(n)}_{\epsilon\epsilon} = \langle \epsilon^n(x_1)\epsilon^n(x_1+r)\rangle$$

for two values of the exponent $n = 1.5$ and $n = 2$. In the inertial range we have

$$R^{(n)}_{\epsilon\epsilon} = \langle \epsilon^n(x_1)\epsilon^n(x_1+r)\rangle \sim \langle \epsilon \rangle^{2n}(r/L)^{-\mu n}$$

It is well known that for the lognormal law $\mu_n = n(2n-1)\mu$, and for the β-model $\mu_n = \mu(2n-1)$ (see, for example, Monin and Yaglom [1967], Antonia, Satyaprakash, and Hussain [1982]). Thus, we obtain $\mu_{1.5} = 3\mu$, $\mu_2 = 6\mu$ for the lognormal law and $\mu_{1.5} = 2\mu$, $\mu_2 = 3\mu$ for the β-model.

Presented in Figs. 4.16 and 4.17 is a comparison of the measured ratios $\mu_{1.5}/\mu$ and μ_2/μ in different flows with the two aforementioned theories. It is clear that the β-model agrees with the experimental data better than the lognormal law which, as shown above (Fig. 4.14), describes the moments $\langle v_1^n \rangle$ well.

7. Verification of the hypothesis on the statistical independence of fine- and large-scale motions and formula (4.13). The hypothesis referred to in the heading of this section occupies a central place in the theory for the concentration PDF developed in Chapter 3. For the dynamic problem, when the large-scale motion is characterized by the velocity field and the fine-scale

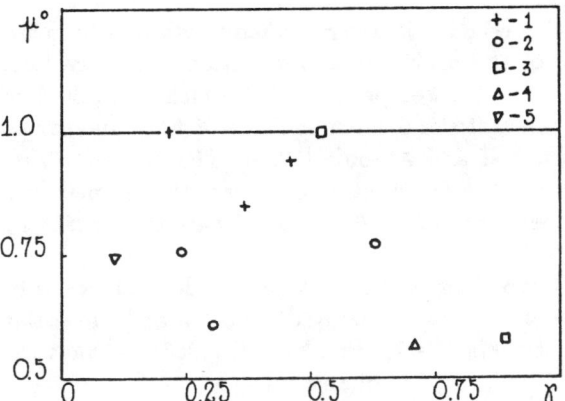

Figure 4.15 Comparison of constant μ in the correlation function of energy dissipation and the exponent of $q(6)$ of the sixth-order structure function from data by Kuznetsov, Praskovskii, and Sabel'nikov. Points 1 – 5 correspond to the flows in table 5. The horizontal straight line corresponds to formula (4.15), $\mu = 2 - q(6)$, $\mu^0 = \mu[2-q(6)]$.

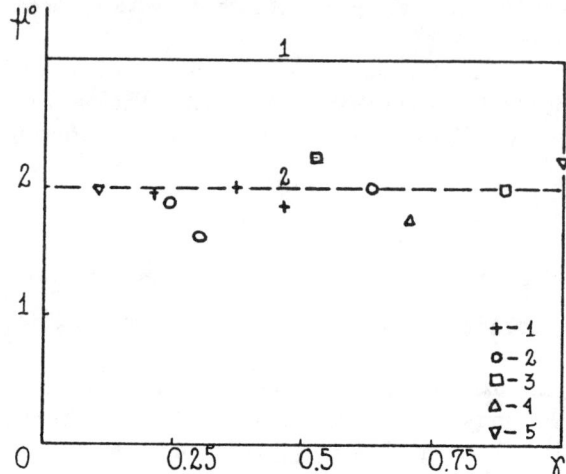

Figure 4.16 The exponent in the correlation function of energy dissipation $R_{\epsilon\epsilon}^{(1,5)}$ from data by Kuznetsov, Praskovskii, and Sabel'nikov. Points 1 – 5 correspond to the flow numbers in table 5. The solid horizontal straight line corresponds to the lognormal law $\mu_{1,5} = 3\mu$, the dashed straight line represents the β-models $\mu_{1,5} = 2\mu$, $\mu^0 = \mu_{1,5}/\mu$.

motion by the energy dissipation field, this hypothesis enables the conditional density of the joint PDF of the velocity and energy dissipation in the turbulent fluid $P_t(u_1, \epsilon)$ to be represented by the product of the conditional probability distribution density of velocity $P_t(u_1)$ and energy dissipation $P_t(\epsilon)$, i.e.,

$$P_t(u_1, \epsilon) = P_t(u_1) P_t(\epsilon) \tag{4.36}$$

The experimental verification of this relation is a very difficult task due to insufficient accuracy of measurements of the probability distributions for large fluctuation amplitudes of ϵ and u_1. For the temperature field such a method of verification of the hypothesis of the statistical independence of fine- and large-scale motions was adopted by Anselmet and Antonia [1985]. The measurements conducted by them of the joint probability distribution density of the temperature and scalar dissipation in the plane jet for $x_1/d = 40$, $\text{Re}_\lambda \approx 160$ are in satisfactory agreement with this hypothesis.

A simpler way of testing the hypothesis on the statistical independence is to measure the conditionally averaged moments of energy dissipation and the scalar dissipation (Kuznetsov and Rashchupkin [1977], Praskovskii [1983], Kuznetsov, Praskovskii, and Sabel'nikov [1984a, b], i.e., the quantities

$$\langle \epsilon^n \rangle_{t, u_1} = \int_0^\infty \epsilon^n P_t(\epsilon | u_1) d\epsilon$$

where $P_t(\epsilon | u_1)$ is the conditional probability distribution density in the turbulent fluid for a given value of the velocity u_1. These moments can be measured with far smaller errors than the distribution densities themselves. If relation (4.36) applies and thereby, $P_t(\epsilon|u_1) = P_t(\epsilon)$, then the conditionally averaged moments $\langle \epsilon^n \rangle_{t,u_1}$ are independent of u_1, i.e.,

$$\langle \epsilon^n \rangle_{t,u_1} = \langle \epsilon^n \rangle_t \tag{4.37}$$

The moments $\langle \epsilon^n \rangle_{t,u_1}$ for $n = 1, 2, 3$ were measured at various points of the five flows indicated in table 5. The intermittency factor was varied in the range $0.1 < \gamma < 1.0$. A notable systematic variation of the quantity $\langle \epsilon^n \rangle_{t,u_1}$ as a function of velocity u_1 was not detected within a wide range of fluctuation amplitudes. With increasing order of the moment n the random spread of the results of measurements of the ratio $\langle \epsilon^n \rangle_{t,u_1}/\langle \epsilon^n \rangle_t$ relative to unity rises but the experimental data generally agree satisfactorily with formula (4.37). It is significant that this result is true even at very moderate values of Re_λ.

The importance of identification of the turbulent fluid is evident from Figs. 4.18 and 4.19, which show the results of measurements of the moments $\langle \epsilon^n \rangle_{t,u_1}$ in the mixing layer and in a plane wake behind a cylinder at the points where

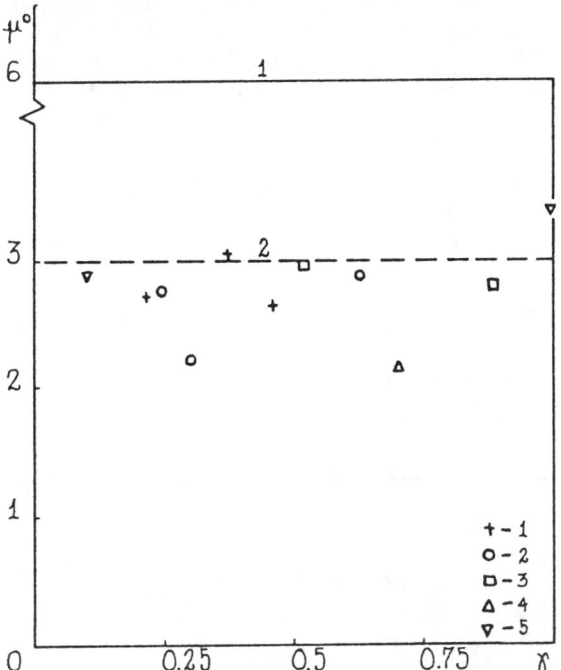

Figure 4.17 The exponent in the correlation function of energy dissipation $R_{\epsilon\epsilon}^{(2)}$ from data by Kuznetsov, Praskovskii, and Sabel'nikov. Points 1 – 5 correspond to the flow numbers in table 5. The solid horizontal straight line corresponds to the lognormal law $\mu_2 = 6\mu$, the dashed straight line represents the β-models $\mu_2 = 3\mu$, $\mu_0 = \mu_2/\mu$.

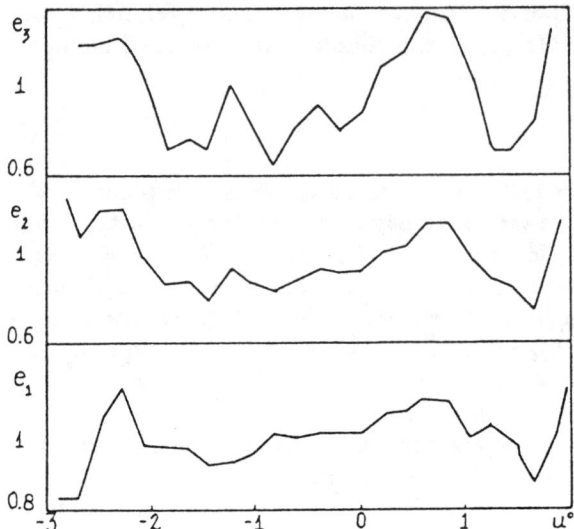

Figure 4.18 Conditionally averaged moments of energy dissipation from data by Kuznetsov, Praskovskii, and Sabel'nikov. The measurements were conducted in the mixing layer. $y/x = 0$, $\langle(u_1 - \langle u_1 \rangle)^2\rangle^{1/2}/\langle u_1 \rangle = 0.21$, $\text{Re}_\lambda = 1660$, $L = 1.21$ m, $\eta = 0.21$ mm, $\gamma = 0.89$, $u^\circ = (u_1 - \langle u_1 \rangle)/\sqrt{\langle(u_1 - \langle u_1 \rangle)^2\rangle}$, $e_n = \langle \epsilon^n \rangle_{u_1}/\langle \epsilon \rangle^n$, $n = 1, 2, 3$.

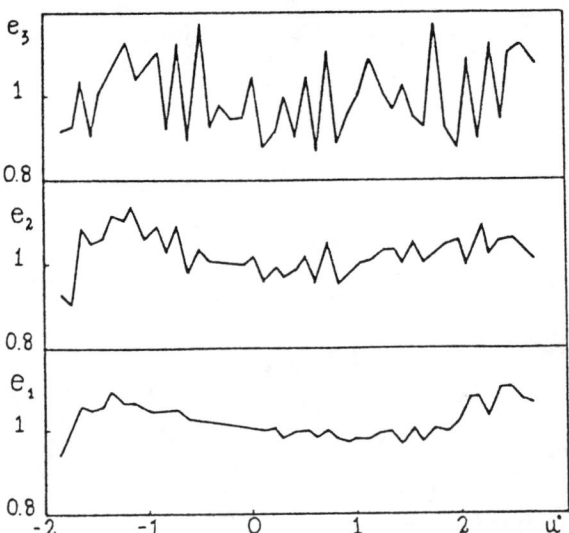

Figure 4.19 Conditionally averaged moments of energy dissipation from data by Kuznetsov, Praskovskii, and Sabel'nikov. The measurements were conducted in a plane wake behind a cylinder. $y/d = 0$, $\langle(u_1 - \langle u_1 \rangle)^2\rangle^{1/2}/\langle u_1 \rangle = 0.069$, $\text{Re}_\lambda = 196$, $L = 0.135$ m, $\eta = 0.22$ mm, $\gamma \approx 1$. The notations are the same as in Fig. 4.18.

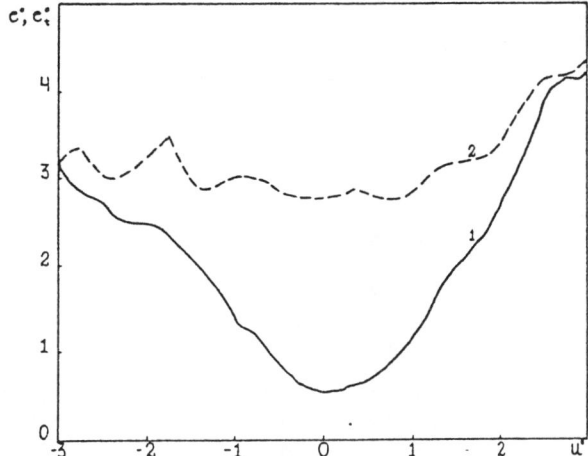

Figure 4.20 Conditionally averaged energy dissipation from data by Kuznetsov, Praskovskii, and Sabel'nikov. The measurements were conducted in a three-dimensional wake. $y/d = 1.23$, $\mathrm{Re}_\lambda = 403$, $\gamma = 0.244$, $L = 0.66$ m, $\eta = 0.38$ mm, $u^0 = (u_1 - \langle u_1 \rangle)/\sqrt{\langle (u_1 - \langle u_1 \rangle)^2 \rangle}$, 1) $e^0 = \langle \epsilon \rangle_{u_1}/\langle \epsilon \rangle$, 2) $e_t^0 = \langle \epsilon \rangle_{u_1,t}/\langle \epsilon \rangle$.

intermittency is insignificant (in the mixing layer $\gamma = 0.89$, in the plane wake $\gamma \approx 1$), and Fig. 4.20 which shows the conditionally averaged dissipation $\langle \epsilon \rangle_{t,u_1}$ in a three-dimensional wake at a point where $\gamma = 0.214$, i.e., where intermittency is significant. Figure 4.20 also shows the experimental dependence of conditionally averaged dissipation $\langle \epsilon \rangle_{u_1}$ which is obtained in the same flow without the presence of intermittency. It can be seen that, firstly, $\langle \epsilon \rangle_{t,u_1} \geq \langle \epsilon \rangle_{u_1}$ is in complete agreement with (3.10) and, secondly, the conditional mean $\langle \epsilon \rangle_{u_1}$ in contrast to $\langle \epsilon \rangle_{t,u_1}$ depends strongly on the velocity level at which averaging is carried out. This result proves that the presence of turbulent fluid is a fundamental basis in the formulation of the hypothesis on the statistical independence of fine- and large-scale motions in sheared flows.

Let us now go to the results of verification of hypothesis (4.13) for conditionally averaged energy dissipation at a specified difference of velocities v_1. The quantity $\epsilon_{11} = \nu(\partial u_1/\partial x_1)^2$, which was conditionally averaged at a constant value of v_1 was measured at different points of the flows. Similar measurements enable only qualitative confirmation of relation (4.13) to be obtained, since its quantitative verification requires the simultaneous measurement of all three velocity components and, if the assumption of local isotropy is not accepted, a large number of spatial velocity derivatives.

The character of the dependence of moment $\langle \epsilon_{11} \rangle_{v_1}$ on the difference of velocities v_1 was found to be identical at all points. As an example, Fig. 4.21 shows the results of measurements of $\langle \epsilon_{11} \rangle_{v_1}$ in the mixing layer (flow 3). It is evident that increasing $|v_1|$ leads to a large increase of $\langle \epsilon_{11} \rangle_{v_1}$. The rate of this increase rises

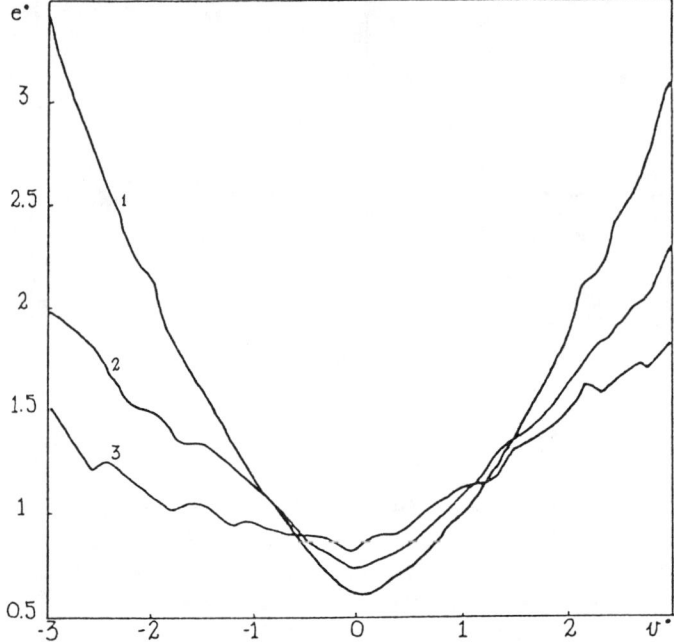

Figure 4.21 Dependence of the conditionally averaged energy dissipation on the difference of velocities from data by Kuznetsov, Praskovskii, and Sabel'nikov. The measurements were conducted in the mixing layer. *1)* $r/L = 0.01$, *2)* $r/L = 0.051$, *3)* $r/L = 0.26$, $e^0 = \langle \epsilon \rangle_v / \langle \epsilon \rangle$, $v^0 = v_1 / \sqrt{\langle v_1^2 \rangle}$. The experimental conditions were the same as in Fig. 4.18.

with decreasing distance r between the points. These results are in qualitative agreement with formula (4.13).

In conclusion, let us consider the way these results can be interpreted. To this end, it should be born in mind that the Kolmogorov-Obukhov theory is asymptotic (Re $\to \infty$, $r/L \to 0$), and the statistics of fine-scale fluctuations are determined not only by the structure function D_{11}, but also by an entire aggregate of quantities of the form $\langle v^n \rangle$, $\langle \epsilon(x_1)\epsilon(x_1+r) \rangle$, ... Therefore, the nonuniversality of constant μ means that either the Kolmogorov-Obukhov theory does not yield asymptotically accurate formulas for all characteristics of the fine-scale fluctuations, or the Reynolds number in the tests conducted was not sufficiently high. As already indicated, the tests were conducted at very high Reynolds numbers $Re_\lambda \sim 10^3$ and the variation of Re_λ by an order of magnitude with an accuracy within the errors of measurements did not have an effect on the results. Therefore, it appears that the Kolmogorov-Obukhov theory is not asymptotically exact, although one cannot of course totally exclude the possibility that its conclusions may be true at Reynolds numbers far exceeding the values at which experiments were conducted.

§4.6 THEORY OF INERTIAL INTERVAL AND THE PROBLEM OF TURBULENCE MODELING

The solution of the questions considered in the present chapter has exceptionally important ramifications for the theory of turbulence. Indeed, although the main features of turbulent flow are determined by the large-scale, energy-carrying vortices, the theory of turbulence cannot be limited to the study of only those vortices since their evolution is dependent on energy dissipation which is effected in the fine-scale vortices. Hence, the volume of information required for suitably exact description of flow under high Reynolds numbers is very large. We use the relationships obtained above for estimating this volume when solving the Navier-Stokes equations in a three-dimensional domain ω during time T.

Such an estimate is of considerable interest in view of the development of powerful computers which can integrate the three-dimensional nonstationary Navier-Stokes equations. Let us analyze the accuracy with which numerical integration can be carried out. For this purpose we select n reference points in a domain $\omega \times T$ which enable linear interpolation at any point during the calculation of the velocity field to be carried out. When selecting the position of these points, one must take into consideration that energy dissipation fluctuates strongly and, therefore, the local time and spatial scales of the velocity field vary significantly, i.e., the grid must be substantially nonuniform. It is obvious that the spatial grid scale cannot strongly exceed the quantity $\eta = \nu^{3/4}\epsilon^{-1/4}$ and the temporal scale the quantity $\tau = \sqrt{\nu/\epsilon}$, where ϵ is the true value of dissipation at the given point. Therefore, the number of reference points must not be less than

$$n = \int_0^T \int_\omega \eta^{-3}\tau^{-1} d^3x\, dt \qquad (4.38)$$

As an example, we consider a homogeneous stationary turbulence assuming that there is an external force compensating energy dissipation. Then, from (4.38) we obtain

$$n \sim L^3 T \nu^{-11/4} \langle \epsilon^{5/4} \rangle \qquad (4.39)$$

where L is a characteristic dimension of domain ω (usually it can be assumed that it is of the order of the integral turbulence scale).

Before we continue, it is useful to note that the analyzed approach enables only pseudorealizations whose statistical characteristics coincide with the statistical realization characteristics (see, for example, Lee [1982]), to be obtained. This conclusion is based on the notion that the finite difference approximation introduces an error which can be regarded as a perturbation of the velocity field. As a consequence of flow instability, this perturbation rises exponentially with time and, since perturbations with characteristic scales larger than the Kolmogorov scale are unstable, the time constants in the exponential law are of the order of τ, i.e., very small. For this reason the error of the finite difference approximation at the reference time must be of the order $\exp(-T/\tau)$. Consequently, to obtain a realization of the velocity field that is close to the true value the minimum distance between the nodes of the three-dimensional grid must be of the order $\eta \exp(-T/\tau)$, and the minimum time step of the order $\tau \exp(-T/\tau)$. Hence, it is evident that obtaining an accurate realization of the velocity field is associated with the processing of unrealistically large amount of information.

In view of this situation, it can be assumed that formula (4.39) yields a satisfactory estimate of n only when describing the statistical characteristics of the velocity field. It should be emphasized that this is only an assumption which is based on a number of intuitive considerations. The most important of these considerations is the following:

Perturbations caused by the errors of the finite difference approximation lead to the numerical solution at some points (bifurcation points) appearing to "skip" from one realization to another and, since almost all realizations possess identical properties, this circumstance does not change the statistical characteristics (in the work by Lee [1982] references can be found to the theorem on the equivalency of the means under consideration for some classes of dynamic systems).

Let us now proceed to further estimates. By using the results presented in §4.3 we have

$$\langle \epsilon^{5/4} \rangle = \gamma \langle \epsilon^{5/4} \rangle_t = \gamma \langle \langle \epsilon^{5/4} \rangle_{v,t} \rangle_t$$

$$= \gamma \left\langle \left(\frac{v^3}{2} \right)^{5/4} \left(\frac{vr}{\nu} \right)^{x(5/2)-5/4} \right\rangle_t \tag{4.40}$$

It follows from the approximate formula (4.22) that $x(5/2) - 5/4 \approx 15\mu/128 \approx 0.01$. This means that when estimating the value of n, dissipation fluctuations can be ignored (when seeking the location of the reference points this is not admissible). Then, assuming that formula (4.40) yields a correct order of magnitude of the results when $r \sim L$, we obtain

$$\langle \epsilon^{5/4} \rangle \sim (u^3/r)^{5/4}, \qquad u^2 = \langle (u_k - \langle u_k \rangle)^2 \rangle$$

Therefore, (4.39) assumes the form

$$n \sim \frac{uT}{L} \text{Re}^{11/4}, \qquad \text{Re} = \frac{uL}{\nu} \tag{4.41}$$

Hence it follows that the amount of information required for the description of turbulence increases rapidly with increasing Reynolds number. (The number of reference points at each time layer in domain ω is of the order $\text{Re}^{9/4}$, which agrees with the estimate by Landau and Lifshit [1954]). Therefore it is imperative to look for ways to obtain approximate solutions.

One of these ways can be based on the principle of self-similarity of turbulence with respect to Reynolds number. In this approach, to describe the flow for large Reynolds number Re_1, one can use the results of the numerical solution of the Navier-Stokes equations in which Reynolds number Re_2 is relatively small, i.e.,

$\text{Re}_1 \gg \text{Re}_2 \gg 1$. When computing the velocity, this approach leads to an absolute error of the order of the Kolmogorov velocity, i.e., a relative error of the order $\epsilon r = \text{Re}_2^{-1/4} - \text{Re}_1^{-1/4} \approx \text{Re}_2^{-1/4}$. Eliminating Re_2 from the formula $\epsilon r \sim \text{Re}_2^{-1/4}$ and $n \sim \text{Re}_2^{11/4}$, yields

$$n \sim \frac{uT}{L}\epsilon r^{-11} \qquad (4.42)$$

The second method considers the velocity field partially averaged with respect to the three-dimensional domains with the characteristic dimension ℓ which satisfies the condition $L \gg \ell \gg \nu^{3/4}\langle\epsilon\rangle^{-1/4}$. This approach is applied to the near-wall turbulence models in which an equation similar to the Navier-Stokes equation is formulated for the description of a partially averaged velocity field. Equations obtained from the Navier-Stokes equation by replacing the coefficient of molecular viscosity by the microturbulent viscosity ν_t are used in the simplest case. Moreover, it is assumed that ν_t is universally linked with the vortex characteristics having dimensions of the order ℓ, for example

$$\nu_t = C_\nu \ell^2 \sqrt{\left(\frac{\partial u_k^{(\ell)}}{\partial x_j} + \frac{\partial u_j^{(\ell)}}{\partial x_k}\right)^2} \qquad (4.43)$$

where C_ν is the universal constant, $u^{(\ell)}$ is the partially averaged velocity field. It is natural to assume that, just as in the previous case, when estimating n, dissipation fluctuations can be ignored. Since the spatial scale of the field $u^{(\ell)}$ is of the order ℓ, and temporal scale is of the order $(\ell^2/\langle\epsilon\rangle)^{1/3}$, the same considerations which were used when obtaining estimate (4.41) yield

$$u \sim \frac{uT}{L}\left(\frac{L}{\ell}\right)^{11/3} \qquad (4.44)$$

The approach under consideration yields an absolute error of the order $(\langle\epsilon\rangle\ell)^{1/3}$, i.e., a relative error of the order $\epsilon r \sim (\ell/L)^{1/3}$. By using this ratio in formula (4.44), we again arrive at estimate (4.42). Thus, there is no fundamental difference between the two approaches.

It is evident that the amount of operations increases rapidly with increases in the required accuracy of description. Consequently, it is necessary to somehow decrease the error arising as a result of the inaccurate description of fine-scale fluctuations. This is attainable only if there is some universal connection between the statistical characteristics of large- and fine-scale parts of the turbulence spectrum.

These investigations showed that these connections are presently known only very approximately (the most important constants C and μ vary roughly by two times). For this reason, the constants entering in grid turbulence models cannot be universal.

We explain the conclusion drawn on the example of constant C_ν in formula (4.43). From physical considerations it is apparent that $\nu_t \sim \ell\sqrt{e_\ell}$, where e_ℓ is the energy of the vortices with a dimension of the order ℓ, i.e., $e_\ell \sim C(\langle \epsilon \rangle \ell)^{2/3}$ and $\nu_t \sim \sqrt{C} \langle \epsilon \rangle^{1/3} \ell^{4/3}$. Now the mean value of expression (4.43) is of the order $C_\nu \langle \epsilon \rangle^{1/3} \ell^{4/3}$, i.e., $C_\nu \sim \sqrt{C}$. Since the Kolmogorov constant is nonuniversal, then constant C_ν is also nonuniversal. The nonuniversality of the latter was demonstrated by Ferziger [1985] who stated that the value of C_ν must be varied by approximately two for the best correlation of the calculated results and experimented data obtained in different flows.

Hence it follows that one of the most vital tasks is to develop a reliable theory of the inertial interval of turbulence spectrum. Generally speaking, since there is a direct interaction between large- and fine-scale characteristics of turbulence, this problem might not even have a solution. Such an interaction is, however, found to be quite weak since constants k and $q(2) - 2/3$ characterizing the fluctuations of energy dissipation are extremely small. Therefore, it can be assumed that the details of such an interaction are not very important and, consequently, that there are fairly simple methods to describe the fine-scale structure of turbulence.

The smallness of constants k and $q(2) - 2/3$ merits special comment since when describing the flow of viscous fluids small constants arise. In view of this, on the face of it, an unexpected analogy is revealed between the theory of local homogeneous turbulence and the asymptotic theory of separation of the laminar boundary layer in which the velocity field is expanded into a series of very small Reynolds numbers. Furthermore, as will be seen below, in both cases the characteristic scales of length and velocity are identically dependent on Reynolds number.

Before we explain this last conclusion, it is useful to make the following comments. First of all, it should be remembered that as a result of internal intermittency, regions of large velocity gradients are generated in the turbulent flow and that these have small volume (see §1.1). One of the simplest geometrical interpretations of such regions is of tangential discontinuities in the ideal fluid or viscous mixing layers between plane parallel flows in a viscous fluid. The flow structure in such layers is described by the boundary layer equations, i.e., in a first very rough approximation, it can be assumed that there exists a certain analogy between the boundary layer on a solid surface and the region with large dissipation in a turbulent flow.

It is generally known that mixing layers are unstable, and in the linear theory of boundary layer stability on a solid surface, the same scales of length and velocity arise as in the theory of separation (Zhuk and Ryzhov [1980]). Consequently, it can be assumed that with the growth of instability of the region with high dissipation in the turbulent flow, the same phenomena arise as in the separation of the laminar boundary layer on a solid surface.

Let us turn our attention in this regard to the theory of separation of the boundary layer on a solid surface (Sychev, Ruban, Sychev and Korolev [1987]). We shall assume that the determining parameter is the Reynolds number Re to boundary layer thickness before the separation point. Then, the dependence of the characteristic scales of the length and velocity of the number Re in the vicinity of the separation point has the form

$$\Delta x \sim \text{Re}^{-3/4}, \quad \Delta y \sim \text{Re}^{-5/4}$$

$$\Delta u \sim \text{Re}^{-1/4}, \quad \Delta v \sim \text{Re}^{-3/4} \qquad (4.45)$$

where u, v are longitudinal and transverse components of velocity, x, u are longitudinal and transverse coordinates. Hence it follows that the scale Δx varies with the Reynolds number just as the Kolmogorov scale of length, and the scale Δu varies as the Kolmogorov velocity.

It is also evident from (4.45) that of all the velocity derivatives $\partial u/\partial y$ has the greatest value: $\partial u/\partial y \sim \text{Re}$. If the analogy being developed is valid, then this conclusion enables the character of function $q(n)$ when $n \to \infty$ to be determined. It can be shown that if function q is linear for large n, i.e., $q \sim \alpha n$, then the velocity distribution of the velocity difference at two points belonging to the inertial range are finite, i.e., there exists a large upper value of the velocity difference v_m, and upon variation of r this value varies as $v_m \sim r^\alpha$ (Kuznetsov and Sabel'nikov [1981a]). In this case, function $x(n)$ characterizing the effect of Reynolds number Re on the fluctuations of the velocity derivatives (see §4.4, formula (4.21)) has the form $x(n) = (\alpha - 1)n/(\alpha + 1)$ when $n \to \infty$, i.e., the probability distribution of the derivatives is finite and the maximum value of the derivative with Re changes as $\text{Re}^{(\alpha-1)/(\alpha+1)}$. This result is an indication that the maximum value of the derivative is of the order Re and, consequently, $\alpha = 0$. Thus, $q = \text{const}$ when $n \to \infty$, i.e., v_m is of the order of the energy carrying vortices.

In concluding this discussion of the problems associated with the local structure of turbulence, we once again emphasize that whether a qualitative leap (break-through) in the description of turbulence will occur is largely dependent on the extent of universality of the motion in the fine-scale spectrum and on whether it is feasible to develop a sufficiently accurate and reliable theory of the inertial interval. The solution of these problems is a challenge to both experimentalists and theoreticians.

CHAPTER
FIVE

TURBULENT DIFFUSION COMBUSTION

The burning of nonpremixed gases in turbulent flow is widely used in diverse technical devices (industrial furnaces, burners, gas turbine combustion chambers, and so on). The principles of the theory of this process were set in the works by Burke and Schumann [1928], Shvab [1948], Zel'dovich [1949], Hawthorne, Wedell, and Hottel [1949]. One of the main ideas of the theory was proposed by Burke and Schumann. It is based on the assumption that the combustion process is limited by the mixing process of the fuel and the oxidant. The criterion of validity of this assumption can be obtained from the work of Zel'dovich [1949], whose results are discussed below. In accordance with this assumption, the rates of all chemical reactions can be regarded as infinitely high and, hence, the composition and temperature in the diffusion flame are thermodynamically in equilibrium. This assumption reduces the problem to the description of the field of the inert (nonreacting) contaminant. Indeed, assuming that all the coefficients of molecular transport are equal*, the equations of diffusion of the fuel and oxidant can be written in the form

$$\Lambda(c_f) = -\rho W_f, \qquad \Lambda(c_o) = -\rho W_o$$

$$\Lambda = \rho \frac{\partial}{\partial t} + \rho u \nabla - \nabla(D\rho\nabla) \qquad (5.1)$$

where ρ is density, u is velocity, x is the coordinate, D is the coefficient of molecular diffusion, W is the rate of reaction, c is the concentration, subscripts f and o refer to the fuel and oxidant, Λ is the differential operator corresponding to the equation of diffusion, and t is time.

If the chemical reaction is a one stage process, then the following link exists between quantities W_f and W_o : $W_o = \text{St } W_f$, where St is the stoichiometric coefficient (the number of grams of oxidant required for the complete combustion of a gram of fuel). By virtue of the linearity of operator Λ the reaction rate can be excluded from formula (5.1) leading to the relation (St $c_f - c_o$) = 0. Since

---

*When describing turbulent combustion this is not regarded as a strong limitation owing to the self-similarity of turbulent flows with respect to Reynolds number.

Λ const $\cdot F$) = const $\cdot \Lambda(F)$, Λ (const) = 0, then it is convenient to introduce the quantity

$$z = \frac{\text{St } c_f - c_o + 1}{1 + \text{St}}, \qquad \Lambda(z) = 0 \tag{5.2}$$

which is equal to 1 in a flow of pure fuel and 0 in a flow of pure oxidant. In the absence of combustion it yields the concentration of the fuel, and upon combustion it yields the element concentration of the fuel.

Since thermodynamic equilibrium is usually shifted towards the combustion products, the concentrations of intermediate substances are small, and the condition of thermodynamic equilibrium is approximately written as $c_f c_o = 0$, i.e., the reaction takes place so rapidly that the fuel and oxidant cannot simultaneously exist at each point. From (5.2) we then deduce

$$c_f = 0 \left(z < z_s = \frac{1}{1 + \text{St}} \right), \qquad c_f = \frac{z - z_s}{1 - z_s} (z > z_s) \tag{5.3}$$

It is evident from relation (5.3) that an infinitely thin zone of chemical reactions, located at the isoscalar surface $z = z_s$ on which $c_f = c_o = 0$, corresponds to the assumption made. It is easy to show that on the surface under consideration the flux of fuel and oxidant are in a stoichiometric ratio. Thus, the determination of the characteristics of a diffusion flame is reduced to the description of the concentration of the inert contaminant.

Proceeding from sufficiently general assumptions, it can be shown (§5.1) that this conclusion is also true in the case when the number of reactions and reacting substances is arbitrary. Moreover, formula (5.2) must be generalized, i.e., concentration z is understood to mean a quantity which is obtained as a result of the following operation. Let us consider some point of the flow and mentally direct all the chemical reactions in the reverse direction. We shall continue this process until all atoms in the fuel are assembled into the initial molecule. The fuel concentration thus obtained will be equal to z.

The specific features of diffusion combustion in the turbulent flow were first investigated by Hawthorne, Wedell, and Hottel [1949]. The main problem which arises in the case under consideration is associated with the fact that, as follows from (5.3), the fuel concentration c_f is nonlinearly connected with the concentration of the inert contaminant. Therefore, when finding completeness of combustion the nonlinear dependence $c_f(z)$ must be averaged, for which one must know the concentration PDF of the contaminant z. The need for such averaging is evident from purely physical considerations. The point is that the conclusions drawn in the work by Burke and Schumann [1928] are also true when describing combustion in a turbulent flow, i.e., it can be assumed that the reaction zone is very thin. Owing to random oscillations of the velocity the flame front also randomly oscillates. Hence, at each point of the flow either an excess fuel or excess oxidant will be observed. For this reason, for the same value of $\langle z \rangle$ the amplitude of fluctuations of the flame front also rises with increasing amplitude of concentration fluctuations which leads to a decrease in mean completeness of combustion. Hence, it follows that for the description of the effectiveness of the combustion process, one must have information on the concentration PDF.

The analysis presented above shows that the investigation of the characteristics of turbulent diffusion combustion is reduced to the description of the concentration of the inert contaminant. Therefore, for the solution of the problem under consideration it is natural to enlist the methods developed in the theory of turbulence. These methods are presently being developed intensively (Launder and Spalding [1972], Baev, Golovichev, and Yasakov [1976], Kuznetsov, Lebedev, Sekundov, and Smirnova [1977a, b, 1980], Bywater [1980], Vulis, Ershin, and Yarin [1963], Zimont and Meshcheryakov [1974], Zimont, Meshcheryakov, and Sabel'nikov [1978, 1981, 1983], Alber and Batt [1974, 1976], Borgi [1980], Gromov, Larin, and Levin [1984], Meshcheryakov and Sabel'nikov [1984b] and others). One or another semi-empirical turbulence model, which allows the calculations of quantities $\langle z \rangle$ and $\sigma^2 = \langle (z - \langle z \rangle)^2 \rangle$, is used in these methods. When averaging various nonlinear dependences, it is assumed that the concentration PDF z is universally connected with the quantities $\langle z \rangle$ and σ, i.e., $P = \sigma^{-1} F[(z - \langle z \rangle)/\sigma]$. The form of function F is selected, as a rule, from more or less arbitrary considerations.

The main problems that arise when using these methods are associated with the description of the specific gas dynamic features of diffusion combustion which are caused by a large decrease in density. For example, it is generally known that the penetration of a submerged diffusion flame is considerably higher than the penetration of a nonburning jet of the same gas (Hawthorne, Wedell, and Hottel [1949], Kremer [1966]). Calculations and experiments (Kuznetsov, Lebedev, Sekundov, and Smirnova [1977b], Buriko and Lebedev [1980]) show that two different gas dynamic effects appear in diffusion combustion. The first is connected with the decrease of residence time in the flame. For example, with combustion in a channel the velocity increases, and upon combustion in a free flame the flow velocity decreases less than in a nonburning jet. This effect leads to the inhibition of mixing. The second effect arises only upon combustion in a confined space and is caused by the fact that under the action of the pressure gradient the heavy fluid (oxidant) and light fluid (combustion products) accelerate differently. Therefore, the transverse velocity gradient increases, leading to additional turbulence of the flow and improved mixing. Since both phenomena have opposite effects on the characteristics of mixing, then both a decrease (Buriko and Lebedev [1980]) and an increase (Klyachko and Strokin [1969]) in flame length can take place under different conditions.

The next step in the development of diffusion combustion was made by Zel'dovich [1949], who took into account the effect of the rate of chemical reactions on the structure of the flame. It was established by Zel'dovich that the thickness of the reaction zone, as a rule, is significantly smaller than the characteristic dimensions of the problem. This is due to the fact that chemical reactions, encountered in combustion processes, occur with high rates only at high temperatures. Therefore, at distances removed from the flame front, i.e., from surface $z = z_s$, the reaction rate is strongly reduced. The indicated state enables consideration of the reaction zone as a particular boundary layer whose thickness is much less than its radius of curvature. The analysis of the characteristics of such a boundary layer indicates that at large activation energy of the leading reactions, the distributions of the temperature and concentrations at the flame front differ little from the distributions that are obtained for an infinitely high reaction rate. Therefore, the flow of the fuel to the flame front Q_f is easily estimated from the solution of the problem based on the assumption of infinite reaction rate, i.e., $Q_f = (1 - z_s)^{-1} \rho D |\partial z / \partial n|$ where n is the normal to the reaction zone. This formula follows from (5.3).

This flow plays a decisive role in the theory, since it is established in the work by Zel'dovich [1949] that steady combustion is possible only if $Q_f < Q_{fc}$, where Q_{fc} coincides, in order of magnitude, with the flow of the fuel in a normal flame front. It should be recalled that such a front is referred to as a plane reaction zone which is propagating across the stationary mixture of fuel and oxidant. This front differs from the diffusion front by the fact that both combustible components are located on one side of the reaction zone. The velocity of propagation of this front u_n depends on the characteristic time of the chemical reaction τ_c and the coefficients of molecular transport, i.e., $u_n \sim \sqrt{D/\tau_c}$. The use of quantity u_n is very convenient, since the detailed kinetics are often unknown and velocity u_n is easily measured in experiments. Quantity u_n is dependent on mixture composition and on the initial temperature and pressure. In the current case Q_{fc} must be estimated from the value of u_n in a stoichiometric mixture of the same initial temperature which would have been observed upon mixing of fuel and oxidant without reaction. Hence, it is evident that $Q_{fc} = u_n(z_s) z_s \rho(z_s)$.

Thus, condition $Q_f < Q_{fc}$ acquires the form

$$\left| D \frac{\partial z}{\partial n} \right| < z_s (1 - z_s) u_n(z_s)$$

From the theory of Zel'dovich [1949] it follows that if this condition is implemented, then the combustion process takes place just as at infinitely large reaction rate. Otherwise, flameout takes place.

The assumption that the characteristics of diffusion combustion depend weakly on the reaction rate has become widespread, since it enables the solution of a number of practically important problems; for example, finding the flame length. However, a number of new problems have arisen lately in connection with the need for further enhancement of the combustion process, increase in its efficiency, and reduction of the emission of toxic substances and so on, problems which require substantial refinement of the theory.

Let us consider an actual example — the problem of reducing the concentration of nitrogen oxides. In many cases (in particular, in the combustion chambers of gas turbines) the concentration of nitrogen oxides is one to two orders of magnitude lower than the equilibrium concentration. Therefore, their emission is substantially dependent on the rate of chemical reaction. The rate of oxidation of nitrogen changes sharply with varying temperature (Zel'dovich, Sadovnikov, and Frank-Kamenetskii [1947]). Therefore, even weak thermodynamic nonequilibrium of the main reactions and small heat losses (for example, as a result of radiation) can strongly influence the emission of nitrogen oxides (Bowman [1973], Sarofim and Pohl [1973], Sigal [1977]).

Thus, the general objective is evident and is associated with the analysis of the effect of turbulence on the progress of chemical reactions under the conditions when the principal features of the process (for example, flame length) are independent of the rate of chemical processes. The present chapter is devoted to the solution of this problem.

The methods developed in the chapter are of sufficiently general character. However, it was found convenient to expound their essence on the example of solving a specific problem — the description of the concentration field of nitrogen oxides in a submerged diffusion plume. The convenience of such a method is associated with the need for experimental verification of the main assumptions and conclusions of the theory, and the submerged free flame is best studied experimentally. Accordingly, discussed below is the combustion taking place upon the efflux of propane or hydrogen with velocity u_0 from a round nozzle of diameter d vertically upwards into stationary air. The pressure and temperature are standard. The characteristic values of the determining parameters: $d = 3 - 6$ mm, $u_0 = 10 - 30$ m/sec for propane; $d = 0.5 - 6$ mm, $u_0 = 100 - 800$ m/sec for hydrogen.

The adopted discussion plan in this chapter takes into account that the solution can be divided into different stages. First (§5.1), it is natural to analyze the case when all chemical reactions leading to the oxidation of the fuel are in equilibrium. Then (§5.2), the formation of nitrogen oxides is considered on the assumption that their concentration is much lower than the equilibrium concentration and all the other concentrations are in equilibrium. A general problem arises here which is related to the need for averaging the rates of chemical reactions. Heat losses by radiation are analyzed in §5.3, and the effect of turbulence on the progress of all chemical reactions, including the reactions of fuel oxidation, is analyzed in §5.4. The developed methods are used in §5.5 for the description of nitrogen oxide formation. Finally, in §5.6 an attempt is made to develop a more general theory in which the effect of the chemical reactions on the main characteristics of the plume is not downplayed.

The relation of the above problems with the tasks considered in the preceding chapters is quite distinctly evident. Indeed, the rate of nitrogen oxidation is strongly dependent on temperature. Therefore, when determining the mean reaction rate, it is necessary to know the temperature PDF. A similar problem is also encountered when estimating plume radiation. Finally, when investigating the effect of turbulence on the deviation from thermodynamic equilibrium, one must know, first of all, the local turbulence structure, since the thickness of the zone of chemical reactions in diffusion combustion is very small and, hence, the internal structure of these zones is mainly dependent on the characteristics of the more small-scale vortices.

§5.1 GAS DYNAMIC EFFECTS IN TURBULENT DIFFUSION COMBUSTION

In the present section the combustion of nonpremixed gases is considered under the assumption that the rates of chemical reactions are infinitely high. In this section, as well as everywhere hereafter, it is assumed that the Mach number is small, Reynolds number is low, the coefficients of molecular transport are equal, and, if there are walls, heat loss through them is negligibly small. In this and the next sections it is also assumed that radiation heat losses are absent. One more slightly restrictive assumption is convenient to formulate later on. As shown by Bilger [1976], Kuznetsov, Lebedev, Sekundov, and Smirnova [1977a], these assumptions enable the reduction of the computation of the combustion characteristics to the investigation of the field of element mass fraction of the fuel z, i.e., exclude chemical reactions from consideration.

Let M substance take part in the chemical reactions. Then, the transport equation can be written as $\Lambda(c_\ell) = W_\ell \rho$. Here c_0 is enthalpy, $c_\ell (\ell = 1, \ldots, M)$ are mass concentrations, W_0 is the rate of heat release, $W_\ell (\ell = 1, \ldots, M)$ is the rate of formation (absorption) of substance ℓ. Let all substances be composed of n different atoms (or non-disintegrating group of atoms). It is obvious that the total mass of the atoms of each type (or nondisintegrating group of atoms) is conserved. Conserved also is the total energy which is equal to the sum of the enthalpy and energy of chemical bonds at a small Mach number. The laws of conservation can be written as

$$\sum_{\ell=0}^{M} A_\ell^i W_\ell = 0 \qquad i = 0, 1, \ldots n$$

Here, A_ℓ^0 is the heat of formation of substance ℓ, $A_\ell^i (\ell \geq 1, i \geq 1)$ is the mass fraction of the atoms with number i in a substance with number ℓ. Using this condition by virtue of the linearity of operator Λ, ρW_1 can be excluded from the transport equation $\Lambda(c_\ell) = \rho W_\ell$. Moreover, we obtain $n + 1$ equations of the type

$$\Lambda(\varphi_i) = 0, \qquad \varphi_i = \sum_{\ell=0}^{M} A_\ell^i c_\ell$$

The characteristics of the state which is in equilibrium thermodynamically are completely determined by quantities φ_i. Let us now introduce an assumption which is not very onerous to the effect that the initial conditions for quantities φ_i are similar (in the simplest case this condition means that the initial distribution of the temperature and concentration of the fuel and oxidant are similar). Then, φ_i are linear functions of z which is a consequence of the linearity of operator Λ and the similarity of the initial and boundary conditions. Hence, thermodynamic analysis allows the expression of the temperature and concentration of all substances in terms of z, where $\Lambda(z) = 0$, $z = 1$ in the fuel stream, $z = 0$ in the oxidant stream.

It should be noted now that if the Mach number is small, then when calculating the density the pressure can be assumed constant†. Then, this procedure yields the density ρ in terms of z only.

Thus, the problem is reduced to the search for the concentration PDF of the inert contaminant. Let us consider how this problem is solved in the theory of turbulence. The averaged equations of motion and diffusion are usually used for this purpose. Reynolds stresses entering these equations and the fluxes of the substances are expressed in terms of the gradients of the mean velocity and mean concentration and the coefficient of turbulent transport. The difference of all the theories (and such theories are abundant) is in the methods of computing the coefficient of turbulent transport.

A model proposed by Sekundov [1971] and refined by Kuznetsov, Lebedev, Sekundov, and Smirnova [1977a, b, 1980] is used below. The characteristic features of other models, as well as more elaborate discussion of the principal ideas of the semi-empirical theory of turbulence, can be found in the survey of Ginevskii et al. [1978].

The model includes the equations for the energy of turbulence, turbulent viscosity, and concentration fluctuations. In this model, just as in all the other semi-empirical theories, a very large number of considerably arbitrary assumptions are used when deriving all these equations. It is expedient here to dwell on them in detail. We should just point out that the equations obtained in any model ought to be regarded as a hypothesis whose validity can be established only by comparing the results of the calculation with the experimental data. Let us present the final system of equations written in the boundary layer approximation for an axisymmetric flow

$$\langle\rho\rangle\langle u\rangle\frac{\partial\langle u\rangle}{\partial x} + \langle\rho\rangle v_0 \frac{\partial\langle u\rangle}{\partial y}$$

$$= \frac{1}{y}\frac{1}{\partial y}\left[\langle\rho\rangle\nu_t y \frac{\partial\langle u\rangle}{\partial y}\right] - \frac{\partial\langle p\rangle}{\partial x} - (\langle\rho\rangle - \rho_0)g$$

$$\frac{\partial\langle\rho\rangle\langle u\rangle}{\partial x} + \frac{1}{y}\frac{\partial}{\partial y}\langle\rho\rangle y v_0 = 0, \qquad v_0 = \frac{\langle\rho v\rangle}{\langle\rho\rangle}$$

$$\langle\rho\rangle\langle u\rangle\frac{\partial\langle z\rangle}{\partial x} + \langle\rho\rangle v_0\frac{\partial\langle z\rangle}{\partial y} = \frac{1}{y}\frac{\partial}{\partial y}\left[\langle\rho\rangle\frac{\nu_t}{\text{Sc}} y \frac{\partial\langle z\rangle}{\partial y}\right]$$

$$\langle\rho\rangle\langle u\rangle\frac{\partial e}{\partial x} + \langle\rho\rangle v_0 \frac{\partial e}{\partial y} = \frac{1}{y}\frac{\partial}{\partial y}\left[\langle\rho\rangle k_2 \nu_t y \frac{\partial e}{\partial y}\right]$$

$$+\langle\rho\rangle\nu_t\left(\frac{\partial\langle u\rangle}{\partial y}\right)^2 - \beta_2\left[1 + \frac{\nu_t}{e}\left|\frac{\partial\langle u\rangle}{\partial y}\right|\right]\frac{\langle\rho\rangle e^2}{\nu_t}$$

†In the equations of motion the pressure must be accounted for at all Mach numbers.

$$\langle\rho\rangle\langle u\rangle\frac{\partial \nu_t}{\partial x} + \langle\rho\rangle v_0 \frac{\partial \nu_t}{\partial y}$$

$$= \frac{1}{y}\frac{\partial}{\partial y}\left[\langle\rho\rangle k_3 \nu_t y \frac{\partial \nu_t}{\partial y}\right] + \beta_3 \frac{\langle\rho\rangle \nu_t^2}{e}\left(\frac{\partial\langle u\rangle}{\partial y}\right)^2 - \beta_4 \langle\rho\rangle \nu_t \left|\frac{\partial\langle u\rangle}{\partial y}\right|$$

$$\langle\rho\rangle\langle u\rangle\frac{\partial \sigma^2}{\partial x} + \langle\rho\rangle v_0 \frac{\partial \sigma^2}{\partial y} = \frac{1}{y}\frac{\partial}{\partial y}\left[\langle\rho\rangle k_1 \nu_t \frac{\partial \sigma^2}{\partial y}\right]$$

$$+ 2\langle\rho\rangle\nu_t \mathrm{Sc}^{-1}\left(\frac{\partial\langle z\rangle}{\partial y}\right)^2 - \beta_1\left[1 + \frac{\nu_t}{e}\left|\frac{\partial\langle u\rangle}{\partial y}\right|\right]\frac{\langle\rho\rangle e \sigma^2}{\nu_t} \quad (5.4)$$

Here, x is the longitudinal coordinate, y is the radial coordinate, u is the longitudinal velocity, v is the transverse velocity, ν_t is the coefficient of turbulent viscosity, $e = 1/2\,\langle(u - \langle u\rangle)^2\rangle$ is the turbulence energy, $\sigma^2 = \langle z^2\rangle = \langle z\rangle^2$ is the variance of concentration fluctuations, p is pressure, g is the gravitational acceleration, ρ_0 is the density of the surrounding air (the last component in the first equation in (5.4) is taken into account only when calculating a free jet or a plume). The values of the empirical constants κ_1, κ_2, κ_3, Sc, β_1, β_2, β_3, β_4 are selected so as to describe correctly isothermal mixing. When describing combustion their values remain unchanged. In all the calculations presented below it is assumed that $\kappa_1 = \kappa_2 = 1.2$, $\kappa_3 = 2.4$, $\beta_1 = \beta_4 = 0.14$, $\beta_2 = 0.07$, $\beta_3 = 0.9$, Sc $= 0.8$.

The first equation in (5.4) is the equation of motion written in the boundary layer approximation. Buoyancy forces are taken into account. Here, as well as in the entire present chapter, plumes forming from the fuel flowing vertically upwards are considered. The second relation is the continuity equation, and the third is the equation of turbulent diffusion. The next two equations for turbulent kinetic energy and the turbulent viscosity coefficient describe the adopted turbulence model. These equations, just like the last relation in (5.4) which yields the variance of concentration fluctuations are constructed according to a known scheme reflecting the role of the processes of convection (the left sides of the equations), turbulent diffusion (the first terms on the right sides of the equations), generation (second terms on the right sides of the equations) and dissipation (the last terms on the right sides of the equations). Quantity $\beta_1[1 + \nu_t e^{-1}|\partial\langle u\rangle/\partial y|]e\sigma^2 \nu_t^{-1}$, appearing in the last relation in (5.4), is equal to twice the value of the scalar dissipation $\langle N\rangle = \langle D(\partial z/\partial x_\ell)^2\rangle$, which will later play a very important role.

Dependence $\rho(z)$ is either averaged by means of formulas (3.56), (3.57), or it is assumed that $\langle\rho\rangle = \rho(\langle z\rangle)$. Both methods of calculation yield results which differ in the cases under consideration by no more than 5%.

As an illustration of the validity of system (5.4) for the description of turbulent diffusion combustion we shall cite several examples. The first example — submerged free diffusion plume forming from the flow of a jet of fuel vertically upwards from a round nozzle of diameter d with velocity u_0.

Figure 5.1 shows the dependence of the length of hydrogen plume on Froude number $2u_0^2/(gd) = \mathrm{Fr}$ which is selected as a parameter characterizing buoyancy

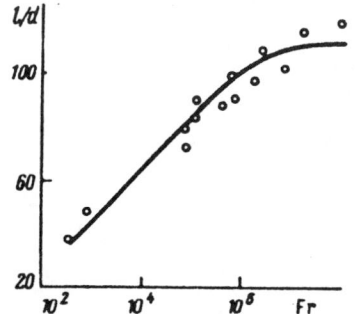

Figure 5.1 Comparison of the calculated length of submerged diffusion hydrogen plume at different Froude numbers with experimental data by Bilger and Beck [1974]. $d = 1.5 - 6.4$ mm, $u_0 = 100 - 200$ m/sec; the symbols are experimental data, the curve — the calculated results.

forces, since the ratio of the densities of air ρ_0 and the combustion products $\rho(z_s)$ in the given case is unchanged. The plume length is understood to mean the distance from the nozzle to that point on the jet axis where equality $\langle z \rangle = z_s = 1/(1 + St)$ is fulfilled. As already indicated, the calculations are very weakly dependent on whether the effect of the fluctuations of concentration z on the mean value of density is taken into account or not. Therefore, the good agreement of the theoretical and experimental data, seen in Fig. 5.1, is an indication that accounting for the concentration fluctuations in the calculation of the gas dynamic structure of flow and the geometrical configuration of the plume is not a prerequisite. In the present case, it is also not necessary to account for the kinetics of the chemical reactions to conduct such calculations.

The effect of concentration fluctuations is pronounced when calculating the mean concentrations of the reacting substances which follows from Fig. 5.2. These data are obtained by Buriko and Kuznetsov [1978] in measurements conducted on the axis of a submerged free propane plume. Measured directly in the tests was quantity c_7^0 which is the mean volumetric concentration of CO_2 in a "dry" sample (i.e., the water condensing at normal temperature was removed). Therefore, quantity $\langle c_7^0 \rangle$ is plotted along the ordinate axis in Fig. 5.2. Here and later on, concentrations are numbered in the following order H_2, OH, O, H, O_2, H_2O, CO_2, CO, N_2, NO. When calculating $c_7(z)$ the results of previously conducted thermodynamic calculations were used, and the dependence $c_7^0(z)$ obtained by this means was averaged with the aid of formulas (3.56), (3.57).

Note that in the experiments gas samples were extracted and their composition analyzed later. This method yields Favre averaging, i.e., quantities of the type $\langle \rho c_\ell \rangle / \langle \rho \rangle$. Henceforth, wherever concentrations are calculated the symbol $\langle \ \rangle$ will correspond to such averaging.

Figure 5.2 shows that the fluctuations of z have noticeable effect on the mean concentration of CO_2, since the maximum value of $\langle c_7^0 \rangle$ is 60% less than the maximum value of the thermodynamic equilibrium of CO_2 concentration.

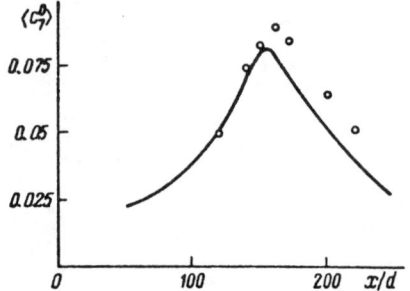

Figure 5.2 Comparison of the calculated mean volumetric concentration of carbon dioxide on the axis of a submerged diffusion propane plume with experimental data by Buriko and Kuznetsov [1978]. $d = 3$ mm, $u_0 = 25$ m/sec; the fuel flows vertically upwards; the symbols are experimental data; the curve — calculated results.

The next graph (Fig. 5.3) indicates that the effect of the concentration fluctuations can be exceptionally strong. This graph generalizes the results of measurements of the concentrations of hydrocarbons in a submerged free plume of chemically pure propane[‡] (contaminant content 0.03%). The absolute error of data is equal to 10^{-4}. All hydrocarbons with number of atoms C from 1 to 4 are found in the samples, i.e., in addition to oxidation, hydrocarbon pyrolysis takes place.

Analysis of the experimental data showed that pyrolysis and oxidation interact very weakly. When processing the data the equivalent concentration of propane was introduced

$$c_f = \sum_i \frac{44}{36 + 3n_i} c_i$$

where c_i is the mass concentration of hydrocarbons, n_i is the ratio of the number of atoms H and C in the i-th hydrocarbon, and summing is over all values of the index i which correspond to the hydrocarbons. This value is maintained in all reactions in which oxygen-containing compounds do not take part; in the absence of the processes of pyrolysis it is equal to the concentration of propane.

It was assumed during processing that oxidation takes place only on the stoichiometric surface $z = z_s$, and the rate of this process is infinitely high, i.e., formula (5.3) is valid. Quantity c_f, which is determined by this formula, was averaged by means of relations (3.56), (3.57) (the solid line in Fig. 5.3). Quantities $\langle c_f \rangle_t = \langle c_f \rangle / \gamma$, $\langle z \rangle_t = \langle z \rangle / \gamma$ are plotted along the axes of the graph. These quantities are obtained by averaging over the turbulent fluid (γ is the intermittency factor, subscript t corresponds to conditional averaging over the turbulent fluid). Quantities $\langle c_f \rangle$ and $\langle z \rangle$ were measured and the intermittency factor γ calculated from (3.56), (3.57), (5.4).

Such calculation showed that the intensity of concentration fluctuations $\sigma/\langle z \rangle$ on the jet axis is higher than 0.55, i.e., more than in an isothermal jet. Analysis

---

[‡]These data were obtained by Yu. Ya. Buriko and V. R. Kuznetsov.

of system (5.4), conducted in the works by Kuznetsov, Lebedev, Sekundov, and Smirnova [1977a, b, 1980], shows that this effect is caused by buoyancy forces (in the absence of buoyancy forces, the intensity of velocity and concentration fluctuations is the same as in an isothermal jet; accounting for buoyancy forces leads to considerable increase in the intensity of concentration fluctuations but does not lead to an increase in the intensity of velocity fluctuations). This effect was observed in tests by Kotsovinos [1977] (buoyant jet without combustion) and Gengembre, Cambray, Karmed, and Bellet [1984] (buoyant jet with combustion).

It follows from computations that $\gamma < 1$ at all points at which measurements were made. Therefore, from (3.56), (3.57) it follows that $P_t(z)$, and hence $\langle c_f \rangle_t$, depends only on $\langle z \rangle_t$. From the graph in Fig. 5.3 it is evident that the theoretical and experimental data agree well. Since logarithmic coordinates are not very convenient when analyzing the accuracy of calculations, one of the radial distributions $\langle c_f \rangle$ is presented in ordinary coordinates in Fig. 5.4.

The data enable the indirect establishment of the accuracy with which the theory developed in Chapter 3 describes concentration PDF in the turbulent fluid

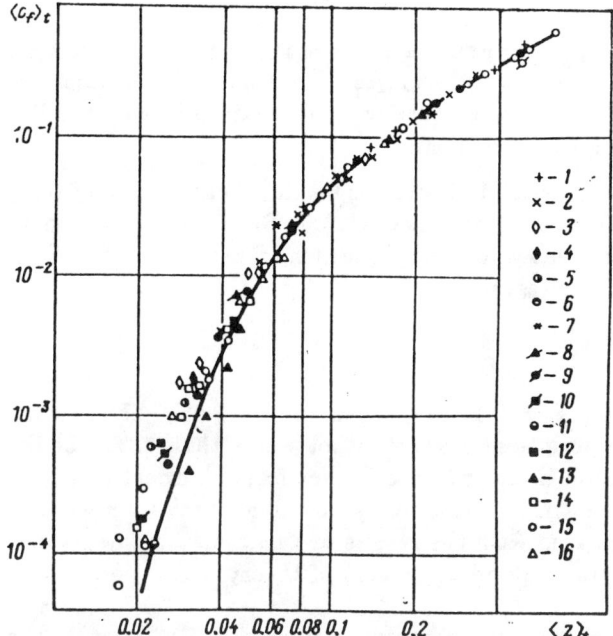

Figure 5.3 Equivalent concentration of propane in submerged diffusion plume from data by Buriko and Kuznetsov. I. $d = 3$ mm, $u_0 = 19.8$ m/sec. *1)* $x/d = 40$; *2)* $x/d = 80$; *3)* $x/d = 120$; *4)* $x/d = 160$; *5)* $x/d = 200$; *6)* $x/d = 240$. II. $d = 6$ mm, $u_0 = 10.7$ m/sec; *7)* $x/d = 40$; *8)* $x/d = 80$; *9)* $x/d = 120$; *10)* $x/d = 160$. III. $d = 3$ mm. *11)* $u_0 = 19.8$ m/sec; *12)* $u_0 = 14.2$ m/sec; *13)* $u_0 = 28.2$ m/sec. IV. $d = 6$ mm. *14)* $u_0 = 7.3$ m/sec; *15)* $u_0 = 10.7$ m/sec; *16)* $u_0 = 14.9$ m/sec. Symbols *1 - 10* correspond to radial profiles. Symbols *11 - 16* correspond to axial profiles.

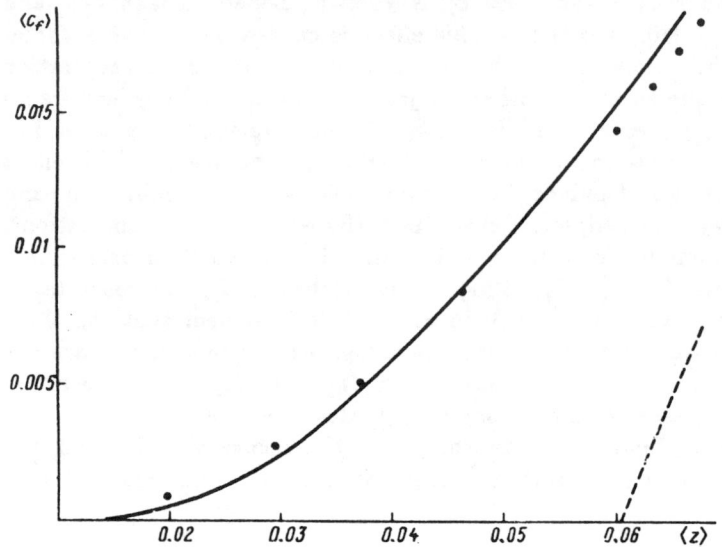

Figure 5.4 Equivalent concentration of propane at one of the sections of submerged diffusion plume. $x/d = 169$, $d = 3$ mm, $u_0 = 19.8$ m/sec; the symbols are the results of measurements; the solid line — calculation accounting for fluctuations; the dashed line — calculation without accounting for fluctuations.

$P_t(z)$. Comparison of the experimental data and the results of calculation of $\langle c_f \rangle_t$ shows that $\langle z \rangle_t > 0.03$, the error does not exceed 12%. Such a contribution to the value of $\langle c_f \rangle_t$ yields fluctuations whose amplitudes is higher than $3\langle z \rangle_t$. This contribution is described by the integral

$$j = \int_{3\langle z \rangle_t}^{\infty} c_f(z) P_t dz$$

whose computation shows that $j = 0.12 \langle c_f \rangle_t$ when $\langle z \rangle_t = 0.03$. Thus, when $z/\langle z \rangle_t < 3$ the accuracy of description of $P_t(z)$ is not worse than 12%. In the indicated range $P_t(z)$ changes by three orders and, therefore, its direct measurement is unlikely at the present time. Accordingly, let us go back to Fig. 3.10 in which function P_t was compared with the results of direct measurements in a narrower range of values of $z/\langle z \rangle_t$. Good agreement of theory and experiment is also evident in this case.

A number of important conclusions can be drawn from Fig. 5.4. First, in the region $\langle z \rangle < z_s$ the quasilaminar model (fluctuations absent) leads to gross errors, so $c_f = 0$ when $z < z_s$. Second, it is very important to describe correctly the form of PDFs, since small values of $\langle c_f \rangle_t$ are determined by fluctuations with very large amplitude. For example, the use of the normal law with the same values of $\langle z \rangle$ and σ when $\langle z \rangle_t = 0.03$ reduces $\langle c_f \rangle$ by more than an order of magnitude. Third, in the present case, the accuracy requirement for the turbulence model rises strongly (it is seen from Fig. 5.3 that $\langle c_f \rangle_t$ is strongly dependent on $\langle z \rangle_t$

and, hence, small errors in the calculation of $\langle z \rangle$ greatly influence the value of $\langle c_f \rangle$).

In connection with the last conclusion, we recall that the model ensures an accuracy of calculation of $\langle z \rangle$ of the order of 30%. This accuracy is inadequate for $\langle c_f \rangle$. Therefore, to improve this accuracy when analyzing the experimental data obtained in some section, the results of calculation of γ and $\langle z \rangle$ were not in a section with the same value of x/d, but in a section with the same value of the axial concentration $\langle z \rangle$. Such a technique is, apparently, fully justified when analysing the accuracy of the theory describing the concentration PDF. Indeed, Fig. 5.3 depicts the results obtained for a very large variation of d, u_0, the position of the point at which measurements are taken, and the measured quantity itself. Hence, the good agreement of the theoretical and experimental data in Fig. 5.3 cannot be accidental.

Let us now present another example from the work by Buriko and Lebedev [1980]. Studied was the diffusion combustion in a cylindrical channel of diameter d_1 in which air is supplied with velocity u_1. Propane was supplied with velocity u_0 through a central nozzle of diameter $d (d_1/d = 13.3)$. The results of the measurements and the calculations based on formulas (5.4) are given in Fig. 5.5. It is seen that at the very end of the plume mixing with combustion accelerates as a result of creation of an additional velocity shear (air and the combustion products accelerate differently owing to the difference in density and pressure gradient). The model correctly describes this effect qualitatively; however, its quantitative agreement with the experiment is unsatisfactory. Calculations based on the widely used "$k-\epsilon$" model and on model (5.4) yielded very slightly differing results. This is an indication that some important effects are being neglected. We

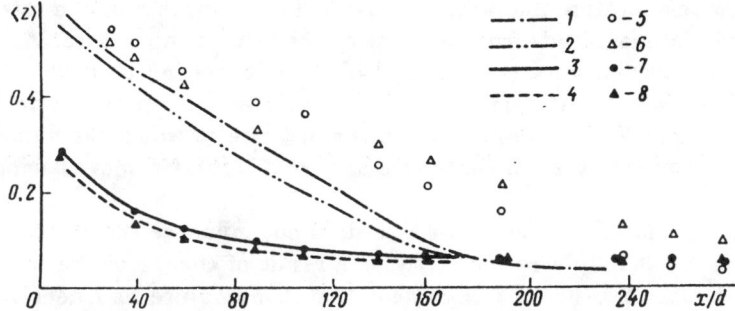

Figure 5.5 Element mass fraction of the fuel (propane) on the axis of diffusion plume in a channel according to the tests and calculations by Buriko and Lebedev [1980]. *1)* $u_1/u_0 = 0.32$, plume calculation; *2)* $u_1/u_0 = 0.16$, plume calculation; *3)* $u_1/u_0 = 0.32$, calculation of isothermal jet; *4)* $u_1/u_0 = 0.16$, calculation of isothermal jet; *5)* $u_1/u_0 = 0.32$, experiment with combustion; *6)* $u_1/u_0 = 0.16$, experiment with combustion; *7)* $u_1/u_0 = 0.32$, experiment without combustion; *8)* $u_1/u_0 = 0.16$, experiment without combustion; $u_0 = 16$ m/sec, $d = 3$ mm.

emphasize that such effects are important only for combustion in a channel and are insignificant in a submerged plume.

In conclusion to this section, let us make a number of comments. It is frequently necessary to average various nonlinear dependencies in the calculations. The results of such averaging depend on the character of nonlinearity and quantity $\langle z \rangle$. A whole series of quantities exist during the calculation of which fluctuations either can be totally ignored (for example, density), or can account for fluctuations without yielding too large a correction (for example, concentration of CO_2). In the latter case, as calculations show almost any reasonable model for PDF is suitable, i.e., it is important only to account for the existence of fluctuations. There are also, however, a number of quantities for whose computation one must exactly know the form of the PDF and intensity of fluctuations (concentration of carbon monoxide). An important role is also played by the mean composition at which one or another quantity is considered. For example, when $\langle z \rangle > z_s$, the effect of the fluctuations on the mean concentration of hydrocarbons is not very large, and already when $\langle z \rangle = 0.5z_s$ it has an essential value (Fig. 5.4).

We should also note that turbulence models should be used with great caution, since their accuracy is not always satisfactory.

§5.2 EFFECT OF TEMPERATURE AND CONCENTRATION FLUCTUATIONS ON THE MEAN REACTION RATE

Accounting for the effect of temperature and concentration on the mean rate of chemical reaction is of great practical interest. This conclusion is clearly illustrated on the example of nitrogen oxidation reaction whose rate is strongly dependent on temperature. Estimates show that at a mean temperature of 2000 K and relatively low temperature fluctuations (say, 10%), the mean reaction rate can differ by an order of magnitude from the rate at the mean temperature. Another example — the ignition of a cold jet of hydrogen (coflowing or near the wall) being discharged into hot stream of air. It is shown that temperature and concentration fluctuations lead to an increase in the distance at which the flame front appears by approximately 2 – 3 times (Kuznetsov [1972b], Gromov, Larin, and Levin [1984]).

It should be noted that turbulent on its own does not influence the reaction rate. This influence is felt only very indirectly as a result of change of the conditions of heat and mass transfer. It is evident that temperature and density fluctuations are caused by the fluctuations of the flame front. As a result of this, the reaction rate at a fixed point in the flow attains either very large or very small values which naturally alter the mean reaction rate. In this case, the reaction rate in the coordinate system connected with the flame front can stay the same as in laminar flow.

The analyzed problem was, apparently, first touched upon by Zel'dovich [1949] who noted that as a result of the strong dependence of the reaction rate on temperature the mean reaction rate must strongly differ from the reaction rate at mean temperature. The first attempt to quantitatively analyze the effect of fluctuations on the mean reaction rate was undertaken by Vulis [1960]. The dependence of the reaction rate on concentration was not taken into consideration in this work, as a result of which a conclusion is made to the effect that temperature fluctuations always lead to an increase in the mean rate compared with the reaction rate computed from the mean values of temperature and concentrations. However, as a rule, fluctuations lead to the opposite effect. In many cases the field of temperature and concentration are similar, i.e., the fluctuations of these quantities are strongly connected and can be expressed in terms of z. Accounting for this state leads to the fact that the dependence of the reaction rate on temperature has a maximum at some value $z = z_m$. It is evident that at points of the flow where $\langle z \rangle = z_m$ any concentration fluctuation leads to the reduction of the reaction rate (Kuznetsov [1969]). This effect is manifested most strongly at the edge of the jet or plume, i.e., in the region where intermittency is substantial (Kuznetsov [1972b]). Note also that at those points where $\langle z \rangle$ strongly differs from z_m, concentration fluctuations lead to an increase of the mean reaction rate. A more detailed discussion of the question under consideration is included in the book by Kompaniets, Ovsyannikov, and Polak [1979].

In the general case, averaging of the rate of chemical reaction is a very difficult task, since chemical reactions on their own can strongly affect the temperature and concentration PDF. One example of such a strong effect (turbulent combustion of a homogeneous combustible mixture) is considered in Chapter 1. Hence, it cannot be assumed that the temperature and concentration PDFs in a turbulent flow with chemical reactions have universal form. In this sense, the process of formation of nitrogen oxides is a happy exception. As already mentioned, the concentration of nitrogen oxides in many cases is far lower than the equilibrium concentration and, therefore, the rate of nitrogen oxidation is very weakly dependent on the concentration of the final product. Furthermore, the reaction under consideration has little effect on the temperature and concentration of all substances with the exception of NO. In the present section, we shall assume that the reactions between all substances, with the exception of NO, are so fast that their concentrations are in equilibrium. Then, the rate of nitrogen oxidation depends only on z. Indeed, let us consider the scheme of nitrogen oxidation established by Zel'dovich et al. [1947]

$$N + O_2 = NO + O, \qquad N_2 + O = NO + N \tag{5.5}$$

The first reaction in (5.5) occurs much faster than the second. Therefore, as shown by Zel'dovich, Sadovnikov, and Frank-Kamenetskii [1947], the rate of nitrogen oxidation has the form

$$W_{10} = 2k_{10} M_{10} M_3^{-1} M_9^{-1} c_3 c_9 \tag{5.6}$$

if the concentration of NO is substantially lower than the equilibrium concentration. Here, W_{10} enters into the equation of diffusion of nitrogen oxides in the form

$$\Lambda(c_{10}) = \rho W_{10} \tag{5.7}$$

where c_3, c_9, c_{10} are mass concentrations of O, N_2, NO; k_{10} is the constant of the rate of the second reaction in (5.5); $M_\alpha (\alpha = 1 - 10)$ are molecular weights of the substances in the mixture. Heretofore, the rate of formation of various substances have subscripts corresponding to their number. Quantity k_{10} has the form

$$k_{10} = k_{10}^0 \exp(-E/RT), \quad \text{where} \quad k_{10}^0 = 1.1 \cdot 10^{14} \text{cm}^3/(c \cdot \text{mole})$$

$E = 3.1 \times 10^5$ J/mole, R is the universal gas constant. If it is assumed that heat losses by radiation are absent and all reactions (with the exception of the reactions of nitrogen oxidation) are in equilibrium, then quantities c_3, c_9, and T are dependent only on z. Thereby, W_{10} can be found from thermodynamic calculations and for its averaging it suffices to know only the concentration PDF of the inert contaminant.

Thermodynamic analysis shows that the dependence of $W_{10}(z)$ has a sharp maximum when $z = z_s$ as a result of which the following asymptotic formula (Kuznetsov [1969]) is true

$$\langle \rho W_{10} \rangle = \int \rho W_{10} P(z) dz = w P(z_s), \quad w = \int \rho W_{10} dz \quad (5.8)$$

Estimates indicate that this formula yields fully acceptable accuracy when $\sigma \geq \Delta z$, where Δz is the range of the values of z in which W_{10} exceeds one half of its maximum value ($\Delta z = 10^{-2}$ for the combustion of propane, $\Delta z = 5 \times 10^{-3}$ for the combustion of hydrogen). Analysis of the results of numerical computations showed that the formulated condition was always fulfilled.

Formula (5.8) simplifies the investigation significantly, since only parameter w is featured. The physical meaning of relation (5.8) is fairly obvious if one takes into account that in accordance with (1.21) $P(z_s)$ characterizes the volume enclosed between surfaces $z = z_s$ and $z = z_s + dz$. Calculation shows that upon combustion under normal conditions we have $w = 2.8 \times 10^{-8} g/(\text{cm}^3 \text{ sec})$ for propane, $w = 4.9 \times 10^{-8} g/(\text{cm}^3 \text{ sec})$ for hydrogen.

These results enable the analysis of the means by which concentration fluctuations affect the formation of nitrogen oxides. Let us average (5.7) and assume that the coefficient of turbulent diffusion of nitrogen oxides coincides with the coefficient of turbulent diffusion of the inert contaminant (the discussion of this assumption is given below in §5.5). We deduce

$$\langle \rho \rangle \langle u \rangle \frac{\partial \langle c_{10} \rangle}{\partial x} + \langle \rho \rangle v_0 \frac{\partial \langle c_{10} \rangle}{\partial y}$$

$$= \frac{1}{y} \frac{\partial}{\partial y} \left[\frac{\nu_t}{\text{Sc}} \langle \rho \rangle y \frac{\partial \langle c_{10} \rangle}{\partial y} \right] + \langle \rho W_{10} \rangle \quad (5.9)$$

Equation (5.9) was solved numerically together with equations (5.4) with two assumptions. In the first case, it was assumed that the fluctuations are absent, i.e., $\langle \rho W_{10} \rangle = \rho(\langle z \rangle) W_{10}(\langle z \rangle)$. In the second case, fluctuations were accounted for, i.e., formulas (3.56), (3.57) were used when averaging. The results of the calculations are depicted in Fig. 5.6.

Comparison of curves *1* and *2* shows that concentration fluctuations lead to a strong reduction in the concentration of nitrogen oxides (almost by three times).

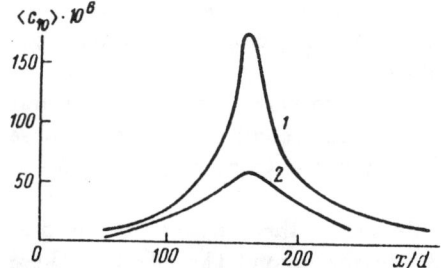

Figure 5.6 Effect of fluctuations on the mean NO concentrations on the axis of submerged diffusion plume of propane. *1*) calculation without accounting for fluctuations, *2*) calculation with the fluctuations accounted for. In the calculations conditions were specified corresponding to the experiments by Buriko and Kuznetsov [1978].

§5.3 EFFECT OF RADIATION ON THE CHARACTERISTICS OF TURBULENT DIFFUSION COMBUSTION

The conditions of laboratory tests, which are used to verify the theory, heat losses by radiation can comprise 25% (Markstein [1975], Buriko and Kuznetsov [1978, 1980]). Even in this case the characteristics of the plume (with the exception of its temperature) are very weakly dependent on heat losses by radiation. This conclusion is evident from Fig. 5.2 in which the results of the calculations in which radiation was not taken into account and the experimental data obtained under the conditions when heat losses by radiation amounted to 20 – 25% are compared (Buriko and Kuznetsov [1978, 1980]).

The effect under consideration, nonetheless, plays an important role when investigating the formation of nitrogen oxides (Sigal [1977]), since the rate of nitrogen oxidation is strongly dependent on temperature. In particular, estimates show that when heat losses by radiation increase by 1%, quantity w decreases by 30%. Hence, it follows that the effect of radiation on the oxidation of nitrogen is substantial in almost all practically important cases. The purpose of the investigation conducted below is to develop a universal method of calculation and to estimate approximately the effect of radiation on the formation of nitrogen oxides under the conditions of laboratory tests.

When analyzing a free diffusion plume there are a number of circumstances which significantly simplify the calculation of heat losses by radiation. The first is associated with the absence of walls, i.e., the radiated energy is irreversibly lost. The second circumstance is caused by the fact that under conditions characteristic for the majority of tests, the approximation of optically thin layer can be used, i.e., it can be assumed that all the molecules or carbon particles radiate independently from each other and radiation is not absorbed. The usefulness of such an approximation can be established from the following estimate. At normal pressure the radiation path length in the combustion products of a stoichiometric mixture is a maximum and equal to 20 cm. Therefore, in all cases, whenever the thickness of the plume is less than 20 cm, the approximation of the optically thin layer is applicable. Finally, a third simplifying circumstance is associated with the small level of losses such that the radiation flux density can be estimated from the temperature computed for adiabatic conditions.

Accordingly, the investigation will be conducted in three stages. In the first stage heat losses by radiation are computed. In the second the effect of these losses on the flame temperature is found. Finally, in the third stage the rate of nitrogen oxidation is computed.

Accordingly, the investigation will be conducted in three stages. In the first stage heat losses by radiation are computed. In the second the effect of these losses on the flame temperature is found. Finally, in the third stage the rate of nitrogen oxidation is computed.

The method described below was developed by Buriko and Kuznetsov [1983a, b]. It is of approximate character and is based on the analysis of the relative heat losses which are determined by formula

$$q(x) = \frac{\int_0^x \int_0^\infty \langle I \rangle y\, dy\, dx}{Q_0 \int_0^x \int_0^\infty \langle \rho W_f \rangle y\, dy\, dx} \tag{5.10}$$

where I is the radiation flux density, W_f is the rate of oxidation of the fuel, Q_0 is the quantity of heat released upon combustion of one gram of fuel. The numerator in (5.10) yields the energy which is radiated by the volume of the plume enclosed between the initial and present sections, and the denominator yields the energy released in this volume.

The method is based on the following technique. Let us conduct a thermodynamic analysis in which the initial enthalpy decreases by the amount $q(x)Q(z)$, where $Q(z) = Q_0 z$ when $z < z_s$, $Q(z) = Q_0(1-z)/(1-z_s)$, i.e., $Q(z)$ is the amount of heat released during the adiabatic combustion of the mixture in which the concentration of the fuel is equal to z.

Let us elucidate the possible inaccuracies of such an approach. Owing to the random character of the process, the flame front can be observed at different points in the same section. Moreover heat losses, strictly speaking, are dependent on the point at which the flame front is located. In the analysis this circumstance is ignored (the relative level of heat losses is determined so that only the dependence on one coordinate x is taken into account). This assumption can be indirectly substantiated by the experimental data in Chapters 1 and 3, where it was shown that the statistical character of concentration in the turbulent fluid changes weakly across the section, i.e., these characteristics are approximately homogeneous inside the fluctuating jet boundary at each section. Since the position of the flame front is determined by z, and this field is statistically homogeneous in the given section, the fluctuations of the flame front can be disregarded. Another inaccuracy of the method is connected with the fact that heat losses at each given point are of random character, since the temperature and concentration distributions on each surface $z = $ const are also of random character. This circumstance is not taken into account since the results of the analysis are dependent only on quantity $q(x)Q(z)$ which is not random when $z = $ const. Rigorous substantiation of these assumptions is not possible. Therefore, the accuracy of the method is assessed below indirectly by analysing the mean temperature field.

Let us now proceed to the first stage of the calculations. We first find the denominator in (5.10). In the diffusion approximation of the concentration of fuel c_f there is some function z. Let us substitute this function into (5.1) and make use of the diffusion equation of the passive contaminant $\Lambda(z) = 0$. We have (Bilger [1976, 1977], Kuznetsov [1979a])

$$\rho W_f = \rho N \frac{d^2 c_f}{dz^2}, \qquad N = D(\nabla z)^2 \tag{5.11}$$

Using (5.3), we derive $d^2 c_f/dz^2 = (1 - z_s)^{-1}\delta(z - z_s)$, where δ is delta function. Let us substitute this relation into (5.11), average, and make use of the hypothesis discussed in Chapter 3 concerning the concentration fluctuations and scalar dissipation in the turbulent fluid being statistically independent. Then, we obtain

$$\langle \rho W_f \rangle = (1 - z_s)^{-1} \langle N \rangle_t \rho(z_s) \gamma P_t(z_s)$$
$$= (1 - z_s)^{-1} \langle N \rangle \rho(z_s) P_t(z_s) \tag{5.12}$$

where $\langle N \rangle_t$ is the mean value of the scalar dissipation in the turbulent fluid, P_t is the concentration probability density in the turbulent fluid. Formula (5.12) was derived by Bilger [1976] (for $\gamma = l$) on the implicitly adopted assumption of statistical independence of z and N.

Let us now estimate the radiation flux I. Consider first the combustion of hydrogen. In this case, radiation is caused by water vapor. Let us make use of the experimental data presented in the book by Mikheev [1949], in which the results of investigation of radiative heat transfer are given between a hemisphere filled with water vapor and the central element of its base. For sufficiently small radius of the sphere, i.e., in the approximation of an optically thin layer, the effective emissivity factor of water vapor can be approximated by expression $\epsilon_6 = \beta_6(T) p_6 \ell_0$, where ℓ_0 is the radius of sphere, p_6 is the partial water vapor pressure, T is the temperature, and β_6 can be approximated by the expression

$$\beta_6 = b_{61} - b_{62}T, \ b_{61} = 2 \cdot 10^{-7} \text{cm}^{-1}\text{Pa}^{-1}, \ b_{62} = 6.4 \cdot 10^{-11} \text{cm}^{-1}\text{Pa}^{-1}\text{grad}^{-1}$$

Therefore, in accordance with the data by Mikheev [1949] we obtain

$$I_6 = I = 4\beta_6(T) p_6 \sigma_0 T^4 \tag{5.13}$$

where σ_0 is the Stephan-Boltzmann constant. Quantities p_6 and T appearing in (5.13) can be found from thermodynamic analysis on the assumption that radiation has little effect on the temperature and composition of the combustion products. Thus, parameter q can be calculated from formulas (3.56), (3.57), (5.10), (5.12), (5.13).

Let us now consider the combustion of propane. In this case, in addition to water vapor, carbon dioxide and soot particles also radiate. The radiation energy of carbon dioxide can be described by a formula similar to relation (5.13)

$$I_7 = 4\beta_7(T)p_7\sigma_0 T^4, \qquad \beta_7 = b_{71} - b_{72}T$$

$$b_{71} = 4.6 \cdot 10^{-7} \text{cm}^{-1}\text{Pa}^{-1}, \qquad b_{72} = 1.3 \cdot 10^{-10} \text{cm}^{-1}\text{Pa}^{-1}\text{grad}^{-1} \tag{5.14}$$

where subscript 7 corresponds to CO_2. These formulas are also obtained from analysis of the experimental data by Mikheev [1949]. The total radiation by water vapor and carbon dioxide is found from formula $I = I_6 + I_7$. As expected, calculations in which the radiation by soot particles is not taken into account yield lower value of $\langle I \rangle$. As an example we cite the calculation conducted for a propane plume ($d = 0.3$ cm, $u_0 = 25$ m/sec). In this case, the total radiation of the plume is equal to $q(\infty) = 0.15$, whereas the experimental data by Buriko and Kuznetsov [1978] indicate that $q(\infty) = 0.25$, i.e., approximately 40% of the energy lost is radiated by soot particles.

Since the kinetics of soot formation in the diffusion combustion of propane is practically unstudied, the parameter q is conveniently computed from the semi-empirical formula

$$I = \mu(I_6 + I_7) \tag{5.15}$$

where coefficient μ is selected so that the calculated value of I would agree with the experimentally measured value. This coefficient in all calculations was assumed equal to $\mu = 1.67$. The temperature and concentration of carbon dioxide c_7 and water c_6 are found as a result of thermodynamic analysis. Formula (5.15) is averaged with the aid of relations (3.56), (3.57), parameter q is found from (5.10), (5.12) and from the averaged relation (5.15).

To examine the validity of the proposed method, we present the results of calculation of the mean temperature fields in a diffusion propane plume. As was already mentioned, for the determination of the temperature, one must use the results of thermodynamic analyses from which dependence $T(z, q)$ can be derived. This dependence must be averaged with the help of formulas (3.56), (3.57). Figure 5.7, taken from the work by Buriko and Kuznetsov [1983b], shows a comparison of the results of calculated and experimental data. It is seen that the data and the results of calculation agree well. Analogous results were also obtained for a hydrogen plume.

Two circumstances stand out. Firstly, upon combustion of propane the maximum temperature in the diffusion plume is 750 K lower than the theoretical temperature, i.e., lower than the temperature of adiabatic combustion of a stoichiometric mixture $T_s = 2260$ K (normal initial conditions). Secondly, the maximum temperature in the given section of the plume is almost constant with increasing distance from the nozzle.

Analysis of the results of the calculations shows that the last effect is caused by the interaction of two processes. On the one hand, as we move away from the nozzle the position of the averaged surface of the flame front (i.e., surface $\langle z \rangle = z_s$) approaches the axis of the plume, i.e., it is displaced to the region where the intensity of concentration pulsations $\sigma/\langle z \rangle$ is low. This circumstance leads to an increase in the mean temperature. On the other hand, as we move away

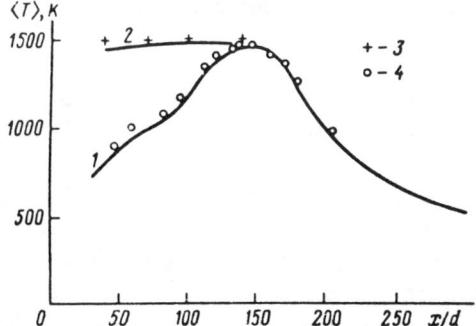

Figure 5.7 Comparison of the calculated values of the mean temperature in a submerged diffusion plume of propane with the experimental data by Buriko and Kuznetsov [1980]. $d = 3$ mm, $u_0 = 25$ m/sec; *1*) calculation of mean temperature on the axis, *2*) calculation of maximum mean temperature in various sections, *3*) measurement of maximum mean temperature in various sections, *4*) measurement of mean temperature on axis of plume.

from the nozzle, radiation increases. This increase is associated with the fact that during the mixing process the distance between two closely located surfaces $z = z_1$ and $z = z_1 + ds$ increases (see §3.8), i.e., the radiating volume and, hence, heat losses increase. Both effects are mutually compensated. On the whole, radiation and concentration fluctuations affect the mean temperature roughly equally.

It is useful to note that concentration fluctuations affect only the mean flame temperature and do not act on the actual flame temperature, i.e., the temperature on the surface $z = z_s$. Therefore, owing to the increase of radiation with increasing distance from the nozzle, the temperature of the flame decreases monotonically.

The method proposed above easily enables taking into account the effect of flame radiation on the rate of nitrogen oxidation. For this purpose, one must conduct a thermodynamic analysis in which the initial enthalpy decreases by the amount $qQ(z)$. Such an analysis yields dependencies $T(z,q)$, $c_3(z,q)$. Substituting $T(z,q)$ and $c_3(z,q)$ into (5.6), we find $W_{10}(z,q)$ and, hence, compute the quantity $w(q)$. Following this, we integrate equations (5.4), (5.9) in which quantity $\langle \rho W_{10} \rangle$ is given, as before, by formula (5.8). Parameter w appearing in (5.8) is dependent on quantity q, which is computed by the methods developed above.

This analysis indicates that for the combustion of a propane jet ($d = 0.3$ cm, $u_0 = 25$ m/sec), the maximum concentration of nitrogen oxides on the axis of the plume is equal to 9×10^{-6}. Comparing this number with the results given in Fig. 5.6, we conclude that, in this case, radiation leads to a roughly six-fold decrease in the concentration of NO. It is also evident that the computed value of the maximum concentration of NO on the axis of the plume is substantially lower than the measured value by Buriko and Kuznetsov [1980] $\langle c_{10} \rangle = 3.6 \times 10^{-5}$. This result is not surprising, since, as indicated at the beginning of the chapter, the concentration of monatomic oxygen can be larger than the equilibrium concentration, and the influence of turbulence on the kinetics of the main chemical reactions is not yet taken into account.

§5.4 EFFECT OF THE RATE OF CHEMICAL REACTIONS ON TURBULENT DIFFUSION COMBUSTION

As generally known, the kinetics of oxidation of the majority of fuels are of chain character, i.e., the process takes place in several stages with many intermediate substances formed at each stage. Investigation of the oxidation kinetics in turbulent flows can be considerably simplified if one takes into account that the chemical reactions take place at high rates with moderate deviations from the thermodynamic equilibrium. Therefore, the conversion of the substances takes place in very narrow spatial zones. The specific feature of turbulence is manifested only in the fact that such zones are contorted and are displaced in space in a random manner.

The problem under consideration is characterized by the following features. Just as with the majority of strongly nonlinear systems, chemical reactions in some range of the values of the determining parameters can be very sensitive to the conditions under which mixing takes place. Since the process is of multistage character, this means that several regions, in which the oxidation mechanisms strongly differ, can appear in a diffusion flame. Moreover, the various regions are separated by narrow zones in which sharp changes in the mechanism of the reactions takes place. Another feature is that since the thickness of such zones is small, the conditions of internal mixing are determined only by the local characteristics of turbulence, i.e., by energy dissipation ϵ, by scalar dissipation N, and by the coefficients of molecular transport. As mentioned in Chapter 4, the amplitude of fluctuations of quantity ϵ far exceeds its mean value. Analogous assertions are true also for scalar dissipation. A complete analysis of the problem under consideration is hardly likely at the present time. Therefore, we shall later consider a number of examples illustrating these general considerations. We shall discuss below the solution of the equations of transfer of heat and mass in various regions of the flame. It will be shown that, in a whole series of cases one can find either asymptotically accurate solutions connecting the concentration of the reacting substances with the local nonaveraged turbulent characteristics, or can reduce the solution to the integration of the diffusion equation without sources with boundary conditions that are dependent on the local characteristics of turbulence and on the rate of chemical reactions. Since the PDFs of quantities ϵ and N are dependent on Reynolds number (see Chapter 4), one of the important problems is to explain how the processes of molecular transport affect the progress of chemical reactions in a developed turbulent flow.

Very Small Deviations from Thermodynamic Equilibrium. As indicated in §5.1, at an infinitely high rate of chemical reactions the composition and temperature in the diffusion flame can be easily computed from the conditions of thermodynamic equilibrium. It can be shown that this approach yields the first term of the asymptotic expansion of the solutions of the transport equations into a series with respect to some large parameter (Damköhler number). In the notation adopted in §5.1, the transport equations are written as

$$\Lambda(c_m) = \rho W_m, \qquad \Lambda = \rho \frac{\partial}{\partial t} + \rho u \nabla - \nabla(D\rho\nabla)$$

Let us reduce these equations to a dimensionless form by refining all length scales to the turbulence scale L, all velocity scales to $\sqrt{e}$ (e is the turbulence energy), all reaction rates to the constant of the rate of the most rapid reaction K. The system under consideration then acquires the form

$$\Lambda_0(c_m) = G_e \rho W_m^{(0)}, \qquad G_e = \frac{KL}{\sqrt{e}} \qquad (5.16)$$

where Λ_0 is the dimensionless operator of Λ, $W_m^{(0)}$ are dimensionless reaction rates. It is evident that when $G_e \to \infty$, the solution is found from the system $W_m^{(0)} = 0$, i.e., there is a state of thermodynamic equilibrium in which c_m is dependent only on z. Let us denote this solution by $d_m^{(e)}$ and seek small corrections in the form of series

$$c_m = c_m^{(e)}(z) + \frac{1}{G_e} c_m^{(1)} + \frac{1}{G_e^2} c_m^{(2)} + \ldots$$

Substitute this series into (5.16) and make use of the diffusion equation $\Lambda_0(z) = 0$. For the sake of clarity we return to the initial dimensional variables. It can be shown that the second term of expansion $(c_m^{(1)})$ satisfies the linear system of equations

$$N \frac{d^2 c_m^{(e)}}{dz^2} + \frac{\partial W_m}{\partial c_n} c_n^{(1)} = 0 \qquad (5.17)$$

where matrix $\partial W_m/\partial c_n$ is computed for $c_n = c_n^{(e)}$, i.e., its elements are known from the thermodynamic analysis of function z. Hence, it follows that the small corrections to the thermodynamic equilibrium are dependent only on the element concentration of the fuel and scalar dissipation. It is also seen that $c_m^{(1)}$ is linearly dependent on N. Since the mean value of the scalar dissipation is independent of Reynolds number, then quantity $\langle c_m^{(1)} \rangle$ is also independent of the processes of molecular transport which justifies the assumption above concerning the equality of the molecular transport coefficients. It should be noted that the fluctuation amplitude of quantities $c_m^{(1)}$, generally speaking, is dependent on Reynolds number in all turbulent flows, since the amplitude of scalar dissipation fluctuations, just as the amplitude of energy dissipation fluctuations, is dependent on Reynolds number (see Chapter 4).

Unfortunately the region of applicability of relation (5.17) is limited (the Damköhler number must be very large), since the rates of various reactions differ usually by several orders of magnitudes. Moreover, the characteristics of the process are determined by the slowest reactions, and det $(\partial W_m/\partial c_n)$ by the fastest reactions. This property of the system of kinematic equations is well known (the so-called stiff differential system) as it creates great difficulties in numerical calculations (see, for example, Baev, Golovichev, and Yasakov [1976]).

Effect of the Rate of Chemical Reactions on Moderate Deviations from Thermodynamic Equilibrium. Let us analyze the process of combustion of hydrogen or hydrocarbons. Consider the following kinetic scheme

$$H + O_2 = OH + O, \qquad H_2 + OH = H_2O + H \tag{5.18}$$

$$2OH = H_2O + O, \qquad CO + OH = CO_2 + H \tag{5.19}$$

$$H + OH + M = H_2O + M, \quad H + O + M = OH + M, \quad 2H + M = H_2 + M \tag{5.20}$$

where symbol M corresponds to the molecule of any substance in the mixture. In consequence of the works of Kaskan [1958] and Ramshaw [1980], we will show that this scheme correctly reflects the main kinetic behavior of the combustion of mixtures that are nearly stoichiometric with moderate deviations from thermodynamic equilibrium (the quantitative formulation of these conditions is conveniently presented later).

Let us first of all note that thermodynamic equilibrium is not reached without completing the reactions (5.20) even if the most general kinetic scheme is considered. This circumstance plays a decisive role, since reactions (5.18), (5.19) near equilibrium take place with a rate far exceeding reaction (5.20). This conclusion is based on the estimate in which the rates of the direct reactions in (5.18) – (5.20) were computed from the thermodynamically equilibrium values of temperature and concentration. It is established that the rate of the slowest reaction in (5.18), (5.19) in the adiabatic combustion of a stoichiometric mixture for the most unfavorable estimate[§] is one and a half times higher than the rate of the fastest reaction in (5.20).

Hence, it follows that the conversion of the substances is limited by reactions (5.20), and reactions (5.18), (5.19) are close to equilibrium. This conclusion is true even when the most general kinetic scheme is considered. It will be shown below that the conditions which follow from the equality of the rates of direct and reverse reactions in (5.18), (5.19) are found to be adequate for the unambiguous description of the rates of the reactions in (5.20). Moreover, it must be assumed that the substances which are not enumerated in (5.18) – (5.20) do not participate in the chemical conversions. Such an assumption contributes only a small inaccuracy, since the thermodynamic analysis indicates that equilibrium concentrations of the substances, not enumerated in (5.18) – (5.20), are much less than the smallest of the concentrations of the substances taking part in reactions (5.18) – (5.20). The indicated considerations justify the use of a strongly simplified scheme described by reactions (5.18) – (5.20).

[§]The uncertainty of the estimates is caused by inaccurate measurements of the constants of the rates of the chemical reactions.

Let us now proceed to the consequences of the analysis conducted above. Consider first the combustion of hydrogen. From (5.18), (5.20) it is evident that the process is described by seven variables (six concentrations and the temperature). The law of conservation of energy and the laws of conservation of atoms O and H enable finding three relations between seven variables. As is evident from §5.1, quantity z enters into these relations. As independent variables we select the concentrations of H_2, OH, O, H, denoting them c_1, c_2, c_3, c_4, respectively.

The conditions following from the detailed equilibrium reactions (5.18) yield three more relations, i.e., only one independent variable remains. The selection of this variable is facilitated with the aid of a technique which is well known in chemical kinetics (Kaskan [1958], Ramshaw [1980]). We select such a linear combination from $c_1 - c_4$ so that the fastest reactions (5.18) would not affect its distribution. For this purpose, equations $\Lambda(c_n) = \rho W_n$ are multiplied by arbitrary numbers, the resulting relations are added, and the numbers selected so that the rapid reactions of (5.18) would be absent in the final relations. Using this technique, we obtain

$$\Lambda(c) = \rho W, \qquad c = c_1 + \frac{1}{17} c_2 + \frac{1}{8} c_3 + 3c_4$$

$$W = -4(W_1 - W_{-1} + W_2 - W_{-2} + W_3 - W_{-3}) \tag{5.21}$$

where subscripts $1 - 3$ in the symbols W_n correspond to the ordinal number of the reaction in (5.20), the positive subscripts correspond to the direct reactions and the negative subscripts to the reverse reactions. The rates of these reactions depend on $c_1, \ldots, c_4$ and z. The conditions of equilibrium of reactions (5.18) yield three algebraic relations. These relations enable expressing the quantity W, which will be referred to henceforth as the effective oxidation rate, in terms of c and z only. The dependence of the temperature and concentration of the remaining substances on c and z can be computed beforehand.

As an example of such a calculation Figs. 5.8 and 5.9 show the dependences of the concentration of monatomic oxygen c_3 and the temperature on quantity c (normal conditions). Note that when $z = z_s$, the concentration of monatomic oxygen differs strongly from the equilibrium concentration ($c_3^{(e)} = 2.3 \times 10^{-4}$).

Thus, the problem is reduced to the solution of one equation (5.21). It is found that the source appearing in this equation can be approximated by a simple formula (Bilger [1980a])

$$W = k(c - c^{(e)})^2 \tag{5.22}$$

where k is a constant. The verification of this formula is presented in Fig. 5.10. The vertical sections give the range of variation of the results of the calculation of which the constants of the rates of chemical reactions presented by Jensen and Jones [1978] were used. The value of k computed on the basis of these data were used. The value of k computed on the basis of these data is equal to $k = 1.4 \times 10^6$ sec^{-1}; the calculation based on the constants of the reaction

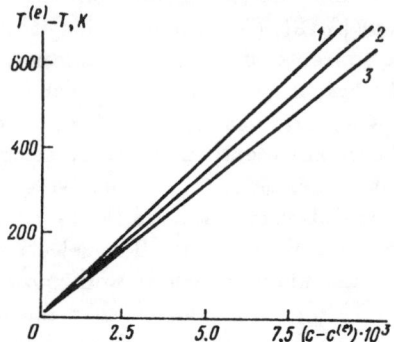

Figure 5.8 The effect of deviations from thermodynamic equilibrium on the combustion temperature of hydrogen in air. *1)* $z = 0.02$ (lean mixture), *2)* $z = 0.028$ (stoichiometric mixture), *3)* $z = 0.036$ (rich mixture).

rates given by Jenkins, Yumlu, and Spalding [1966] yields a value one order of magnitude larger: $k = 1.4 \times 10^7$ sec^{-1} (Buriko and Kuznetsov [1983a]). For comparison purposes we present the data by Bilger [1980a]: $k = 1.6 \times 10^6$ sec^{-1}.

As will be evident later, this calculation accuracy can be regarded as satisfactory, since deviation from equilibrium turn out to be proportional to $k^{-1/3}$. Note that as the deviations from the equilibrium increase, i.e., with increasing $c - c^{(e)}$, the ratio of the rates of the trimolecular reactions in (5.20) and bimolecular reactions in (5.18), (5.19) increases, i.e., in the final analysis the condition of validity of formula (5.22) is violated. In the stoichiometric mixture the rate of the fastest reaction in (5.20) is compared with the rate of the slowest reaction in (5.18), (5.19) when $\Delta c = c - c^{(e)} \approx 10^{-2}$, i.e., when $\Delta c > 10^{-2}$ formula (5.22) is inapplicable. Note also that the value of k is weakly dependent on the level of

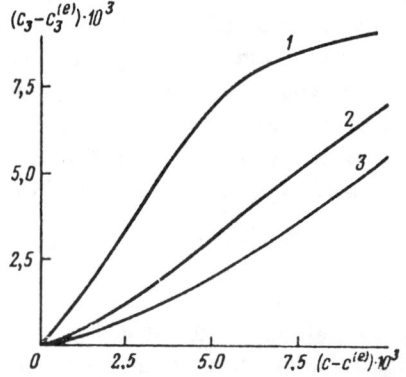

Figure 5.9 The effect of deviations from thermodynamic equilibrium on the concentration of monatomic oxygen in the combustion of hydrogen in air. *1)* $z = 0.02$ (lean mixture), *2)* $z = 0.028$ (stoichiometric mixture), *3)* $z = 0.036$ (rich mixture).

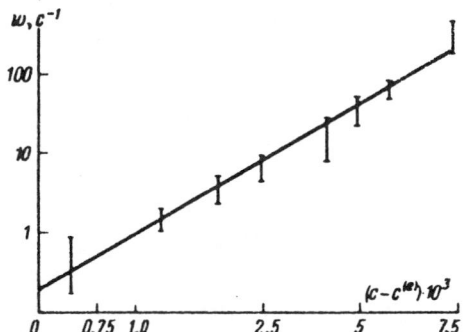

Figure 5.10 Calculation of the effective oxidation rate of hydrogen in air. The vertical sections correspond to the range of variation of W when z varies within the bounds $z = 0.02 - 0.036$ and q within the bounds $q = 0 - 0.2$.

heat losses by radiation, since W is determined by trimolecular reactions in (5.20) whose rate changes little with temperature.

Let us dwell now on dependence $c^{(e)}(z)$. The equilibrium values of the reduced concentration $c^{(e)}$ is depicted in Fig. 5.11. It is seen that $c \approx c_1^{(e)}$ (c_1 is the concentration of H_2) and, as a first approximation, the equilibrium values of the reduced concentration and the concentration of H_2 are described by formula (5.3) which is true for one-stage irreversible reaction. The indicated circumstance is caused by the fact that the equilibrium of the gross reaction $H_2 + 1/2\, O_2 = H_2O$ is strongly shifted towards the formation of combustion products (in the stoichiometric mixture we have $c_6^{(e)}/(c_1^{(e)}\sqrt{c_5^{(e)}}) = 3600$, where c_6, c_5 are the concentrations of H_2O and O_2). Therefore, quantity $c^{(e)}$ is conveniently represented in the form (Kuznetsov [1982a])

$$c^{(e)} = \frac{1}{\kappa}\varphi(s), \qquad s = \kappa k_0(z - z_s) \tag{5.23}$$

where $\varphi \to 0$ when $s \to -\infty$, $d\varphi/ds \to 1$ when $s \to \infty$, k_0 is a constant ($k_0 \approx 1$ for hydrogen). It is seen that when $k \to \infty$, formulas (5.3) and (5.23) coincide. Henceforth, it will be assumed everywhere that $k \gg 1$.

This method can also be used when describing the combustion of hydrocarbons. We should just note the final results obtained for propane by Buriko and Kuznetsov [1983a]. The reduced concentration acquires the form

$$c = c_1 + \frac{1}{17}c_2 + \frac{1}{8}c_3 + 3c_4 + \frac{1}{14}c_8$$

i.e., in the formula, by means of which the value of c is determined, a superfluous term appeared proportional to c_8 is the concentration of CO. The equilibrium values of the reduced concentration is depicted in Fig. 5.12, from which it is evident that formula (5.23) remains true if we assume $k_0 = 0.42$. Formula (5.22) undergoes some change which is evident from Fig. 5.13, which is borrowed from

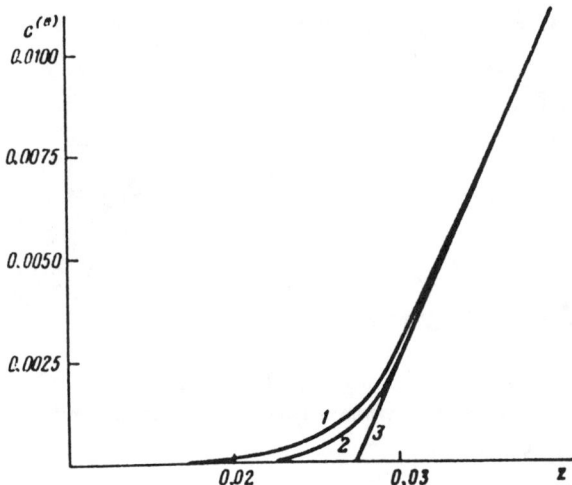

Figure 5.11 Calculation of equilibrium values of the reduced concentration upon combustion of hydrogen in air. *1)* $c^{(e)}$, *2)* $c_1^{(e)}$, *3)* dependence (5.3).

the work by Buriko and Kuznetsov [1983a]. The constants of the rates of chemical reactions are taken from the work by Jenkins, Yumlu, and Spalding [1966]. The solid lines in Fig. 5.13 correspond to the dependence

$$W = k(c - c^{(e)})^m \tag{5.24}$$

where $m = 2.4$, $k = 6.3 \times 10^7$ sec^{-1}. Calculations based on the data by Jensen and Jones [1978] show that $m = 2.4$, $k = 6.3 \times 10^6$ sec^{-1}. The presented approximations above are valid when $z = 0.045 - 0.08$ and $\Delta c = c - c^{(e)} < 5 \times 10^{-3}$.

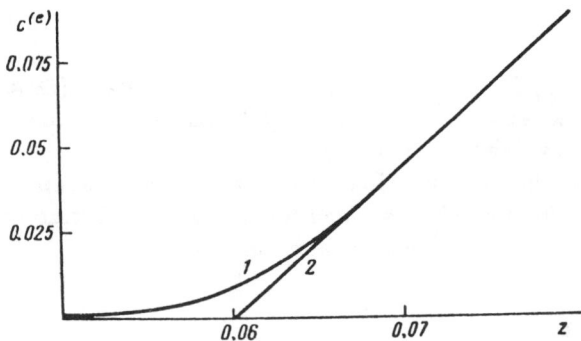

Figure 5.12 Calculation of equilibrium values of the reduced concentration upon combustion of propane in air. *1)* $c^{(e)}$, *2)* piecewise-linear approximation corresponding to one-stage irreversible reaction.

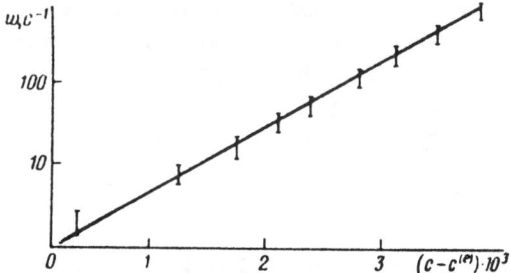

Figure 5.13 Calculation of the effective oxidation rate of propane in air. The vertical lines correspond with the range of variation of W when z is varied within the bounds $z = 0.045 - 0.08$ and q within the bounds $q = 0 - 0.25$.

Let us now proceed to the solution of equation (5.21). This solution, evidently, gives the distribution of $c(x,t)$ which enables finding the temperature and concentrations of all substances from algebraic relations. We make use of the following considerations (Kuznetsov [1982a]). We shall first assume that the zone of the chemical reactions has very small thickness. From this assumption we obtain the solution and then derive the conditions of its validity.

Let the zone of reactions be localized near the contorted, randomly shifting surface $z = z_s$ and let its thickness δ_c be much less than the Kolmogorov scale $\eta = \nu^{3/4}\langle\epsilon\rangle^{-1/4}$. We go to a coordinate system moving with some point on the surface $z(x,t) = z_s$. When $\delta_c \ll \eta$, the radius of curvature of the reaction zone is much larger than its thickness as a result of which the derivatives in the directions of the tangents to surface $z = z_s$ can be ignored. Furthermore, in the coordinate system under consideration the characteristics are determined only by the quantities ν and $\langle\epsilon\rangle$. Therefore, estimates $\partial/\partial n \sim 1/\delta_c, \partial/\partial t \sim (\nu/\langle\epsilon\rangle)^{-1/2}$ are true (n is the normal to surface $z = z_s$). Hence, when $\delta_c \ll \eta$, equation (5.21) acquires the form (2.18). The left side of (2.18) is much less than the first term on the right side if $\delta_c \ll \eta$. We also take into account that since δ_c is small, then the reduced concentration c is also small in the reaction zone. Therefore, there is no sense in accounting for the effect of temperature on ρD. Thus, from (2.18), (5.24) we obtain

$$D \frac{dc^2}{dn^2} - k[c - c^{(e)}]^m = 0 \tag{5.25}$$

This equation is true for $\delta_c \ll \eta$. The indicated condition is still incomplete (δ_c is unknown). Therefore, we solve equation (5.25), find δ_c, and thereby obtain the condition for the validity of the results in explicit form.

Note that, since $\delta_c \ll \eta$, the distribution of the element fuel concentration in the reaction zone can be approximately described by expression $z = z_s + n\, dz/dn\, (dz/dn = \text{const})$. Then (5.25) assumes the form

$$N \frac{d^2 c}{dz^2} - k[c - c^{(e)}(z)]^m = 0, \quad N = D\left(\frac{\partial z}{\partial n}\right)^2 = D\left(\frac{\partial z}{\partial x_k}\right)^2 \tag{5.26}$$

Hence there follows an important conclusion: the structure of the reaction zone is determined by a single fluid dynamic characteristic — the scalar dissipation N. This conclusion was made by Kuznetsov [1977b] from the analysis of the equation for the concentration PDF of a reacting contaminant.

Let us introduce the new variables and parameters

$$s = \kappa k_0 n \frac{dz}{dn}, \qquad \psi = \kappa c, \qquad g_0 = \frac{k}{N} k_0^{-2} \kappa^{-m-1}$$

where dz/dn is computed on surface $z = z_s$. Then, from (5.23), (5.25) we obtain

$$\frac{d^2\psi}{ds^2} - g_0[\psi = \varphi(s)]^m = 0 \tag{5.27}$$

Let us investigate two limiting cases: $\langle g_0 \rangle \gg 1$ and $\langle g_0 \rangle \ll 1$. In the first case we seek a solution in the form of the asymptotic series $\psi = \varphi + \varphi_1 g_0^{-1/m} + \ldots$. Then, we find $\varphi_1 = (\varphi'')^{1/m}$ and so on. Thus,

$$c = c^{(e)}(z) + \left(\frac{1}{G_0} \frac{d^2 c^{(e)}}{dz^2}\right)^{1/m} + \ldots, \qquad G_0 = \frac{k}{N}$$

For $m = 2$ this formula was obtained by Bilger [1980a] by another method. The formula is true only if $|c - c^{(e)}| \ll c^{(e)}$ and, hence, its application to the question under consideration here is very limited.

Let us consider the other limiting case $\langle g_0 \rangle \ll 1$ which is analyzed by Kuznetsov [1982a] and Buriko and Kuznetsov [1983a]. This case corresponds to the limit G_0 = const, $k \to \infty$. Then, function φ acquires the form $\varphi = s\theta(s)$, where $\theta = 0$ when $s < 0$, $\theta = 1$ when $s > 0$. Physically this means that the thickness of the reaction zone is much larger than the width of the zone in which the deviations from dependence $\varphi = s\theta(s)$ is significant. In this case, equation (5.27) is written as

$$\frac{d^2\psi}{ds^2} - g_0[\psi - s\theta(s)]^m = 0 \tag{5.28}$$

This equation has one solution which is smooth and bounded for all s

$$c = c^{(e)}(z) + k_0 G^{-1/(1+m)} \left(\frac{2}{m+1}\right)^{1/(1-m)}$$

$$\times \left\{\frac{m-1}{2}\left[|z - z_s|G^{1/(1+m)} + 2^{(2m-1)/(m+1)}\frac{(m+1)^{1/(m+1)}}{m-1}\right]\right\}^{2/(1-m)}$$

$$G = \frac{k_0^{m-1} k}{N} \tag{5.29}$$

The special case of this relation $(m - 2)$ is given by Kuznetsov [1982a]. The solution at arbitrary m is shown by Buriko and Kuznetsov [1983a].

Let us consider the conditions of applicability of formula (5.29). One of these conditions ($\langle g_0 \rangle \ll 1$) is inessential. Indeed (5.29) describes incorrectly only the small deviations from equilibrium ($|c - c^{(e)}| \ll c^e$), since an approximate form of dependence $\varphi(s)$ is used. However, the correct description of small deviations from the equilibrium is not of particular interest. The second condition is $\delta_c \ll \eta$. On the basis of formula (5.29) it is natural to determine the thickness of the reaction zone from relation $|z(\delta_c) - z_s| G^{1/(1+m)} = 1$. We further have $|z(\delta_c) - z_s| = \delta_c |\partial z/\partial n| = \delta_c \sqrt{N/D}$. Hence, we obtain $\delta_c = \sqrt{D/N} G^{-1/(1+m)}$. By comparing δ_c and η we conclude that when $\nu = D$, the condition of validity of formula (5.29) is

$$\beta = \frac{\delta_c}{\eta} = \langle \epsilon \rangle^{1/4} \nu^{-1/4} k^{-1/(1+m)} (\langle N \rangle k_0^2)^{(1-m)/2(1+m)} \ll 1 \tag{5.30}$$

We now show that when $\beta \gg 1$ and $L \gg \delta_c$, formula (5.29) yields a correct order of magnitude of c, i.e., condition (5.30) is also inessential. Since the integral turbulence scale L is much larger then δ_c and the Kolmogorov scale η is much smaller than δ_c, dimension δ_c then belongs to the inertial range.

It is obvious that the effect of vortices with a dimension much larger than δ_c is inessential as before, and the effect of vortices with small dimensions leads, as a first approximation, to the intensification of the transport processes inside the reaction zone. In order to take into account this effect, we make use of a technique proposed by Kolmogorov [1962a, b] and Obukhov [1962]. Let us consider a cube with a side of the order δ_c and the center at point x. Quantities u, c, ρ, z are averaged over this cube and the results of the partial averaging is denoted $u(x, t, \delta_c)$, $c(x, t, \delta_c)$, $\rho(x, t, \delta_c)$, $z(x, t, \delta_c)$. The reaction zone is understood to mean the region located near the surface $z(x, t, \delta_c) = z_s$ and having a thickness of the order δ_c. Since it is assumed that the effect of the vortices with a dimension less than δ_c is reduced to the intensification of transport processes only inside the reaction zone, it can be considered that relations (5.21), (5.23), (5.24) describe the distribution of $c(x, t, \delta_c)$ if the quantities appearing in them u, ρ, c, z, D are replaced by the partially averaged quantities, i.e., by the quantities $u(x, t, \delta_c)$, $\rho(x, t, \delta_c)$, $c(x, t, \delta_c)$, $z(x, t, \delta_c)$, D_δ, where D_δ is the coefficient of turbulent diffusion resulting from vortices with dimension δ_c and less. The adopted assumption is valid only for estimating orders of magnitudes.

The idea of describing the vortex-induced intensification of mixing when the scale of the vortices does not exceed the characteristic dimension of the zone, with the aid of the coefficient of diffusion D_δ goes back to Richardson [1926]. In the theory of combustion this idea was mainly utilized for the analysis of turbulent combustion of a homogeneous mixture [Prudnikov et al. [1971], Zimont and Sabel'nikov [1975b], Zimont [1979]; see also the references indicated in the beginning of §1.4). It should be remembered also that Richardson's idea lies at the basis of various estimate turbulence models which were mentioned in the introduction and in Chapter 4.

Note that δ_c and, hence, D_δ are not known beforehand and, therefore, the analysis of the equations for partially averaged quantities is necessary. For this purpose, let us proceed to the coordinate system moving with some point on surface $z(x, t, \delta_c) = z_s$, thereby excluding the effect of vortices with a dimension far larger than δ_c. We take into account that all quantities in the region with a dimension of the order δ_c are dependent only on $\langle N \rangle_\delta$, $\langle \epsilon \rangle_\delta$, δ_c, since dimension δ_c belongs to the inertial range (see Chapter 4). Here, $\langle N \rangle_\delta$ and $\langle \epsilon \rangle_\delta$ are the values of the scalar dissipation and energy dissipation which are averaged over a region with the characteristic dimension δ_c. From dimensional considerations we have $D_\delta \sim \langle \epsilon \rangle_\delta^{1/3} \delta_c^{4/3}$. The velocity of the medium in the new system of coordinates is of the order $(\langle \epsilon \rangle_\delta \delta_c)^{1/3}$, the spatial scale of its variation is, by definition, equal to δ_c, and the temporal scale is of the order $\langle \epsilon \rangle_\delta^{-1/3} \delta_c^{2/3}$. Therefore, the following estimate applies

$$u(x, t, \delta_c) = (\langle \epsilon \rangle_\delta \delta_c)^{1/3} \mathbf{v}(y, \tau)$$

$$y = \frac{x}{\delta_c}, \qquad \tau = t \langle \epsilon \rangle_\delta^{1/3} \delta_c^{-2/3}$$

where $\mathbf{v}$ is a random function which is dependent on y and τ.

An analogous formula also applies for the density, i.e., $\rho(x, t, \delta_c) = \rho(y, \tau)$. The quantity $z(x, t, \delta_c)$ is represented in the form

$$z(x, t, \delta) = z_s + \langle N \rangle_\delta^{1/2} \delta_c^{1/3} \langle \epsilon \rangle_\delta^{-1/6} \psi(y, \tau)$$

where ψ is a random function which is dependent only on y and τ. The value of c is represented in the form

$$c = c_\delta \chi(y, \tau)$$

where c_δ is some as yet unknown scale of the variation of the reduced concentration. Let us replace u, c, ρ, z, D in relations (5.21) – (5.24) with $u(x, t, \delta_c)$, $c(x, t, \delta_c)$, $\rho(x, t, \delta_c)$, $z(x, t, \delta_c)$, D_δ, go to new variables y, τ, and make use of the estimates obtained above. When $m = 2$ (combustion of hydrogen), we have

$$\rho \frac{\partial \chi}{\partial t} + \rho v_k \frac{\partial \chi}{\partial y_k} = \frac{\partial}{\partial y_k} \rho \frac{\partial \chi}{\partial y_k} - k^0 [\chi - \chi^0 \psi \theta(\psi)]^2 \rho$$

$$k^0 = k \delta_c^{2/3} \langle \epsilon \rangle_\delta^{-1/3} c_\delta, \qquad \chi^0 = N_\delta^{1/2} \delta_c^{1/3} c_\delta^{-1} \langle \epsilon \rangle_\delta^{-1/6} \qquad (5.31)$$

By definition, quantities $\mathbf{v}$, χ, ψ, ρ are dependent only on y, τ which is possible only when both criteria k^0 and χ^0, appearing in this relation, are of the order of unity. Therefore, we have two conditions for the determination of two unknown quantities c_δ and δ_c. Thus,

$$c_\delta \sim \left(\frac{\langle N \rangle_\delta}{k} \right)^{1/3}, \qquad \delta_c \sim \langle \epsilon \rangle_\delta^{1/2} k^{-1} \langle N \rangle_\delta^{-1/2}$$

These formulas were obtained by Kuznetsov [1982a].

By comparing the first of these relations with formula (5.29) ($z = z_s$, $m = 2$), we arrive at the conclusion that in both cases the order of the deviations from thermodynamic equilibrium is identical. Therefore, as a first approximation, relation (5.29) is also valid when condition (5.30) is not fulfilled. Two stipulations ought to be made here. First, if condition (5.30) is fulfilled, then the valid, i.e., nonaveraged, value of the scalar dissipation N enters into (5.29). And if $\beta \gg 1$, i.e., $L \gg \delta_c \gg \eta$, then (5.29) must feature the scalar dissipation averaged over a domain with a small dimension of the order of δ_c. This conclusion is based only on the quantities averaged over such a domain enter into equation (5.31). Second, the estimates of quantities δ_c and c_δ are valid only when $L \gg \delta_c$.

Let us analyze formula (5.29) and carry out a number of estimates. This relation enables linking the temperature and concentration of all substances with the element mass fraction of the fuel. Thereby, our task is reduced to the description of the field of concentration of the inert contaminant. Featured in (5.29) are the nonaveraged values of the parameters, i.e., within the limits of applicability of the formulas for the value of c the description of the effect of all the details of turbulence on the chemical processes is given as a first approximation. It is evident that two random parameters z and N enter (5.29). The fluctuations of the first parameter are caused by large-scale vortices and, hence, these fluctuations lead to the transfer of the reaction zone as a whole without changing its internal structure. The fluctuations of the scalar dissipation are determined by the small-scale vortices. These fluctuations lead to a change in the internal structure of the reaction zone.

From the standpoint of practical applications, the computation of the mean values from various functions of type $F(c)$ is of greatest interest. For this purpose, we make use of the hypothesis already discussed in Chapter 3 concerning the fluctuations of the concentration of the inert contaminant and the scalar dissipation being statistically independent in the turbulent fluid. Let us first analyze the magnitude of the effect of the fluctuations of scalar dissipation. It follows from (5.29) that $c \sim N^{1/3}$ ($m = 2$). Therefore, as a quantitative characteristic of this influence we can take relation $I_c = \langle N^{1/3} \rangle / \langle N \rangle^{1/3}$.

We assume in the estimates that the PDFs of the scalar dissipation and energy dissipation are described by the same law. This assumption is confirmed experimentally by Sreenivasan, Danh, and Antonia [1977]. Consequently, formula (4.23) is valid also in the case under consideration. The value of μ in this formula, generally speaking, must be different. Using the results of the cited work, we have $\mu = 0.36$. Thus, we obtain $\langle N^n \rangle_t \approx \langle N \rangle_t^n (L/\eta)^{\mu n(n-1)/2} \sim \langle N \rangle_t^n \mathrm{Re}^{3\mu n(n-1)/8}$, where $\mathrm{Re} = \sqrt{\epsilon} L/\nu$. Since viscosity must be computed from the temperature of the combustion products, the value of Re under usual conditions rarely exceeds 10^3. Then $I_c \sim 0.81$, i.e., the fluctuations of the scalar dissipation in a given case can be neglected and, when computing different mean values, it can be assumed that $N = \langle N \rangle_t$, where $\langle N \rangle_t$ is the dissipation averaged over the turbulent fluid (Kuznetsov [1982a]).

From this estimate it follows that the deviations from the equilibrium are determined by the parameter $G = k k_0^{m-1}/\langle N \rangle_t$, which characterizes the ratio of substance supply time to the flame front to the time of the chemical reaction. The value of this parameter is usually large. As an example, let us consider the combustion of a hydrogen jet discharging from a nozzle of diameter $d = 0.05$ cm with a velocity $u_0 = 880$ m/sec. Calculations based on the system of equations (5.4) show that when $\langle z \rangle = z_s$ and $y = 0$, the scalar dissipation is equal to $\langle N \rangle_t = 0.08$ sec^{-1} and, hence, when $k = 1.4 \times 10^6$ sec^{-1}, we obtain $G = 1.7 \times 10^7$. Then, from (5.29) we have $\Delta c = c - c^{(e)} = 2.7 \times 10^{-3}$ and from the data in Figs. 5.8, 5.9 we conclude that the flame temperature T_s decreases by 200 K and the concentration of monatomic oxygen is 6 times higher than the equilibrium concentration (for $z = z_s$).

In conclusion, we once again note that formula (5.29) is valid only for a limited range of values of z and only when $|c - c^{(e)}|$ is not too large. For hydrogen the formula is true when $0.02 < z < 0.036$, and for propane when $0.045 < z < 0.08$. Thus, if the composition differs significantly from the stoichiometric, then the solution of (5.29) becomes invalid. Furthermore, with increasing scalar dissipation $\langle N \rangle_t$, the value $|c - c^{(e)}|$ rises, and at some value of $\langle N \rangle_t$ the rate of the slowest reaction in (5.18) or (5.19) becomes less than the rate of the fastest reaction in (5.20), i.e., the validity of the assumption on which relation (5.24) is obtained is violated. For hydrogen this takes place when $c - c^{(e)} \approx 10^{-2}$, and for propane when $c - c^{(e)} \approx 5 \times 10^{-3}$. Thus, at very high mixing rates formula (5.29) is also invalid.

These limitations can be relaxed, since based on the arguments which were used upon derivation of (5.26), a system of equations of a more general form is obtained (Peters [1984])

$$N \frac{d^2 c_\ell}{dz^2} - W_\ell = 0 \qquad (5.32)$$

In the representation of this system it is assumed that all the consequences of the laws of conservation of heat and mass, i.e., some independent variables are excluded and, hence, W_ℓ are known functions of c_ℓ and z. When deriving (5.32) it is not assumed that reactions (5.18), (5.19) progress with a rate much greater than reactions (5.20). The main condition of applicability of (5.32) has the previous form, i.e., the integral turbulent scale L must be much larger than the thickness of the zone of chemical reactions δ_c. In the general case, the method of calculating δ_c cannot be identified beforehand and, consequently, system (5.32) must be first integrated and then its applicability condition analyzed.

Let us consider the physical meaning of condition $L \gg \delta_c$. It evidently follows from this condition that the temperature and concentration gradients change in the reaction zone very sharply. Therefore, system (5.32) is true only in the region of sharp variation of the gradients of quantity c_ℓ. Outside this region another approach to the solution of transport equations is required. This approach is convenient to analyze in §5.6.

Condition for the Existence of the Turbulent Diffusion Flame. It is evident from physical considerations that system (5.32) can have solutions describing the combustion process. Such solutions are obtained at very large values of parameter N. Then, in (5.32) component W_ℓ can be neglected, and the solution has the form $c_\ell = A_\ell z + B_\ell (A_\ell, B_\ell$ are constants), i.e., fields c_ℓ and z are similar (only mixing takes place). Thus, the process of combustion is possible only when $N < N_{\text{cr}}$ where parameter N_{cr} can be exactly computed when solving system (5.32). This problem has not been solved as yet, and therefore we limit ourselves to the rough estimate of quantity N_{cr}.

Such an estimate can be obtained from the theory by Zel'dovich [1949], Zel'dovich and Frank-Kamenetskii [1938a, b]. It is established in the latter work that there exists a definite link between the constants of the rates of chemical reactions (these constants appear in the quantities W_ℓ) and the velocity of normal flame propagation u_n. Such an approach significantly simplifies the problem, since the characteristics of the chemical kinetics can, as a first approximation, be described with the help of just one quantity u_n whose value is known from experiments.

Another simplification is indicated in the work by Zel'dovich [1949], wherein it is established that the process of diffusion combustion is stable if the flow of fuel Q_f towards the flame front is less than some critical value Q_{fc}, and Q_{fc} coincides in order of magnitude with the flow of fuel towards a normal flame front propagating in a stoichiometric mixture. As will be seen later, comparison of quantities Q_f and Q_{fc} enables estimation of the parameter N_{cr}.

Indeed, the flow of fuel toward any diffusion flame when $z_s \ll 1$ is equal to

$$Q_f = \rho^{(b)} D^{(b)} \left| \frac{\partial z}{\partial n} \right| = \rho^{(b)} \sqrt{N D^{(b)}}, \qquad z_s \ll 1$$

where superscript b refers to the combustion temperature, and the superscript 0 to be used later refers to the initial temperature, n is the normal to the flame surface. When deriving this relation formula (5.3) is used. The flow of fuel in a normal flame propagating in a stoichiometric mixture is obviously equal to $Q_{fc} = \rho^{(0)} u_{ns} z_s$, where u_{ns} is the velocity of normal flame propagation in a stoichiometric mixture. Assuming that $\rho \sim T^{-1}$, $D \sim T^2$, the condition for the existence of the diffusion flame $Q_f < Q_{fc}$ is reduced to the form

$$N < N_{\text{cr}} = \frac{u_{ns}^2 z_s^2}{D^{(0)}} \tag{5.33}$$

where superscript 0 corresponds to the initial temperature.

A similar result was obtained by Peters and Williams [1981].

Formula (5.33) is convenient for the analysis of turbulent combustion, since the scalar dissipation N is one of the most important quantities determining the mixing process and its mean value appears practically in all turbulence models.

It must be emphasized that the nonaveraged value of the scalar dissipation appears in (5.33). In a number of cases this might, apparently, have important significance, since, as already mentioned, it follows from this that at high Reynolds numbers fluctuations of N which far exceed the mean value of the scalar dissipation are observed with high probability. Such fluctuations are of local character in the sense that large values of N are observed in the regions with small characteristic dimension. Therefore, "holes" must appear in the flame front, i.e., condition (5.33) in some regions of the flame front is not fulfilled and combustion in them is terminated. The appearance of these holes can easily be observed in the core regions of the diffusion flame.

Let us now estimate the value of N_{cr}. Note that in real cases the model used above yields only a qualitatively accurate result. This is due to two causes. First, the activation energies of the leading reactions are not sufficiently high. This leads to the fact that in a number of cases upon variation of mixture composition, i.e., the values z, the normal velocity of flame propagation reaches a maximum not at $z = z_s$, i.e., in the stoichiometric mixture whose combustion temperature is a maximum, but at some other value $z = z_m$. In order to account for this situation the quantity Q_{fc} is, naturally, estimated in a mixture with maximum value of u_m. Therefore, in (5.33) one must replace u_{ns} with u_{nm} and z_s with z_m, where u_{nm} is the maximum value of the velocity of normal flame propagation.

Second, (5.33) contains some uncertainty associated with differences in the coefficients of diffusion of the fuel and oxidant. Therefore, all the arguments presented above are valid not only for the fuel, but also for the oxidant. In the latter case relation (5.33) is also derived. However, this relation will feature the coefficient of diffusion of the oxidant in place of the coefficient of diffusion of the fuel. It is natural to select the largest of the coefficients in these estimates. Thereby the most "rigid" condition of the existence of the flame is selected. Hence, by using for the estimation of z_m, u_{nm}, and $D^{(0)}$ the experimental data presented by Lewis and Elbe [1961], Bretshnaider [1966], Dubovkin [1961], we conclude that under normal conditions in the combustion of hydrogen the parameter N_{cr} is equal to 200 sec^{-1}, and in the combustion of propane N_{cr} is equal to 30 sec^{-1}. With increasing pressure p or initial temperature $T^{(0)}$, the values of N_{cr} increase roughly as $(T^{(0)})^2 p^{0.7}$. In conclusion we note that formula (5.33) gives only the condition of existence of a stabilized diffusion flame. It can be shown that the indicated limitation is very important.

§5.5 FORMATION OF NITROGEN OXIDES IN TURBULENT DIFFUSION COMBUSTION

The main purpose of the present section is to verify the theoretical concepts developed above when investigating the formation of nitrogen oxides. Therefore, we shall make use of the best-studied experimental case — submerged diffusion plume. The theory of formation of nitrogen oxides in such a plume was developed by Buriko and Kuznetsov [1983a]. Such an investigation is also of direct practical interest.

Let us first consider the formation of NO under the conditions when radiation is low and the temperature and concentration of monatomic oxygen are in thermodynamic equilibrium. Let us consider the combustion of hydrogen. In this case the principal investigations are conducted at very high flow rates and the effect of buoyancy forces is small. Then, parameter w which is determined by formula (5.8) is independent of x, y, d, u_0, equation (5.9) which yields the mean concentration NO ($\langle c_{10} \rangle$) is linear, and the coefficients in this equation $\langle u \rangle$, v_0, ν_t, $P(z_s)$ are functions of only x, y, u_0, d. Hence

$$\langle c_{10} \rangle = w\tau F\left(\frac{x}{d}, \frac{y}{d}\right)$$

where F is the dimensionless function, $\tau = d/u_0$ (this quantity is tentatively referred to later as the residence time).

This conclusion is not confirmed by the data presented in Fig. 5.14 which show that quantity $c^0 = \langle c_{10} \rangle^{(m)}/\tau(\langle c_{10} \rangle^{(m)}$ is the maximum concentration of NO on the axis) decreases, as it should, with increasing τ. Hence, it is clear that the effects considered in §§5.3, 5.4 and which are not taken into account when deriving the relation $\langle c_{10} \rangle = w\tau F$ have fundamental significance. In the case under consideration the relation is valid, i.e., chemical nonequilibrium is determined by parameter τ. The effect of radiation is characterized by parameter q which, as follows from (5.10), (5.12)–(5.15), can be written as $q \sim \tau$. Hence it follows that c^0 is a universal function of τ.

This conclusion was reached by Buriko and Kuznetsov [1978] and, as can be seen from Fig. 5.14, is well supported by experiments. Let us consider the quantitative aspect of the problem. The formulas derived above give one the opportunity to compute the emission of nitrogen oxides. In the first stage of the calculation, system (5.4) is not used and only dependence $w(G, q)$ is found. For this purpose one specifies the values of z, G, q and solves the algebraic relations following from the laws of conservation of energy and mass of atoms H, O, and C and the conditions of equilibrium of reactions (5.18) and (5.19). Moreover, formula

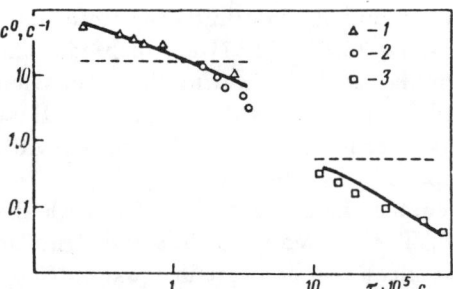

Figure 5.14 Comparison of the calculated maximum averaged concentration of NO on the axis of a submerged diffusion plume with the experimental data by different authors. *1*) hydrogen, Lavoie and Schlader [1974]; *2*) hydrogen, Bilger and Beck [1974]; *3*) propane, Buriko and Kuznetsov [1978]. Solid lines — calculation in which radiation and flame nonequilibrium are taken into account; dashed lines — calculation for adiabatic flame which is in thermodynamic equilibrium.

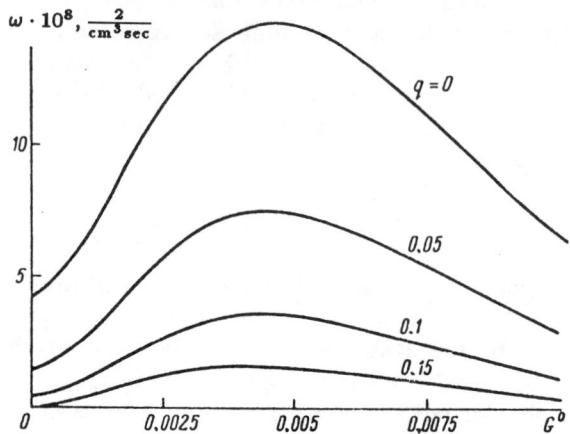

Figure 5.15 Effect of nonequilibrium and radiation in the diffusion flame on the effective oxidation rate of nitrogen in the combustion of hydrogen in air. $G^0 = G^{-1/(m+1)}, m = 2$.

(5.29) is used, and the role of radiation is taken into account by decreasing the initial enthalpy by the amount $qQ(z)$ ($Q = Q_0 z/z_s$ for $z < z_s$, $Q = Q_0(1-z)/(1-z_s)$, where Q_0 is the heat of combustion of one gram of fuel). This analysis enables finding the dependence of the oxidation rate of nitrogen from formula (5.6) as a function of z, G, q and determining function $w(G, q)$ from relation (5.8).

The results of the calculations are depicted in Figs. 5.15, 5.16, which show that the deviations from equilibrium can increase the oxidation rate of nitrogen by several times.

In the final stage the system of equations (5.4), (5.9) is numerically solved. The last equation in (5.4) describes the distribution of concentration fluctuations. One of the components in this equation, namely the one which is proportional to constant β_1, yields the value $2\langle N \rangle$. In order to find the intermittency factor and, consequently, quantity $\langle N \rangle_t = \langle N \rangle/\gamma$, formulas (3.56), (3.57) are utilized. The quantities $\langle z \rangle$ and σ entering into these formulas are found during the calculation of system (5.4). Parameter G is determined from the known value of $\langle N \rangle_t$. Heat losses by radiation in the combustion of hydrogen are given by formulas (5.10), (5.12), (5.13), and in the combustion of propane by formulas (5.10), (5.12), (5.14), (5.15). The computed values of G and q enable the computation of w with the aid of the data presented in Figs. 5.15, 5.16. The derived value of w and formulas (3.56), (3.57) permit the computation of the mean reaction rate in equation (5.9). The numerical integration of the latter allows obtaining the distribution of the mean concentration of NO with ease.

Let us now consider the results of numerical calculations. All calculations were carried out on the basis of the rate constants for the chemical reactions from the work by Jenkins, Yumlu, and Spalding [1966], i.e., $k = 1.4 \times 10^7$ sec^{-1} for hydrogen, $k = 6.3 \times 10^7$ sec^{-1} for propane. The results of the computation of the maximum concentration of NO on the axis of the plume are depicted in

Fig. 5.14, where $c^0 = \langle c_{10} \rangle^{(m)}/\tau$, $\langle c_{10} \rangle^{(m)}$ is the maximum concentration of NO on the plume axis. Figure 5.14 shows that the results of the calculations agree satisfactorily with the experimental data.

A comparison of the calculated distributions of NO with the experimental data by Lavoie and Schlader [1974] is presented in Figs. 5.17 – 5.19 (the symbols are experimental data, the curves are the calculated results, $d = 2.2$ mm, $u_0 = 200$ m/sec). Figure 5.20 shows the dependence of the calculated dimensionless flow rate of nitrogen oxides Q^0 on the distance from the nozzle. Q^0 is determined by the relation

$$Q^0 = \frac{4 \int_0^\infty \langle \rho \rangle \langle u \rangle \langle c_{10} \rangle y \, dy}{\rho_f u_0 d^2}$$

where ρ_f is the density of fuel. Since the profiles of $\langle \rho \rangle$, $\langle u \rangle$ in the tests by Lavoie and Schlader [1974] were not measured, the computed profiles of $\langle \rho \rangle$, $\langle u \rangle$ and the profiles of $\langle c_{10} \rangle$, measured by them, were used. From Figs. 5.17, 5.18 it is seen that the axial and radial distribution of $\langle c_{10} \rangle$ at small distances from the nozzle is not described satisfactorily by the present theory. At large distances from the nozzle the agreement between the theoretical and experimental data is good. Figure 5.20 also shows that the consumption of nitrogen oxides can be accurately calculated at all distances from the nozzle. A more detailed analysis of the causes behind the discrepancies indicated above will be presented somewhat later. Analogous calculations were also conducted for a submerged diffusion plume of propane. A comparison of the calculated distributions of NO with the experimental data by Buriko and Kuznetsov [1978] is presented in Figs. 5.21 – 5.24 (the symbols are for the experimental data, the curves are for the calculated results).

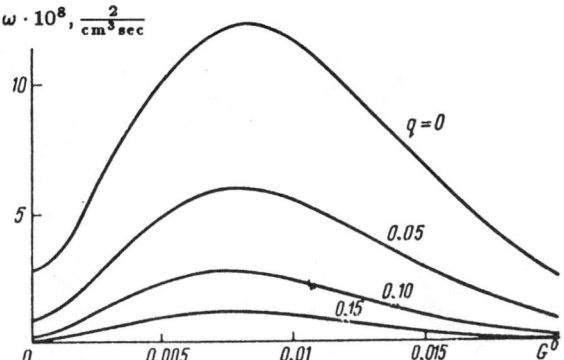

Figure 5.16 Effect of nonequilibrium and radiation of the diffusion flame on the effective oxidation rate of nitrogen in the combustion of propane in air. $G^0 = G^{-1/(m+1)}, m = 2.4$.

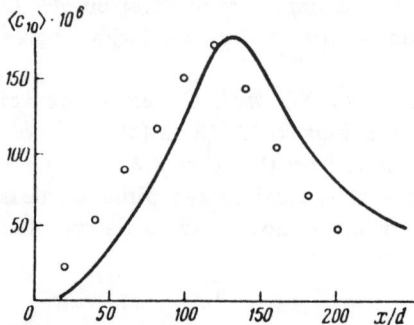

Figure 5.17 Comparison of the calculated mean concentration of NO on the axis of a submerged diffusion plume of hydrogen with the experimental data by Lavoie and Schlader [1974].

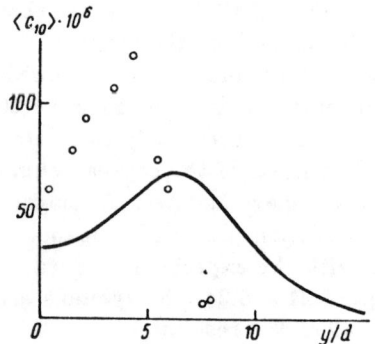

Figure 5.18 Comparison of the calculated profile of the mean concentration of NO in a submerged diffusion plume of hydrogen at section $x/d = 40$ with the experimental data by Lavoie and Schlader [1974].

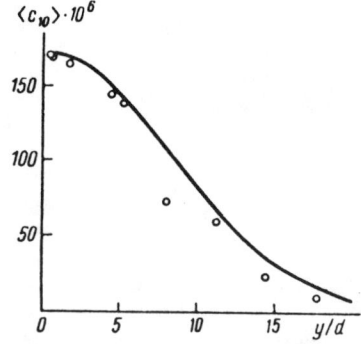

Figure 5.19 Comparison of the calculated profile of the mean concentration of NO in a submerged diffusion plume of hydrogen at section $x/d = 160$ with the experimental data by Lavoie and Schlader [1974].

TURBULENT DIFFUSION COMBUSTION 229

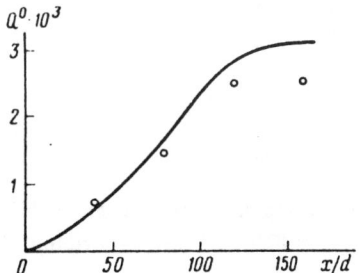

Figure 5.20 Comparison of the calculated flow rate of nitrogen oxides at various sections of a submerged diffusion plume of hydrogen with the experimental data by Lavoie and Schlader [1974].

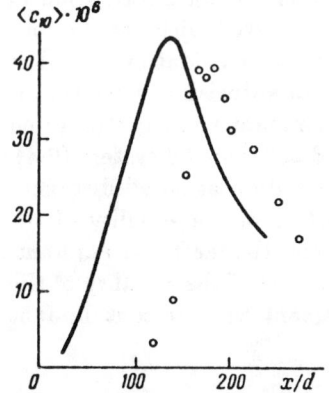

Figure 5.21 Comparison of the calculated mean concentration of NO on the axis of a submerged diffusion plume of propane with the experimental data by Buriko and Kuznetsov [1978]. $d = 3$ mm, $u_0 = 25$ m/sec.

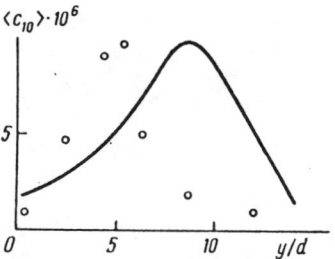

Figure 5.22 Comparison of the calculated profile of the mean concentration of NO in a submerged diffusion plume of propane at section $x/d = 22$ with the experimental data by Buriko and Kuznetsov [1978].

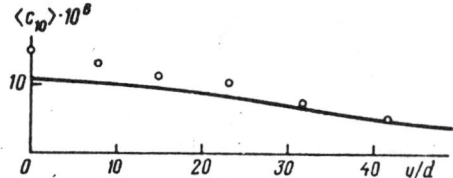

Figure 5.23 Comparison of the calculated profile of the mean concentration of NO in a submerged diffusion plume of propane at section $x/d = 270$ with the experimental data by Buriko and Kuznetsov [1978].

It is evident that all the conclusions from analyzing the combustion of hydrogen are also valid for the oxidation of propane.

Let us now consider the causes of the low accuracy of calculation of nitrogen oxide concentration at small distances from the nozzle. We first of all note that the theory in all cases yields a correct description of the total formation of nitrogen oxides, i.e., the quantity $\int_0^\infty \langle \rho W_{10} \rangle y \, dy$ is calculated with satisfactory accuracy. Hence, it follows that the mean rate of nitrogen oxidation $\langle W_{10} \rangle$ is correctly computed at all points of the plume. This is not surprising, since W_{10} is dependent only on the characteristics of the inert contaminant and they, as noted at the beginning of the chapter, are described with good accuracy by system (5.4). Hence, the reason behind the low accuracy of the calculation at small distances from the nozzle can be attributed only to the assumption of the equality of the coefficients of turbulence transport of nitrogen oxides and the inert contaminant.

Doubts were raised about the validity of the assumption of the equality of the coefficients of turbulent transport of the inert contaminant and the contaminant

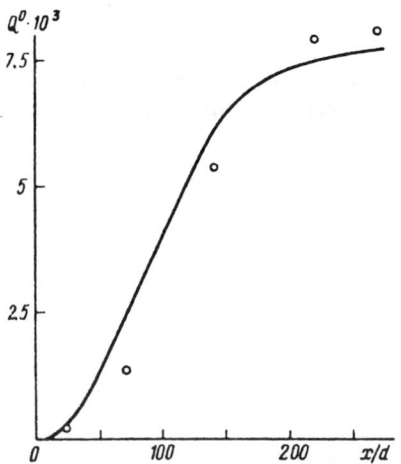

Figure 5.24 Comparison of the calculated flow rate of nitrogen oxides at various sections of a submerged diffusion plume of propane with the experimental data by Buriko and Kuznetsov [1978].

taking part in the chemical reaction in the works by Chang [1970], Vilyunov and Dik [1976] in which semi-empirical methods were developed for the description of the effect of chemical reactions on the laws of turbulent transport. Since the validity of the assumptions made in these papers is not directly tested, it is useful to indicate the direct proof of the importance of the effect under consideration. Following the example of the work by Kuznetsov [1979b], we consider the turbulent diffusion of the fuel. We assume that the reaction is single-step and irreversible and its rate is infinitely high, i.e., formula (5.3) is true. Then, the flow rate of the fuel is given by the relation

$$\langle c_f u \rangle = \int c_f(z) u P(u,z) d^3u\, dz = \int c_f(z) \langle u \rangle_z P(z) dz$$

where $P(u,z)$ the PDF of the velocity and concentration, and the conditionally averaged velocity of the medium $\langle u \rangle_z$ is determined by formula (3.16) in which the flow of the inert contaminant and the variance of concentration fluctuations σ^2 are featured. This formula allows computing the coefficient of turbulent diffusion of the fuel which, by definition, is equal to

$$-\frac{\langle c_f v \rangle - \langle c_f \rangle \langle v \rangle}{\partial \langle c_f \rangle / \partial y}$$

The calculation, based on formulas (3.16), (3.56), shows that: 1) the coefficient of turbulent diffusion of the fuel and the inert contaminant are noticeably different; 2) these differences increase when concentration fluctuations increase (Kuznetsov [1979b]. Thus, the chemical reaction really affects the process of turbulent transport of the reacting contaminant.

In the case under consideration this effect is nonessential, since the transport processes do not change the total flow rate of nitrogen oxides, and the source in equation (5.9) is independent of c_{10}. If such a dependence becomes significant (which, for example, takes place at large τ when the value of c_{10} approaches equilibrium), the total flow rate of nitrogen oxides will be dependent on their coefficient of turbulent diffusion.

In conclusion to this section we note that in the majority of the cases considered above, condition (5.30) was not fulfilled, i.e., strictly speaking, formula (5.29) yields only the correct order of magnitude. Nonetheless, considering the good agreement between calculations and experimental data, it can be assumed that formula (5.29) is true even when condition (5.30) is not fulfilled.

CHAPTER
SIX
TURBULENT COMBUSTION OF A HOMOGENEOUS MIXTURE

The combustion of a homogeneous mixture, just like the combustion of non-premixed gases, is used in a number of technical devices (internal combustion engines, afterburners of gas turbines and so on). In contrast to diffusion combustion, the combustion of a homogeneous mixture is studied with some difficulty, since the rate of chemical reactions substantially influences the characteristics of the process. This situation leads to the creation of a whole series of nontrivial effects whose significance has been understood only lately. The purpose of this chapter is to identify these effects by analyzing experimental data and obtaining a criterial description of the process.

§6.1 MAIN PROBLEMS

The theory of turbulent combustion of a homogeneous mixture adopts the main concepts and ideas from the theory of propagation of a plane (normal) flame front in a stationary mixture of fuel and oxidant. Therefore, before we state the main problems arising when investigating turbulent combustion, it is useful to consider the means for solving the problem of normal flame propagation (Zel'dovich and Frank-Kamenetskii [1938a, b]).

Let us assume that 1) one-stage reaction takes place between the fuel and oxidant, i.e., the process is described by three variables T, c_0, c_f: the temperature, concentrations of the oxidant and fuel, respectively; 2) the coefficients of molecular transport are equal and, hence, concentrations c_0 and c_f can be expressed in terms of temperature T owing to the similarity of the equations of diffusion and heat transfer. Therefore, the description of the process is reduced to the solution of a single equation

$$\rho^{(0)} u_n \frac{dc}{dn} = \frac{d}{dn} a\rho \frac{dc}{dn} + \rho W(c), \qquad c = \frac{T - T^{(0)}}{T^{(b)} - T^{(0)}} \tag{6.1}$$

where ρ is density, W is the rate of heat release, a is the thermal diffusivity, c is the dimensionless temperature, u_n is the velocity of normal flame propagation, superscripts 0 and b refer to the fresh mixture and combustion products, n is the direction normal to the flame front. The direction of the normal is determined so that $c = 0$ when $n = -\infty$ (fresh mixture), $c = 1$ when $n = \infty$ (combustion products). The boundary-value problem for equation (6.1), i.e., equality $c(-\infty) = 0$, $c(\infty) = 1$ enables the determination of u_n.

The method of solution is based on the following considerations. Usually, the rate of chemical reaction is strongly dependent on temperature, i.e., $W \sim \exp(-E/(RT))$, where parameter $E/(RT^{(b)})$ is much larger than unity (E is the activation energy, R is the universal gas constant). Therefore, two regions can be identified in the flame-thermal zone in which $0 < c < 1 - \delta c$, where $\delta c \sim RT^{(b)}/E \ll 1$, and the zone of chemical reactions in which $1 - \delta c < c < 1$. In the first zone the component in equation (6.1) containing the source is inessential, and in the second zone the convective component can be ignored which enables finding u_n in an explicit form.

The details of the procedure are given in the book by Zel'dovich, Barenblatt, Librovich, and Makhviladze [1980]. Therefore, we indicate only those results required below. In order to simplify the final relations, we assume that $\rho = \rho^{(0)} T^{(0)}/T, a = a^{(0)} (T/T^{(0)})^2$ (such an assumption enables describing the real dependences $\rho(T)$ and $a(T)$ sufficiently accurately). The first result is to the effect that in the range $(E/(RT^{(b)}) \to \infty)$ the thickness of the reaction zone tends to zero. Consequently, it must be regarded as a surface on which the temperature is continuous and its derivative undergoes a discontinuity, i.e.

$$\left[\frac{\partial c}{\partial n}\right]_{n=+0} = 0, \qquad \left[\frac{\partial c}{\partial n}\right]_{n=-0} = \frac{a^{(0)}\beta}{u_n}, \qquad \beta = \frac{T^{(b)}}{T^{(0)}} \tag{6.2}$$

The second result is formula

$$u_n = \sqrt{2a^{(0)} \int_0^1 W\, dc}, \qquad \frac{E}{RT^{(b)}} \gg 1 \tag{6.3}$$

i.e., the rate of the normal flame propagation is dependent on the coefficient of molecular transport and rate of chemical reactions.

Formula (6.3) reveals the possibility of the approximate description of the problem under arbitrary kinetics of chemical reactions, since it enables considering, instead of a large number of inexactly known parameters characterizing the rates of individual reactions, one quantity u_n which is easily and sufficiently accurately measured in experiments. The quantity u_n depends on the pressure p, on the initial temperature $T^{(0)}$, and on the excess air coefficient in the fresh charge $\alpha^{(0)}$. As an example, curve 1 in Fig. 6.1 is presented showing the experimental dependence $u_n(\alpha^{(0)})$ for gasoline-air mixture (Talantov [1975]). The data of the tests are described, to a first approximation, by the dependence

$$u_n = \int (\alpha^{(0)})(T^{(0)})^2 p^{-0.2} \tag{6.4}$$

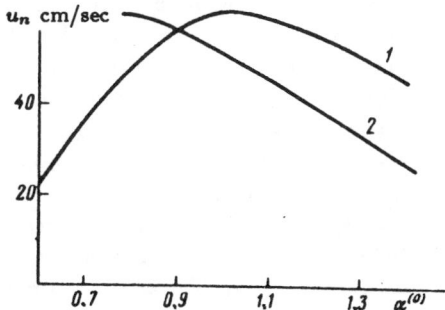

Figure 6.1 Comparison of the velocities of normal propagation of a laminar flame obtained in experiments with a plane flame (curve *1*) and calculated from the composition at the leading edge (curve *2*).

which approximates the results of measurements in gasoline-air mixtures. Function $f(\alpha^{(0)})$ can be found in Fig. 6.1.

According to these observations it will, henceforth, be assumed everywhere that $E/(RT^{(b)}) \gg 1$. This means that formula (6.2) is valid not only for a plane, but also for an arbitrary flame front. Hence, it follows that the characteristics of the process are described by the heat equation without sources. It is important that three boundary conditions be given: two appear in formulas (6.2), and one is set at an infinite distance from the reaction zone ($c = 0$). An additional boundary condition determines the position of the flame front. Thus, in addition to fluid dynamic characteristics, the process depends on $a^{(0)}$ and u_n. Hence, the rate of chemical reactions affects only the value of u_n. Since u_n is dependent on the rate of chemical reactions and on molecular transport, the effect of chemical kinetics alone is described by just one parameter, namely $\tau_c = a^{(0)}/u_n^2$. It is often convenient to consider the thickness of the thermal zone of the flame front δ_n which, according to (6.2), is determined as

$$\delta_n = \frac{a^{(0)}\beta}{u_n} \tag{6.5}$$

Let us indicate the order of the values of u_n and δ_n under normal conditions. For the majority of hydrocarbons and air mixtures (methane, propane, and so on) when the composition varies the maximum value of u_n is reached at $\alpha = \alpha_{nm}$, where α_{nm} is close to unity. In this case $u_n \approx 40$ cm/sec, $\delta_n = 0.4$ mm ($\beta = 7.5$, $a^{(0)} = 0.2$ cm^2/sec). For a hydrogen-air mixture $\alpha_{nm} \approx 0.6$, $u_n(\alpha_{nm}) \approx 250$ cm/sec, $\delta_n = 0.09$ mm ($\beta = 8$, $a^{(0)} = 0.3$ cm^2/sec). The estimates of quantities u_n and δ_n are based on the data presented in the works by Dubovkin [1961], Karpov and Severin [1977]. The calculation of quantity $a^{(0)}$ is based on the technique described by Bretshnaider [1966].

Let us now consider combustion in a turbulent flow. The main information in this case is obtained from measurements of the analogs of quantities u_n and δ_n corresponding to the velocity of propagation of the turbulent flame u_t and the length of the combustion zone δ_t. These concepts are not defined, however, as precisely as in the theory of laminar combustion. It should be remembered that quantity u_n characterizes the specific flux rate of the fresh mixture into the flame front and is equal to the ratio of the volume flow rate of the mixture to its surface area. Such a surface can be determined from equality $c = c_0 = $ const. As the estimates made above indicate, the thickness of the flame front δ_n is small in comparison with the characteristic dimension of the problem. Consequently, the areas of different isotherms $c = c_0$ differ slightly from each other. Quantity δ_t in the turbulent flow is always of the order of the characteristic dimension of the problem and, therefore, the areas of the averaged isotherm $\langle c \rangle = c_0 = $ const significantly differ.

Since there is no one preferred isotherm, the concept of the velocity of propagation of the turbulent flame cannot be defined strictly. In spite of this, the study of the effect of various parameters on the introduced value of u_t by any one method can be very useful. As will be seen later, such a study reveals relatively simply a whole series of very complex affects and qualitatively describes their interaction. Therefore, it is suggested in the present chapter that the main information on turbulent combustion be given only by quantity u_t.

Henceforth, mainly two groups of experiments are used. The first group of tests were conducted by Talantov with coworkers ([Talantov [1975], Yankovskii and Talantov [1969], Kuzin, Yankovskii, Apollonov, and Talantov [1972], Kuzin and Talantov [1977], Golubev, Yankovskii, Postnov, and Talantov [1973]). Combustion in a channel of square cross section with side d was investigated. The flame was stabilized by recesses located on opposite walls of the channel. The composition of the gasoline-air mixture, its initial temperature, pressure, inlet velocity into the channel, and the value of d were varied in the experiments. The velocity of propagation of the turbulent flame u_t was determined from the position of the forward boundary of the flame which was found by photographing the plume (it is important to bear in mind that one of the averaged isotherms $\langle c \rangle = $ const which is adjacent to the fresh mixture is referred to as the frontal boundary; the exact value of the constant is not known).

The second group of tests were conducted by Karpov and Severin [1977, 1980]. A spherical bomb was used with blades on the walls to make the mixture turbulent. The composition and type of mixture and the fluctuating velocity (normal temperature and pressure) were varied. The mixture was ignited at the center of the bomb and the dependence of the pressure on time was recorded. This enabled determining the derivative of the volume of the burned mixture with respect to time. This derivative was ascribed to the surface of a sphere whose volume is equal to the volume of the combustion products. Thus, somewhat differing definitions of u_t were used in the two groups.

Let us proceed to the analysis of these and also some other experiments. We shall first consider the effect on u_t of the two most important characteristics of turbulence — fluctuating velocity and integral scale

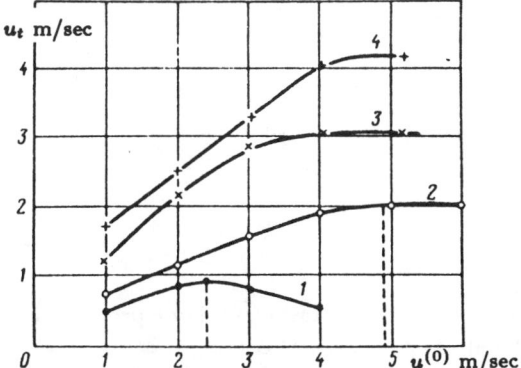

Figure 6.2 Effect of velocity fluctuations on the velocity of propagation of the turbulent flame from data by Karpov and Severin [1977], $T^{(0)} = 293$ K, $p = 0.1$ MPa. *1)* propane-air mixture, $\alpha = 0.6$; *2)* propane-air mixture, $\alpha = 0.9$; *3)* hydrogen-air mixture, $\alpha = 3$; *4)* hydrogen-air mixture, $\alpha = 2$.

$$u^{(0)} = \sqrt{\frac{1}{3}\langle(u^{(0)} - \langle u^{(0)}\rangle)^2\rangle}, \qquad L = \frac{(u^{(0)})^3}{\langle\epsilon\rangle}$$

where $\langle\epsilon\rangle$ is the energy dissipation. Figure 6.2 depicts the results of experiments by Karpov and Severin [1977]. The value of $u^{(0)}$ — the fluctuating velocity in the bomb in the absence of combustion — is plotted along the abscissa. It is evident that when the fluctuating velocity is increased, the velocity of flame propagation first increases and then remains unchanged or decreases. Note that in the majority of tests only the first section of the curves presented in this figure are observed (the section in which u_t increases with increasing $u^{(0)}$). In particular, only such dependences were revealed in the tests by Talantov and coworkers.

The effect of the turbulence scale L is illustrated in Fig. 6.3 which is from the work by Yankovskii and Talantov [1969]. It is seen that when d is increased (and hence quantity L which is proportional to the side of the channel d), the velocity of flame propagation increases. On the other hand, it is reported by Ballal and Lefebvre [1975], Ballal [1979] that when L is increased the velocity of flame propagation rises when $u^{(0)}/u_n < 2$ and decreases when $u^{(0)}/u_n > 3$. Henceforth, $\langle u_1 \rangle$ is the mean flow velocity.

These graphs indicate that both turbulence characteristics $u^{(0)}$ and L of the mixture entering the flame affect the velocity of flame propagation. These characteristics must change in the reaction zone as a result of decrease of density. This effect is illustrated in Fig. 6.4 in which the experimental results by Suzuki, Oba, Hirano, and Tsuji [1979] are presented. The combustion of propane-air mixture was investigated ($\alpha = 0.9$). A grid allowing the turbulence energy to be varied was mounted inside a tube from which the mixture discharged into the stationary surrounding air. The solid lines in the figure depict the lines of equal luminosity. The dashed line indicates the boundaries of the core of the isothermal jet. It is evident that the forward boundary of the flame is located much closer to the

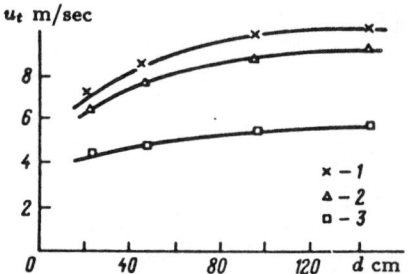

Figure 6.3 The effect of channel size on the velocity of turbulent flame propagation from data by Yankovskii and Talantov [1969]. *1)* $\alpha = 0.85$, *2)* $\alpha = 1$, *3)* $\alpha = 1.4$; $p = 0.1$ MPa, $T^{(0)} = 443$ K, $\langle u_1 \rangle = 50$ m/sec, gasoline-air mixture.

nozzle than the inner boundary of the core of the isothermal jet. As a result of change of density this effect must cause a strong restructuring of the whole flame.

The experiments indirectly indicate that the fluctuating velocity in the flame $u^{(b)}$ rises sharply. Indeed, it is clear from physical considerations that the flame cannot move relative to the medium with a velocity which is higher than the fluctuating velocity. Therefore, in estimating u_t, one can give an indirect estimate of the fluctuating velocity. For the conditions of the experiments whose results are illustrated in Fig. 6.4, we find from the position of the forward boundary of the flame that $u_t = 2.2$ m/sec, i.e., $u_t/u^{(0)} = 8$. Analogous data presented in the work under consideration indicate that $u_t/u^{(0)} = 14.5$ when $u^{(0)} = 9$ cm/sec. Assuming that $u_t \sim u^{(b)}$, it can be concluded that $u^{(b)} \gg u^{(0)}$. Large values of quantity $u_t/u^{(0)}$ were also recorded by Talantov and coworkers. Estimates show that the value of $u_t/u^{(0)}$ in these tests reached 5 (just as for the developed flow in

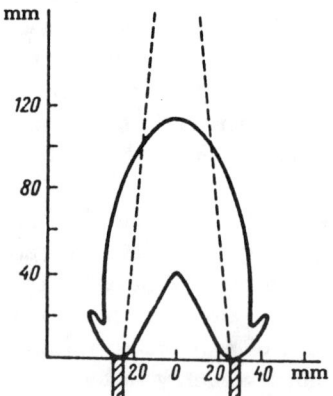

Figure 6.4 Flame contours upon combustion of propane-air mixture in a Bunsen burner from data by Suzuki, Oba, Hirano, and Tsuji [1979]. $d = 5.4$ cm, $\langle u_1 \rangle = 4.5$ m/sec, $u^{(0)} = 0.28$ m/sec, $\alpha = 0.9$.

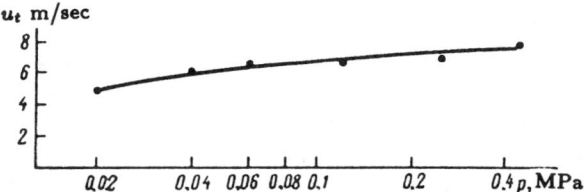

Figure 6.5 Effect of pressure on the velocity of turbulent flame propagation in a channel from data by Golubev, Yankovskii, Postnov, and Talantov [1973]. $\alpha = 1$, $T^{(0)} = 523$ K, $d = 5$ cm, $\langle u_1 \rangle = 50$ m/sec, gasoline-air mixture.

a channel, it is assumed that turbulence intensity in these tests in the incoming flow amounts to 5%). Direct measurements of the fluctuating velocity in the flame conducted by Ballal [1970] also indicate that $u^{(b)} > u^{(0)}$.

Thus, we conclude that the change of density affects the combustion characteristics substantially. This effect is described by parameter $\beta = T^{(b)}/T^{(0)}$.

The effect of pressure p on the velocity of flame propagation is illustrated in Fig. 6.5 where the data by Golubev, Yankovskii, Postnov, and Talantov [1973] are presented. It is evident that u_t increases with increasing p. Analogous results were obtained by Doroshenko and Nikitskii [1960], Khramtsov [1960]. Owing to the principle of self-similarity of turbulent flows with respect to Reynolds number, the change of p does not affect $u^{(0)}$ and L. Therefore, the previous graph is an indication of the effect of the rate of chemical reactions. The characteristic reaction time $\tau_c = a^{(0)}/u_n^2$, as seen from (6.4), decreases with increasing pressure ($a^{(0)} \sim p^{-1}, u_n \sim p^{-0.2}$). In accordance with the increase of the rate of reaction u_t also increases. This effect, however, cannot be described with the aid of quantity u_n alone. Indeed, all theories starting from Damköhler [1940] and Shchelkin [1943], predict the rise of u_t with increasing u_n which is not in agreement with the data presented in Fig. 6.5 and formula (6.4) (u_n decreases and u_t increases with increasing p). Therefore, in addition to u_n, the coefficient of molecular transport must also be featured in any theory. If it is assumed that the principle of self-similarity with respect to Reynolds number is valid when describing turbulent combustion, then it follows that the determining parameter is only some combination of parameters u_n and $a^{(0)}$, i.e., quantity $\tau_c = a^{(0)}/u_n^2$ which, as already mentioned, is independent of the processes of molecular transport and characterizes only the chemical reactions. Then, taking into account that quantities $u^{(0)}$, L, β are also determining parameters, we arrive at a formula of the type

$$u_t = u^{(0)} \varphi_1(\beta, \text{Mi}), \qquad \text{Mi} = \frac{u^{(0)} a^{(0)}}{L u_n^2} \qquad (6.6)$$

The Mihelson parameter Mi was introduced by Dunskii when analyzing flame stabilization by high-drag bodies (the results of this work are included in the books by Raushenbakh et al. [1964] and Shchetinkov [1965]). It was later encountered in the works by Klimov [1977b], Baev and Tret'yakov [1968, 1972, 1977]. In all these works somewhat different definition of quantity Mi was used ($u^{(0)}$ was replaced with the mean velocity, and the turbulence scale was replaced with a characteristic geometric dimension).

Note that formula (6.6) describes the results of all the calculations based on the semi-empirical theories of turbulence in which the principle of self-similarity with respect to Reynolds number is used. Such an approach to the solution of the problem is frequently used at the present time. As an example we can cite the works by Vulis, Ershin, and Yarin [1963], Launder and Spalding [1972], Librovich and Lisitsyn [1975], Baev, Golovichev, and Yasakov [1976], Vilyunov and Dik [1976], Bray and Libby [1976], Bray and Moss [1977a, b], Bray [1980], and others.

If one rejects the principle of self-similarity with respect to Reynolds number, i.e., one assumes that in addition to the characteristic chemical time τ_c the coefficients of molecular transport also have an influence, for example, $a^{(0)}$, then the generalized dependence (6.6) has the form

$$u_t = u^{(0)} \varphi_2 \left(\beta, \text{Mi}, \frac{u^{(0)}}{u_n} \right) \qquad (6.7)$$

A special case of this dependence is formula

$$u_t = u^{(0)} \varphi_3 \left(\beta, \frac{u^{(0)}}{u_n} \right) \qquad (6.8)$$

which follows from the flamelet combustion model (Damköhler [1940], Shchelkin [1943]) (see §1.4). It is assumed in this model that combustion takes place in a thin contorted flame front. Since its thickness is much less than the scale of fluid dynamic nonhomogeneity, it can be assumed that the flame is a surface which moves relative to the medium with velocity u_n. The characteristics of such a surface are dependent only on u_n, i.e., $a^{(0)}$ is not included in the set of determining parameters and formula (6.8) is then obtained.

On the face of it, formula (6.7) yields the most general form of the dependence of u_t on the determining parameters, since the ratios of the various coefficients of transport for the given fuel and oxidant are constant and the dependence of these coefficients on temperature can, as a first approximation, be characterized by parameter β. This conclusion, however, is not confirmed by the experimental results presented in Figs. 6.6, 6.7.

For the elucidation of these conclusions we note that the change in $\alpha^{(0)}$ affects all three parameters in (6.7). When the composition of gasoline-air mixtures is varied, all three parameters reach extreme values if $\alpha^{(0)} \approx \alpha_{nm}$. In the meantime, it is evident from Fig. 6.6 that in this case u_t has a maximum when $\alpha^{(0)} = \alpha_{tm}$, where $\alpha_{tm} < \alpha_{nm}$ (α_{nm} is the excess air coefficient in a mixture with maximum value of u_n). In the combustion of hydrogen an opposite effect is observed: $\alpha_{tm} > \alpha_{nm}$ (Fig. 6.7).

There are only two possible explanations of the indicated behavior: 1) either quantity u_n is not the only parameter characterizing chemical kinetics; 2) or as a result of the differences in the coefficients of molecular transport in the flame the composition and, consequently, the value of u_n computed from the initial excess

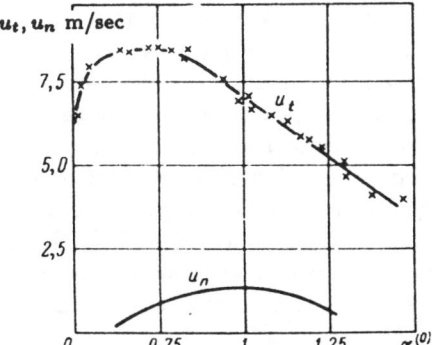

Figure 6.6 Effect of composition on the velocity of turbulent flame propagation in gasoline-air mixtures from data by Kuzin, Yankovskii, Apollonov, and Talantov [1972]. $d = 5$ cm, $\langle u_1 \rangle = 50$ m/sec, $p = 0.1$ MPa, $T^{(0)} = 493$ K.

air coefficient incorrectly reflects the role of chemical effects. There are a number of factors indicating that the second explanation is the most probable.

Indeed, it should be noted that the anomalous influence of the composition on combustion characteristics is observed not only in turbulent flow, but also in laminar flow. For example, analysis of the experimental data presented by Lewis and Elbe [1961] indicates that when the composition changes the quenching diameter, the minimum spark ignition energy, and the critical velocity gradient when the flame travels in a tube reach extreme values at excess air coefficients close to α_{tm}. On the basis of these data Baev and Tret'yakov [1968, 1977] developed a description of the process of turbulent combustion in which instead of quantity u_n the results of measurements of the critical velocity gradient in laminar flame

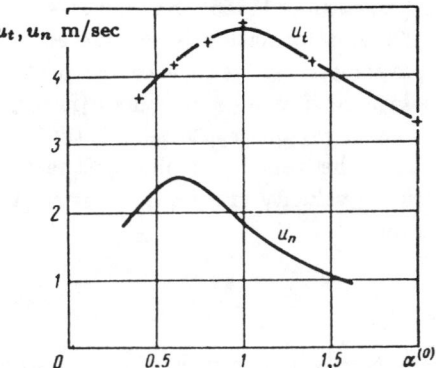

Figure 6.7 Effect of composition on the velocity of turbulent flame propagation in gasoline-air mixtures from data by Karpov and Severin [1977]. $L = 1$ cm, $u^{(0)} = 3$ m/sec, $p = 0.1$ MPa, $T^{(0)} = 293$ K. Presented in the original are the dependences of u_t on $u^{(0)}$ for variable values of α.

propagation is used. A similar approach was used later by Buriko and Kuznetsov [1976] in which data obtained from measurements of the quenching diameter and minimum spark ignition energy were included for the same purpose.

Such an approach enabled a significantly more accurate generalization of the experimental data and led simultaneously to the appearance of a whole series of theoretical problems. These problems had already become apparent in the work by Buriko and Kuznetsov [1976], wherein it was shown that the anomalous influence of the composition on the characteristics of turbulent combustion was caused by differences in the coefficients of molecular diffusion of the fuel D_f and oxidant D_0. In particular, $D_f < D_o$ for gasoline-air mixture and $D_f > D_0$ for hydrogen-air mixture. Respectively, $\alpha_{tm} < \alpha_{nm}$ in the first case and $\alpha_{tm} > \alpha_{nm}$ in the second (Figs. 6.6, 6.7).

Thus, analysis of experimental data enables the formulation of two main questions: 1) why the characteristics of turbulent combustion are dependent on the processes of molecular transport despite the small thickness of the flame front, i.e., there exist two independent determining parameters $a^{(0)}$ and u_n, 2) why the differences in the coefficients of molecular diffusion are important (such differences, as generally known, do not lead to a change in the mean composition of the combustion products)? The next section is devoted to the analysis of these questions.

§6.2 MECHANISM OF TURBULENT COMBUSTION

As noted in §1.4, the available experimental data indicate that turbulent combustion has flamelet character, i.e., the fresh mixture and products of combustion are separated by a narrow boundary layer in which a sharp rise in temperature takes place. This layer is contorted as a result of velocity fluctuations of different scales. The effect of turbulence on its internal structure is determined by a parameter known from the theory of flame stretching (Kovasznay [1956], Klimov [1963]) in which a homogeneous, vortex-free, laminar flame deformation is considered.

In order to analyze the role of such a parameter, let us turn to the problem (Klimov [1963]) in which two reaction zones located in planes $x_1 = \pm x_0$ (in region $|x_1| < x_0$ combustion products are found, and in region $|x_1| > x_0$ reactants are found) are considered. We shall assume that the velocity profile of the form $u_2 = gx_2$ (g is a positive constant characterizing velocity gradient), $u_1 = u_1(x_1)$, where u_1 is found from the continuity equation

$$\frac{d\rho u_1}{dx_1} + \rho(x_1)\frac{du_2}{dx_2} = 0 \tag{6.9}$$

The streamlines of such a flow are schematically shown in Fig. 6.8, in which the dashed lines indicate the position of the reaction zone. The solution of the problem under consideration, obtained on the assumption that $a = D_o = D_f \sim T^2$, $\rho \sim T^{-1}$, shows that the Klimov-Kovasznay parameter $K = a^{(0)}g/u_n^2$ plays a most important role.

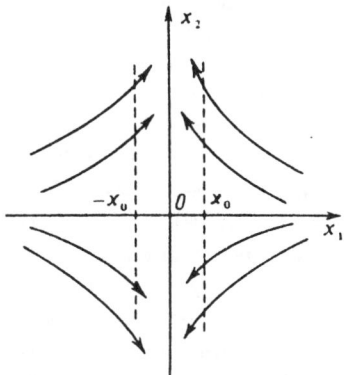

Figure 6.8 Schematic diagram of flow and position of flame front under homogeneous deformation.

When K ≪ 1 the two flame fronts are located far apart ($x_0 \gg \delta_n$) and the temperature profile in them differs from the temperature profile in the normal flame. With increasing K this profile is contorted, the distance between both fronts decreases, and when $K > \pi/2$ combustion becomes impossible.

Thus, parameter K characterizes the effect of fluid dynamic deformations on the internal structure of the flame and, hence, can be used for the analysis of the effect of the small-scale part of the turbulence spectrum on the combustion process. It must be taken into account that the velocity gradient g is a random quantity. Therefore, in further estimates we shall assume that $g = \sqrt{\langle g^2 \rangle} = \sqrt{\langle \epsilon \rangle/(15\nu^{(0)})}$. The last equality follows from the local isotropy of turbulence (Batchelor [1953]). Then, taking into consideration the definition of the turbulence scale L we obtain

$$K = \frac{a^{(0)}}{u_n^2} \sqrt{\frac{(u^{(0)})^3}{15\nu^{(0)} L}}, \qquad L = \frac{(u^{(0)})^3}{\langle \epsilon \rangle} \qquad (6.10)$$

When $a^{(0)} = \nu^{(0)}$ this criterion is proportional to the ratio of the squared Kolmogorov velocity $(\nu^{(0)}\langle \epsilon \rangle)^{1/4}$ and the velocity of normal flame propagation, or proportional to the ratio of the squared thickness of the normal flame and the Kolmogorov scale $\eta = (\nu^{(0)})^{3/4}\langle \epsilon \rangle^{-1/4}$. It is also evident that $K \sim \sqrt{Mi} u^{(0)}/u_n$, i.e., quantity K is a combination of the parameters introduced in §6.1.

Let us estimate quantity K under the conditions of the tests by Talantov and coworkers by making use of the data presented by Townsend [1956]

$$u^{(0)} = 0.05 \langle u_1 \rangle, \qquad L = 0.2d \qquad (6.11)$$

where $\langle u_1 \rangle$ is the mean flow velocity, d is the diameter of the channel. Using the graph presented in Fig. 6.1 and formulas (6.4), (6.11), we conclude that the maximum and minimum values of K in these tests are not large $K = 10^{-2}$ ($\alpha^{(0)} = 1$, $T^{(0)} = 793$ K, $p = 0.1$ MPa, $\langle u_1 \rangle = 30$ m/sec, $d = 5$ cm) and $K = 0.8$ ($\alpha^{(0)} = 1.4$, $T^{(0)} = 393$ K, $p = 0.1$ MPa, $\langle u_1 \rangle = 100$ m/sec, $d = 5$ cm).

Nevertheless, from the results of these tests it is evident that the value of α_{tm} (excess air coefficient at which u_t is a maximum) remains unchanged when all parameters are varied. Thus, the effect of the differences in the coefficients of molecular transport is manifested even in the cases when it can be expected that the effect of turbulence on the internal structure of the flame front is small.

It must be noted here, however, that these estimates were based on the characteristics of turbulence in the incoming flow. In the flame these characteristics can change which is indicated by the following considerations. When $K \ll 1$, the minimum scale of velocity fluctuations in the reacting is larger than the thickness of the flame front δ_n. This indicates that the flame can be regarded as locally planar. In gas dynamic approximation ($a \to 0$) such a flame is unstable relative to the perturbations of any wavelength (Landau [1944]). Consideration of the effects resulting from viscosity and heat conduction is presented in "Nonsteady flame propagation" edited by Markstein [1968], by Istratov and Librovich [1966a, b]. It is shown in these works that the harmonic perturbations with wavelength $\ell > \ell_{cr} \sim \delta_n$ are unstable, while perturbations with wavelength $\ell < \ell_{cr}$ are stable. These conclusions were confirmed experimentally by Petersen and Emmons [1961], who investigated the stability of a flame that is stabilized by oscillating wire. Analyzing these data reveals that

$$\ell_{cr} = c_1 \delta_n, \qquad c_1 \approx 2 \tag{6.12}$$

Flame instability causes turbulence (Landau and Lifshits [1954]). It is important that, first of all, the large-scale perturbations appear. Indeed, as shown by Landau [1944], for an infinitely thin flame front, the amplitude A of the perturbation with wavelength ℓ increases with time t according to the law

$$A = A_0 \exp\left(\frac{t}{\tau_i}\right), \qquad \tau_i = \frac{\ell}{u_n} \psi(\beta), \qquad \psi = \frac{\beta}{\beta+1}\left[\sqrt{\beta + 1 - \frac{1}{\beta}} - 1\right] \tag{6.13}$$

i.e., the amplitude increases faster the smaller the wavelength. For a flame front of a finite thickness, one more determining parameter ℓ_{cr} emerges and, hence, from dimensional considerations we obtain

$$\tau_i = \frac{\ell}{u_n} \varphi_4\left(\beta, \frac{\ell}{\ell_{cr}}\right)$$

where $\varphi_4 > 0$ when $\ell/\ell_{cr} > 1$ and $\varphi_4 < 0$ when $\ell/\ell_{cr} < 1$. It follows from these inequalities that the quantity τ_i is a minimum when $\ell = \ell_{cr}\varphi_5(\beta)$. Since $\ell_{cr} \sim \delta_n$, then the perturbations that are comparable with the thickness of the normal flame increase the most rapidly.

Such perturbations cause strong contortion of the flame as a result of which the effects caused by the differences in thermal diffusivity a, coefficients of diffusion of the fuel D_f and oxidant D_0 become important (Zel'dovich [1944], Lewis and Elbe [1961], Eckhaus [1961]). Note that at large distances from the combustion zone in the reactants and in combustion products, the heat and mass fluxes are determined exclusively by convection. Therefore, for a flame which is in the

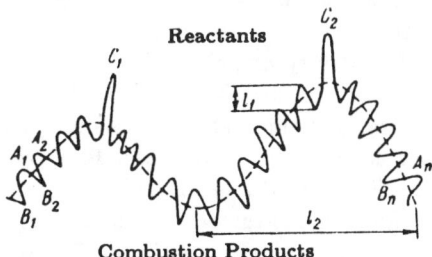

Figure 6.9 Qualitative structure of the turbulent flame.

mean plane it follows from the laws of conservation of energy and mass that the composition and temperature in the reaction zone are on the average unchanged. Therefore, the systematic effect of the differences in the coefficients of molecular transport on the combustion process is possible only when flame propagation is determined by some selected points on which the structure of the reaction zone (for example, its curvature) is unchanged. Following Zel'dovich and Frank-Kamenetskii [1947], for these points, we can consider the leading points, i.e., those points which have penetrated farthest into the fresh mixture (points $A_1, A_2, \ldots, C_1, C_2, \ldots$ in Fig. 6.9).

It is seen from Fig. 6.9 that in the vicinity of the leading points the curvature of the flame has the same sign. Therefore, when $D_o \neq D_f$ the composition of the mixture in the vicinity of these points changes systematically which, in its turn, leads to the systematic change of u_n. In the vicinity of the points under consideration, an analogy can be made with the combustion in a spherical flame inside which combustion products are found. On the basis of this analogy, it is shown that in the combustion of gasoline ($D_o > D_f$) the composition in the reaction zone is weakened (Buriko and Kuznetsov [1976], Kuzin and Talantov [1977]). In the combustion of hydrogen ($D_o < D_f$) the composition in the reaction zone is enriched (Buriko and Kuznetsov [1976]). Therefore the maximum values of u_t and u_n are attained at different excess air coefficients α_{tm} and α_{nm}. In particular, in the combustion of gasoline, the excess air coefficient in the reaction zone will be equal to α_{nm} if the excess air coefficient in the initial mixture is less than α_{nm}. Consequently, $\alpha_{tm} < \alpha_{nm}$ in the combustion of gasoline. In the combustion of hydrogen, on the contrary, the inequality $\alpha_{tm} > \alpha_{nm}$ must be satisfied.

In order to conclude this qualitative analysis of the behavior of flame propagation, one more effect which has not been considered in detail before must be indicated. Analysis of experimental data (for example, the tests mentioned earlier by Talantov and coworkers) shows that α_{tm} changes very weakly when the regime parameters are varied over a wide range of values. Hence, it can be assumed that in the vicinity of the leading points the flame front structure is universal (for example, its curvature is the same). This hypothesis was first formulated in Baev and Tret'yakov [1968, 1977] on the basis of the concept that the most intensive velocity fluctuations thrust out tongues of flame deep into the reactants, and the forward motion of such tongues continues until a critical regime for the propagation of the flame arises near the leading points.

To elucidate let us turn to Fig. 6.9. In the vicinity of points C_1, C_2, ... the structure of the flame tongue is analogous to the structure of the plane layer of the products of combustion with both sides being surrounded by reactants. The analysis conducted above showed that such a layer exists only when $K < \pi/2$, i.e., sufficiently strong fluid dynamic deformations lead to termination of flame propagation. A similar effect arises when the flame is contorted. For example, in the vicinity of points A_1, A_2, ... in Fig. 6.9 an analogy can be drawn between the structure of the flame tongue and the structure of a stationary spherical flame inside which combustion products are located. It is shown by Buriko and Kuznetsov [1976] that the radius of such a flame r cannot be less than the thickness of the normal flame δ_n. In both cases the regime in which $K = \pi/2$ or $r = \delta_n$ is referred to as critical, i.e., flame propagation becomes impossible.

Summing the results of this analysis, we arrive at the conclusion that the propagation of the turbulent flame is determined by the following processes: 1) by the leading points; 2) by flame instability; 3) by the appearance in the vicinity of the leading points of a regime which is critical for flame propagation; 4) by the change in the composition and temperature in the leading points as a result of the differences in the coefficients of molecular transport. This scheme was formulated in the work by Kuznetsov [1982b].

As an example of this work let us investigate with more detail some "elementary" processes affecting combustion and analyze their interaction. First of all, note that in addition to the three earlier indicated parameters, at least two more new ones should also be considered: D_o/D_f and a/D_o. Despite the fact that these parameters are constant for the same fuel and oxidant, the analysis of their effect on the velocity of flame propagation is necessary. Since it is assumed that u_t is determined by the motion of the leading points and that the structure of the flame near these points is universal, the effect of the differences in the coefficients of molecular transport, as a first approximation, can be taken into account with the aid of the following example.

We shall assume that the dependence of the composition and temperature of combustion on the excess air coefficient in the initial mixture $\alpha^{(0)}$ near the leading points can be computed by considering the critical regimes which arise either upon the contortion of a spherical flame or under homogeneous deformation of a plane layer of combustion products. Such a calculation is conducted in §6.3. It yields formulas connecting the excess air coefficient in the leading point $\alpha^{(b)}$ with the initial excess air coefficient $\alpha^{(0)}$. In these formulas the coefficients of molecular transport are featured. It is then assumed that quantity u_n, which appears in all the relations that will be derived later, must be computed not from the initial composition but from the composition of the leading points.

Such a procedure enables us to limit ourselves to the analysis of dependence (6.7) in which only three parameters appear. The next step is the investigation of two last cases: $K \ll 1$ and $K \gg 1$. The first is considered in §§6.5, 6.6, and the second in §6.7. In the first case, instead of quantities β, Mi, $u^{(0)}/u_n$ it is convenient to consider the equivalent system of criteria β, K, $a^{(0)}/(u_n L)$. The significance of the last two parameters is small. However, only one of them (K) can be excluded from consideration.

In order to explain this conclusion, it should be remembered that parameter K describes the effect of fluid dynamic deformations on the internal flame structure and, in accordance with formula (6.10), it is computed from the gradient of the fluctuating velocity in the initial reactants. Since an additional small-scale turbulence is formed in the flame, the definition of criterion K ought to be reconsidered. Accordingly, we introduce quantity $K^{(b)} = g^{(b)} a^{(0)}/u_n^2$, where $g^{(b)}$ is the characteristic value of the velocity gradient in the flame. Moreover, symbol K corresponds as before to the quantity determined by formula (6.10). The characteristics of flame-generated turbulence are dependent on u_n, β, and ℓ_{cr}. When $K \ll 1$, the minimum scale of the perturbations in the flow into the flame is much larger than ℓ_{cr} and, hence, the characteristics of turbulence in the reactants cannot be included in the set of determining parameters. Consequently, $g^{(b)} = u_n F(\beta)/\ell_{cr}$, i.e., $K^{(b)}$ is dependent only on β. This means that if $K \ll 1$, then when K is varied the degree of deformation and contortion of the flame does not change, i.e., parameter K is not a determinant.

The parameter $a^{(0)}/(u_n L)$ cannot be excluded from consideration since, otherwise, the dependence of u_t on the scale and pressure cannot be explained (Fig. 6.3, 6.5).

Thus, two determining parameters remain: β and $a^{(0)}/(u_n L)$. Analysis of the characteristics of flame generated turbulent (§§6.5, 6.6) enables combining both parameters into one. This parameter is used in §6.6 for analyzing the experimental data.

§6.3 EFFECT OF DIFFERENCES IN THE COEFFICIENTS OF MOLECULAR TRANSPORT ON THE FLAME STRUCTURE AT THE LEADING POINTS

On the example of the program formulated at the end of §6.2, let us consider two model problems for finding u_n at the leading points. In one of these problems the critical deformation of a plane layer of combustion products is analyzed, and in the other the critical propagation regime of a spherical flame is analyzed. The formulation of the first problem is due to Klimov [1963, 1977a], who provides a solution for $a = D_o = D_f$. A somewhat more general case ($a \neq D_o = D_f$) is discussed by Gremyachkin and Istratov [1972]. The solution of the first problem, obtained by Kuznetsov and Sabel'nikov [1977] when $a \neq D_o \neq D_f$ is given below.

Assume that the specific heat is constant, the density ρ and coefficient of transport are dependent only on the temperature and pressure, $\rho \sim 1/T$, $a \sim T^2$, $D_o \sim T^2$, $D_f \sim T^2$. We shall also assume that the Zel'dovich - Frank-Kamenetskii approximation is valid, i.e., the zone of chemical reactions is a surface on which the derivatives of temperature T, concentrations of the fuel c_f and oxidant c_o have a discontinuity (this assumption was discussed at the beginning of §6.1). The schematic diagram of the flow is shown in Fig. 6.8 in which two reaction zones (fresh mixture is found in the region $|x_1| > x_0$) are depicted by the dashed lines. The component of the velocity of the medium parallel to the reaction zone has the form $u_2 = gx_2 (g > 0)$, and the normal component depends only on x_1 and is determined by equation (6.9). Outside the reaction zone the main equations have the form

$$\rho u_1 \frac{dT}{dx_1} = \frac{d}{dx_1} a\rho \frac{dT}{dx_1}$$

$$\rho u_1 \frac{dc_f}{dx_1} = \frac{d}{dx_1} D_f \rho \frac{dc_f}{dx_1}$$

$$\rho u_1 \frac{dc_o}{dx_1} = \frac{d}{dx_1} D_o \rho \frac{dc_o}{dx_1}, \qquad |x_1| > x_0 \tag{6.14}$$

Let us introduce the Dorodnitsyn variable (see, for example, Schlichting [1960])

$$\xi = \frac{\int_0^{x_1} \rho x_1 dx_1}{\rho^{(0)}}$$

Taking into account that $\rho \sim 1/T$, $D_o \sim D_f \sim a \sim T^2$, from (6.9), (6.14) we obtain

$$-g\xi \frac{dT}{d\xi} = a^{(0)} \frac{d^2 T}{d\xi^2}$$

$$-g\xi \frac{dc_f}{d\xi} = D_f^{(0)} \frac{d^2 c_f}{d\xi^2}$$

$$-g\xi \frac{dc_o}{d\xi} = D_o^{(0)} \frac{d^2 c_o}{d\xi^2} \tag{6.15}$$

The superscript 0 here and superscript b used later correspond, as before, to the states in the reactants and in the reaction zone respectively.

Let us, for the sake of definiteness, assume all the fuel in the reaction zone is burned. Then, the boundary conditions assume the form

$$c_o = c_o^{(0)}, \qquad c_f = c_f^{(0)}, \qquad T = T^{(0)}, \qquad |x_1| = |\xi| = \infty$$

$$c_f = 0, \qquad \text{St } D_f^{(0)} \left|\frac{dc_f}{d\xi}\right| = D_o^{(0)} \left|\frac{dc_o}{d\xi}\right|$$

$$Q D_f^{(0)} \left|\frac{dc_f}{d\xi}\right| = a^{(0)} \left|\frac{dT}{d\xi}\right|, \qquad |\xi| = \xi_0 \tag{6.16}$$

Here, $\pm\xi_0$ are coordinates of the reaction zone, St is the stoichiometric coefficient, $Q = (1 + \text{St})\, T_s$ is the calorific value of the fuel divided by the specific heat, T_s is the temperature of the adiabatic combustion of the stoichiometric mixture. The first three conditions in (6.16) yield the characteristics in the reactants and the last two follow from the laws of conservation of mass and energy. Coordinate ξ_0 can be found from the additional condition (6.2) connecting the heat flux from the reaction zone which is subjected to deformation with the heat flux in the normal flame front. In the given case, this condition must link quantities ξ_0 and g. This connection, however, is not of particular interest, since the critical regime in which $\xi_0 \to 0$ is being investigated. Thus, a solution is sought for problem (6.15), (6.16) when the magnitude of g is so large that both reaction zones merge into a single zone. This problem is easily solved. We have

$$T^{(b)} = T^{(0)} + T_s(1 + \text{St})c_f^{(0)}\sqrt{\frac{D_f^{(0)}}{a^{(0)}}}$$

$$c_o^{(b)} = c_o^{(0)} - \text{St}\, c_f^{(0)}\sqrt{\frac{D_f^{(0)}}{D_o^{(0)}}}, \qquad \alpha^{(b)} > 1 \qquad (6.17)$$

where $\alpha^{(b)}$ is the excess air coefficient in the reaction zone.

If all the oxidant is consumed in the reaction zone, then we similarly obtain

$$T^{(b)} = T^{(0)} + T_s\left(1 + \frac{1}{\text{St}}\right) c_o^{(0)}\sqrt{\frac{D_o^{(0)}}{a^{(0)}}}$$

$$c_f^{(b)} = c_f^{(0)} - \frac{1}{\text{St}} c_o^{(0)}\sqrt{\frac{D_o^{(0)}}{D_f^{(0)}}}, \qquad \alpha^{(b)} < 1 \qquad (6.18)$$

The last relations in (6.17), (6.18) determine the excess air coefficient in the reaction zone. Let $\alpha^{(b)} > 1$. The second formula in (6.17) determines the concentration of the oxidant following the completion of the reaction. This concentration is, evidently, equal to $\text{St}\, (\alpha^{(b)} - 1)/(1 + \alpha^{(b)}\, \text{St})$. A similar relation can be found from (6.18) when $\alpha^{(b)} < 1$. Taking into account that usually $\text{St} \gg 1$, we have

$$\alpha^{(b)} = \alpha^{(0)}\sqrt{\frac{D_o^{(0)}}{D_f^{(0)}}}, \qquad \alpha^{(b)} > 1$$

$$\alpha^{(b)} = \frac{\alpha^{(0)}\sqrt{D_f^{(0)}/D_o^{(0)}}}{\alpha^{(0)}\left(\sqrt{D_f^{(0)}/D_o^{(0)}} - 1\right) + \sqrt{D_f^{(0)}/D_o^{(0)}}}, \qquad \alpha^{(b)} < 1 \qquad (6.19)$$

Hence, it is evident that when $D_o^{(0)} > D_f^{(0)}$, the composition in the reaction zone is reduced compared with the composition of the reactants. This case corresponds to the combustion of the majority of hydrocarbons (propane, gasoline, and so on). In the combustion of hydrogen and methane we have $D_o^{(0)} < D_f^{(0)}$, i.e., the composition in the reaction zone is, on the contrary, enriched.

From the first relations in (6.17), (6.18) it follows that, generally speaking, the combustion temperature in the reaction zone does not coincide with the temperature of adiabatic combustion of a mixture with the same composition as in the reaction zone. In particular, for hydrogen we have $a^{(0)} > D_o^{(0)}$, and the temperature in the reaction zone is lower than the temperature of adiabatic combustion of a mixture with an excess air coefficient equal to $\alpha^{(b)}$. For the majority of hydrocarbons it can be assumed with acceptable accuracy that $a^{(0)} = D_o^{(0)}$ and, hence, the temperature in the reaction zone coincides with the temperature in the normal flame propagating in a mixture whose excess air coefficient is equal to $\alpha^{(b)}$. Therefore, for the determination of u_n we can use the experiments in which a plane flame in a flow without deformation was investigated. Moreover, only the composition ought to be recalculated from formula (6.19). This procedure is inapplicable for hydrogen. However, even in the case under consideration formula (6.19) is valid for estimates since the differences between D_o and D_f are larger than the differences between D_o and a.

Let us now analyze the quantitative aspect of the problem. Since flame propagation is determined by the leading points, and these points are found in the critical regime, then upon variation of $\alpha^{(0)}$ the maximum value of u_t is reached when $\alpha^{(0)} = \alpha_{tm}$, i.e., when quantity $u_n(\alpha^{(b)})$ becomes maximum. Using (6.19) we conclude that the link between α_{tm} and α_{nm} is described by expression $\alpha_{tm} = \alpha_{nm}\sqrt{D_f^{(0)}/D_o^{(0)}}$. For gasoline-air mixture we have $\alpha_{nm} = 1$ (Talantov [1975]), $D_o^{(0)} = 0.19$ cm^2/sec (Dubovkin [1961]), i.e., $\alpha_{tm} = 0.7$ which agrees well with the experimental data of Kuzin, Yankovskii, Apollonov, and Talantov [1972], Golubev, Yankovskii, Postnov, and Talantov [1973], Yankovskii and Talantov [1969]. For hydrogen-air mixtures $\alpha_{nm} = 0.6$ (Karpov and Severin [1977]), $D_f^{(0)} = 0.6$ cm^2/sec, $D_o^{(0)} = 0.2$ cm^2/sec (Bretshnaider [1966]) i.e., $\alpha_{tm} = 1$ which agrees well with the experimental data by Karpov and Severin [1977]. The relation found between α_{nm} and α_{tm} correlates qualitatively with the data obtained on the combustion of ethane $(D_o > D_f)$ and methane $(D_o < D_f)$. As established by Karpov and Severin [1980], $\alpha_{tm} < \alpha_{nm}$ in the first case and $\alpha_{tm} > \alpha_{nm}$ in the second. However, the differences between D_o and D_f for these fuels and, accordingly, between α_{tm} and α_{nm} are not large, which makes qualitative comparisons difficult.

The effect of the differences in the coefficients of diffusion of the fuel and oxidant on the velocity of normal flame propagation is illustrated in Fig. 6.1. Curve 1 corresponds to the experimental data for plane flame presented by Talantov [1975]. Curve 2 is obtained by transforming curve 1, i.e., the composition in the leading points was recalculated by formula (6.19). It is seen that the effect of the differences in the coefficients of molecular transfer is very significant.

The analysis conducted above pertains to the case when the formation of the critical regime is caused by stretching of the flame. In an exactly similar manner we can investigate another case when the appearance of the critical regime is caused by flame contortion. As already indicated an analogy can be drawn in this case between combustion in the vicinity of the leading points and combustion in a stationary spherical flame surrounding combustion products. This problem is considered in the work by Buriko and Kuznetsov [1976]. It is shown that in the critical regime (the flow power is equal to zero, the radius of the reaction zone is minimum) formulas (6.19) are retained if the quantities D_o, D_f, a entering into them are replaced with D_o^2, D_f^2, a^2, i.e., the effect under consideration is expressed even more strongly. Since formulas (6.19) yield values of α_{tm} that agree with experiments, it can be assumed that it is flame stretching that leads to the critical conditions.

§6.4 SPECTRAL REPRESENTATION OF THE VELOCITY OF FLAME PROPAGATION

As repeatedly indicated, turbulence and, hence, turbulent combustion are essentially multiscale processes. For the description of this feature a structure function and its Fourier transform, called spectral energy density, is introduced. Upon analysis of the combustion process it is useful to introduce a similar quantity which would characterize the role of contortions of the flame with various scales.

The first of such an attempt apparently belongs to Povinelli and Fuchs [1962], who assumed that the flame front surface x_1, (x_2, x_3, t) is single valued and simply connected and quantity x_1 is a stationary random function of x_2 and x_3. This assumption enabled use of a mathematical technique, which was developed in the theory of turbulence, to describe the process. However, in the general case such an approach is essentially unsuitable, since the flame front is not a simply connected and single valued surface. Intuitive considerations of the spectral representation of the velocity of flame propagation were used by Kuznetsov [1977a], Klimov [1977a], Zimont and Sabel'nikov [1975b], Zimont [1979] for the development of various turbulent flame models.

A rigorous definition is given by Kuznetsov [1980]. It is based on averaging of random quantities over regions with finite volume as introduced by Kolmogorov [1962a, b] and Obukhov [1962]. Such a procedure has already been employed in §5.4. It enables smoothing considerably the small-scale details of the process. As the example of this approach, let us introduce quantity

$$U(\ell) = \max[\overline{W}(\ell)], \qquad \overline{W} = \frac{1}{4\rho^{(0)}\ell^2} \int_\omega \rho W d^3 x \qquad (6.20)$$

where W is the rate of heat release, ω is a cube with the center at point x and a side of 2ℓ. Since the size of the cube is finite, $\overline{W}$ is a random function which is dependent on t, x, ℓ. Quantity U is the maximum value of $\overline{W}$ in all realizations of the process. Therefore, U is a nonrandom function.

Let us explain the meaning of these quantities. It is apparent that upon integration of the rate of heat release over a region with dimension ℓ, all the details of the process dependent on perturbations whose dimension is much less than ℓ, are excluded from consideration. For example, if there exists in the flame only two perturbations with strongly differing scales $\ell_2 \gg \ell_1$, then the structure of the reaction zone has the form depicted schematically in Fig. 6.9. If $\ell_2 \gg \ell \gg \ell_1$, then quantity $\overline{W}$, which is regarded as a coordinate function, is different from zero only in a narrow range with the characteristic dimension ℓ. This region is found near the surface $\overline{W}(x,t) = U(x,\ell)$ which is represented in Fig. 6.9 by the dashed line. The surface under consideration is referred to as a partially smoothed flame surface. Owing to (6.2) we obtain from (6.20)

$$W = \frac{u_n S^{(b)}(\ell)}{4\ell^2} \tag{6.21}$$

where $S^{(b)}(\ell)$ is the area of that part of the reaction zone which is enclosed in cube ω (the amount of heat released in cube ω is equal to the area of the reaction zone multiplied by the heat flux from this zone).

Let us consider two limiting cases: $\ell \to 0$ and $\ell \to \infty$. In the first zone the reactions can be regarded as plane. Therefore, $\overline{W}$ is maximum if the reaction zone is a diagonal plane of the cube. Hence, from (6.21) we obtain

$$U(\ell) = u_n \sqrt{2}, \qquad \ell \to 0 \tag{6.22}$$

Now let $\ell \to \infty$. Consider a flame which is, on the average, planar and stationary moving in direction 1. Since the process is statistically homogeneous in directions 1 and 3, then

$$\langle W \rho \rangle = \lim_{\ell \to \infty} (2\ell)^{-2} \int_{-1}^{\ell} \int_{-1}^{\ell} \rho W \, dx_2 dx_3$$

$$\overline{W} = \frac{1}{\rho^{(0)}} \lim_{\ell \to \infty} \int_{-1}^{\ell} \langle W \rho \rangle dx_1 \tag{6.23}$$

Let us average equation (2.1) and integrate the relation over all x_1. Taking into account that $c = 1$, $\langle u \rangle = \beta u_t$ in the region occupied by combustion products, we obtain from (6.23)

$$U(\infty) = u_t \tag{6.24}$$

Note now that for a flame which is, on the average, planar, the integral $\int_\omega \rho W d^3 x$ increases with increasing ℓ not more slowly than ℓ^2 (this follows from statistical homogeneity in the directions 2 and 3). Hence, U is a nondiminishing function of ℓ. Thus, from (6.22), (6.24) we see that with increasing ℓ function $U(\ell)$ rises monotonically from a value of the order of u_n to u_t. Hence, $U(\ell)$ describes the contribution made to the velocity of flame propagation by all the perturbations whose dimensions do not exceed ℓ in order of magnitude. This function characterizes also that amount of reactants which is converted per unit time by a unit surface area of the partially smoothed flame. The derivative $\partial U/\partial \ell$ is, evidently, non-negative. It describes the contribution made to the combustion rate by the perturbations with dimensions of the order ℓ. This quantity is, naturally, referred to as the spectral representation of the velocity of flame propagation.

The definitions presented above can be used in arbitrary and nonhomogeneous turbulent flow. In this case quantity $U(\ell)$ will depend on ℓ and x. It must be taken into account that when analyzing combustion in a bounded flow, there exists some maximum possible flame perturbation scale L_m. For example, in combustion in a tube of diameter d, $L_m = d/2$. Similarly, in a Bunsen burner we obtain $L_m = d/2$, d is the diameter of the burner (the flame cannot penetrate the boundary of the combustible mixture jet, and the characteristic dimension of this jet is of the order of d). Therefore formula (6.24) assumes the form $u_t = U(L_m)$. Note also that if $K \ll 1$, then there are no wrinkles on the flame front with a dimension less than ℓ_{cr} and, therefore, from formula (6.22) we obtain the estimate $U(\ell_{\mathrm{cr}}) \sim u_n$. Thus from (6.22), (6.24) we have

$$U(\ell_{\mathrm{cr}}) \sim u_n, \qquad U(L_m) \sim u_t \tag{6.25}$$

§6.5 EFFECT OF FLAME INSTABILITY ON THE TURBULENT COMBUSTION OF A HOMOGENEOUS MIXTURE

As already indicated, the plane flame front is unstable. Two mechanisms of instability are possible. The first is caused only by the decrease in density upon combustion (Landau [1944]). The second is associated with the difference in the thermal diffusivity and in the diffusion coefficients (Zel'dovich and Barenblatt [1959], Barenblatt, Zel'dovich, and Istratov [1962]). When $a < D$ this mechanism causes flame instability even if $\beta = 1$. In the present book only fluid dynamic (thermal) flame instability will be considered, since this describes the main features of the problem.

At the present time, the behavior of a flame within the framework of the linear theory of stability is well studied. The main results of this theory are stated in the book "Nonsteady flame propagation" edited by Markstein [1968] and in the review by Istratov and Librovich [1966b].

It is established that the plane flame is unstable with respect to harmonic perturbations with a period (scale) ℓ which satisfies the condition $\ell > \ell_{\mathrm{cr}}$. It is essential that quantity ℓ_{cr} be very small. However, one often observes in experiments a stable laminar plume with a characteristic dimension far greater than ℓ_{cr}. The stability of such a plume is apparently associated with two circumstances. First, an important role can be played by the global flame configuration. For example, a spherical flame is more stable than a plane flame (Istratov and Librovich [1966c]). Second, the development of flame instability can be considerably dependent on the amplitude of the initial perturbations. The strong effect of the amplitude of such perturbations is quite clearly observed when analyzing transitions of turbulence in an incompressible fluid. For example, in a tube with smooth walls laminar flow can be observed up to a Reynolds number of 2×10^5, i.e., by two orders of magnitude higher than the critical value (see, for example, Schlichting [1960]). The effect of the initial perturbations during combustion on the development of instability is, apparently, expressed even more strongly. As an example we cite the experiments, already mentioned, by Susuki, Oba, Hirano, and Tsuji [1979]. The estimates carried out in §6.1 showed that the fluctuating velocity in the reactants in this case is less than u_n and much less than u_t ($u_t = 1.4$ m/sec when $u^{(0)} = 9$ cm/sec, $u_t = 2.2$ m/sec when $u^{(0)} = 28$ cm/sec, $u_n = 40$ cm/sec). It is evident from such estimates that the change of the fluctuating velocity, which is small compared with u_t, leads to a large change in the position of the front flame boundary.

A natural question arises concerning the character of the flow and combustion following loss of stability of the flame front. It is well known that in incompressible fluid flows, the appearance of turbulence following loss of stability is far from being immediate. With increasing Reynolds number a whole series of ordered stationary and nonstationary regimes appear and only when Reynolds number becomes sufficiently high do stochastic velocity fluctuations appear. Similarly, following loss of stability in the flame, ordered regimes of flow and combustion can also be formed, for example, cellular flames (see the book "Nonsteady flame propagation" edited by Markstein [1968]).

It can be assumed that in this case the amplitude and character of the initial perturbations also play an important role. To elucidate let us turn to tests by Ballal [1979], in which the combustion of propane-air mixture was investigated in a channel with a grid at the entrance. In these tests the effect of flow acceleration during combustion is excluded, since the channel walls were divergent. The

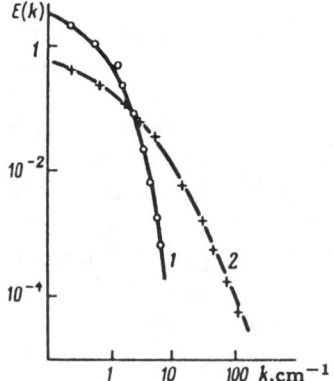

Figure 6.10 Spectral density of turbulence energy in a flow behind a turbulence grid from data by Ballal [1979]. The units of measurements along the ordinate axis are arbitrary. $p = 0.02$ MPa, $T^{(0)} = 293$ K; *1*) with combustion, *2*) without combustion.

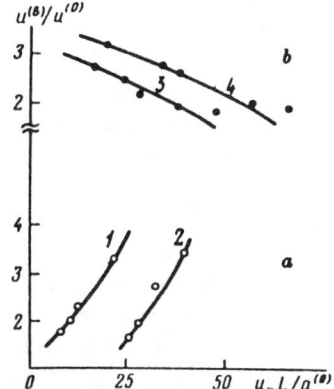

Figure 6.11 Effect of the scale of the initial turbulence in the incoming flow on the energy generated by the turbulent flame from data by Ballal [1979]. a) 1) $u^{(0)}/u_n = 0.6$, 2) $u^{(0)}/u_n = 1.5$; b) 3) $u^{(0)}/u_n = 3.3$, 4) $u^{(0)}/u_n = 7.6$.

mixture was ignited by a pilot flame. The turbulence energy and its spectrum were measured on the flame axis by means of laser Doppler anemometer.

The initial perturbations were random, since the flow beyond the grid was turbulent. From the experimental data presented in the work, it follows that the flame is turbulent, and the turbulence energy in the flame is greater than in the incoming stream. The first conclusion is evident from Fig. 6.10 which shows the dependence of the spectral energy density E on the wave number k. It is seen that the spectrum of velocity pulsations is continuous, i.e., there are no discrete perturbations in the flame. Two regions can be identified — short-wave and long-wave regions. In the long-wave region ($k < k_{cr}$) the spectral turbulence energy density in the flame is not greater than in the incoming stream (k_{cr} is defined as the abscissa of the intersection of curves 1 and 2 in Fig. 6.10). In the short-wave region ($k > k_{cr}$), an opposite picture is observed which is associated with the great increase in the viscosity of the combustion products. Simple estimations indicate that the values of quantities k_{cr} and ℓ_{cr} practically coincide (formula (6.12) is used). Based on these estimates it can be assumed that the increase in the spectral energy density in the long-wave region is associated with the instability of the flame. Illustrated in Fig. 6.11a is the effect of the turbulence scale L on the ratio of fluctuating velocities in the flame and in the incoming stream. It is evident that $u^{(b)} > u^{(0)}$, and when scale L is increased, the turbulence energy in the flame increases* as expected from the preliminary analysis conducted in §6.2.

Thus, it can be assumed that when small random perturbations act on the flame, a turbulent regime of flow and combustion appears. On the basis of the analogy with the appearance of turbulence in flows of imcompressible fluid, we

---

*In the work under consideration it is also established that for sufficiently large values of $u^{(0)}/u_n$ the turbulence energy in the flame drops with increasing L (see Fig. 6.11b). The results of these experiments are conveniently explained in §6.7.

also assume that the characteristics of the turbulence generated by the flame are independent of the amplitude of the initial perturbations if $L_m/\ell_{cr} \to \infty$. It will be shown below that the turbulence energy generated by the flame increases without restriction when $L_m/\ell_{cr} \to \infty$. The rate of this rise is however very small. Therefore, this assumption does not contradict the remarks made earlier concerning the effect of the initial perturbations on the development of flame instability.

Let us elucidate the course of further discussions. We first consider a flame, which is plane on the average, propagating in unbounded flow $(L_m = \infty)$. Let the coefficients of molecular transport be equal, and the turbulence energy in the fresh mixture very small $(u^{(0)} \ll u_n)$. For the analysis of the nonlinear stage of development of flame instability, we use the spectral representation of the velocity of flame propagation, introduced in §6.4. Let us formulate the similarity hypothesis for this representation and then show how the results can be used to describe the general case.

The process under consideration is nonstationary. Indeed, let us assume the opposite, i.e., assume the velocity of flame propagation u_t and combustion zone thickness δ_t tend to constant values when $t \to \infty$. Let us consider stability with respect to very large-scale perturbations $\ell \gg \delta_t$. Then, the flame can be regarded as a discontinuity of density and, hence, all the corollaries of the linear theory of stability, developed by Landau [1944], are valid. In this case, it suffices to replace u_n with u_t in all the formulas of this theory. Therefore, the amplitude of the perturbations under consideration increases with time which leads to the increase of u_t and δ_t. However, the nonstationary state of the process affects only the characteristics of large-scale perturbations whose amplitudes, in accordance with the linear theory of stability, increases more slowly, the larger the length of the perturbation. With increasing time the amplitude of the perturbation with a fixed wavelength (scale) must reach the stationary level as a result of nonlinear effects.

This discussion enables identifying the parameters on which function $U(\ell)$, derived in the previous section, is dependent. It is evident that quantities u_n, $a^{(0)}$, ℓ, β, x_1 (x_1 is measured from the isotherm $\langle c \rangle = 1/2$) must be among these parameters. When $t \to \infty$ time is not one of these parameters, since by definition $U(\ell)$ is independent of nonstationary large-scale perturbations. When the initial perturbation energy is low, the process is totally determined by the flame-generated turbulence. The characteristics of this turbulence must depend only on the function under study $U(\ell)$. Hence, the above parameters exhaust the list of quantities on which function $U(\ell)$ is dependent. Therefore, from dimensional considerations we obtain

$$U = u_n V\left(\beta, \frac{u_n x_1}{a^{(0)}}, \frac{u_n \ell}{a^{(0)}}\right), \qquad t \to \infty, \qquad |x_1| \lesssim \delta_t, \qquad \ell \ll \delta_t$$

where V is some dimensionless function.

Since $\delta_t \to \infty$ when $t \to \infty$, then it is always possible to identify the perturbations with such a scale that $\ell_{cr} \ll \ell \ll \delta_t$. This region of scales will play the same role as the logarithmic layer in boundary layer theory or as the inertial range in the theory of locally homogeneous turbulence. Just as in these theories, it can be assumed in the present case that the characteristics of the perturbations with scale $\ell \gg \ell_{cr}$ are not directly dependent on the processes of molecular transport. This assumption allows finding the spectral representation of the combustion velocity within two arbitrary functions of β. The search for these functions is based on the concepts of the theory of logarithmic boundary layer (see, for example, Monin and Yaglom [1965]), i.e., it is assumed that the spectral representation of the combustion rate obtained in the region of intermediate scales "joins" smoothly with the spectral representation in the region of large ($\ell \sim \delta_t$) and small ($\ell \sim \ell_{cr}$) scales. For the determination of the form of $U(\ell)$ in these regions, the results of the linear theory of stability are used.

Thus, let us consider the region $\ell_{cr} \ll \ell \ll \delta_t$. The quantity $\Delta U = U(L) - U(\ell)$ characterizes the contribution to the combustion rate by the perturbations whose scale changes from ℓ to L.

Assume that the interaction of the perturbations with scales that do not greatly differ plays the main role. Hence, it follows that ΔU is independent of $a^{(0)}$.

Since ΔU is independent of $a^{(0)}$, then

$$V\left(\beta, \frac{u_n x_1}{a^{(0)}}, \frac{u_n L}{a^{(0)}}\right) - V\left(\beta, \frac{u_n x_1}{a^{(0)}}, \frac{u_n \ell}{a^{(0)}}\right) = G\left(\beta, \frac{x_1}{L}, \frac{\ell}{L}\right) \tag{6.26}$$

where G is some dimensionless function. Let us differentiate this relation twice with respect to L and ℓ. We obtain

$$\xi \frac{\partial^2 G}{\partial \xi^2} + \frac{\partial G}{\partial \xi} + \eta \frac{\partial^2 G}{\partial \xi \partial \eta} = 0, \qquad \xi = \frac{\ell}{L}, \qquad \eta = \frac{x_1}{L} \tag{6.27}$$

The solution of this equation has the form $G = \eta^n f_n(\xi)$. If $n \neq 0$, then $|G| = \infty$ either when $x_1 = 0$, or when $|x_1| = \infty$ which has no physical meaning. Therefore, $n = 0$, i.e., G is independent of η and, hence, $\partial G/\partial \eta = 0$ must be substituted into (6.27). By integrating this equation we find

$$G = -\Phi \ln \xi + B$$

where Φ and B are dependent only on β. Let us differentiate (6.26) with respect to ℓ. We obtain

$$\frac{\partial}{\partial \ell} V\left(\beta, \frac{u_n x_1}{a^{(0)}}, \frac{u_n \ell}{a^{(0)}}\right) = \frac{\Phi(\beta)}{\ell}$$

By integrating this relation we find

$$V = \Phi(\beta) \ln \left(\frac{u_n \ell}{a^{(0)}}\right) + C\left(\beta, \frac{u_n x_1}{a^{(0)}}\right) \tag{6.28}$$

where C is the integration "constant". Assume that this distribution "joins" smoothly with the representation of the combustion rate in the region $\ell \sim \ell_{\mathrm{cr}}$. This means that formula (6.28) yields the correct order for $\ell \sim \ell_{\mathrm{cr}}$. Since all perturbations with scales $\ell \lesssim \ell_{\mathrm{cr}}$ are stable, then in cube ω with a dimension of the order ℓ_{cr} the reaction zone can be regarded as plane. Then from the first estimate in (6.25) and formulas (6.12), (6.28) we obtain

$$U(\ell) = u_n \Phi(\beta) \ln \frac{u_n \ell}{c_1 a^{(0)} \beta} + c_2 u_n \tag{6.29}$$

where c_2 is some constant.

As is seen from (6.29), U is independent of x_1, i.e. the local structure of the reaction zone and, hence, the local structure of the flame-generated turbulence are statistically homogeneous.

In order to find function $\Phi(\beta)$ let us consider the nonstationary, large-scale perturbations with scales of the order of δ_t and more. As indicated, when describing such perturbations it is permissible to regard the flame as thin. Therefore, the linear theory of stability can be used assuming that the nonlinear effects yield small corrections. We also use a number of phenomenological concepts developed by Zel'dovich [1966], who provides an equation describing the development of discrete disturbances at the flame front. In the present case, one ought to account for a continuous perturbation spectrum. It should be remembered that the contribution to combustion by perturbations with a scale of the order of ℓ is described by quantity $\partial U(\ell)/\partial \ell$. In order to obtain the geometric characteristics of the flame from this quantity, it must be divided by u_n, i.e., one must consider the quantity $\kappa = u_n^{-1} \partial U/\partial \ell$ which, evidently, characterizes the amplitude of perturbations with dimension ℓ. Let us now deduce the phenomenological equation for quantity κ.

It is evident that when $t = $ const and $\ell \to \infty$, quantity κ is described by the linear theory of stability. Within the framework of this theory the amplitude of the perturbation increases according to (6.13). Therefore, it can be assumed that the equation for κ in the linear approximation has the form

$$\frac{\partial \kappa}{\partial t} = c_4 \psi(\beta) \frac{u_n \kappa}{\ell} \tag{6.30}$$

where c_4 is a constant of the order of unity. In this approximation (6.30) coincides exactly with the relation obtained by Zel'dovich [1966].

Note now that when $\ell \to \infty$ the flame can be regarded as a discontinuity of the density. The velocity of motion of this discontinuity relative to the medium is not, however, equal to u_n, since the small-scale perturbations developed up to the moment under consideration have changed its internal structure. Hence, when $\ell \to \infty$, u_n in (6.30) must be replaced with u_t. Similarly, it can be assumed that when $\ell \to \delta_t$, u_n in (6.30) must be replaced with $U(\ell)$. Then, (6.30) in the linear approximation acquires the form

$$\frac{\partial \kappa}{\partial t} = c_4 \psi(\beta) \frac{U(\ell) \kappa}{\ell}$$

Let us now take into account the small nonlinear corrections assuming that they are proportional to κ^2. We then obtain

$$\frac{\partial \kappa}{\partial t} = c_4 \psi(\beta) \frac{U(\ell)\kappa}{\ell} - \lambda \kappa^2 \tag{6.31}$$

where λ is independent of κ.

In order to ascertain parameters affecting the quantity λ, we consider the mechanism stabilizing the perturbation amplitude. One of the possible mechanisms associated with the formation of angular points is pointed out by Zel'dovich [1966]. These points $(B_1, B_2, \ldots$ Fig. 6.9) are formed in the part of the combustion zone which is adjacent to the products of combustion. The existence of such points (break lines in the three-dimensional case) are clearly seen in the flame shadow-graphs given by Palm-Leis and Strehlow [1969]. Let us illustrate the important significance of this mechanism on the example of propagation of a flame, which is on the average plane, in a medium which always stays stationary. We shall assume that 1) the initial position of the flame is described by a single-valued smooth function; 2) each point of the flame moves normal to its surface with the same velocity.

This problem was analyzed by Kuznetsov [1975], Vilyunov and Dik [1975]. In the first work it is shown from the exact solution that there are two periods of flame evolution. During the first period its surface remains smooth, and the mean flame area unchanged. In the second period angular points appear and the mean flame surface begins to decrease.

The mechanism under consideration is purely of mechanical character and, hence, it can be assumed that the rate of decrease of the flame surface, which is determined by quantity λ, is independent of β. When $\ell \gg \ell_{cr}$, the coefficient of molecular transport (quantity a characterizes small-scale perturbations which interact weakly with large-scale fluctuations) is not one of the determining parameters. Hence, λ depends only on ℓ and $U(\ell)$. Then, from dimensional considerations we conclude that $\lambda \sim U(\ell)$, i.e., (6.31) assumes the form

$$\frac{\partial \kappa}{\partial t} = c_4 \psi(\beta) \frac{U(\ell)\kappa}{\ell} - c_5 U(\ell) \kappa^2$$

where c_5 is a constant. The stationary solution of this equation is $\kappa = c_3 \psi/\ell (c_3 = c_4/c_5)$, and since $\kappa = u_n^{-1} \partial U/\partial \ell$, then $U = c_3 \psi u_n \ln \ell +$ const. By comparing this formula with relation (6.29), we conclude that $\Phi = \psi$, i.e.,

$$U(\ell) = c_3 \psi(\beta) \ln \frac{u_n \ell}{c_1 a^{(0)} \beta} + c_2 u_n \tag{6.32}$$

Formula (6.32), derived for a flame which is plane on the average, is also valid in the general case if $\ell \ll L_m$, and the turbulence energy in the reactants is sufficiently small. Hence, it can be assumed that relation (6.32) also yields the correct order of magnitude of $u(\ell)$ when $\ell \sim L_m$. Then, from the second formula in (6.25) we obtain the estimate

$$u_t \sim c_3 \psi(\beta) u_n \ln \frac{u_n L_m}{c_1 a^{(0)} \beta} \tag{6.33}$$

where L_m is the maximum scale of perturbations (see §6.4). Since relation (6.32) is true when $\ell \gg \ell_{cr}$, the small component $c_2 u_n$ is, henceforth, ignored.

Let us now consider the characteristics of the flame-generated turbulence. We introduce the conditionally averaged structure function

$$D_{uu}^{(b)}(\ell) = \frac{1}{3} \langle [u(x) - u(x + \ell)]^2 \rangle_b$$

where subscript b means that averaging is effected under the condition that $\overline{W} = U$, i.e., only those cases when the flame fronts are most densely "packed" inside cube ω. Function $D_{uu}^{(b)}$ characterizes the contribution made to the turbulent energy by the velocity fluctuations whose scale does not exceed ℓ. It is readily seen that $D_{uu}^{(b)}(\infty) \sim (u^{(b)})^2$, since velocity fluctuations at two infinitely distant points are statistically independent. Owing to the absence of other parameters with the dimension of velocity[†] it can be assumed that $D_{uu}^{(b)} \sim U^2$ when $\ell \gg \ell_{cr}$, i.e.,

$$D_{uu}^{(b)} = c_6 u_n^2 \psi^2(\beta) \ln^2 \frac{u_n \ell}{c_1 a^{(0)} \beta} \tag{6.34}$$

Just as in formula (6.33), relation (6.34) is true when $\ell_{cr} \ll \ell \ll L_m$, and when $\ell_{cr} < \ell < L_m$ it yields the correct order of the values of the structure function (it is, of course, assumed here that the turbulence energy in the reactants is sufficiently small). From (6.34) it is evident that when $\ell \ll L_m$ the structure of the flame-generated turbulence is statistically homogeneous and isotropic.

Formulas (6.32) – (6.34) were obtained by Kuznetsov [1980]. Following this work, we now consider how flame instability affects combustion in a flow with an arbitrary initial turbulence energy. For this purpose, we compare quantity $D_{uu}^{(b)}$ with the structure function of velocity pulsations in the incoming flow into the flame $D_{uu}^{(0)} = \frac{1}{3} \langle [u(x) - u(x + \ell)]^2 \rangle$. It is evident that if condition $D_{uu}^{(b)}(\ell) \gg D_{uu}^{(0)}(\ell)$, is fulfilled for some values of ℓ, then flame instability substantially affects the perturbations with the scale under consideration.

Comparing both structure functions, one ought to keep in mind that combustion usually takes place in a limited volume so that the scale of flame perturbations is always limited. For example, in combustion in a tube, this scale cannot exceed the diameter. Note also that in turbulent flows the integral turbulence scale is always of the order of the characteristic dimension of the system, i.e., $L \sim L_m$. Therefore, from the definition of the structure function follow the estimates $D_{uu}^{(b)}(L_m) \sim (u^{(b)})^2$, $D_{uu}^{(b)}(L_m) \sim (u^{(0)})^2$[‡] ($u^{(0)}$ is the fluctuating velocity in the incoming flow into the flame).

[†]Quantity u_n cannot be a determining parameter, since it characterizes only those flame perturbations and, hence, those velocity perturbation whose scale is comparable with ℓ_{cr}.

[‡]Note that if the velocity fluctuating at two points are statistically independent, when $D_{uu}^{(0)} = 2(u^{(0)})^2$. This equality, which is strictly true when $L \to \infty$, can be used for the approximate estimate of $u^{(0)}$ if the distance between the points is of the order of the integral scale.

From the comparison of quantities $u^{(b)}$ and $u^{(0)}$ the following parameter follows

$$\mu = \frac{u^{(b)}}{u^{(0)}} = \frac{u_n}{u^{(0)}} \psi(\beta) \ln \frac{u_n L_m}{c_1 a^{(0)} \beta} \qquad (6.35)$$

As already mentioned at the end of §6.2, the combustion process is described by criteria β, K, $a^{(0)}/(u_n L)$, where K is given by formula (6.10). Usually β does not change much ($\beta = 5-8$) and, therefore, it can be assumed that β = const. Let us also take into account that $L \sim L_m$. Then, from (6.10), (6.35) it follows that instead of quantity $a^{(0)}/(u_n L)$ and K we can select another system of independent parameters: μ and K. Of greatest practical interest is the investigation of just such an independent variation of μ and K for which parameters $u^{(0)}/u_n$ and $u^{(0)} L/a^{(0)}$ remain much larger than unity (of course, these criteria are dependent on μ and K). It can be shown that for such a variation of quantities μ and K the following conditions are always fulfilled: $\mu \ll 1$ if $K \geq 1$, and $K \ll 1$ if $\mu \gtrsim 1$. Hence, one must study only three cases: 1) $\mu \gg 1$, $K \ll 1$; 2) $\mu \ll 1$, $K \ll 1$; 3) $\mu \ll 1$, $K > 1$. The set task is solved by comparing dependences $D_{uu}^{(0)}(\ell)$ and $D_{uu}^{(b)}(\ell)$, where formula (6.34) is, as before, considered to be valid.

Let us consider the first case. By definition of quantity μ, condition $D_{uu}^{(0)}(L_m) \ll D_{uu}^{(b)}(L_m)$ is valid. An analogous condition $D_{uu}^{(0)}(\ell) \ll D_{uu}^{(b)}(\ell)$ is fulfilled if $\ell < L_m$, since when ℓ varies function $D_{uu}^{(b)}(\ell)$ changes much more weakly than $D_{uu}^{(0)}(\ell)$. For example, from (6.34) we have $D_{uu}^{(b)} \sim \ln^2 \ell$, and from the Kolmogorov theory [1941] we conclude that $D_{uu}^{(0)} \sim \ell^{2/3}$ (Fig. 6.12, a). Hence, it follows that flame instability significantly influences the entire turbulence spectrum in the combustion zone, and turbulence in the incoming flow is inessential, i.e., all the results of the analysis carried out above are true.

Let us consider another extreme case ($\mu \ll 1$, $K > 1$). From the definition of parameter K it follows that when $K > 1$ the following estimates are valid: $\eta < \delta_n, v^2 \sim D_{uu}^{(0)}(\eta) > u_n^2$ (δ_n is the thickness of the normal flame front, η is the Kolmogorov scale, v is the Kolmogorov velocity). With increasing K these estimates gain in magnitude. On the other hand, from (6.34) it follows that $D_{uu}^{(b)}(\ell) \sim u_n^2$ if $\ell \sim \ell_{cr}$ (but $\ell > \ell_{cr}$). Since $\ell_{cr} \sim \delta_n$, then $D_{uu}^{(b)}(\ell) < D_{uu}^{(b)}(\ell)$ when $\ell \sim \ell_{cr}$ (but $\ell > \ell_{cr}$). As can be seen from Fig. 6.12c this estimate gains in

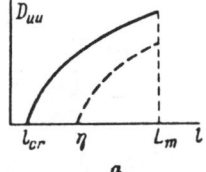
a

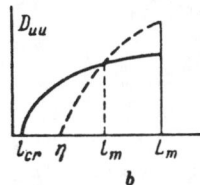
b

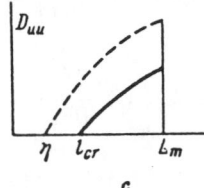
c

Figure 6.12 Qualitative character of the dependence of the structure functions on distance. a) $\mu \gg 1$, $K \ll 1$; b) $\mu \sim 1$, $K \ll 1$; c) $\mu \ll 1$, $K \gtrsim 1$; in all three cases the solid line is $D_{uu}^{(b)}$, the dashed line is $D_{uu}^{(0)}$.

magnitude with increasing ℓ ($D_{uu}^{(b)}$ increases more slowly than $D_{uu}^{(0)}$). Thus, flame instability does not play a role in the entire turbulence spectrum.

Finally, in the intermediate range ($\mu \ll 1$, K $\ll$ 1) it is evident from the definition of parameter μ that flame instability does not play a role in the large-scale part of the spectrum ($\ell \sim L_m$). In the small-scale part of the spectrum this role can be large, since when $\ell \to \ell_{cr}$ function $D_{uu}^{(b)}$ decreases more slowly than $D_{uu}^{(0)}$ (Fig. 6.12b). The boundary between both regions is determined by condition $\ell = \ell_m$ where ℓ_m is the root of equation $D_{uu}^{(0)}(\ell_m) = D_{uu}^{(b)}(\ell_m)$.

Thus, when K $\lesssim$ 1, flame instability can play an important role. It is essential in this case that the characteristics of the process are dependent on the linear scale of the system. Indeed, from (6.35) it follows that turbulence energy in the combustion zone tends to infinity if quantity L_m/δ_n, which appears under the integral sign in (6.35), increases unrestricted. The rate of such an increase, however, is very weak and, therefore, the effect under consideration was ignored for a long time in the theory of turbulent combustion and in the known theories of flame autoturbulization. For example, in the works by Karlowitz, Denniston, and Wells [1951], Kozachenko [1960], Kuzin and Talantov [1977] relations were obtained from which it follows that the ratio of the energies of turbulence in the combustion zone and in the incoming flow is a quantity which is dependent only on β and independent of the scale of the system.

§6.6 CRITERIAL DESCRIPTION OF TURBULENT COMBUSTION OF A HOMOGENEOUS MIXTURE

As estimates show, the main part of the experimental investigations referred to in §6.2 (for example, Talantov and coworkers) is conducted for K < 1, i.e., when flame instability can play an important role. It is this part of the investigations that is considered in the present section. Note that in addition to flame instability other gas dynamic effects can also have important significance. Therefore, prior to proceeding to the analysis of the experimental data, it is useful to consider the equation for turbulence energy in a flow with variable density (Monin and Yaglom [1965]). In the stationary flow in the absence of mass forces at high Reynolds numbers this equation assumes the form

$$\frac{\partial}{\partial x_k}\left[e\langle u_k\rangle + \frac{1}{2}\langle \rho u'_k u'^2_s\rangle + \langle p' u'_k\rangle\right] = -\langle \rho\rangle\langle \epsilon\rangle + \left\langle p'\frac{\partial u'_k}{\partial x_k}\right\rangle$$

$$-\langle \rho u'_k u'_s\rangle\frac{\partial \langle u_k\rangle}{\partial x_s} + \langle \rho u'_k\rangle\langle u_s\rangle\frac{\partial \langle u_k\rangle}{\partial x_s} \qquad (6.36)$$

where $e = 1/2\langle \rho u'^2_k\rangle$ is the turbulence energy, $p' = p - \langle p\rangle$, $u' = u - \langle u\rangle$.

The three terms which appear on the left side of (6.36) characterize convection and turbulent diffusion; they just redistribute the turbulence energy inside the combustion zone without changing the total energy of the fluctuating motion. The first term on the right side describes viscous dissipation, the second term describes the work done by the pressure fluctuations upon expansion or compression of the medium. The third and fourth terms characterise the interaction of fluctuating and averaged motions.

For a flame which is on the average plane in an unbounded flow, the third and fourth terms cause a decrease in the turbulence energy in the flame (Librovich and Lisitsyn [1975], Bray and Libby [1976], Libby and Bray [1981]). This conclusion is qualitatively apparent from the theory of rapid vortex-free deformation of a compressible fluid (Ribner and Tucker [1953]; see also Ievlev [1975], Sabel'nikov [1975]). The relations presented in these works indicate that upon rapid expansion of the medium the fluctuation energy decreases. If the combustion of a homogeneous mixture in a tube is considered, then one more effect arises which is caused by the fact that as a result of the pressure gradients the fluids with different density (reactants and combustion products) accelerate differently. This state leads to an increase of mean velocity shift which causes a rise in turbulence energy (see Chapter 5).

Thus, if one considers a flame which propagates in an unbounded flow, then only correlation of pressure fluctuations and velocity divergence can cause an increase in turbulence energy upon combustion. Hence, from the theory developed in §6.5, it follows that the correlation characterizes the role of flame instability.

Quantity $\langle p' \text{div } u' \rangle$ plays a special role in the equation of energy which follows from the following reasonings. First, in subsonic flows it is significant only upon combustion of a homogeneous mixture. This conclusion is drawn in the work by Kuznetsov [1979a] in which it is shown that 1) $\langle p' \text{div } u' \rangle = 0$ under nonisothermal mixing in the absence of reactions, 2) $\langle p' \text{div } u' \rangle / (\langle \rho \rangle \langle \epsilon \rangle) \sim 10^{-2}$ in diffusion combustion; 3) $\langle p' \text{div } u' \rangle / (\langle \rho \rangle \langle \epsilon \rangle) \sim 1$ in combustion of a homogeneous mixture. Second, from comparison of (6.35) and (6.36) it can be surmised that due to the correlation of the pressure fluctuations and velocity divergence, turbulence energy in the combustion zone tends to infinity if $L/\ell_{\text{cr}} \to \infty$.

Bearing in mind the indicated consideration we proceed to the analysis of the experimental data. Let us consider first the tests in which turbulence characteristics in the combustion zone were measured directly by laser Doppler anemometer. In the work by Ballal [1979] already discussed, it is established that in all investigated regimes the turbulence energy in the combustion zone is higher than in the incoming flow (Fig. 6.11a, b). As already indicated, in these tests the effect of the mean pressure gradient was excluded and, therefore, additional shift of the mean velocity did not occur. From 6.11a it is evident that if the energy perturbations in the fresh mixture is small ($u^{(0)} < 2u_n$) the increase of their scale leads to an increase in fluctuating velocity in the combustion zone. This conclusion agrees with formula (6.35). A decrease in turbulence energy, which is seen from Fig. 6.11b ($u^{(0)} > 2u_n$), does not contradict the developed theory as will be shown in §6.8. Unfortunately, a quantitative comparison of the theory and experiment in the present case is not feasible, since the maximum value of the turbulence scale L exceeded quantity ℓ_{cr} only by three times and the condition of applicability of formula (6.35) has the form $\ln(L/\ell_{\text{cr}}) \gg 1$.

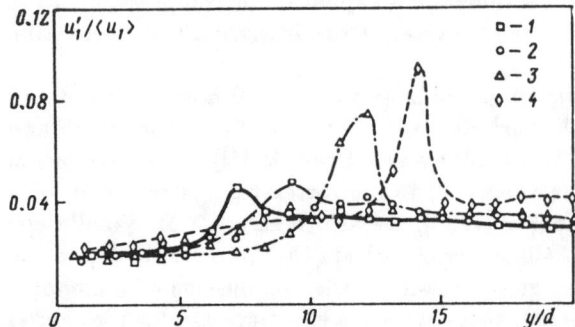

Figure 6.13 Profiles of root-mean-square fluctuations of the longitudinal velocity upon combustion of a homogeneous mixture of ethylene with air behind turbulence-inducing grid in a channel from data by Bill, Namer, and Talbot [1981]. Test conditions are the same as in Fig. 1.23. *1)* $x_1 = 7$ cm, *2)* $x_1 = 7.5$ cm, *3)* $x_1 = 8$ cm, *4)* $x_1 = 8.5$ cm, x_1 is the distance downstream from the grid, $u_1' = \sqrt{\langle(u_1 - \langle u_1 \rangle)^2\rangle}$, $\langle u_1 \rangle$ is the flow velocity behind the grid.

An increase in turbulence energy in the combustion zone was also recorded in the works by Bill, Namer, and Talbot [1981], Chang and Ng [1983]. The initial conditions in both tests were roughly the same (a thin flame-stabilizing wire is placed in the homogeneous turbulent flow). The results obtained in the first work are depicted in Fig. 6.13. It is evident that the turbulence energy in the flame increases.

In the second work turbulence intensity outside the combustion zone amounted to 5%. Moreover, the maximum intensity of fluctuations registered in the combustion zone amounted to: for the transverse velocity component 16%, for the longitudinal component 10%, i.e., turbulence energy rose by almost one order of magnitude. Even more interesting results were obtained when measuring Reynolds stresses. It was established that turbulent viscosity at every point is negative, i.e., fluctuation energy transforms into an averaged motion (and, hence, the increase in energy can be caused only by flame instability, i.e., by correlation $\langle p' \text{div } u'\rangle$). Should these results be confirmed by further experimental investigations, it would necessitate a cardinal change of the principles of constructing the semi-empirical theory of turbulence which are used when describing the combustion of a homogeneous mixture.

Let us now consider the results of measurements of the velocity of turbulent flame propagation. It is established in a number of experiments that u_t far exceeds $u^{(0)}$. The tests, already mentioned, by Suzuki, Oba, Hirano, and Tsuji [1979] (Fig. 6.4) can be cited as an example. Formula (6.33) enables the estimation of the order of the values of u_t in these tests. Note that two experimental constants c_1 and c_3 enter into (6.33). The first is determined by formula (6.12), and the value of the second is unknown. In the theory constructed with the utilization of dimensional considerations the values of the empirical constants must be of the order of unity (provided that the theory is correctly constructed). Therefore, it is natural to assume that $c_3 = 1$. Then, in the conditions of the analyzed tests ($u_n = 40$ cm/sec, $\beta = 7.6$, $a^{(0)} = 0.2$ cm^2, $L_m = d/2$, where d is the diameter of the burner) we obtain from (6.33) $u_t = 2.3$ m/sec, which agrees well with the results of measurements ($u_t = 2.2$ m/sec when $u^{(0)} = 28$ cm/sec).

In order to analyze the results of the other tests in which the velocity of turbulent flame propagation u_t was measured, it should be remembered that the development of perturbations with dimension $\ell \lesssim \ell_m$, as established in §6.5, is determined by flame instability and when $\ell \ll \ell_m$ the contribution of these perturbations to the velocity of flame propagation is given by formula (6.32) (ℓ_m is the root of equation $D_{uu}^{(b)}(\ell_m) = D_{uu}^{(0)}(\ell_m)$, where $D_{uu}^{(b)}$ is given by formula (6.34), and $D_{uu}^{(0)}$ is the structure function of velocity pulsation in the incoming flow). Flame instability has no effect on large-scale ($\ell > \ell_m$) flame perturbations.

Hence, one can relatively easily estimate the effect of flame instability on the combustion process when K $\lesssim$ 1. For this purpose, we make use of the procedure described in §6.3, namely all perturbations with scale ℓ_m and less are smoothed on the flame surface. It is evident that the specific rate of conversion of the substance into a partially smoothed flame surface is equal to $U(\ell_m)$, where $U(\ell_m)$ is given by formula (6.32). It is obvious that the characteristics of a partially smoothed flame surface cannot be directly dependent on u_n or on the normal flame front. By definition of this surface, flame instability cannot also directly affect its characteristics. The effect of the above enumerated factors are described by the single quantity $U(\ell_m)$.

Thus, by analyzing the effect of large-scale perturbations ($\ell > \ell_m$) on the motion of a partially smoothed flame surface, we arrive at the statement of the problem known from the theory of frontal combustion. The only difference is that the velocity of normal flame propagation u_n must be replaced with the considerably larger quantity $U(\ell_m)$ and all perturbations with scales less than ℓ_m must be excluded from the turbulence spectrum. When $\ell_m \ll L$ the energy of such perturbations is much lower than the turbulence energy $(u^{(0)})^2$. Therefore, just as in the theory of frontal combustion, quantity u_t is dependent only on β and on the two parameters $u^{(0)}$ and $U(\ell_m)$, i.e., we obtain formula $u_t = u^{(0)}\varphi[\beta, u^{(0)}/U(\ell_m)]$, which is analogous to formula (6.8). Practically, parameter $u^{(0)}/U(\ell_m)$ is inconvenient. Therefore, it is expedient to use the equivalent parameter μ. It can be assumed that when K $\lesssim$ 1 the velocity of turbulent flame propagation is described by the formula (Kuznetsov [1982b])

$$u_t = u^{(0)} F(\mu), \qquad \mu = \frac{u_n}{u^{(0)}} \psi(\beta) \ln \frac{u_n L_m}{c_1 a^{(0)} \beta} \tag{6.37}$$

where L_m is the maximum scale of perturbations.

Let us find the form of this function in two limiting cases: $\mu \gg 1$ and $\mu \ll 1$. In the first case, flame instability plays a determining role in all regions of the turbulence spectrum. Therefore, quantity u_t is given by formula (6.33) from which it follows that $F = \text{const} \cdot \mu$ when $\mu \gg 1$. Let us now analyze the second case $\mu \ll 1$, i.e., when flame instability does not affect the turbulence energy. In order to determine the form of function $F(\mu)$ we consider a specific example

— the combustion of a homogeneous mixture in the wake of a flame stabilizer. As already mentioned in Chapter 1, one of the most characteristic features of such a flow is intermittency, i.e., fluctuation of the boundary between the turbulent and nonturbulent fluids. In the nonturbulent fluid turbulence energy is small and the exchange between the two fluids is of unidirectional character (fluid particles cannot leave the turbulent fluid). Since the fluctuating velocity outside the boundaries of the turbulent fluid attenuates rapidly, the flame cannot spread beyond these boundaries.

It is readily seen that when $K \lesssim 1$ the flame tends to approach the boundary of the turbulent fluid. Indeed, it follows from (3.108) that the velocity of the medium relative to the boundary is of the order $(\langle\epsilon\rangle\nu)^{1/4}$. If $K \lesssim 1$, then this velocity is less than the velocity of motion of the flame relative to the medium u_n. Hence follows the assertion made. Thus, when $K \lesssim 1$ the boundary of the turbulent fluid is also simultaneously the flame front (Kuznetsov [1982b]).

Hence, it is obvious that when $K \lesssim 1$ and $\mu \ll 1$ the entire substance reaching the wake burns. Since the rate of wake expansion is of the order of the pulsating velocity, then $u_t \sim u^{(0)}$, i.e., F is independent of μ when $\mu \ll 1$ and $K \lesssim 1$.

This conclusion is also true when $\mu \gg 1$ when flame instability is substantial. But in this case, the fluctuating velocity is unknown.

The conclusions made are confirmed when the experimental data by Talantov and coworkers are processed (the conditions of these tests and the corresponding references are indicated in §6.1). In the tests under consideration the regime parameters were varied within a very wide range of values: $\alpha^{(0)} = 0.8 - 1.4$, $T^{(0)} = 393 - 793$ K, $p = 0.04 - 0.45$ MPa, $d = 2.5 - 15$ cm, $\langle u_1 \rangle = 30 - 140$ m/sec. Upon processing it was assumed that $L_m = d/2$, $a^{(0)} = 0.2$ cm^2/sec when $T^{(0)} = 300$ K and $p = 0.1$ MPa, $a^{(0)} \sim [T^{(0)}]^2/p$. When computing $u^{(0)}$ formula (6.11) was used. Taken into account was the change of composition in the leading points, i.e., according to formula (6.19) it was assumed that $\alpha^{(b)} = 1.4\alpha^{(0)}(\alpha^{(b)} > 1)$. The results of the calculation of $u_n(\alpha^{(b)})$ for one of the regimes is shown in Fig. 6.1 as curve 2. For the regimes in which $T^{(0)}$ and p were varied formula (6.4) was used when calculating u_n.

The results of analyzing the experimental data are depicted in Fig. 6.14. It is evident that the experimental points are grouped around a single curve corresponding to the interpolation formula $u_t = u^{(0)}(c_7 + c_3\mu)$ which correctly describes the dependence of u_t on μ in two limiting cases: $\mu \gg 1$ and $\mu \ll 1$. The empirical constants c_7 and c_3 are of the order of unity ($c_3 = 1, c_7 = 1.4$). Thus, parameter μ allows generalization of the experimental data with a large variation of the determining parameters, and the theory describes correctly the form of the critical dependence in the limiting case.

Let us now briefly analyze the effect of various parameters on the quantity u_t. The effect of turbulence energy in the reactants requires no special comment. It is also obvious that the characteristics of the effect of composition on u_t result from the differences in the coefficients of molecular transport. The results obtained in §6.3 enable finding the excess air coefficient at which u_t is a maximum. Increasing the scale of turbulence causes a rise in u_t which is connected with the strengthened

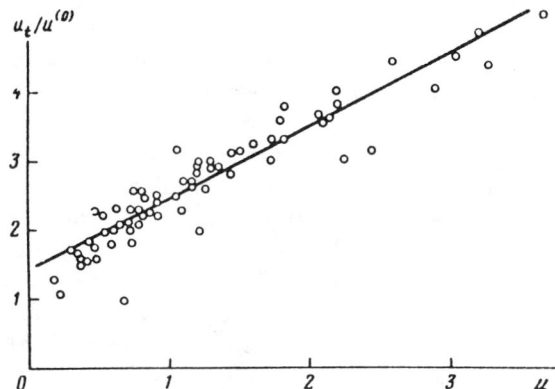

Figure 6.14 Critical analysis of measurements of the velocity of turbulent flame propagation in tests described in the book by Talantov [1975]. $\mu = (u_n/u^{(0)})\psi(\beta)\ln(u_n L_m)/(c_1 a^{(0)}\beta)$.

role of flame instability and flow turbulence. The effect of pressure turns out to be very nontrivial. It is easily seen that with increasing pressure quantity $U(L_m)$ increases and then it decreases. Such dependence is caused by the interaction of two competing factors. On the one hand, with increasing pressure the role of flame instability increases (ℓ_{cr} decreases) which leads to the increase of its surface. On the other hand, with increasing pressure u_t decreases, i.e., the specific rate of substance conversion on this surface decreases. Estimates show that maximum is reached at very high pressures. Thus, for example, when $\alpha^{(0)} = 1$, $T^{(0)} = 373$ K in a tube of 5 cm diameter, the maximum value of $U(L_m)$ for gasoline-air mixture and, hence, u_t are attained at a pressure of 0.6 MPa. Test results at such high pressures are unknown. Therefore, in all experiments the velocity of flame propagation increases with increasing pressure.

§6.7 MAXIMUM POSSIBLE INTENSITY OF THE FLOW IN TURBULENT COMBUSTION OF A HOMOGENEOUS MIXTURE

Considered in this section is the combustion of a homogeneous mixture characterizing condition $K \gtrsim 1$ (see formula (6.10)), i.e., it is assumed that the thickness of the normal flame δ_n is greater than the smallest (Kolmogorov) turbulence scale $\eta = \nu^{3/4}\langle\epsilon\rangle^{-1/4}$. From the practical viewpoint it is of interest to study combustion in devices of considerably large dimension in which the integral turbulence scale is much larger than δ_n. Since Reynolds number is usually very high, this means that this case is realized in such devices in which turbulence energy attains large values. Since turbulence intensity $u^{(0)}/\langle u_1\rangle$ changes, as a rule, within moderate limits, combustion with large flow rates of the reactants is considered. Analysis of such a regime is of considerable practical interest in view of the concern about the ultimately possible heat release of the combustion process.

It should be important to bear in mind that $K \gtrsim 1$, flame instability does not practically affect the turbulence spectrum. Therefore, the system of three determining parameters indicated in §6.2 must be reconsidered. Strictly speaking, the dependence on parameter β is maintained even in the case under consideration, since, in addition to flame instability, other gas dynamic effects discussed in §6.6 can play an important role. These effects also exist in the diffusion flame and, as evident from §5.1, their influence is described only by quantity β. Thus, they are independent of scale factors, i.e., of the quantity L_m/δ_n appearing in the definition of criteria μ (6.35). Hence, when $L_m/\delta_n \to \infty$ and $K > 1$, the turbulence energy in the combustion zone cannot tend to infinity, i.e., when $K > 1$ the gas dynamic effects do not essentially alter the fluid dynamic structure of the flow. Thus, within the framework of approximate theory the dependence on parameter β can be neglected. According to this, distinction will not, henceforward, be made between the fluctuating velocity in the reactants and in the combustion products, and quantities $u^{(0)}$ and $u^{(b)}$ are denoted by one symbol u. Consequently, two parameters Mi and K are significant and their definitions are given in (6.6) and (6.10), respectively.

In order to illustrate the specific features of the regimes in which $K > 1$, let us go back to the analysis of combustion in the wake of a stabilizer. It was shown in §6.6 that when $K \lesssim 1$ the flame front is located near the boundary of the turbulent fluid. When $K > 1$ the velocity of motion of this boundary relative to the medium, i.e., the Kolmogorov velocity, becomes larger than u_n (parameter K is proportional to the ratio of the squares of these velocities). Therefore, the unburned mixture "intrudes" into the wake. Such an intrusion takes place everywhere, since the velocity of the boundary of the turbulent fluid is a random quantity. Nonetheless, the angle of inclination of the average isotherm becomes less than the expansion angle of the wake.

Let us consider what will happen when parameter K is increased. Obviously, the expansion angle of the wake stays unchanged. The mean value of the velocity of the turbulent fluid boundary will increase compared with u_n. Hence, the flow rate of the reactants penetrating these boundaries increases, and the angle of inclination of the averaged isotherm decreases. This means that with increasing u the velocity of turbulent flame propagation no longer changes proportionally to the fluctuating velocity, i.e., $u_t/u \to 0$ when $u \to \infty$.

It can be assumed that this feature is of general character. Indeed, as already indicated in Chapter 1, intermittency is regarded as a general property of all turbulent flows, i.e., turbulence is always made up from individual "pieces" of vortex and potential fluids. Therefore, the reasoning used when analyzing combustion in the wake of the stabilizer are also true when analyzing combustion in each "piece" of the vortex fluid. Hence, it can be assumed that if the mixture is burned inside an individual "piece" of the vortex fluid, then if $K \lesssim 1$ the flame front approaches the boundaries of this "piece". And if $K \gtrsim 1$, then the flame will be pushed away from these boundaries deep into the turbulent fluid. This state of affairs must prevent flame propagation.

This conclusion is clearly illustrated by the tests of Karpov and Severin [1977]. Figure 6.2 shows that there exist two combustion regimes. In the first $(u < u_{cr})$ the combustion rate is proportional to u. In the second $(u > u_{cr})$ with increasing fluctuating velocity the combustion rate is either unchanged or even decreases with increasing u. The boundary between the two regimes is traced quite clearly (in Fig. 6.2 the ordinates of points $u = u_{cr}$ are denoted by vertical dashed lines). Simple estimates indicate that change of regimes in these tests occurs at $K \approx 1$ with an accuracy of 20%. In these estimates it was assumed that $\langle \epsilon \rangle = u^3/L, u^{(0)} = a^{(0)}$, the difference between the coefficients of molecular transport was taken into account, i.e., formulas (6.19) were used, and quantity $\langle \epsilon \rangle$ entering into the definition of parameter K was calculated from quantity u_{cr}. The value of the turbulence scale was $(L = 1 \text{ cm})$.

Thus, when $K \gtrsim 1$ the heat release of the combustion process does not increase with increasing fluctuating velocity. Experiments by Karpov and Severin [1977], at the same time, show that the combustion process in the regime under consideration is stable and flameout does not occur (see Fig. 6.2).

As will become clear below, in order to explain this combustion feature, processes of two types must be taken into account. Firstly, the macrostructure of the flame must be analyzed (i.e., its global geometric configuration). Secondly, the microstructure of the flow must be considered. In the latter the internal intermittency of turbulence plays a large role. The important role of the internal intermittency of turbulent combustion was, apparently, first emphasized by Khomyak [1970, 79].

To illustrate the role of flame configuration we shall first consider a flame, which is on the average plane, propagating in an unbounded flow with arbitrary turbulence characteristics. Following the work of Kuznetsov and Sabel'nikov [1977], we can show that, generally speaking, the reaction zone is a continuous surface.

In accordance with the statement of the problem which is traditional for the theory of combustion, we assume that $a = D_o = D_f$. In this case the rate of heat release W is dependent only on the dimensionless temperature. This dependence is, evidently, continuous and quantity W reaches a maximum at some value c, say $c = c_m$. Let us select an arbitrary continuous line extending from the fresh mixture, i.e., in the region where $c = 0$, to the combustion products, i.e., in the region where $c = 1$. The solution of the heat equation (2.1) is a continuous function. Therefore, the dimensionless temperature along the line under consideration continuously changes from $c = 0$ to $c = 1$. Hence, value $c = c_m$ is reached on this line at least once. At this point the maximum rate of heat release is the same as in the normal flame front. Since an arbitrary line is selected, it hence follows that there are no "holes" in the flame.

It is pertinent, however, to note that, strictly speaking, the reasoning presented above is true only if $a = D_o = D_f$. Otherwise, W is dependent on several variables. Moreover, the temperature and concentration can change so that as a result of heat losses the limits of flame propagation are reached. In the combustion of a mixture with a composition close to the stoichiometric, these limits are not, however, reached, since it follows from formula (6.19) that the change in composition in the reaction zone is not too big. Thus, on the average, a plane flame exists for any turbulent characteristics, and its surface is described by a continuous function.

These conclusions refute the viewpoint expressed by Ballal and Lefebvre [1975], Andrews, Bradley, and Lwakawamba [1975], according to which "holes" can be formed in the flame.

We should note that "holes" cannot form only in the combustion of a homogeneous mixture. In diffusion combustion the reaction rate is a function of several variables and the analysis presented above cannot be used.

Let us now consider the case when the characteristic dimension of the region occupied by the combustion products is limited. As an example we can cite the combustion in the wake of a high-drag body or the homogeneous deformation of a plane layer of combustion products which has already been considered in §§6.2, 6.3. From these discussions it follows that the flame surface is also described by a continuous function. However, the flame itself does not exist under all conditions. For example, at sufficiently high velocities of the incoming flow the flame is separated from the high-drag body, and at a sufficiently large velocity gradient the plane layer of combustion products does not exist.

Let us now analyze the effect of small-scale fluctuations on the combustion process. In this case, it is necessary to take into account that as a result of internal intermittency the energy of the small-scale part of the turbulence spectrum in space is distributed extremely nonuniformly (see Chapters 1, 4). These peculiarities indicate that the definitions of parameters (6.10) must be reconsidered.

For this purpose, let us consider a cube ω with center at point x and side 2ℓ. We introduce quantity

$$K(\ell) = \frac{a^{(0)}}{u_n^2} \left[\frac{1}{8\ell^3} \left(\int_\omega \frac{\partial u_1}{\partial x_j} d^3x \int_\omega \frac{\partial u_j}{\partial x_i} d^3x \right)^{1/2} \right] \tag{6.38}$$

i.e., parameter K is calculated from the velocity gradient averaged over cube ω. This value $K(\ell)$ is a random function of the coordinates and of time. For the parameter defined by formula (6.10) we retain the old symbol K without indicating the dimension of the region ω.

Let us now select dimension ℓ which satisfies condition $L \gg \ell \gg a/u_n$ and identify in the flow two types of regions so that $K(\ell) < 1$ in the regions of the first type and $K(\ell) > 1$ in the regions of the second type. If the statistical characteristics of the small-scale part of the turbulence spectrum possessed ordinary properties (for example, the velocity gradient probability distribution is close to the normal), then the regions of the first type with $K \gg 1$ would have practically not existed. This case, however, is not characteristic for turbulence at very high Reynolds numbers. As was shown in Chapters 1 and 4, the PDF of the absolute value of the velocity difference is close to the lognormal, i.e., small-scale fluctuations with very large and very small amplitudes are highly probable in the flow. Hence, even when $K \gg 1$ there are extended regions in the flow wherein the turbulence energy is small, i.e., $K(\ell) < 1$.

It is evident that the combustion in similar circumstances takes place in accordance with the flamelet mechanism discussed in the preceding section.

Let us analyse what happens to the flame in the regions in which velocity gradients are too large. In the Zel'dovich-Frank-Kamenetskii approximation the zone of the chemical reaction is still the surface on which the temperature is constant (the difference in the coefficients of molecular transport here are not taken into account for the sake of simplicity). It is obvious that the temperature gradient on the surface of the reaction zone can be found from the solution of the heat equation if the law of motion of this zone is known (both boundary conditions are specified — at infinity and on a known surface). This law is found from the additional condition (6.2), which is always true in the Zel'dovich-Frank-Kamenetskii approximation.

Let us analyse the consequence of such a statement of the problem. To this end we shall temporarily disregard condition (6.2). In the region under consideration when $K(\ell) \gg 1$ contortions appear on the surface of the reaction zone with a scale much larger than $\delta_n \sim a/u_n$. This means that the temperature gradient in the reaction zone is, generally speaking, much higher than the value predicted by formula (6.2), i.e., the conductive heat removal is too large. Since condition (6.2), i.e., the conductive heat removal is too large. Since condition (6.2) is fulfilled, then the conductive heat losses must be compensated by convective heat addition. Hence, it follows that the reaction zone begins to move relative to the medium with a velocity much higher than u_n. Since holes are not formed in the flame, it is rapidly forced out from the region of the second type into the region of the first type (Kuznetsov [1982b]).

The existence of the effect under consideration was, apparently, demonstrated in the experiments by Khomyak and Yarosinskii [1982] in which the combustion in a vertical tube was studied. The tube was filled with a mixture and a strong turbulent region was generated in the upper part by means of a fan. The mixture was burned in the lower part of the tube where velocity fluctuations were small. It was established that combustion ceases when the flame reaches the strongly turbulent region.

It follows from the analysis that when $K \gg 1$, as a result of internal intermittency, the flame front is almost always found in the regions of the first type, i.e., in the region where $K(\ell) \lesssim 1$. Thus, the combustion is of flamelet character always.

This conclusion is indirectly confirmed by tests investigating the conditions for flameout in the wake of high-drag bodies. Prior to analyzing the results of such experiments, let us find the parameter for flameout. It is well known that flame stabilization is determined by the processes taking place in the zone of reverse flows. The characteristic linear scale of this zone is proportional to the transverse dimension of the high-drag body d. Then, from the above analysis it follows that in the zone of reverse flows the flame can exist only under the condition $K(d) \lesssim 1$. Since the integral turbulence scale L is proportional to d, then

$$\frac{1}{d^3}\left(\int_\omega \frac{\partial u_i}{\partial x_j}d^3x \int_\omega \frac{\partial u_j}{\partial x_i}d^3x\right)^{1/2} \sim \frac{u}{L}, \quad L \sim d$$

where ω is a cube with disk $d/2$. Then, from (6.38) we obtain that K $(d) \sim$ Mi, where parameter Mi is given by formula (6.6). Thus, the condition of flameout has the form Mi ~ 1.[§]

This condition was first obtained by Dunskii when analyzing the experimental data (a review of the results of his work is included in the book by Shchetinkov [1965], Raushenbakh et al. [1964]). It was also used by Baev and Tret'yakov [1977] and Klimov [1977b] for the description of flameout in the wake of high-drag bodies. In all these works the definition of parameter Mi included quantities $\langle u_1 \rangle$ and d ($\langle u_1 \rangle$ is the velocity of the incoming mixture) instead of the fluctuating velocity u and the turbulence scale L, respectively. Since $u \sim \langle u_1 \rangle$, $L \sim d$, the indicated difference is inessential.

It was shown by Dunskii that when the flameout occurs the values of parameter Mi can be greatly different, in particular, if we are comparing the regimes in which the combustion of rich and lean mixtures with identical u_n is investigated. This shortcoming was overcome by Baev and Tret'yakov [1968, 1972, 1977]. In these works complex $a/u_n^2 = \tau_c$ appearing in the definition of parameter Mi (6.6) was found from experiments in which the critical velocity gradient was measured during the propagation of a laminar flame in a tube. As noted in §6.2, in such tests effects that are caused by the differences in the coefficients of molecular transform are observed. Baev and Tret'yakov [1972, 1977] showed that the use of τ_c enables a significantly better generalization of the experimental data. Hence, it is evident that the effect of the differences in the coefficients of molecular transport is manifested even with flameout. The considerable influence of this effect on flame stabilization was apparently first noted by Zukoski and Marble [1955].

To illustrate let us consider two values of the excess air coefficient in the initial mixture $\alpha_1^{(0)} < \alpha_2^{(0)}$ so that $u_n(\alpha_1^{(0)}) = u_n(\alpha_2^{(0)})$ and $u_0(\alpha_2^{(0)})$ during flameout in the wake of a stabilizer. It is well known that $u_0(\alpha_1^{(0)}) > u_0(\alpha_2^{(0)})$ in the combustion of propane-air mixtures $(D_f < D_o)$ (Shchetinkov [1965]). In the combustion of methane-air mixtures $(D_f > D_o)$ an opposite effect is observed (Volodina and Andreev [1958]).

For confirmation of this conclusion, Fig. 6.15 shows the results of analyzing the experiments by Dunskii in the form of the dependence of Mi on Re. In these experiments, the excess air coefficient was varied within the bounds $\alpha^{(0)} = 0.6$ - 1.4, the diameter of the stabilizer was varied from $d = 5$ mm to $d = 11$ mm, and the velocity of the incoming flow of propane-air mixture was $\langle u_1 \rangle = 20 - 140$ m/sec. It is assumed in the calculations that $L = d$, $u = 0.3 \langle u_1 \rangle$. The effect of the differences in the coefficients of molecular transport was taken into account, i.e., the value of u_n was computed from the composition which is given by formula (6.19).

It is seen from Fig. 6.15 that when flameout occurs, parameter Mi is of the order of unity. Moreover, the maximum value of Mi is not more than twice the

---

[§]Note that K $\gg 1$ for Mi ~ 1 and high Reynolds number.

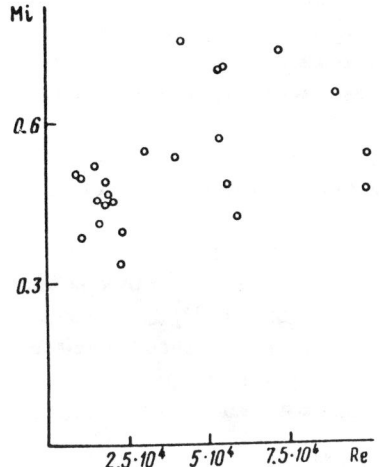

Figure 6.15 Critical analysis of the results of measurements of flame stabilization process in the wake of high-drag bodies conducted by Dunskii (these results are presented in the books by Raushenbakh et al. [1974] and Shchetinkov [1965]). $\mathrm{Mi} = ua^{(0)}/(du_n^2)$, $\mathrm{Re} = \langle u_1 \rangle d/\nu^{(0)}$.

minimum value. The small "scatter" of the values of parameter Mi is caused by accounting for the differences in the coefficients of molecular transport. Analyzing of the same data conducted by Dunskii without accounting for these differences leads to a situation where the maximum value of parameter Mi differs from the minimum value by more than an order of magnitude.

Thus, the differences in the coefficients of molecular transport affect the combustion process even in the most heat-intense regime. Indeed, otherwise (the characteristic scale of the change of the instantaneous temperature field is of the order of the integral turbulence scale) from the principle of self-similarity of turbulent flows with respect to Reynolds number it would have followed that the differences in the coefficients of molecular transport do not affect the process of flame stabilization. Hence, it follows that even in the most stressed state the combustion process takes place in accordance with the flamelet.

These concepts enable explaining the results of the tests, already discussed, by Ballal and Lefebvre [1975] and Ballal [1979]. Mechanisms which do not, on the face of it, agree with the concepts developed in §§6.5, 6.6 were established in these tests, namely, it was shown that when $u \gtrsim 3u_n$, the velocity of flame propagation and turbulence energy in the reaction zone increased with increasing turbulence scale. The experiments under consideration have two specific features. First, the combustion in the flow behind a grid, i.e., in a flow in which turbulence degenerated, was investigated. Second, the effect of sufficiently small-scale turbulence $(L/\delta_n = Lu_n/(a^{(0)}\beta) < 6)$ was studied. Estimates made by Ballal and Lefebvre [1975] indicate that in the conditions of their tests $\eta < \delta_n$ when $u \gtrsim 3u_n$. Hence, it is evident that condition $K \gtrsim 1$ is fulfilled at the measurement points. This condition, however, cannot be fulfilled for the whole flame, since turbulence behind the grid degenerates. Since $\langle \epsilon \rangle \sim 1/L$, then with increasing turbulence scale the rate of this degeneration decreases. Hence, the characteristic dimension of the region in which condition $K \gtrsim 1$ is fulfilled increases with increasing turbulent scale.

Therefore, with increasing scale the values of parameter $K(\ell)$ and the turbulence energy, averaged over the entire combustion zone, increase which lead to a decrease in u_t.

Analogous considerations are valid for the analysis of the effect of turbulence scale on the energy of perturbations in the combustion zone. It is obvious that with increasing L the average volume of the regions at which $K(\ell) \lesssim 1$ decrease. As already noted, in these regions flame instability is important. Since the relative volume of such regions decreases, a decrease in the energy of flame-generated perturbations takes place. The latter causes a drop in the total energy of turbulence in the flame when L is increased.

Let us now determine the maximum heat intensity of the process of turbulent combustion of a homogeneous mixture. We first estimate the rate of the chemical reaction for $K \lesssim 1$, $\mu \ll 1$. As was shown in §6.6, in this regime we have $u_t \sim u$. Hence, we obtain $\langle W \rangle \sim u_t/\delta_t \leq u/L$ (the last inequality follows from the fact that the thickness of the turbulent flame δ_t cannot be less than the turbulence scale L). This estimate, as a first approximation, is also true when $K \sim 1$ which is clear from the above analysis of the experiments by Karpov and Severin [1977]. Since $\langle \epsilon \rangle \sim u^3/L$, then $u \sim L^{1/3} u_n^{4/3} (a^{(0)})^{-1/3}$ when $K \sim 1$, i.e., $\langle W \rangle \sim u_n^{4/3} L^{-2/3} (a^{(0)})^{-1/3}$ when $K \sim 1$. Since when $K \sim 1$ the limit heat intensity of the combustion process is reached, we obtain

$$\langle W \rangle \leq W_m = u_n^{4/3} L^{-2/3} (a^{(0)})^{-1/3} \tag{6.39}$$

Hence, it follows that the ratio of the mean reaction rate in the turbulent flame to the reaction rate averaged over the thickness of the normal flame, i.e., to quantity $u_n^2/a^{(0)}$ does not exceed $(u_n L/a^{(0)})^{-2/3}$. This ratio is usually very small (of the order of several percent).

Summing up the results of this analysis, we conclude that the combustion process of a homogeneous mixture in a flow with very large turbulent energy possesses a number of particular features. 1) When $K \gtrsim 1$ and $Mi < 1$, the increase of the turbulence energy leads neither to the intensification of the combustion process nor to flameout. 2) There exists a limit heat intensity of the combustion process which is determined by the reaction rate, the coefficients of molecular transport, and the turbulence scale. This heat intensity is usually much lower than the heat intensity of the normal flame. 3) Flameout occurs when $Mi \sim 1.4$. Combustion takes place in accordance with the flamelet model even in the most stressed regimes.

§6.8 FLAME STABILIZATION BY BLUNT BODIES

Analysis conducted in the preceding section showed that immediately before flame blow-out, i.e., in the most critical regime, the combustion of a homogeneous mixture takes place in accordance with the frontal mechanism and at flame blow-out the value of parameter Mi is approximately constant (Fig. 6.15). In the meantime a closer analysis of this picture shows that a noticeable scatter is observed in the points obtained in different regimes. Therefore, the need arises to verify these considerations. This need is also caused since experimental investigation of the process understudy reveals a number of nontrivial facts which, on the face of it, contradict common sense.

Indeed, traditional concepts of flame stabilization are based on the experimental data presented in Figs. 6.16, 6.17. Figure 6.16 shows the experiments by Dunskii with propane-air mixture, and Fig. 6.17 shows experiments with hydrogen (Gorbatko, Kuznetsov, and Lipatov [1986]). Experiments with hydrogen in the latter work show that the region of stable combustion does not always have the

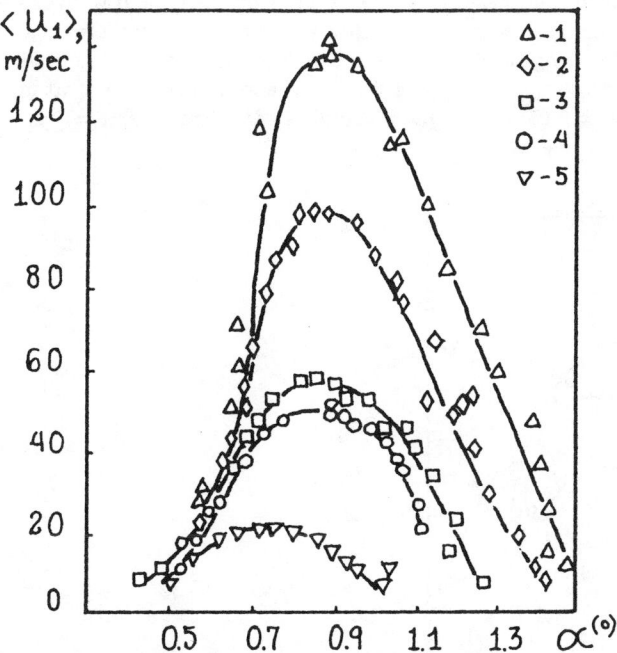

Figure 6.16 The dependence of velocity at flame blow-out $\langle u_1 \rangle$ with a stabilizer on the excess air factor in the incoming flow $\alpha^{(0)}$ from data by Dunskii (these data are presented in the book by Raushenbakh et al. [1974] and Shchetinkov [1965]). The combustion of propane-air mixture behind a cone with an apex angle of 30° mounted in the exit section of an 18 mm diameter tube, $T^{(0)} = 293$ K, $p = 0.1$ MPa, 1) $d = 11$ mm, 2) $d = 8.7$ mm, 3) $d = 5.7$ mm, 4) $d = 4.5$ mm, 5) $d = 3$ mm.

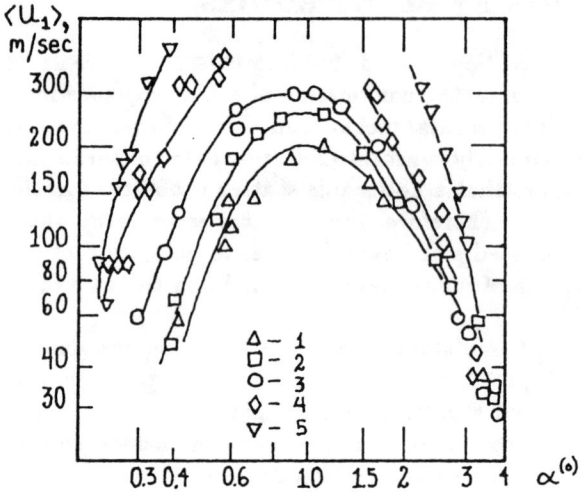

Figure 6.17 The dependence of the velocity at flame blow-out with a stabilizer on the excess air factor in the incoming flow $\alpha^{(0)}$ from data by Gorbatko, Kuznetsov, and Lipatov [1986]. Combustion of hydrogen-air mixture in the wake behind a cylinder with spherical head section mounted coaxially with the flow in the exit section of a 40 mm diameter tube, $T^{(0)} = 288$ K, $p = 0.05$ MPa, *1*) $d = 3$ mm, *2*) $d = 4$ mm, *3*) $d = 6$ mm, *4*) $d = 9$ mm, *5*) $d = 16$ mm.

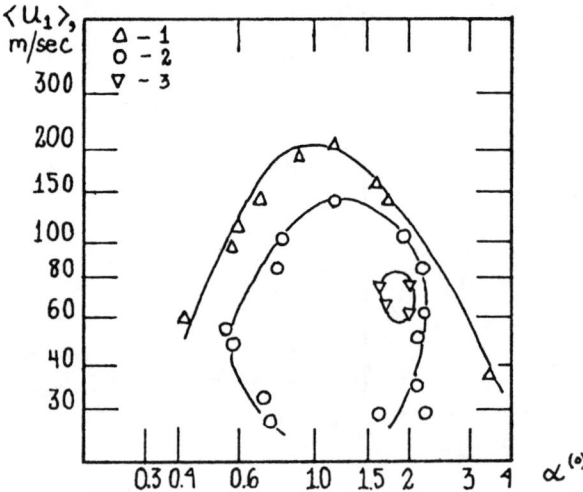

Figure 6.18 Regions of stable combustion of hydrogen-air mixtures at various pressures from data by Gorbatko, Kuznetsov, and Lipatov [1986]. Test conditions are the same as in Fig. 6.17, $T^{(0)} = 288$ K, $d = 3$ mm, *1*) $p = 0.05$ MPa, *2*) $p = 0.034$ MPa, *3*) $p = 0.029$ MPa.

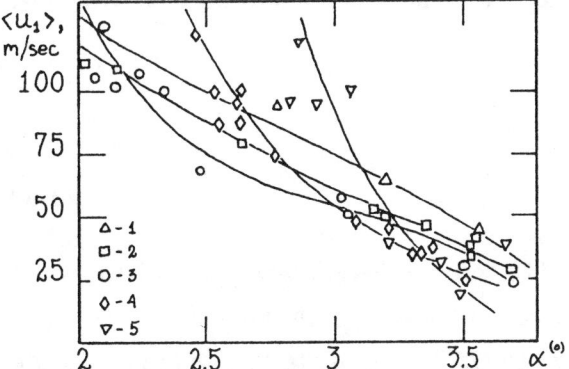

Figure 6.19 Regions of stable combustion of hydrogen-air mixtures at large excess air factors from data by Gorbatko, Kuznetsov, and Lipatov [1986]. Test conditions and notations are the same as in Fig. 6.17.

form depicted in Figs. 6.16, 6.17. In particular, it can be seen from Fig. 6.18 that when the pressure is decreased, the region of stable combustion becomes closed, and from Fig. 6.19 it follows that with increasing diameter of the stabilizer, the velocity at flame blow-out can decrease.

An explanation of these phenomena and verification of the flame stabilization criteria is given by Gorbatko, Kuznetsov, and Lipatov [1986] whose analysis showed that such a verification requires: a) taking into account the effect of Reynolds numbers and heat release on the structure of the flow in the wake behind the stabilizer, b) taking into account that the change of the composition in the reaction zone caused by the differences in the coefficients of molecular transport.

It has been established by trial and error that the best of the experimental data is obtained if parameters Mi and Re are calculated with respect to the complex parameter $\langle u_1 \rangle / \sqrt{\beta}$, where $\langle u_1 \rangle$ is the velocity at flame blow-out, β is the ratio of densities of the fresh mixture and combustion products, i.e.

$$\mathrm{Mi} = \frac{a^{(0)} \langle u_1 \rangle}{\sqrt{\beta}\, u_n^2 d}, \qquad \mathrm{Re} = \frac{\langle u_1 \rangle d}{\sqrt{\beta}\, \nu^{(b)}}$$

where $a^{(0)}$ is the thermal diffusivity of the fresh mixture, $\nu^{(b)}$ is the kinetic viscosity of the combustion products.

The description of the effect of the changes in the coefficient of molecular transport was based on the results obtained in §6.5, where it was shown that in the combustion of hydrocarbons one must take into consideration only the effects caused by selective diffusion and that the change of composition in the reaction zone is determined by the parameter $D_f^{(0)}/D_o^{(0)} = \psi$. Furthermore, the connection between the composition in the incoming flow and in the reaction zone is given by formula (6.19). Since it was established that the deformation of the flame (curvature, extension) plays a specific role, an attempt was made to generalize formula (6.19) in the form

$$\alpha^{(b)} = \frac{\alpha^{(0)}(1+\text{St}) + 1 - \psi}{1 + \psi \text{St}}, \qquad \alpha^{(b)} > 1$$

$$\alpha^{(b)} = \frac{\alpha^{(0)}(1+\psi \text{St})}{\psi(1+\text{St}) + (\psi - 1)\alpha^{(0)}\text{St}}, \qquad \alpha^{(b)} < 1 \qquad (6.40)$$

In contrast to formula (6.19), when deriving these relations it was not assumed that $\text{St} \gg 1$ and the ratio $\sqrt{D_f^{(0)}/D_o^{(0)}}$ is replaced by the symbol ψ. It is readily seen that when $\text{St} \to \infty$ and $\psi = \sqrt{D_f^{(0)}/D_o^{(0)}}$, formulas (6.40) and (6.19) coincide. As noted in §6.5, formulas (6.19) are also valid for the description of a curved flame when the diffusion coefficients are replaced by their squares, i.e., it is assumed that $\psi = D_f^{(0)}/D_o^{(0)}$. Thus, formulas (6.40) describe both extreme cases corresponding to the curving and stretching of the flame.

In using relations (6.40) it was assumed that quantity ψ is variable and a function of only one parameter

$$\frac{u_n d}{a^{(0)}} = u_n^0$$

which is expressed in terms of parameters Mi and Re $(u_n^0) = \sqrt{\text{Mi}/\text{Re}}$ with the equality of all the coefficients of molecular transport. Since Mi is a function of Re at flame blow-out, it is thereby assumed that depending on Re the role of selective diffusion is different.

The dependence $\psi(u_n^0)$ can be found from the experiments in which combustion of both rich and lean mixtures is investigated. Indeed, let us assume that the dependence of $\langle u_1 \rangle$ on $\alpha^{(0)}$ is known from the experiments and, correspondingly, the dependence $\langle u_1 \rangle / \sqrt{\beta} = F(\alpha^{(0)})$. Let us select some value $\alpha_1^{(0)} > 1$ which will be considered as a variable parameter. We find the quantity $\alpha_2^{(0)} < 1$ satisfying the condition

$$F(\alpha_1^{(0)}) = F(\alpha_2^{(0)}) \qquad (6.41)$$

i.e., roughly speaking, we compare rich and lean mixtures with equal rates of flame blow-out. The values of u_n must be the same in such mixtures. More precisely, the following equality must be satisfied

$$u_n(\alpha_1^{(b)}) = u_n(\alpha_2^{(b)}) \qquad (6.42)$$

The two formulas in (6.40) and the relations (6.41) and (6.42) yield a system of four equations with four unknowns $\alpha_2^{(0)}$, $\alpha_1^{(b)}$, $\alpha_2^{(b)}$ and ψ. Its solution (conducted numerically) enables the connection between ψ and $\alpha_1^{(0)}$, i.e., find the composition of the reaction zone to be found and thereby the value of u_n entering the parameter Mi to be calculated.

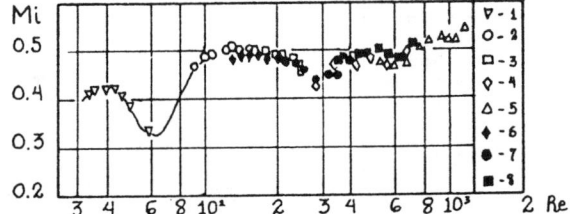

Figure 6.20 The dependence between parameter Mi and Re at flame blow-out in propane-air mixture from data by Dunskii and methane-air mixture from data by Volodina and Andreev [1958]. $T^{(0)} = 293$ K, $p = 0.1$ MPa; propane: *1)* $d = 3$ mm, *2)* $d = 4.5$ mm, *3)* $d = 5.7$ mm, *4)* $d = 8.7$ mm, *5)* $d = 11$ mm; methane: *6)* $d = 5$ mm, *7)* $d = 7$ mm, *8)* $d = 9$ mm.

The effectiveness of the indicated procedure for processing the experimental data is demonstrated in Fig. 6.20 and 6.21. The data for propane were obtained by Dunskii. These are depicted in Fig. 6.16. The data for methane were obtained by Volodina and Andreev [1958] in similar conditions (a cone with an apex angle of 70° placed in the section of 18 mm diameter tube). The quantity $\nu^{(b)}$ was calculated as a function of the combustion temperature of the mixture with an excess air factor equal to $\alpha^{(0)}$. As was expected, selective diffusion makes the reaction zone lean in the combustion of propane and rich in the combustion of methane (it can be seen from Fig. 6.21 that $\psi < 1$ in the first case and $\psi > 1$ in the second case and then it follows from (6.40) that $\alpha^{(b)} > \alpha^{(0)}$ for propane and $\alpha^{(b)} < \alpha^{(0)}$ for methane). With increasing u_n^0, which is accompanied by a simultaneous increase of Re, the role of selective diffusion decreases but, apparently, never totally disappears.

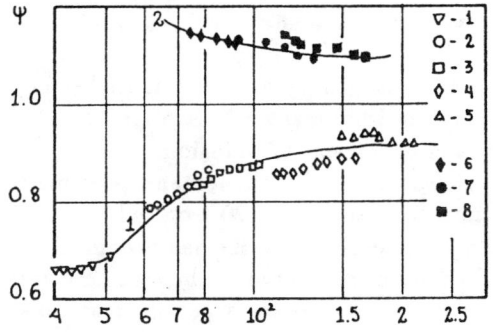

Figure 6.21 Effect of parameter $u_n d/a^{(0)}$ on selective diffusion in the combustion of propane-air mixture from data by Dunskii and methane-air mixture from data by Volodina and Andreev [1958]. The notation is the same as in Fig. 6.20. *1)* propane, *2)* methane.

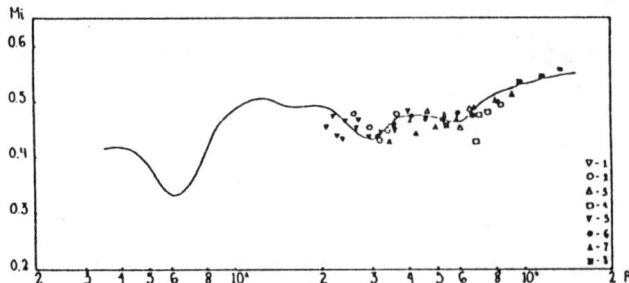

Figure 6.22 Critical processing of the experimental data obtained when investigating the combustion of propane-air mixture from data by Dunskii, and methane-air mixture from data by Volodina and Andreev [1958] at different initial temperatures $T^{(0)}$. The solid line is transferred from Fig. 6.20, $p = 0.1$ MPa; propane: $d = 5.7$ mm, *1)* $T^{(0)} = 293$ K, *2)* $T^{(0)} = 353$ K, *3)* $T^{(0)} = 503$ K, *4)* $T^{(0)} = 624$ K; methane: $d = 7$ mm, *5)* $T^{(0)} = 293$ K, *6)* $T^{(0)} = 373$ K, *7)* $T^{(0)} = 473$ K, *8)* $T^{(0)} = 573$ K.

Despite the opposite character of the effect of selective diffusion on the combustion of propane and methane, the results of the experiments carried out with both fuels are well generalized by a single curve (Fig. 6.20).

Finally, it is evident from Fig. 6.20 that the effect of Re is very noticeable and leads to the creation of several extremum in the dependence Mi (Re). The latter is an indication of repeated rearrangement of the region of reverse currents taking place with changing number Re which is very characteristic for the flows at transition Reynolds numbers.

The results of a series of experiments with propane and methane in which the initial temperature was varied are presented in Fig. 6.22. In this series the parameter β was varied over a much wider range than in the experiments in Fig. 6.20. Since a satisfactory correlation is obtained in this case, it can be assumed that the complex parameter $\langle u_1 \rangle / \sqrt{\beta}$ truly reflects the effect of heat release on flame stabilization.

The data obtained from the combustion of hydrogen were not considered since, in this case, in addition to selective diffusion, an important role is played by the difference between the thermal diffusivity and coefficient of diffusion.

In order to elucidate the qualitative character of the phenomena presented in Fig. 6.18, we assume that $D_f = D_o$ and turn to Fig. 6.20 from which it is evident that when Re = 60 - 200 the dependence of Mi on Re has a nonmonotonic character. It should be noted that viscosity is calculated at the temperature of the combustion products, i.e., its value is 30 - 40 times larger than at normal temperature; therefore, the Reynolds number range under consideration is of practical interest. As already mentioned, the nonmonotonic character of the dependence Mi (Re) is associated with the characteristic features of the transition in the back-flow region. This process is very sensitive to the conditions in the incoming flow, geometry of the blunt body and so on. Therefore, the data presented in Fig. 6.20 are largely dependent on the specific features of the experimental set-up (note that both series of experiments in Figs. 6.20 and 6.22

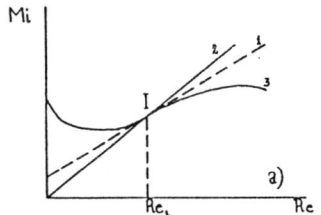
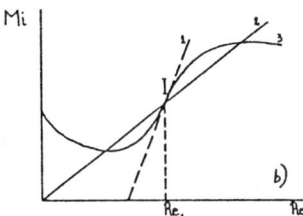

Figure 6.23 Qualitative character of dependence Mi (Re) *1)* tangent to the inflection point, *2)* straight line Mi = hRe, *3)* the curve bounding the region of stable combustion.

were conducted in practically similar conditions). In view of this, it is expedient to assume henceforth that in the range under consideration only the qualitative character of dependence Mi (Re) is preserved.

We shall now attempt to plot the region of stable combustion in the traditional coordinates $\langle u_1 \rangle$, $\alpha^{(0)}$. For this, it is necessary to analyze the pattern of variation of the root of the following equation with the variation of $u_n(\alpha^{(0)})$

$$\text{Mi}(\text{Re}) = h\,\text{Re}, \qquad h = \frac{a^{(0)}\nu^{(b)}}{u_n^2 d^2} \qquad (6.43)$$

As follows from Fig. 6.23a, b, two different cases are possible. In the first (Fig. 6.23a) the following condition is fulfilled

$$\text{Mi}(\text{Re}_1) > \text{Re}_1 \text{Mi}'(\text{Re}_1), \qquad \text{Mi}' = \frac{d\,\text{Mi}}{d\,\text{Re}} \qquad (6.44)$$

where Re_1 is the coordinate of the inflection point I of the dependence Mi (Re). In this case, there is always only one root. If condition (6.44) is not fulfilled, then depending on constant h one or three roots can be realized (Fig. 6.23b). Note that for the data shown in Fig. 6.20 equality Mi (Re_1) = Re_1 Mi' (Re_1) is approximately fulfilled. Therefore, on the basis of the above remark concerning the regimes of stable operation in the range of Re numbers under consideration being dependent on the specific features of the experimental set-up, it can be assumed that both cases are encountered in practice.

It is evident that in the first case the region of stable combustion has the usual shape (Fig. 6.24a). In the second case, two variations are possible. Indeed, the slope of the straight lines Mi = hRe is minimal when $\alpha^{(0)} = \alpha_m^{(0)}$, where $\alpha_m^{(0)}$ is the excess air factor at which u_n is a maximum. Consequently, with the rise

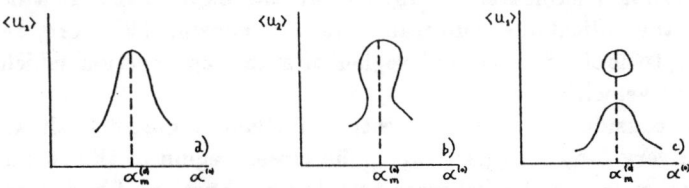

Figure 6.24 Qualitative shape of the regions of stable combustion.

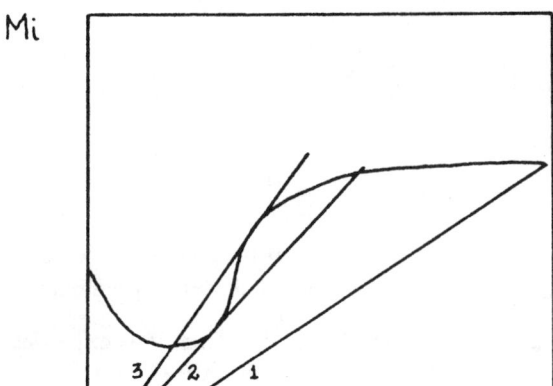

Figure 6.25 The evolution of the roots of equation Mi = hRe 1, 2, 3) successive locations of line Mi = hRe with increasing $\delta\alpha$. The straight lines are tangential to curve Mi = Mi (Re).

of $\delta\alpha = |\alpha^{(0)} - \alpha_m^{(0)}|$, the quantity h can only increase. If $h(\alpha_m^{(0)})$ is sufficiently small, then the evolution of the straight line Mi = hRe with the variation of $\delta\alpha$ has the shape illustrated in Fig. 6.25, i.e., at small $\delta\alpha$ there is only one root. When $h_2 < h < h_3$ there are two roots (h_2, h_3 are the slopes of the curves 2 and 3 in Fig. 6.25 passing through the origin of the coordinate and tangential to the curve Mi = Mi (Re)). With further increase of $\delta\alpha (h > h_3)$ there is again only one root. Therefore, the region of stable combustion has the shape illustrated in Fig. 6.24b.

If condition $h_2 < h(\alpha_m^{(0)}) < h_3$ is satisfied, then when $\alpha = \alpha_m^{(0)}$, there are three roots. When $\delta\alpha$ is increased, the evolution of two roots, corresponding to the largest Mi, yields a closed region of stable combustion in Fig. 6.24c, and the evolution of the third root, corresponding to the smallest Mi, yields a region of stable combustion of the usual shape which is located below the closed region in Fig. 6.24c.

Thus, it can be assumed that the appearance of a closed region of stable combustion in Fig. 6.19 is caused by the rearrangement of the zones of backflow with changing Reynolds numbers. In 1970 when the experiments in this figure were conducted, the indicated circumstance was not known. Therefore, an attempt was not made to look for a second region of stable combustion which would favor low mixture velocities.

In conclusion, let us consider the effect of selective diffusion. Figure 6.19 shows that this effect can be very large. In particular, the closed region of the stable combustion is formed when $\alpha^{(0)} \sim 1.8$, whereas with the combustion of hydrogen the maximum value of u_n is attained when $\alpha^{(0)} = 0.6$. As indicated above,

$\psi = D_f^{(0)}/D_o^{(0)}$ in the combustion of a curved flame. Therefore, we conclude from (6.40) that $\alpha^{(b)} = 0.6$ when $\alpha^{(0)} = 1.8$ $(D_f^{(0)}/D_o^{(0)} \simeq 3)$ i.e., formulas (6.19) and the more exact relations (6.43) predict correctly the change of composition in the reaction zone.

It should also be noted that in the combustion of hydrogen the dependence of ψ on u_n^0 has qualitatively the same form as for methane (in both cases $D_f > D_o$). Naturally, the effect of selective diffusion on the combustion of methane is far less than on the combustion of hydrogen (D_f/D_o in the first case is much less than in the second). This explains the anomalous effect of the diameter on the limits of stable combustion (Fig. 6.18). Indeed, at large $\alpha^{(0)}$ with decreasing d the quantity ψ increases (curve ℓ in Fig. 6.21). Therefore, the composition in the reaction zone becomes leaner and, therefore u_n increases, possibly leading to the expansion of the region of stable combustion despite the decrease in the diameter of the stabilizer.

CHAPTER
SEVEN
TURBULENT COMBUSTION OF PARTIALLY PREMIXED GASES

In the present chapter the combustion of partially premixed gases in a turbulent flow is considered. The need for the analysis of such a process is called for for at least two reasons. Firstly, during the combustion of atomized fuel as a result of phase changes, some of the drops can end up in the region where the composition in the gaseous phase is so lean that the rate of reaction taking place is negligible. Evaporation of such drops leads to the fuel which has not been involved in the reaction appearing in the region with lean composition. Secondly, when burning premixed gases, as already noted in Chapter 5, holes can form in the flame front through which the nonreacted fuel penetrates to the lean regions.

Thus, the investigation of combustion in a flow with strongly variable composition in the presence of unburned fuel and, hence, unburned oxygen in all flame regions is of substantial practical interest. Consequently, in the case under consideration two particular features arise:

1. The combustible components are located on both sides of the reaction zone and are transferred to it from both lean and rich mixtures, i.e., the process is similar to the combustion of nonpremixed gases.

2. The combustible components diffuse to the reaction zone in parallel, i.e., the process is similar to the combustion of a homogeneous mixture.

As will be shown later, this characteristic feature leads to the overall laws of combustion of partially premixed gases differing very greatly from the laws of both diffusion combustion and combustion of a homogeneous mixture. Therefore, the combustion of partially premixed gases ought to be identified as an independent type of combustion. The theoretical investigation of this type of combustion is essentially at a very early stage; therefore, the results expounded below are mainly of a qualitative character.

§7.1 EFFECT OF TURBULENCE ON THE COMBUSTION OF ATOMIZED LIQUID FUEL

It is obvious that two regimes of drop combustion are possible. In the first, each drop is totally surrounded by the flame, the distance from the center of the drop to the flame front is proportional to its diameter d, i.e., very small. In the second regime, the combustion of individual drops is not possible, for example, due to the large curvature of the flame front or high velocity of drop motion relative to the ambient gas. Furthermore, the position of the reaction zone is determined by mixing the air and the fuel evaporating from all the drops, and the distance from any drop to the flame front is not, generally speaking, linked with its diameter. Since the drops are on both sides of the flame front, combustion of partially premixed gases takes place in the latter case.

The purpose of the present section is to elucidate the conditions under which the second regime of drop combustion is realized. As will become evident later, this regime is the more typical, which is seen from the following considerations. Since fuel flow to the flame front increases with decreasing d, only sufficiently large drops are likely to burn. On the other hand, due to the nonstationary character of velocity fluctuations in the turbulent flow, such drops are never totally entrained. The role of this factor is evident when comparing the velocity of flame blow-out from the drop w_m and the fluctuating velocity $u = \sqrt{\langle (u_k - \langle u_k \rangle)^2 \rangle}$. Under normal conditions, even for very large drops ($d = 1$ mm) we have $w_m \lesssim 30$ cm/sec (Agafonova, Gurevich, and Tarasova [1960]). Meanwhile, in the majority of technical devices $u \sim 1 - 10$ m/sec. Therefore, flame blow-out is possible even with low inertia drops.

Consequently, it is natural henceforth to make assumptions which lead to the most optimistic estimates of the possibility of drop combustion. Thereby, the conditions under which the second combustion regime exists will be explained below.

Let us first consider the combustion of a drop which is totally entrained by the flow. We estimate the minimum diameter d_m at which stable combustion is possible. To this end, we make use of the concepts in Chapter 5. These concepts are based on the Zel'dovich theory [1949] from which it follows that the specific flow of fuel q to any flame front does not exceed the specific flow rate of fuel q_m in the normal flame front propagating across a stoichiometric mixture, i.e., $q < q_m = \rho_0 u_n z_s$, where u_n, ρ_0, z_s are the normal velocity of flame propagation, density, and concentration of the fuel in such a mixture. To estimate q we utilize the theory of drop combustion developed on the assumption that the combustion is quasistationary and the reaction rate is infinitely large (Varshavskii [1945], see also Williams [1965]).

Note that the assumption of quasistationary state of the combustion is true when the process is analyzed far from the critical point at which the boundary between the liquid and gas vanishes (for different fractions of kerosene the critical pressure and temperature fluctuates in the ranges 1.8 - 2.5 MPa, 600 - 700 K, respectively). In view of the fact that as we get closer to the critical point the heat of evaporation decreases, the quasistationary theory underestimates the quantity q, i.e., a more optimistic estimate of the possibility of drop combustion is obtained. It is also readily seen that in the first moments following ignition, the flame front is at a smaller distance from the drop than predicted by the quasistationary theory (in the space surrounding the drop there is little fuel vapor). Therefore, the assumption that the process is stationary also underestimates heat flux from the flame front to the drop and, consequently, again yields a more optimistic estimate of the possibility of drop combustion.

Within the framework of the quasistationary theory we have

$$d^2 = d_0^2 - \kappa t, \qquad 2r_s/d = b \tag{7.1}$$

where d, d_0 is the current and initial drop diameter, t is the time, r_s is the distance from the center of the drop to the flame front, κ and b are constants which are weakly dependent on the pressure and temperature of the air. By finding from the first relation in (7.1) the total fuel flow and from the second the flame surface, we obtain

$$q = \kappa \rho_e / 4b^2 d \tag{7.2}$$

where q is the specific fuel flow to the flame front, ρ_e is the density of the liquid. It follows from (7.2) that since $q < q_m = \rho_0 u_n z_s$, then the combustion of a single drop is possible only if

$$d > d_m = \frac{\kappa \rho_e}{4b^2 \rho_0 u_n z_s} \tag{7.3}$$

It is generally known that the first formula in (7.1) describes the measurement results well. In particular, for kerosene calculation and experiments yield $\kappa = 1$ mm^2/sec (Klyachko [1958]). In the meantime, the calculation results of b conducted in various studies are noticeably different: for kerosene we have $b = 13.1$ (Klyachko [1958]), $b = 35$ (Gogos, Sadkhal, Ayyaswamy, Sundararajan [1986]). Then, assuming that under normal conditions $\rho_e = 0.75$ g/cm^3, $u_n = 40$ cm/sec we obtain from (7.3) $d_m = 5 - 30$ μm.

The refinement of this estimate is unlikely to be justified, since the case considered above can be realized only under very artificial conditions. Indeed, let us find the velocity of the medium u_s at the flame front, i.e., for $r = r_s$. From the first formula in (7.1) and the continuity equation we have

$$u \rho r^2 = \frac{1}{16} \kappa \rho_e d$$

where u is the velocity, ρ is the density, r is the distance from the center of the drop. Using the second relation in (7.1) and inequality (7.3) we obtain

$$u_s \leq u_n z_s \rho_0 / \rho_s \tag{7.4}$$

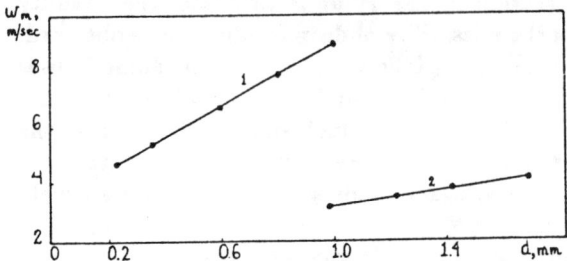

Figure 7.1 Flow velocity at flame blow-out from drops of different diameter from data by Agafonova, Gurevich, and Tarasova [1960]. The combustion of gasoline in air at temperature $T_0 = 1140$ K and normal pressure, 1) freely falling drops, 2) suspended drops.

where ρ_s is the density of the combustion products of the stoichiometric mixture. From (7.4) it follows that $u_s \approx 20$ m/sec in normal conditions, i.e., flow over the drop with even very low velocity must strongly affect the position of the flame front and, hence, its stability.

These estimates are related to the combustion of a single drop. The combustion of an aggregate of drops exhibits a number of characteristic features. Paramount among these is that the temperature of the atmosphere surrounding the given drop is high and its oxygen content low. As evident from the laws of conservation of heat and man, this initial burn out of part of the oxygen cannot change the temperature of the flame front which surrounds the drop. Therefore, chemical reactions in the combustion of both a single drop and an aggregate of drops take place in similar conditions. The only change being that owing to the decrease of oxygen concentration in the surrounding medium, the flame front moves away from the drop surface. For this reason, in the combustion of an aggregate of drops, the first regime of combustion is less probable than in the combustion of a single drop.

Let us now consider the combustion of a single drop in a flow with constant velocity. In order to find the rate of flame blow-out w_m we refer to the experiments by Agafonova, Gurevich, and Tarasova [1960]. From Fig. 7.1 it is evident that the experimental data are described well by the formula

$$w_m = A_1 + B_1 d \tag{7.5}$$

where A_1 and B_1 are independent of d. For the generalization of this formula we make use of a number of nonrigorous theoretical considerations.

Let us first analyze the extreme case $w_m d/\nu \gg 1$, where ν is the kinematic viscosity of the gas. Then, on the front surface of the drop there is a boundary layer whose thickness is a minimum at the front critical point F (Fig. 7.2). At this point the least favorable conditions of combustion are obviously realized. Flame blow-out at this point does not necessarily mean that combustion has been completely disrupted (the flame can partially engulf the drop). However, the data by Agafonova, Gurevich, and Tarasova [1960] indicate that the region of realization of such a regime is quite narrow. In particular, it is seen from Fig. 7.3 that the fraction of the unburnt fuel $1 - \eta_x$ is less by 10% and is independent of the flow velocity w if $w \leq 29$ cm/sec, and when $w > 29$ cm/sec this fraction increases rapidly with increasing w and at $w = w_m = 35$ cm/sec total flame flowout occurs. Hence, it can be assumed that the flame partially engulfs the drop only in a narrow range of velocities $w = 29 - 35$ cm/sec.

Therefore, it is henceforth assumed that flame blow-out at the frontal critical point and of the entire drop can be identified. As will become evident later, such an approach enables us to explain the dependence (7.5) and to find the conditions of combustion stability. The quantitative inaccuracy of this approach can be easily corrected if the empirical constants entering into this condition are determined from the experimental data obtained when investigating complete flame disruption.

It is known (Schlichting [1960]) that the boundary layer in the vicinity of the frontal critical point is determined only by the ratio w/d and not by the quantities w and d each taken separately. We further note that owing to the large values of the stoichiometric factor, the flame is located close to the external boundary layer edge and, therefore, the temperature distribution inside the boundary layer has approximately a universal form: it varies from the boiling temperature on the surface of the drop T_v to the adiabatic combustion temperature of the stoichiometric mixture T_s, the quantities T_v and T_s are weakly dependent on the pressure and temperature. Consequently, the distribution of the thermophysical properties is almost universal so that the processes of molecular transport are characterized by a single quantity, for example, by the viscosity at temperature T_s.

The role of evaporation, including its effect on the velocity within the boundary layer, is described by the parameter $c_p(T_s - T_v)/\ell$, where c_p is the specific

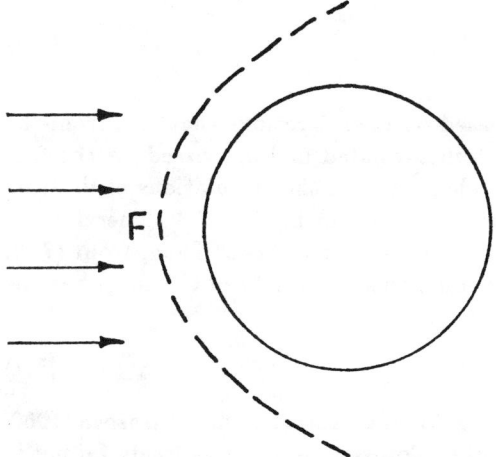

Figure 7.2 Schematic diagram of a drop in a flow. The dashed line represents the conditional boundary layer limit.

290 TURBULENCE IN COMBUSTION

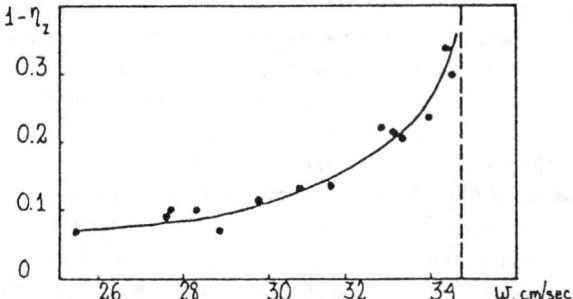

Figure 7.3 Dependence of the completeness of combustion of a single drop on the flow velocity from data by Agafonova, Gurevich, and Tarasova [1960]. Combustion of a suspended drop of gasoline with $d = 1.8$ mm at normal conditions. The dotted line is flame blow-out.

heat at constant pressure, ℓ is the specific heat of evaporation. As evident from the aforementioned, this parameter is roughly constant and will not, therefore, be considered further.

Finally, if follows from Chapter 5 that, as a first approximation, the role of chemical kinetics during the analysis of stability of the diffusion flame is described by the parameter u_n^2/a, where a is the thermal diffusivity which is calculated from the temperature of air T_0, u_n is the maximum value of the velocity of normal flame propagation. From these quantities only one parameter can be compiled $w\,a/u_n^2 d$. This parameter appeared in Chapter 6 when analyzing the limits of combustion stability of a homogeneous mixture. Since its value at flame blow-out must be constant, it hence follows from here that $w_m \sim d$ for very large d which totally agrees with formula (7.5). We thus obtain

$$A_1 = A\,u_n^2/a \tag{7.6}$$

where A is a universal constant.

The estimation of constant B_1 is based on considerations that are, from the theoretical viewpoint, significantly less substantiated and are based on the fact that the value of B_1 is close to u_n as evidenced from the calculations of u_n from formula (6.4) and the results of measurements shown in Fig. 7.3. Therefore, it is assumed that $B_1 = B\,u_n$, where B is a universal constant. Then, from (7.5), (7.6) we conclude that the condition of combustion stability of the drop has the form

$$w < w_m = B\,u_n + A\,u_n^2 d/a \tag{7.7}$$

To find A and B we use the data by Agafonova, Gurevich, and Tarasova [1960] which were obtained by investigating total blow-out from drops freely falling in air (curve 1 in Fig. 7.1). By using (6.4) and assuming that $a \sim T_0^2/p$ (p is the pressure, $a = 0.2$ cm^2/sec, $u_n = 40$ cm/sec under normal conditions, we obtain $A = 0.045$, $B = 0.56$.

The validity of the assumption made above is confirmed by the analysis of a whole series of experimental data. In particular, from a comparison of curves *1* and *2* in Fig. 7.1, it follows that for freely falling drops the quantity w_m is noticeably larger than for suspended drops. It can be assumed that in the latter case u_n decreases due to heat transfer along the suspending element. Despite the fact that such a decrease does not yield to theoretical calculations, it is possible to verify formula (7.7) since the relation (7.5) is true in both cases. It follows from (7.7) that the ratio A_1/B_1^2 is independent of u_n. Analysis of the data presented in Fig. 7.1 confirms this conclusion and indicates that for curves *1* and *2* the values of the ratio A_1/B_1^2 differ by no more than 10% despite the significance of the effect of heat transfer. The latter is evident from the comparison of parameter B_1 which is obtained in both cases. If it is assumed that $B = $ const, then for the suspended drops u_n decreases by 1.9 times.

One more confirmation of the validity of formula (7.7) is obtained from an analysis of the tests in which the effect of the air temperature T_0 on the stability of combustion of suspended drops was investigated. Before we present such an analysis, let us consider the role of natural convection. We estimate the magnitude of the convective velocity w_c on the assumption that the combustion pattern is not significantly disturbed. It is obvious that the characteristic dimension determining the phenomenon under consideration is bd, i.e., the diameter of the sphere at which the flame is located. The role of the buoyancy force is determined by the complex parameter $g(\rho_0 - \rho_s)/\rho_0 \approx g$, where g is the gravitational acceleration, ρ_0 is the air density. Therefore, from considerations of dimensionality we obtain

$$w_c = \sqrt{g b d} \tag{7.8}$$

It is evident, that formula (7.7) is true when $w \gg w_c$.

Taking into account the reservation made, we compare the results of the calculations from formula (7.7) and the experimental data obtained by Agafonova, Gurevich, and Tarasova [1960] for $d = 1.8$ mm and $T_0 = 1140, 873, 673$, and 293 K. The calculated values amount to $w_m = 4.65, 2.75, 1.6$, and 0.27 m/sec and the measured results are $w_m = 4.65, 2.85, 1.7, 0.4 - 0.8$ m/sec. The uncertainty of the last figure is due to the fact that when air moves from bottom to top and top to bottom, different results are obtained, i.e., natural convection plays an important role. That experiments were conducted with suspended drops is taken into account in the calculations. Therefore, in accordance with these estimates the quantity u_n for all T_0 is decreased by 1.9 times. It is evident that at all temperatures, with the exception of the lowest, the correlation of calculation and experiment is quite satisfactory. From (7.7) and (7.8) it is also evident that $w_m \sim w_c$ at $T_0 = 293$ K ($w_c = 50 - 80$ cm/sec when $b = 13 - 35$), i.e., the estimate also indicate the important role played by natural convection.

Thus, this analysis confirms the validity of formula (7.7). Note that this formula was derived on the basis of experimental data obtained for $d \gg d_m$. Therefore, for small d it yields high values of w_m (in particular, since $B > 0$, a physically unrealistic result concerning the possibility of combustion of zero diameter drops follows). Consequently, the condition for its applicability has the form $d \gg d_m$, where d_m is given by formula (7.3). The second condition for the validity of relation (7.7) is $w \gg w_c$; this is obviously of little significance, since the fluctuating velocities are unusually many times larger than w_c.

Let us now estimate the velocity of motion of the drop relative to the gas. If one assumes that the drops do not rotate and the velocity of the medium varies little within distances of the order d, then when $\rho_\ell \gg \rho_0$ we have

$$\frac{dv_k}{dt} = \varphi(\text{Re}) \frac{u_k - v_k}{\tau}, \qquad \tau = \frac{d^2 \rho_\ell}{18\mu}$$

$$\text{Re} = \frac{wd}{\nu}, \qquad w_k = u_k - v_k \qquad (7.9)$$

Here, v_k and u_k are velocities of the drop and gas in the laboratory system of coordinates, μ, ν are dynamic and kinematic viscosity of the gas. Estimates indicate that Re < 100 in all practically important cases and, therefore, $1 < \varphi < 4$ (Schlichting [1960]). As will be seen later, $w \sim 1/\sqrt{\varphi}$, i.e., Stokes formula ($\varphi = 1$) yields no more than twofold error which is quite acceptable. Therefore, it is natural to use Stokes formula and, in order to confirm the estimate $w \sim 1/\sqrt{\varphi}$, to consider a simplified model in which it is assumed that $\varphi = \text{const}$.

Such an approach is quite justified, since from (7.9) it follows that w is a random quantity and, therefore, that there is no clear boundary between the regimes in which the drops burn or just evaporate. Indeed, when $\langle w^2 \rangle \sim w_m^2$, condition (7.7) is fulfilled only at some times, i.e., the drops are successively ignited and extinguished. Consequently, the exact calculation of w is almost useless, and the qualitative laws of the process can be identified by comparing w_m^2 with the results of the estimates of $\langle w^2 \rangle$.

The initial portion of the path of the drop is not considered further, i.e., a statistically stationary solution of equation (7.9) is sought. Thereby, an underestimate of the velocity of motion of the drop relative to the gas is given and, consequently, a more optimistic estimate of the possibility of its combustion is obtained. The following formula is true (Shraiber, Milyutin, and Yatsenko [1980])

$$\langle w^2 \rangle = \int_0^\infty \frac{x^2 E(\omega)}{1 + x^2} d\omega, \qquad x = \frac{\omega \tau}{\varphi} \qquad (7.10)$$

where ω is the frequency, E is the spectral density of the energy of the gas velocity fluctuations along the drop path.

From (7.10) it is apparent that the main contribution to $\langle w^2 \rangle$ is made by the fluctuations with frequencies of the order φ/τ and higher. In typical cases they belong to the inertial interval of the turbulence spectrum. As seen from Chapter 4, E in this case is dependent only on the mean energy dissipation $\langle \epsilon \rangle$ and frequency ω, i.e., (Gurvich and Yurchenko [1980], see also the review by Yaglom [1981])

$$E(\omega) \sim \langle \epsilon \rangle / \omega^2 \qquad (7.11)$$

Then, from (7.10), (7.11) it follows that within a constant multiplier

$$\sqrt{\langle w^2 \rangle} \sim \sqrt{\langle \epsilon \rangle \tau / \varphi} \qquad (7.12)$$

Thus, the velocity of the drop relative to the medium is proportional to the slowly varying multiplier $1/\sqrt{\varphi}$ and its further consideration is inexpedient. The results of measurements of the spectral energy density of the gas velocity fluctuations along the path of the particles are unknown. Hence, the coefficient of proportionality in (7.12) is unknown. Therefore, this is assumed equal to unity in the estimates that are to follow.

It should be noted that there are cases when the frequency ω lies outside the inertial interval. Then, it follows from (7.12) that $\sqrt{\langle w^2 \rangle} > u$ if τ is large, or $\sqrt{\langle w^2 \rangle} < (\nu \langle \epsilon \rangle)^{1/4}$ if τ is small. In the first case, the drops are very passive and it can, therefore, be assumed that the flow velocity around them coincides with the fluctuation velocity of the gas u. In the second case, the value of E when $\omega \gtrsim 1/\tau$ is so small that, as a first approximation, $E = 0$. Thus, the generalization of formula (7.12) assumes the form

$$\sqrt{\langle w^2 \rangle} = u, \qquad \langle \epsilon \rangle \tau > u^2$$

$$\sqrt{\langle w^2 \rangle} = \sqrt{\langle \epsilon \rangle \tau}, \qquad \sqrt{\nu \langle \epsilon \rangle} < \langle \epsilon \rangle \tau < u^2$$

$$\sqrt{\langle w^2 \rangle} = 0, \qquad \langle \epsilon \rangle \tau < \sqrt{\nu \langle \epsilon \rangle} \qquad (7.13)$$

Let us now consider the estimation of viscosity. Since the flame is positioned at the outer edge of the boundary layer, there exists an analogy between a burning drop in a stream and a sphere at temperature T_v in a stream at temperature $T_s \gg T_v$. Consequently, one can use the empirical relation $\mu = \sqrt{\mu(T_v) \mu(T_s)}$ obtained by Kabanov and Klubnikin [1985] when investigating the drag of a cold sphere in a hot stream. Furthermore, it must be born in mind that the calculation of the Kolmogorov velocity $(\nu \langle \epsilon \rangle)^{1/4}$ must be carried out with respect to the viscosity of the medium at a large distance from the drop.

We note that the third formula in (7.13) leads to a more favorable estimate of the possibility of drop combustion, since when $\langle \epsilon \rangle \tau \lesssim \sqrt{\nu \langle \epsilon \rangle}$ drop combustion does not, generally speaking, take place as in a stationary medium. The fact is that in such a medium the flame is at a large distance from the drop and its diameter bd can be larger than the Kolmogorov dimension $\nu^{3/4} \langle \epsilon \rangle^{-1/4}$. At such distances the velocity of the medium changes by the order $(\nu \langle \epsilon \rangle)^{1/4}$ which can be larger than u_n in view of which the action of deformation of the outer flow on the combustion process becomes very substantial.

Let us consider several typical cases. The first corresponds to the conditions of the experiments by Tikhomirov [1958] which were conducted at a pressure $p = 0.1$ MPa, temperature $T_0 = 573$ K and drop diameters $d = 50 - 100$ μm. The regimes of most interest are those in which a high degree of completeness of combustion is achieved at high excess air factor α. This is associated with the fact that at low α propagation of the flame is possible across fairly rich mixture of air and fuel which evaporates but does not burn. Therefore, an unambiguous conclusion about the combustion of a single drop can be made only from the experiments conducted at large α when the mixture of air and vaporized fuel is not hot. The regimes under consideration were indeed observed by Tikhomirov [1958] for $\alpha = 1.2 - 2$ and, hence, for large excess air factor values in the gaseous phase. The fluctuating velocity in all these regimes is small ($u = 0.8 - 2$ m/sec, i.e., $w < 0.8 - 2$ m/sec). Since it follows from (6.4), (7.7) that $w_m = 0.9 - 1$ m/sec ($d = 50 - 100$ μm), it can be concluded that $w \sim w_m$, i.e., the combustion of single drops is possible.

As a second example, we consider the regime of the secondary gas of a gas turbine ($T_0 = 450$ K, $p = 0.3$ MPa). The mean gas velocity in such chambers is of the order of 50 m/sec and, therefore, there will not be a large error if it is assumed that $u = 10$ m/sec. We also assume that the scale of turbulence $L = 5$ cm, i.e., 20% of the diameter of the flame tube just as in an ordinary tube. Thus, using the estimate $\langle \epsilon \rangle = u^3/L$, we obtain $\langle \epsilon \rangle = 2 \times 10^8$ cm^2/sec^3.

Figure 7.4 shows the dependences of $\sqrt{\langle w^2 \rangle}$ and w_m on d. These are only qualitative illustrations, since it follows from (7.7), (7.13) that the slope of the straight line 1, described by the quantity $\sqrt{\langle w^2 \rangle}$, is two orders of magnitude larger than the slope of the straight line 2 corresponding to the dependence $w_m(d)$, i.e., with increasing d the velocity of flow around the drop increases much more rapidly than the growth of the stability margin of combustion. Therefore, either very large ($d > d_1$) or very small ($d < d_2$) drops are burned (Fig. 7.4). Calculations show that the dimension of d_1 is unrealistically large ($d_1 = 6$ mm) and the dimension of d_2 is unrealistically small ($d_2 = 5$ μm) and is, furthermore, comparable with d_m ($d_m = 1.5 - 10$ μm for $b = 13 - 35$). In this case, it follows from (7.13) that drops of dimension of the order d_2 are totally entrained by the flow. This does not mean that they burn just as in a stationary medium, since the diameter of the flame in the stationary medium $bd = 65 - 200$ μm is much larger than the Kolmogorov

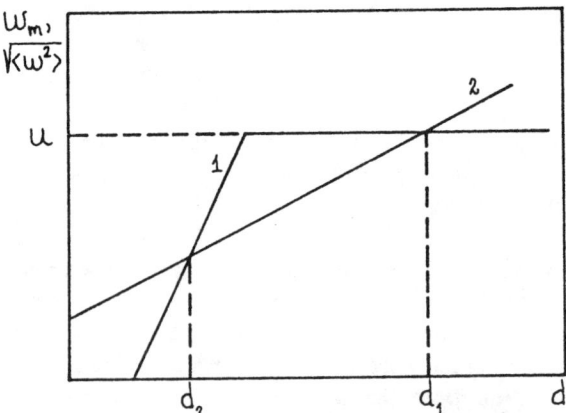

Figure 7.4 Effect of drop diameter on the velocity of flame blow-out and on the root mean square velocity of the flow *1)* $\sqrt{\langle w^2 \rangle}$, *2)* w_m.

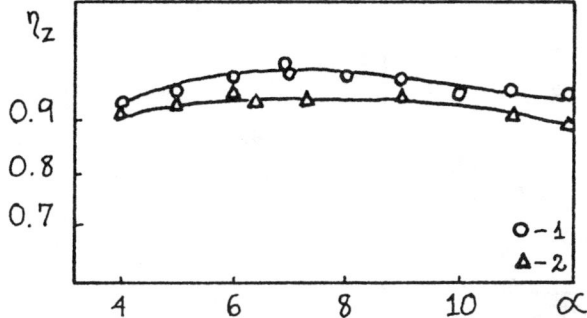

Figure 7.5 Effect of excess air factor on the completeness of combustion in a gas turbine chamber from data by Doroshenko [1960]. *1*) vaporous fuel, *2*) liquid fuel.

scale $\eta = \nu^{3/4}\langle\epsilon\rangle^{-1/4} = 14~\mu m$, and at distances of the order η the velocity of the medium changes by the order $(\nu\langle\epsilon\rangle)^{1/4} = 70$ cm/sec which is comparable with $u_n = 75$ cm/sec.

Thus, in this case combustion of a single drop is not possible, and the behavior of the process is determined by the combustion of the gases. This is apparent from the work by Doroshenko [1960]. As an illustration Fig. 7.5 depicts the results of measurement of completeness of combustion η_z at the exit from the chamber. The initial distributions of fuel concentration are approximately equal. It follows from this figure that the substitution of the liquid fuel by evaporated fuel only slightly increased the effectiveness of the process. In this work one more qualitative proof is obtained of the validity of this conclusion: It is shown that the results of measurements of completeness of combustion at various operating regimes are generalized with the aid of some parameter representing the ratio of the time of the chemical reaction to the time of residence of the gas particle in the combustion chamber. Since this parameter does not include the characteristics of evaporation, it follows that the process is determined by the combustion of the gases.

§7.2 QUALITATIVE STRUCTURE OF THE ZONE OF CHEMICAL REACTIONS DURING THE TURBULENT COMBUSTION OF PARTIALLY PREMIXED GASES

The analysis conducted in §7.1 shows that at sufficiently high turbulence intensity the combustion of single drops is not possible. This circumstance is important, since upon fuel atomization with the aid of an injector the drops can end up in the region with lean composition owing to phase slip. In such regions only drop evaporation takes place and chemical reactions progress very slowly. As noted at the very beginning of the present chapter, owing to the formation of holes in the diffusion flame, lean nonreactive mixtures of air and fuel are possible even with the combustion of initially nonpremixed gases. It should also be noted that even in the absence of holes as a result of the finite nature of the reaction rate, part of

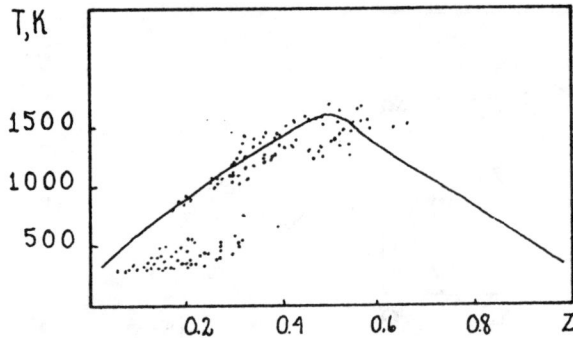

Figure 7.6 Effect of composition on the temperature during the combustion of hydrogen diluted by nitrogen from data by Dibble and Magre [1987]. The solid line — thermodynamic equilibrium temperature, the dots — data obtained upon "instantaneous" measurement of composition and temperature at a point.

the unburned fuel can also pass through the flame front. This effect is shown in Fig. 7.3 from which it follows that at low flow velocities around the drop, when it is totally engulfed by the flame, the completeness of combustion is not equal to unity.

Analogous data were obtained by Dibble and Magre [1987] when investigating the turbulent diffusion flame generated by the flow of a mixture of hydrogen and nitrogen in a coflowing air stream. In this study the composition and temperature were simultaneously measured with the aid of laser Raman spectroscopy with high temporal and spatial resolution. The results of measurements at some times are depicted as points in Fig. 7.6 along whose axes temperature T and the restored concentration of the fuel z are plotted. The scatter of the points is caused by the random character of the process. The presence of points above the solid curve, which corresponds to the thermodynamic equilibrium temperature, indicates a noticeable effect of the differences between the coefficient of molecular diffusion and thermal diffusivity. It is also apparent that when $0 < z < 0.35$, the temperature at some times is exactly equal to the initial temperature, i.e., in the lean part partially burned mixture of hydrogen, nitrogen, and oxygen is observed.

In the region $z > z_1 \approx 0.35$ the reaction progresses at a high rate (all the points are located close to the solid curve which describes the distribution of temperature under thermodynamic equilibrium). In region $z < z_1$ there are very many points at which the temperature is equal to the initial temperature. Thus, two regions can be identified involving high and low reaction rate which are divided by an explicitly pronounced boundary.

This characteristic feature is due to the rate of reaction being strongly dependent on the temperature and, consequently, on the conserved scalar of the fuel z which determines the maximum temperature. The consequences resulting from this feature are analyzed in the work by Kuznetsov [1983], in which a very simplified model of oxidation of hydrocarbons is formulated. The need for the construction of such a model is called for by the kinetic model developed in §5.4 begin valid only for the combustion of mixtures with composition that is not strongly different from the stoichiometric. As we get closer to the region of lean mixtures (decreasing z) the conditions for its applicability become more rigid. On the basis of the study under consideration, we assume that combustion is limited only by the reaction $CO + OH \rightleftharpoons CO_2 + H$, and that the concentrations of radicals OH and H are in thermodynamic equilibrium. Then the system of equations (5.32) acquires the form

$$N \frac{d^2 c_8}{dz^2} = -k c_2^{(e)}(z) \left[c_8 - c_8^{(e)}(z) \right] \tag{7.14}$$

where N is the scalar dissipation, c_2, c_8 is the concentration of OH and CO, k is the reaction rate constant for $CO + OH \rightarrow CO_2 + H$ (this constant is a very weak function of temperature and, therefore, it is henceforth assumed to be independent of z), superscripts e correspond to thermodynamic equilibrium. In writing (7.14) it is taken into account that the concentration of CO_2 is usually many times greater than the concentration of CO (regions $z > z_8$, in which this assumption is violated, will not henceforth be considered).

Thermodynamic calculations show that at normal conditions and $0.5 z_s < z < 0.9 z_s$ the following approximation is true

$$c_2^{(e)} = \text{const} \exp(bz), \qquad c_8^{(e)} = (c_2^{(e)})^n \tag{7.15}$$

where b and n are slowly changing functions of z ($n = 2\text{--}3.5$, $b = 70 - 180$).

Since b is large, i.e., the reaction rate is very strongly dependent on z, two regions can be identified. In the first ($z > z_1$) the reaction rate is very large and, therefore, $c_8 \approx c_8^{(e)}$. In the second region ($z < z_1$) the reaction rate is small and, therefore, it follows from (7.14) that c_8 is linearly dependent on z^*. In order to fine the magnitude of z_1, we shall assume that $b = \text{const}$, $n = \text{const}$, and make substitution of the variables

$$z = z_1 + \frac{1}{b} \varsigma$$

Then, by selecting z_1 from the condition

$$k c_2^{(e)}(z_1) = b^2 N \tag{7.16}$$

we obtain the following equation from (7.14), (7.15)

$$\frac{d^2 c_8}{d\varsigma^2} = e^\varsigma (c_8 - e^{n\varsigma}) \tag{7.17}$$

*As already stated in §5.4, in this region system (5.32) and, hence, model (7.14) are not applicable.

which does not include large or small parameters. Thus, both regions under consideration are separated by a thin boundary layer near the surface $z = z_1$. In this layer the conserved scalar of the fuel varies by a small magnitude $1/b$. It is also easy to see that the location of surface $z = z_1$, which we shall refer to as the freezing front, is a very weak function of N. Indeed, from (7.15), (7.16) we obtain

$$z_1 = \text{const} + \frac{1}{b} \ln N \qquad (7.18)$$

Hence, it is evident that all the characteristic features of the behavior of fuel ending up in the lean regions of the flame ($z < z_1$) are described by the equation of diffusion without reaction. The boundary conditions for this equation are specified on a curved nonstationary surface $z = z_1$. Since it follows from (7.18) that z_1 is a very weak function of N, it can, as a first approximation, be considered as an isoscalar surface, i.e., $z_1 = \text{const}$. The boundary conditions on such a surface are determined from joining the solution of equation (7.17) with the solution of the equation of diffusion without reaction

$$\rho \frac{\partial c_8}{\partial t} + \rho u_k \frac{\partial c_8}{\partial x_k} = \nabla D \rho \nabla c_8$$

It can be shown that they have the form

$$c_8 = c_8^{(e)}(z_1) \sim N^n, \qquad z = z_1 \qquad (7.19)$$

which follows from (7.15), (7.16), i.e., the concentration of CO on surface $z = z_1$ is thermodynamically in equilibrium; the concentration of CO over this surface naturally changes, since although z_1 changes the small, equilibrium concentration of CO is greatly dependent on composition. Since N is a random function of time, then despite the similarity of the equations for c_8 and z, fields c_8 and z are dissimilar and concentration c_8 in region $z < z_1$ is dependent not only on z and N, but also on the previous history of the process. This circumstance is unnoticed in the study by Liu, Bray, and Moss [1984], in which a number of results were obtained based on a vaguely made assumption that system (5.32) is true for all z. Although the concentration of CO in the region $z > z_1$ is close to the equilibrium, it follows from the boundary condition (7.19) that owing to the finite reaction rate, carbon monoxide ends up in the lean part of the flame ($z > z_1$) even if it did not exist there initially. However, since $c_8^{(e)}(z_1)$ is a small quantity, this circumstance does not significantly affect the completeness of combustion. Accordingly, it is natural to ignore this quantity when considering the fate of fuel or products of incomplete combustion that end up in the lean regions due to the specific character of the combustion process (for example, upon liquid atomization).

Further similar analysis will assume that the patterns indicated above are retained even in the general case, i.e., 1) there are two regions separated by thin freezing front $z = z_1$, 2) the chemical reactions in region $z < z_1$ flow with negligibly low rate, 3) the chemical reactions in region $z > z_1$ are so fast that the concentration of the reacting substances are close to thermodynamic equilibrium,

4) the penetration of the unburned fuel from the rich regions $(z > z_1)$ into the lean regions $(z < z_1)$ can be neglected. Furthermore, the quantity z is understood to mean the conserved scalar of the fuel in the gaseous phase. Within the framework of this model the initial stage of the process is not considered and it is assumed that, to start with, there is unburned fuel in the regions $z < z_1$. The initial concentration of the fuel in these regions can be determined by both the characteristic features of the atomizing unit and by the chemical kinetics at the initial stages of mixing (for example, by the formation of holes in the flame). We note that in addition to the freezing front $z = z_1$ located in the lean parts of the flame $(z_1 < z_s)$, there is another freezing front which is located in the rich region $(z > z_s)$. As will be seen later, this state of affairs is significant only in combustion chambers with stoichiometric coefficients less than unity.

Let us illustrate the main features of this model on the example of a combustion chamber at the entrance of which all the fuel and all the air is supplied. Furthermore, we shall assume that its length is large and, therefore, the distribution of z at the exit is practically in equilibrium. Let us consider the structure of surfaces $z = z_1$ and regions $z < z_1$. First of all, we note that since the solution of the equation of diffusion (6.2) is a continuous function of x_k and t, there are no holes in the surface $z = z_1$. Regions $z < z_1$, generally speaking, are not simply connected (Fig. 7.7). Let us now follow the path of an unburned fuel molecule. It should be noted that in the coordinate system moving with the velocity of the medium the equation of diffusion describes random walks of the molecules caused by their collision. During the process of these walks, each molecule quickly "forgets" about its motion in the preceding stages. This means that the velocity vector of the thermal motion of the molecules can have any direction. Since the probability density distribution of the thermal motion of the molecules is Maxwellian, this vector can have an arbitrary absolute magnitude. Therefore, even in a moving medium any continuous function can be the trajectory of the molecule. Of course, each trajectory is realized with different probability.

It is evident that two fundamentally different cases are possible. In the first, the stoichiometric coefficient in the chamber is small and, therefore, $z > z_1$ at its outlet. As seen from Fig. 7.7a, any trajectory crosses the freezing front, i.e., all the fuel burns. In the second case, the stoichiometric coefficient large and, therefore, $z < z_1$ at its outlet. Then, as seen from Fig. 7.7b, there are two types of trajectories starting in the lean regions L: trajectory 1 intersects the freezing front and trajectory 2 does not intersect it. Therefore, for sufficiently large stoichiometric coefficient part of the fuel never burns.

For a more graphic illustration of this conclusion, let us consider an extremely simplified example. Just as in the first sections of Chapter 5, we shall assume that an irreversible one-stage reaction takes place at a rate which is infinite when $z = z_s$ and equal to zero when $z \neq z_s$. We assume that a small amount of fuel is added to the air supplied to the combustion chamber and that this fuel is uniformly distributed in the air flow. Then, the concentration of air and fuel in the lean part of the flame $(z < z_s)$ are identical. Therefore, the fractions of the reacting air and burned fuel coincide (the fuel ending up in the lean part of the

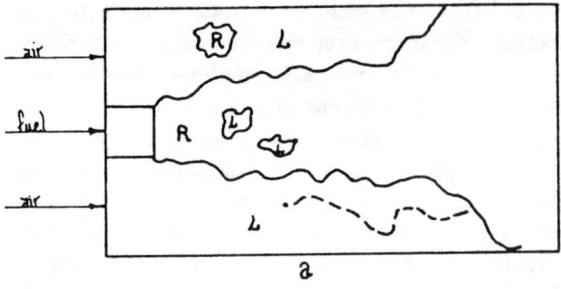

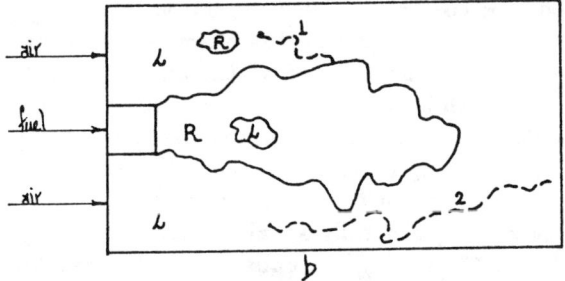

Figure 7.7 Structure of the freezing front (solid lines). The dashed line represents the trajectory of the molecules. R is the regions in which $z > z_1$, L is the regions in which $z < z_1$, a) $z > z_1$ at the exit from the chamber, b) $z < z_1$ at the exit from the chamber. The dashed curves — trajectories of the molecules of the unburned fuel.

flame). From simple equilibrium relations it can be shown that in this case the fuel concentration at the outlet is equal to

$$c_f = c_{f0}\left(1 - \frac{1}{\alpha}\right)$$

where α is the excess air factor, c_{f0} is the initial fuel concentration in the air flow. It hence follows that with increasing α completeness of combustion decreases.

By going to a combustion chamber of finite length we conclude that in the combustion of partially premixed gases, the effectiveness of the process is a nonmonotonic function of α and the optimum regime is reached when $\alpha > 1$. Indeed, with α decreasing from ∞ to 1, the surface $z = z_s$ is stretched and approaches the exit of the combustion chamber. Just as in the combustion of nonpremixed gases, this circumstance degrades the combustion of the fuel in the region $z < z_s$ (as seen from the work by Burk and Shuman [1928], the length of the flame is infinitely large when $\alpha = 1$). On the other hand, increasing the flame surface leads to improvement of the combustion of the fuel ending up in the lean areas. The interaction of these competing processes leads to the maximum effectiveness of the process being attained with $\alpha > 1$. The presence of a similar interaction is distinctively clear from Fig. 7.5.

From this analysis the main principles of the combustion process of liquid fuel in devices with large excess air are deduced. In particular, it is evident that the layout depicted in Fig. 7.7 is unsatisfactory and the solution of the problem is based on the stage-by-stage organization of the combustion process, i.e., on the principle employed in gas turbines. For this purpose two zones are created (Fig. 7.8). All the fuel and a small portion of the air are supplied to the primary zone I. At the outlet of this zone a rich fuel-air mixture is formed and is burned in the secondary zone II by introducing a jet of air.

The length of the primary zone must be sufficiently large for the drops to evaporate since, otherwise, they could end up in lean regions. In view of the fact that a significant amount of time is required for their evaporation, lean $(z < z_1)$ regions $L_1, L_2, \ldots$ in which unburned fuel is found are also formed in the primary zone. These regions are inside the freezing front 1 separating the primary and secondary zones. It appears from Fig. 7.8 that one would think all trajectories originating from regions $L_1, L_2, \ldots$ must intersect the freezing front. This, however, does not mean that all the fuel ending up in the regions under consideration must necessarily burn.

This conclusion becomes clear if one takes into account that regions L_1, $L_2, \ldots$ are moving together with the medium. Indeed, let us consider the sequence of the events that are illustrated in Fig. 7.8a, b, c. Figure 7.8a shows the initial location of the lean region L_1 which is inside the freezing front 1. At some time it appears to be close to the freezing front (Fig. 7.8b). Because of molecular diffusion the barrier between the lean regions L_1 and L disappears and they are joined together (Fig. 7.8c). This process leads to the unburned fuel ending up at the exit of the combustion chamber.

Therefore, the solution of the problem is based on two techniques. Firstly, the excess air factor in the primary zone is selected to be small and the volume of the lean regions is then small. Then these volumes quickly mix with the surrounding medium before reaching the freezing front 1. Secondly, mixing in the primary zone can be intensified, for example, by swirling the flow and generating a zone of reverse flow. Both techniques are used in the combustion chambers of gas turbines.

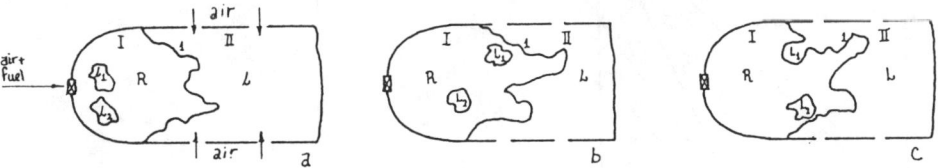

Figure 7.8 Schematic diagram of the combustion chamber of a gas turbine.

§7.3 EFFECT OF CHEMICAL REACTIONS ON THE COMBUSTION OF PARTIALLY PREMIXED GASES

Let us consider the effect of chemical reactions on the combustion of partially premixed gases. For this purpose, we analyze the interaction between mixing and chemical kinetics which arises upon the contact of two-phase homogeneous flows, namely pure air and rich fuel-air mixture. Such a formulation of the problem models the process in the secondary zone of the chamber of the gas turbine where a jet of air is directed toward the rich fuel-air mixture, supplied from the primary zone, through an opening in the flame tube. The main simplification is based on the assumption that there are no fluctuations of the concentration in the fuel-air mixture flow, i.e., the conserved scalar of the fuel z is exactly equal to some number z_0. The variation of this number enables the unusual laws of the process under study to be identified.

For simplicity we assume that the effect of chemical reactions is described by a model, formulated at the end of §5.4, i.e., 1) the conversion of the substances is localized in a narrow zone near the stoichiometric surface $z = z_s$, 2) when condition (5.33) is fulfilled, the reaction does not affect the combustion process, 3) if condition (5.33) is not fulfilled, combustion is not achieved.

Let us consider the $c_f - z$ diagram (Fig. 7.9). In the absence of chemical reactions we have $c_f = z$ (straight line 1). Upon combustion of completely

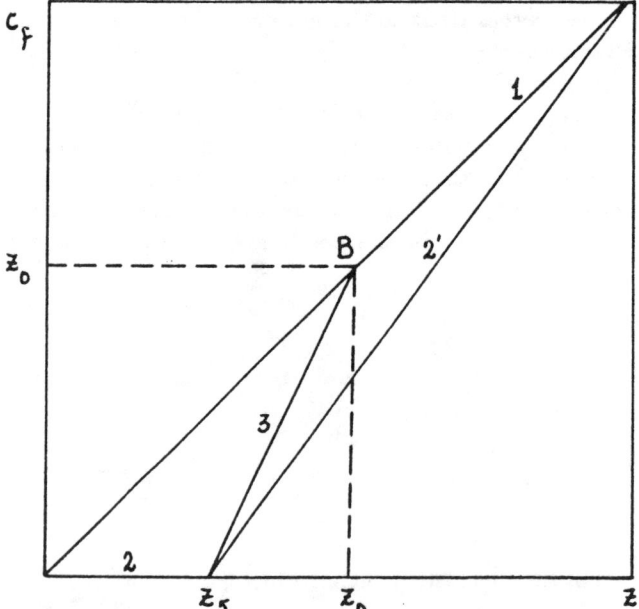

Figure 7.9 $c_f - z$ diagram. The straight line: 1) mixing without reactions, 2 – 2) combustion of completely premixed gases, 3 – 2) combustion of partially premixed gases.

nonpremixed gases relation (5.3), which is depicted by the broken line 2 − 2′, is valid. Upon combustion of partially premixed gases, the initial state of the system corresponds to point B at which $c_f = z_0$. In region $z_0 > z > z_s$, the chemical reaction does not take place, i.e., fields c_f and z are identical and, consequently, dependence $c_f(z)$ is linear. If the rate of reaction is infinitely large when $z = z_s$, then $c_f = 0$ when $z < z_s$. Consequently,

$$c_f = z_0 \frac{z - z_s}{z_0 - z_s}, \quad z > z_s$$

$$c_f = 0, \quad z < z_s \tag{7.20}$$

This dependence is depicted by the broken line 3 − 2.

From (7.20) it is evident that the fuel flow to the flame front is equal to

$$q_f = \frac{z_0}{z_0 - z_s} D \left| \frac{\partial z}{\partial n} \right|$$

where n is the normal to surface $z = z_s$. Therefore, from considerations that are analogous to the ones used in §5.4 when analyzing the stability of combustion of completely premixed gasses, we conclude that stable combustion is possible only if

$$\frac{z_0^2}{(z_0 - z_s)^2} N < N_{\mathrm{cr}}, \quad N = D \left(\frac{\partial z}{\partial n} \right)^2 \tag{7.21}$$

From the linearity of the equation of diffusion it follows that with unchanged geometry of the device and unchanged velocity of the air and mixture, the scalar dissipation N is proportional to z_0^2. We shall also take into account that the scalar dissipation N is proportional to the difference of the velocities of air u_1 and mixture u_2, and with the variation of z_0, and at constant flow rates this difference changes. Hence, combustion stability is determined by the complex parameter

$$S(z_0) = \frac{z_0^4}{(z_0 - z_s)^2} |u_1(z_0) - u_2(z_0)| \tag{7.22}$$

whose value must not exceed some value which depends only on pressure, initial temperature, dimensions and geometry of the combustion chamber.

Let us analyze two extreme cases. In the first $u_1 \gg u_2$. This case corresponds to the combustion in the secondary zone of the combustion chamber of a gas turbine where jets of pure air at high speed are entrained in the mixture. Since z_s is small and α is usually large, a strong variation of z is obtained for weak variation of u_1, i.e., the dependence of u_1 on z_0 can be ignored. Then, from (7.22) we conclude that the minimum value of S is attained when $z_0 = 2z_s$. In the second case the opposite inequality $u_2 \ll u_1$ is fulfilled. Then, from the equation of conservation of fuel flow we obtain $u_2 \sim 1/z_0$ and from (7.22) it follows that S is a minimum of $z_0 = 3z_s$.

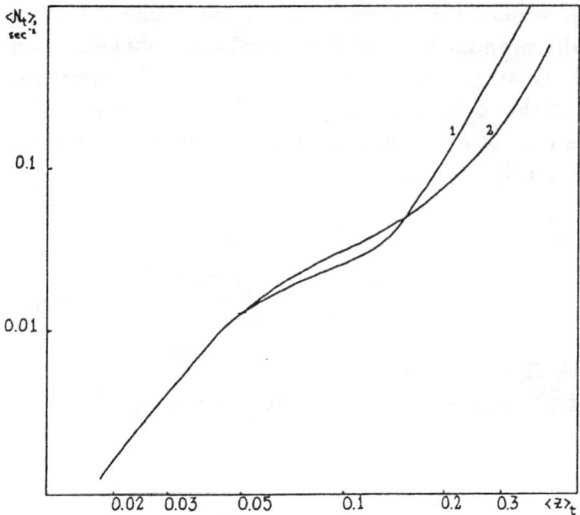

Figure 7.10 Distribution of scalar dissipation at the axis of a submerged diffusion propane flame, *1*) $d = 3$ mm, $u_0 = 19.8$ m/sec, *2*) $d = 6$ mm, $u_0 = 10.7$ m/sec.

It is easy to see that in both cases the minimum value of S is far less than the value of S which corresponds to the combustion of totally nonpremixed gases. For example, for kerosene $(z = 6.35 \times 10^{-2})$ we have $S(2z_s)/S(1) = 6.4 \times 10^{-2}$ in the first case.

Thus, the combustion of partially premixed gases is essentially more stable than the combustion of totally nonpremixed gases and, as seen from these estimates, can be arranged in a flow with a velocity that is an order higher or in a device with a dimension that is an order less than diffusion combustion.

It should be emphasized that this possibility is yet only of a fundamental character. Indeed, the gradient of the conserved scalar of the fuel at the point at which air comes into contact with the rich fuel-air mixture for the first time is equal to infinity, i.e., the condition for the existence of the flame (7.21) is not fulfilled. Therefore, there is a need for the construction of some device for the stabilization of combustion. This device must operate in far more severe conditions, since for constant flow rates of the fuel and air the rich mixture moves with a higher velocity than the velocity at which pure fuel could have moved.

There is one more limitation of the applicability of this method of arranging the combustion process. The fact is that the scalar dissipation which determines the conditions for chemical reactions varies greatly depending on the position of the point under consideration. This characteristic feature is illustrated in Fig. 7.10 which shows the results of calculations of a jet of propane from a nozzle of diameter d with velocity u_0 in a submerged space. These calculations are based on system (5.4) and correspond with the cases considered in Chapter 5. Presented in the figure are the calculation for the axis of the jet. Plotted along the abscissa is the mean concentration in the turbulent fluid $\langle z \rangle_t$ instead of the distance to the nozzle. It is apparent that in the turbulent fluid the mean value of the scalar dissipation $\langle N \rangle_t$ changes by several orders. This feature leads to the fact that at the initial section of mixing a substantial part of the unburned fuel penetrates into the lean section of the jet. Such penetration can be caused either by the appearance of holes in the flame front, or by the mechanism explained in §7.2 with respect to the oxidation of CO. As established in §7.3, in the later stages of mixing part of this fuel diffuses from the lean regions to the reaction zone where it burns and another part leaves the chamber. It is found that with decreasing z_0, the conditions of after-burning of the fuel penetrating the lean regions deteriorate. In order to elucidate this conclusion we shall consider the combustion in the mixing layer between the flows of pure air and the rich fuel-air mixture.

As a result of intermittency there are three regions 1 – 3 (Fig. 7.11) that are separated by randomly oscillating boundaries F_1 and F_2. In region 1 there is pure air, and in region 2 there is the initial mixture. In region 3 (region of turbulent fluid) concentration z changes randomly between zero and z_0. Since the stoichiometric coefficient is usually much larger than unity, then $z_0 \sim 1$ is the freezing front, i.e., surface $z = z_1$ is positioned near the boundary F_1 (Fig. 7.11a). Conversely, when z_0 is small, surface $z = z_1$ is positioned near the boundary F_2 (Fig. 7.11b). In these extreme cases the normal component of the velocity of the medium v relative to surface $z = z_1$ has a definite sign (the normal directed towards the rich region, Fig. 7.11). Indeed, since the trajectory of the fluid (in the hydrodynamic sense) particles cannot depart from the vortex flow present in region 3, the medium always flows into region 3. Therefore, $v > 0$ in the first case and $v < 0$ in the second. Hence it follows that connective motion hinders the transfer of the fuel from surface $z = z_1$ into the lean jet in the first case, and facilitates this transfer in the second. This state of affairs leads with decreasing z_0 to after-burning of the fuel penetrating into the lean part of the flame deteriorates.

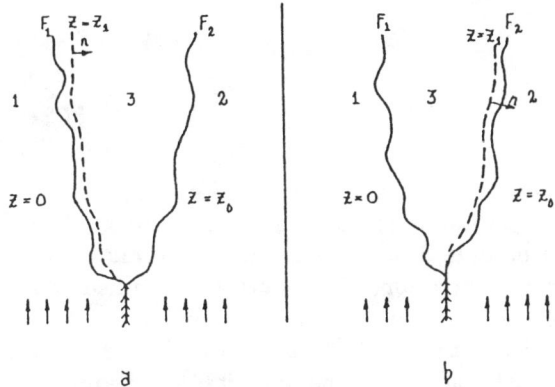

Figure 7.11 Qualitative schematic of the mixing layer.

§7.4 AFTER-BURNING OF THE UNBURNED FUEL PENETRATING INTO THE LEAN PART OF THE FLAME

In considering the after-burning of the fuel penetrating into the lean part of the flame, we shall assume that in region $z < z_1$ reactions do not occur and that the boundary conditions at the freezing front are specified, i.e.

$$c_f = \psi(N), \quad z = z_1$$

The special case of this relation is described by formula (7.19) which is obtained from the approximate model of oxidation of CO.

As evident from §7.3, the correct specification of v, i.e., the velocity of the medium relative to the freezing front, plays an important role.

Just as in §3.4, we change to a system of coordinates moving with a mean velocity, i.e., instead of the evolution along coordinate x_1, we examine the change in time t assuming the flow to be uniform in all directions and the stream unbounded. As already mentioned in §1.3, the relative volume of regions $z < z_1$ is the probability of the quantity z and the change of this volume is determined by the flow rate through surface $z = z_1$, i.e., by quantity v. In order to characterize this flow rate we introduce the parameter

$$B = \lim_{V \to \infty} V^{-1} \int_F \rho v \, dF$$

where V is some volume, F is that part of surface $z = z_1$ which is enclosed inside this volume. We assume the chemical nonequilibrium has little effect on density ρ, i.e., ρ is uniquely expressed in terms of z (see §5.1). By integrating the continuity equation over the region $z < z_1$ we obtain

$$B = -\lim_{V \to \infty} V^{-1} \int_{z < z_1} \frac{\partial \rho}{\partial t} d^3 x = -\frac{\partial}{\partial t} \int_{-\infty}^{z_1} \rho P \, dz \tag{7.23}$$

From (2.15), (7.23), we obtain

$$B = \langle N \rangle_t \left[\frac{\partial \rho P}{\partial z} \right]_{z=z_1} \tag{7.24}$$

here that all the derivatives with respect to three-dimensional coordinates are equal to zero. Hence follow the same qualitative results that were obtained above when analyzing the sign of quantity v. Indeed, let us examine the formulas presented in §3.4. In the case under consideration, they must be modified only slightly, since it was assumed in §3.4 that $\rho = $ const and the maximum concentration equal to 1 and not to z_0. By utilizing the results of the work by Kuznetsov, Lebedev, Sekundov, and Smirnova [1981] and taking the linearity of the equation of diffusion into account, we conclude that in the initial section of mixing ($t \to 0$) formula (3.26) assumes the form

$$\rho P = \gamma \rho P_t = \frac{2}{z_0} \rho(\bar{z}) \exp(-\pi^2 \tau) \sin(\pi \bar{z}/z_0) \sin(\pi z/z_0)$$

$$\tau = z_0^{-2} \int_t^\infty \langle N \rangle_t dt, \qquad \langle N \rangle_t = \frac{a z_0^2}{t}, \qquad 0 < z < z_0$$

$$\bar{z}\rho(\bar{z}) = \langle \rho z \rangle \tag{7.25}$$

where constant a, as shown in §3.4, is equal to π^{-2}.

From (7.24), (7.25) we conclude that $B > 0$ ($v > 0$) for $z_1 < z_0/2$ and $B < 0$ ($v < 0$) for $z_1 > z_0/2$.

Let us consider the quantity

$$\tilde{c}_f = \lim_{V \to \infty} V^{-1} \int_{z<z_1} \rho c_f d^3x$$

which characterizes the fuel concentration in the lean part of the flame and derive an approximate equation for it. Integrating the equation of diffusion over region $z < z_1$ yields the exact relation

$$\frac{\partial \tilde{c}_f}{\partial t} = \varsigma_1 + \varsigma_2, \qquad \varsigma_1 = -\lim_{V \to \infty} V^{-1} \int_F v \rho \psi dF$$

$$\varsigma_2 = \lim_{V \to \infty} V^{-1} \int_F D\rho \frac{\partial c_f}{\partial n} dF \tag{7.26}$$

Component ς_1 does not describe the transport processes, since it causes the change of $\tilde{c}_f$ even when $\tilde{c}_f = \text{const}$ for $z < z_1$ (this change is due to volume $z < z_1$ being variable during the mixing process). The second component ς_2 describes mass transfer between surface $z = z_1$ and the surrounding medium.

The principle assumption which is made later is that v and ψ are considered to be uncorrelated. Then, using (7.24) we obtain

$$\varsigma_1 = \langle \varsigma_1 \rangle = -\langle \psi \rangle B = -\langle \psi \rangle \langle N \rangle_t \left[\frac{\partial \rho P}{\partial z}\right]_{z=z_1} \tag{7.27}$$

We shall also assume that the second component ς_2 can be approximated by the formula which is often encountered in the approximate theory of mass transfer

$$\varsigma_2 = \alpha \tilde{c}_f + \beta \langle \psi \rangle \tag{7.28}$$

where α and β are functions of t. They are found from the following considerations. Let $\psi = \text{const}$ and $c_f = \psi$ in the turbulent fluid. Then $\partial c_f/\partial n = 0$, hence

$$\tilde{c}_f = \langle \psi \rangle \int_{+0}^{z_1} \rho P dz$$

$$\alpha \int_{+0}^{z_1} \rho P dz + \beta = 0 \qquad (7.29)$$

Here, the symbol $+0$ means that the singular components which are included in expression (1.20) for P are not taken into account upon integration. Let us consider another case when $\psi = \text{const}$ and, hence, $c_f = \psi z/z_1$. Then

$$\tilde{c}_f = \langle \psi \rangle z_1^{-1} \int_{+0}^{z_1} z\rho P dz \qquad (7.30)$$

is one of the solutions of equation (7.26). Using this condition and the relation (7.29), we find ς_1 and ς_2. Finally, we obtain

$$\frac{d\tilde{c}_f}{dt} = \langle N \rangle_t \left\{ -\langle \psi \rangle \left[\frac{\partial \rho P}{\partial z}\right]_{z=z_1} + \varsigma \langle \psi \rangle \int_{+0}^{z_1} \rho P dz - \tilde{c}_f \right\}$$

$$\varsigma = \frac{\rho(z_1) P(z_1)}{z_1 \int_{+0}^{z_1} \rho P dz - \int_{+0}^{z_1} z\rho P dz} \qquad (7.31)$$

Let us analyze the solution of this equation in the initial section of mixing and combustion ($t \to 0$). Consider first one special case of boundary conditions on surface $z = z_1$, more specifically we shall assume that if condition (7.21) of the existence of the diffusion flame is fulfilled, then the process is thermodynamically in an equilibrium state, i.e., $\langle \psi \rangle = 0$ and $z_1 = z_s$. Otherwise, we shall assume that mixing takes place without reactions as a consequence of which part of the unburned fuel penetrates into region $z < z_s$. For the sake of simplicity, we shall assume that the velocities of the mixing gases are unchanged. Then, from (7.21), (7.25) we conclude that when $t < t_1$ only mixing takes place, and when $t > t_1$ the process is thermodynamically in nonequilibrium state. Here t_1 is given by the expression

$$t_1 = \frac{a}{N_{\text{cr}}} \frac{z_0^4}{(z_0 - z_s)^2} \qquad (7.32)$$

It is obvious that when $t < t_1$, the amount of fuel penetrating into the lean part of the flame is proportional to t_1 (it is important to bear in mind that the initial mixing section is being considered). Thus

$$\tilde{c}_f \sim t_1, \qquad t = t_1 \qquad (7.33)$$

When $t > t_1$, equation (7.31) assumes the form

$$\frac{d\tilde{c}_f}{dt} = -\varsigma \langle N \rangle_t \tilde{c}_f \qquad (7.34)$$

From (7.33), (7.34) we conclude that

$$\tilde{c}_f \sim t_1 (t/t_1)^{-m}, \qquad m = \frac{\pi^2 \sin s}{s - \sin s} a, \qquad s = \frac{\pi z_1}{z_0} \qquad (7.35)$$

Let us estimate the quantity m for propane ($z_s = 0.06$). Assuming that $z_1 = z_s$, $a = \pi^{-2}$ we obtain $m = 108$ for $z_0 = 1$, i.e., the fuel penetrating into the lean part of the flame at the early stages of the process are very rapidly transferred to the reaction zone. When $z_0 = 2z_s$, $z_1 = z_s$, we have $m = 1.8$, i.e., the relation (7.35) describes a fairly slowly diminishing concentration of the fuel in the lean regions. Therefore, although, as evident from §7.3, scalar dissipation decreases in the last case and, consequently, little fuel penetrates into the lean regions, the after-burning of this fuel is far slower than upon combustion of completely premixed gases.

We note one more nontrivial characteristic feature of the process under consideration for which we analyze the effect of intensification of mixing on the after-burning of the fuel ending up in the lean regions. To this end, we shall formally vary constant a entering the expression for scalar dissipation. Then from (7.32), (7.35) we find that when $t = $ const, concentration $\tilde{c}_f$ is extremal if $\ln(t_1/t) = -1 - 1/m$. This equation can be either without a root or has two roots, i.e., cases are possible when intensification of mixing leads to an increase in the effectiveness of the process. Such behavior is due to the fact that although it is intensification of mixing that leads to the increase of concentration of the fuel ending up in the lean regions in the first stage of the process, the fuel in its second stage is transferred more rapidly to the reaction zone where it burns.

Let us now consider the character of the solutions of the equation (7.31) when $t \to \infty$. In this case, as seen from (3.27), quantity P is described by the normal PDF. Since $P \neq 0$ for all z, the reaction zone is observed at all t. It can be shown that the solution nonetheless does not tend to zero (formally, this follows from all the coefficients in (7.31) tending rapidly to zero when $t \to \infty$). This means that all the fuel entrained in the lean part of the flame cannot burn. The sign of quantity v in this case also plays a very vital role. Indeed, in the concluding stage of mixing, any region $z \neq \langle z \rangle$ shrinks to a point and, therefore, when $z_1 > \langle z \rangle$ the velocity of the medium relative to surface $z = z_1$ is always directed to the lean part of the flame, i.e., the unburned fuel is pushed away from the reaction zone by convective motion. The above conclusion follows from formula (7.24) (in region $z > \langle z \rangle$ the density of the normal PDF is a diminishing function of z).

Thus, in the combustion of partially premixed gases, two regimes can exist. In the first (z_0 is close to unity) the fuel entrained in the lean part of the flame is rapidly transferred back to the reaction zone (m is large). In this regime the previous history does not apparently play a role. In the second regime (z_0 noticeably different from unity), the fuel is slowly carried over to the reaction zone (m is small) and then the previous history of the process becomes significant. Comparison of the results obtained above when analyzing the initial and final sections of mixing and combustion shows that even when $z_0 = 1$, with increasing t (or with increasing distance from the point of contact of the streams of the fuel and oxidant) a gradual transition from the first to the second regime of combustion takes place. A detailed analysis of such a transition has not yet been carried out.

Let us now discuss the results of the experiments by Buriko and Kuznetsov [1986], which were designed to verify the validity of the theoretical ideas formulated above. The conditions of these experiments were discussed in §5.1 when analyzing the dynamics of the concentration of the hydrocarbons in a submerged propane jet. In addition to hydrocarbon concentration, the concentration of CO (c_8) was measured in these experiments. Analysis of the results of these experiments showed that unlike the equivalent concentration of propane, the quantity $\langle c_8 \rangle_t$ is nonuniversally dependent on $\langle z \rangle_t$, i.e., CO is oxidized substantially more slowly and the rate of chemical reactions affects the concentration of CO.

Therefore, the question arises as to whether the first combustion regime is possible, i.e., whether the concentration of CO is a universal function of $\langle z \rangle_t$ and $\langle N \rangle_t$ in which the role of the previous history of the process is insignificant. By a happy coincidence it was found out that with changing d and u_0, the scalar dissipation varies only in the very initial sections of the flame (large values of $\langle z \rangle_t$ in Fig. 7.10) and the scalar dissipation is the same in the main section of the flame (small values of $\langle z \rangle_t$ in Fig. 7.10). Such distribution of scalar dissipation is caused by buoyancy forces which are insignificant for small x_1 and have important significance for large x_1. As already noted in §5.5, $\langle N \rangle \sim 1/\tau$ ($\tau = d/u_0$), if there are no buoyancy forces.

These considerations determined the method of analyzing the experimental data. From the results of measurements in different sections of the flame the position of the line $\langle z \rangle_t = $ const was determined and the distribution of the quantity $\langle c_8 \rangle_t$ on this line analyzed. The quantity $\langle z \rangle_t$ was found from formula $\langle z \rangle_t = \langle z \rangle / \gamma$ where $\langle z \rangle$ was measured and γ calculated by the method discussed in §5.1. The computed value of the scalar dissipation was chosen as the parameter characterizing the position of the point under consideration on line $\langle z \rangle_t = $ const. The results of the analysis are presented in Fig. 7.12.

Analysis shows that all the data are approximated by the dependence $\langle c_8 \rangle_t = \xi(\langle z \rangle_t) + \eta \ln \langle N \rangle_t$, where η is independent of $\langle z \rangle_t$ and ξ is dependent only on $\langle z \rangle_t$, i.e., the change of $\langle N \rangle_t$ exerts the same influence on the results of measurements for all $\langle z \rangle_t$. The apparent change of the slope in Fig. 7.12 is due only to the different scales along the ordinate axis. It is evident that the effect of scalar dissipation is substantial.

Such an effect can be caused by two reasons, namely by chemical reactions or by the change of equilibrium concentration owing to a drop in temperature as a result of radiation heat losses. In order to elucidate which of the reasons is controlling, calculations were conducted in which the mean value of the equilibrium concentration $\langle c_8^{(e)} \rangle_t$ was found. Radiation heat losses were taken into account in the calculations (the method described in §5.3 was used). It is evident from Fig. 7.12 that 1) radiation heat losses exert a weak influence on the mean value of the equilibrium concentration $\langle c_8^{(e)} \rangle_t$; 2) the measured values of $\langle c_8 \rangle_t$ can be both substantially lower and substantially higher than $\langle c_8^{(e)} \rangle_t$. It is hence evident that the rate of chemical reactions in the present case is significant and the effect

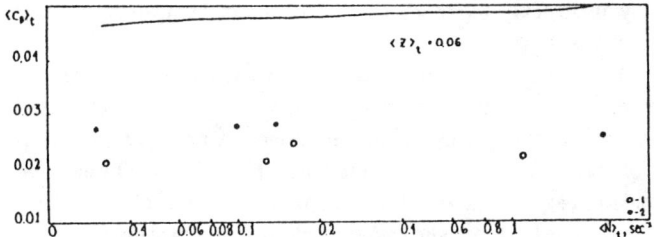

Figure 7.12 Concentration of CO in a submerged diffusion propane flame for various values of $\langle z \rangle_t$. *1)* $d = 3$ mm, $u_0 = 19.8$ m/sec, *2)* $d = 6$ mm, $u_0 = 10.7$ m/sec; the solid lines are the equilibrium concentration of CO.

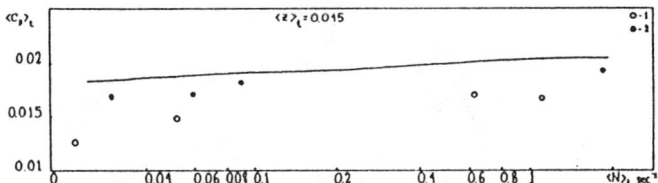

Figure 7.12b

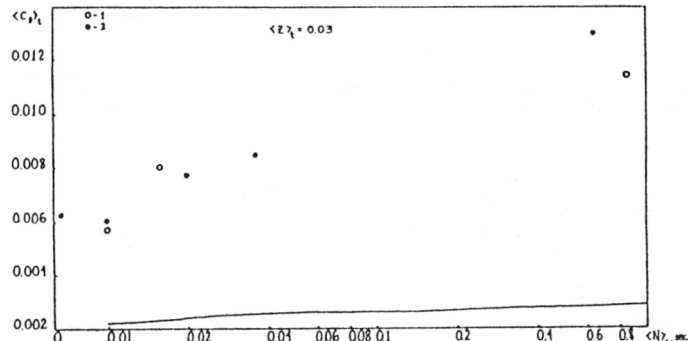

Figure 7.12c

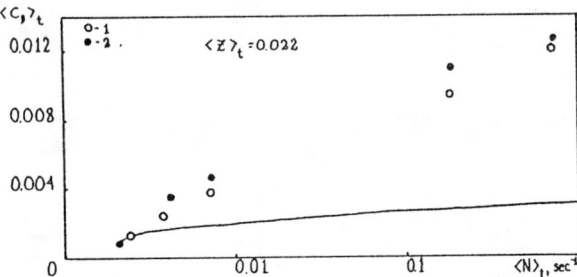

Figure 7.12d

of radiation is apparent only indirectly through the decrease in temperature and slowing of the chemical reactions.

Figure 7.12 also shows that the data obtained in two different regimes are generalized by a single curve. Furthermore, the residence time $\tau = d/u_0$ changed fourfold, the values of the scalar dissipation changed correspondingly at small distances from the nozzle (large $\langle z \rangle_t$) which is seen from Fig. 7.10. Thus, the existence of the first combustion regime, in which the concentration of the reacting substances are considerably dependent on the rate of chemical reactions and in which the previous history of the process plays no role, has been experimentally confirmed.

CONCLUSION

The investigation conducted in this book is based on the equations for the PDFs of various characteristics of turbulence. In essence, this approach means that any fluid dynamic process is studied not in the laboratory, but in a special coordinate system connected with the surfaces on which the value of one or another flow characteristic (concentration, velocity differences, and so on) is constant (isosurfaces). Moreover, the equation for PDF describes some general properties of the transformation of one system of coordinates to another.

Naturally, changing the coordinate system does not eliminate the difficulties of constructing the theory, and an increase in the completeness of the description is attained at the price of excessive mathematical complication (an increase in the number of unknowns)*. Therefore, the approach can be productive only if there is a chance of obtaining some universal PDFs (for example, similar to the Maxwellian distribution), or if the results enable re-evaluating the simpler methods (for example, the methods developed in the so-called semiempirical theories for one-point moments).

It appears that both indicated conditions are fulfilled at least in the problems which are considered in the present book. Apparently, the universality of the statistical characteristics of small-scale turbulence does not require special comment. Therefore, let us consider the degree of universality of the concentration PDF, obtained in Chapter 3. It is evident from the investigation in Chapter 5 that formulas (3.56), (3.57) enable solving a number of practical problems, for example, computing the concentration of nitrogen oxides and hydrocarbons in a diffusion plume. It is important for formula (3.57) to describe not only frequently encountered but also rarely encountered events. This conclusion is evident from the comparison of calculated and measured values of the concentration of propane which is determined in the concluding stage of combustion by the extremely rare appearance of regions with composition that differs strongly from the mean.

However, the main outcome of the investigation is the new approach to the problem of creating a closed theory of turbulence and turbulent combustion. This approach has been devised as a result of analysis of the equations for various PDFs

---

*The difficulties, just as ordinary physical quantities, obey the laws of conservation.

which reflect two phenomena — large-scale fluctuations of isosurfaces (the convective terms in the equation) and small-scale dissipative phenomena taking place in the volume enclosed between two close isosurfaces (the terms proportional to scalar dissipation and energy dissipation). The latter phenomena can be presented as the transfer of momentum and mass between the opposite walls of a contorted plane channel.

The advantages of such an approach are readily seen from an analysis of the concentration field. Let us first consider the transfer of the inert contaminant. In this case, an accurate description can be attained if Reynolds number is considered infinite, i.e., the Kolmogorov scale is equal to zero. Hence, mass transfer between any two closely located isoscalar surfaces is caused by turbulent diffusion, i.e., a statistical description is required. It is important that the characteristics of the process under consideration are universal at all points, since they are determined by the vortices with dimensions in the inertial range.

A totally different picture is observed from the analysis of the concentration field of a reacting contaminant when the Kolmogorov scale can be both greater or smaller than the reaction zone thickness. In this case the transport characteristics between two close isoscalar surfaces are not universal and are dependent on chemical kinetics. Thus, chemical reactions act directly on the dissipative processes. In their turn, dissipative processes influence the large-scale statistics. In particular, such an influence is reflected in the turbulent diffusion of the reacting species. The outcome is clearly confirmed in Chapter 5 where it is shown that the coefficients of turbulent diffusion of nitrogen oxides and the inert contaminant are not the same.

Thus, when constructing a purely statistical theory significant difficulties arise irrespective of whether we consider PDF or its first moments. They are caused by the qualitative character of all closure relations being determined by the type of chemical kinetics, i.e., the theory is essentially of nonuniversal character[†].

The way out of this situation is a combination of statistical and deterministic methods of description. Such an approach is based on the transition to a system of coordinates linked with the isoscalar surface near the reaction zone. The reaction zone is regarded as a very thin plane boundary layer which allows connecting deterministically the parameters describing the fluid dynamic and chemical processes. The characteristics of the isoscalar surface are described statistically.

It is found that in many cases the properties of the isoscalar surface near the reaction zone have a universal form which is not directly dependent on chemical kinetics. For example, in diffusion combustion the reaction zone is near the surface on which the concentration of the inert contaminant is constant. In the combustion of a homogeneous mixture the flame front is, as a rule, situated near the boundary of the turbulent fluid. In all these cases the general theory of turbulent combustion might be applied. The investigation carried out in Chapters 5, 6 is an example.

[†]The presence of such difficulties is graphically described in the article by Spalding [1976]. Evaluating one of his works he wrote: "The results ... did not confirm the expectations: upon modeling turbulence the morning of hope was frequently followed by a day of doubts."

The main conclusion of Chapter 5 is that the structure of the reaction zone in diffusion combustion is dependent only on one fluid dynamic parameter (scalar dissipation). This conclusion lies at the basis of the quantitative theory of nitrogen oxide formation. In Chapter 6 three main combustion regimes of homogeneous mixtures are identified on the basis of this approach; the quantitative description of the effect of flame instability and the differences in the coefficients of molecular transport on the combustion process is given; and a critical description of the velocity of turbulent flame propagation is derived.

Apparently, further improvement of the methods developed in Chapters 5 and 6 would enable reducing the study of combustion to the investigation of the characteristics of turbulence in the absence of chemical reactions.

Let us consider the usefulness of the investigation of the equations for PDFs when solving the last problem. For this purpose, we compare the theory developed above with the widely used semiempirical theories for one-point moments. The principal feature for such models is the assumption that all connections between known and unknown quantities are of a local character. For example, in the "$k-\epsilon$" - model, it is assumed that Reynolds stress at a given point is dependent only on the turbulence energy, on dissipation, and on the gradient of the mean velocity at the same point. The Kolmogorov-Obukhov theory is not used at all in these models. Hence, it is evident that the models for the moments and the equations for PDFs are constructed on different bases, since the Kolmogorov-Obukhov theory is mainly used in the latter case.

As a result an equation of a totally different structure is obtained: the models for the moments do not require the introduction of intermittency, and the equations for PDFs are meaningless without consideration of intermittency. For example, from the equation for concentration PDF it follows that, in the absence of intermittency, the concentration acquires all the values from $-\infty$ to $+\infty$. It is important that intermittency determines the structure of the solutions in all the regions of the flow including those where it has traditionally been regarded as inessential. For example, near the axis or plane of symmetry of jet flows it is established that the less the intermittency factor differs from unity, the less the intensity of concentration fluctuations. The important role of intermittency was also established when analyzing the equation for velocity difference PDF.

This comparison does not yet relate to a shortcoming of the semiempirical theories for relative moments. For example, as seen from (3.57) the characteristics of intermittency can be universally connected with the two first moments of the concentration field and, therefore, the introduction of intermittency does not add new information. Nevertheless, such a shortcoming does indeed exist, since this investigation showed that the characteristics of intermittency are determined by large-scale nonlocal processes. The nonlocal process here is understood to mean a process whose characteristics cannot be described by differential equations with coefficients that are specified in advance.

As an example we can cite equation (3.67) for the concentration PDF. The coefficients in this equation are found from the solution of the boundary-value problem. Without solving this problem, i.e., without accounting for the particular features of the entire flow field in a given section, one cannot draw any conclusions concerning the distribution of the intermittency factor in the vicinity of a point located at the same section.

Equation (4.16) for the PDF of the velocity difference v at two close points at a distance r belonging to the inertial range also has a similar structure. It was established that $\langle v^n \rangle \sim r^{q(n)}$, and function $q(n)$ has singularities. The structure of function $q(n)$ in the vicinity of the singularities reflects the interaction of the turbulent and nonturbulent fluids. Its characteristics are determined by the vortices with dimension of the order of the integral turbulence scale, i.e., again large-scale processes play an important role. Since $q(n)$ is an analytical function, then the structure of the singularity (and, hence, the effect of large-scale processes) affect the behavior of function q for all n.

Thus, intermittency turns out to be an indication of a nonlocal large-scale interaction. Such an interaction is not taken into account in the semiempirical theories for one-point moments. Therefore, the values of the empirical constants in these theories are determined by the type of flow[‡], and the theories themselves do not require the introduction of the intermittency factor.

It is evident that the nonlocal interaction is a very complex problem and, in this sense, the theory based on the equations for PDFs is not any simpler than the theory involving the several first moments of the velocity field. The reason behind all difficulties is evident from Chapter 4 where it is shown that intermittency arises as a result of direct, and not cascade, interaction of vortices with greatly differing scales. Hence, the Kolmogorov-Obukhov theory yields only an approximate description of small-scale fluctuations and this description does not become asymptotically exact if Reynolds number tends to infinity and the scale of the fluctuations under consideration tends to zero (for this it must be additionally assumed that the order of the structure function tends to infinity, i.e., the theory is asymptotically exact only for fluctuations with large amplitude and small scale).

This circumstance is not an indication of any shortcoming of the Kolmogorov-Obukhov theory, since any thorough physical theory cannot be without contradictions and, therefore, it is just a basis for the development of a more accurate and universal description[§]. This investigation shows that the Kolmogorov-Obukhov theory is such a basis, at least, for two reasons.

First, the important role of intermittency and, hence, of the direct interaction between vortices with greatly varying scales follows from the equations for PDFs which are closed on the basis of the Kolmogorov-Obukhov theory, i.e., this theory is capable of predicting effects which were initially not taken into account. Second, in a certain sense, the role of nonlocal processes is not very large.

In order to elucidate the last assertion, it is important to bear in mind that the theory developed in this book enabled either to compute or to estimate a number of constants that are associated with one or another of the characteristics

[‡]It is generally known that for the best description of various flows the values of the empirical constants must be varied.

[§]We recall Hedel's theorem.

of turbulence. This fact on its own is not noteworthy, since the task of any theory is to formulate the most exact and universal method of description and not to compute the constants already measured. Therefore, just one circumstance is worthy of attention here: all constants determining the rate of dissipative processes are small. For example, in formulas (3.31), (4.13) for the conditionally averaged values of the scalar dissipation and energy dissipation two small constants are featured: $a = 1/\pi^2$ and $k \sim 10^{-2}$. Similarly, formula (4.15) describing the dissipation fluctuations includes constant μ which is connected with a small quantity $q_2 = \mu/18$ determining small corrections to the "two thirds" law.

The small value of the latter constant is caused by the fact that in a space of n structure functions $\langle v^n \rangle$ the first singularity of function $q(n)$ ($\langle v^n \rangle \sim r^{q(n)}$) is found in the region $n < -5$, i.e., located at a large distance from those values of n which determine the fluctuation energy ($n = 2$) and dissipation fluctuations ($n = 6$). Since the character of the singularity of function $q(n)$ is determined by intermittency, it then follows that turbulent and nonturbulent fluids interact weakly. Hence, nonlocal large-scale processes causing the formation of intermittency have little effect on the characteristics of small-scale turbulence.

A similar picture is observed when analyzing large-scale turbulence in jet-type flows. Here, the interaction of the turbulent and nonturbulent fluids determines the rate of spread, and this rate, as can be easily shown from integral relations following from the Navier-Stokes equations, characterizes the relative role of dissipative processes. It is generally known that the spreading rate of such flows is small and, therefore, it is not surprising that all the constants connected with dissipative processes are small¶.

Hence, it is obvious that, as a rule, the role of large-scale nonlocal processes is not very large. It is for this reason that the models for the moments yield a valid order of magnitude of the values of the main turbulence characteristics.

The indicated state of affairs yields the confidence that it is possible to find a simple and universal theory of turbulence. In order to analyze the probable mathematical formulation of this theory let us consider several examples that are known in physics. The first example is the Navier-Stokes equation. In this case, the nonlocal character of the transport processes are not taken into account, and the flow is described by differential equations with specified coefficients. The equations of the models for the moments have exactly such a structure. Since they do not ensure sufficient accuracy, such a mathematical formulation is useless.

The second example is heat transfer by radiation. In this case, the process of heat transfer is described by complex relations which feature integrals of various characteristics of the medium over the entire space, i.e., account is taken of the fact that energy radiated in one place can be dispersed and absorbed in other places. In the approximation of an optically thin layer the radiating volume and the wall exchange heat no matter how distant they are from each other, i.e., the process is strongly nonlocal. From the mathematical standpoint a similar ap-

---

¶In flows near walls the relative role of dissipative processes is, apparently, even less (the angle of expansion of the boundary layer is very small, the coefficients of resistance in the channel are small). Note also that even homogeneous turbulence degenerates very slowly.

proach was developed also in the theory of turbulence. The works by Kraichnan [1959, 1974] serve as an example. The spectral turbulence energy density in these works is found from the solution of a system of nonlinear integrodifferential equation. It appears that this mathematical formulation is also useless for describing turbulence, since the role of the nonlocal processes is greatly exaggerated.

The third example is taken from quantum mechanics. Here, the role of the nonlocal processes is neither very large nor very small (the de Broglie wavelength is comparable with an atom). The process is described by a differential equation (Schrödinger wave equation), but the coefficients in this equation (energy levels) are found from the solution of the boundary-value problem i.e., they are not specified beforehand. An analogous description of the problem is obtained also in the present book: concentration PDF satisfies the differential equation in which the coefficients (more precisely, the functions) are also found from the solution of the boundary-value problem. The indicated analogy is not accidental, since the statistical connection (de Broglie wavelength or the integral turbulence scale) in both cases are comparable with the characteristic dimension of the problem. Therefore, it can be assumed that all the equations of the theory of turbulence must have the same character as the Schrödinger equation or the equation for the concentration PDF.

The actual realization of such an approach can be totally diverse depending on which part of the turbulence spectrum is under consideration. For example, in the small-scale part several of the first structure functions do not yield enough information on the local characteristics of turbulence, since the amplitude of dissipation fluctuations significantly exceeds its mean value. Hence, it is necessary to use the equation for the velocity difference PDF. As seen from (4.16), the Kolmogorov-Obukhov theory defines the type of this equation: it is similar to the heat equation, i.e., it features only the first derivative with respect to the time-like coordinate r and the second derivatives with respect to the space-like coordinate v. The principal feature of this equation is that the coefficient for the derivative with respect to the time-like coordinate changes sign. Therefore, the boundary-value problem is set, not only with respect to the space-like coordinate, but also with respect to the time-like coordinate. Equation (3.67) for the concentration PDF $P(z)$ has an exactly analogous structure. The boundary-value problem with respect to the time-like coordinate is redetermined (in both cases the equations feature only the first derivative with respect to this coordinate). The conditions of solvability of such a problem must define the unknown functions entering into the equations initially. In particular, in the equation for $P(z)$ the intensity of concentration pulsations in the turbulent fluid was found just by this means. Similarly, it can be assumed that the conditions of solvability of the boundary-value problem for the velocity difference PDF enable computing constant k and the Kolmogorov constant C which is connected with it.

The application of PDFs for the description of the structure of large-scale turbulence is, apparently, an unjustified complication of the problem, since it is feasible to find a simpler approach leading to an equation of the same structure. To elucidate let us consider jet-type flows. Let us analyze their evolution starting with the section in which the turbulence originated. We shall observe the amplitude of the perturbation with a fixed scale whose value is considerably greater than the initial flow width. It is obvious that at the beginning the amplitude is very small and, hence, the characteristics of such perturbation are described by the linear theory. It is important that in flows of this type the turbulence intensity is small (turbulence intensity is connected with the interaction of turbulent and nonturbulent fluids). Therefore, the increase in amplitude of the large-scale perturbations is described by the linear theory of stability of the profile of the mean velocity (Townsend [1956]).

At the present time numerous theoretical and experimental data are known which establish that the linear theory of stability predicts many features of jet flows (Borisov, Kuznetsov and Sekundov [1985])[‖]. It is readily seen that formulas having the required structure are also derived in this theory. Indeed, the equations of Rayleigh or Orr-Sommerfeld are differential equations in which one of the coefficients (amplification factor) is found from the solution of the boundary-value problem. This, in particular, means that the Reynolds stress, flux of heat and mass at the given point are determined by the entire field of the mean velocity and not just by the gradient of the velocity at the point under consideration. Thus, in the given case, simplifications are possible in which the equations for PDFs are not used.

This approach, however, cannot yield totally closed descriptions of the problem, since the evolution of large-scale oscillations depends on energy dissipation which is determined by large-scale fluctuations. Therefore, this approach must be combined with far more complex methods of study of the small-scale structure of turbulence.

At the present time the means for such a combination are not known and it is evident only that since the main information on the state of turbulent flow is determined by the characteristics of small vortices, the successful realization of this program of investigation is dependent on how accurately and universally one manages to describe the characteristics of turbulence in the inertial and viscous range of its spectrum.

[‖] The attempts to use the linear theory of stability as the starting point for a turbulence theory is, undoubtedly, not new. It suffices to bear in mind the works by Malkus [1965], Gol'dshtik and Shtern [1977], in which mutually exclusive principles of neutral and maximum stability were proposed in the analysis of flows near a wall.

REFERENCES

Abdel-Gayed, R. G. and Bradley, D.
    1981. A two-eddy theory of premixed turbulent flame propagation. Phil. Trans. Roy. Soc., London, A301, No. 1457, pp. 1–25.

Abramowitz, M. and Stegun, I. A.
    1964. Handbook of Mathematical Functions. Applied Mathematics Series. Vol. 55. National Bureau of Standards.

Adrian, R. J.
    1977. On the role of conditional averages in turbulence theory. In: Turbulence in liquids. Ed. by G. K. Patterson and J. L. Zakin. Princeton, Science Press, pp. 323–332.
    1979. Conditional eddies in isotropic turbulence. Phys. Fluids, Vol. 22, No. 11, pp. 2065–2070.

Agafonova, F. A., Gurvich, M. A., and Tarasova, E. F.
    1960. Conditions of combustion stability of single drops of liquid fuel. In: Transactions of the third all-union conference on the theory of combustion, Vol 2. Moscow, Izd-vo AN SSSR, pp. 29–40.

Alber, I. E. and Batt, R. G.
    1974. An analysis of diffusion limited first and second order chemical reactions in a turbulent shear layer. AIAA Paper, No. 593.
    1976. Diffusion-limited chemical reactions in a turbulent shear layer. AIAA J., Vol. 14, No. 1, pp. 70–77.

Alekseev, B. V., Ievlev, V. M., and Kiselev, V. I.
    1976. Application of the equations of probability distributions of fluctuating quantities to the solution of a model Burgers problem. In: Investigations in theoretical and applied physics. Trudy MAI, Iss. 380, pp. 43–52.

Anderson, P., La Rue, J. S., and Libby, P. A.
    1979. Preferential entrainment in a two-dimensional turbulent jet in a moving stream. Phys. Fluids, Vol. 22, No. 10, pp. 1857–1861.

Andrews, G. E., Bradley, D., and Lwakawamba, S. B.
    1975. Turbulence and turbulent flame propagation. Combustion and Flame, Vol. 24, No. 3, pp. 285-304.

Anselmet, F., Gagne, Y., Hopfinger, E. J., and Antonia, R. A.
- 1984. High order velocity structure functions in turbulent shear flows. J. Fluid Mech., Vol. 140, pp. 63–89.

Anselmet, F. and Antonia, R. A.
- 1985. Joint statistics between temperature and its dissipation in a turbulent jet. Phys. Fluids, Vol. 28, No. 4, pp. 1048–1054.

Antonia, R. A.
- 1973. Some small properties of boundary layer turbulence. Phys. Fluids, Vol. 16, No. 8, pp. 1198–1206.
- 1981. Conditional sampling in turbulence measurements. Annual Review of Fluid Mech., Vol. 13, pp. 131-151.

Antonia, R. A., Anselmet, F., and Chambers, A. J.
- 1986. Assessment of local isotropy using measurements in a turbulent plane jet. J. Fluid Mech., Vol. 163, pp. 365–391.

Antonia, R. A. and Bilger, R. W.
- 1976. The heated round jet in a coflowing stream. AIAA J., Vol. 14, No. 11, pp. 1541–1547.

Antonia, R. A., Browne, L. W., and Chambers, A. J.
- 1984. On the spectrum of the transverse derivative of the streamwise velocity in a turbulent flow. Phys. Fluids, Vol. 27, No. 11, pp. 2628–2631.

Antonia, R. A., Chambers, A. J., and Anselmet, F.
- 1984. Fine-scale turbulence measurements in a plane jet. Physicochemical Hydrodynamics, Vol. 5, No. 5–6, pp. 369–382.

Antonia, R. A. and Danh, H. Q.
- 1977. Structure of temperature fluctuation in a turbulent boundary layer. Phys. Fluids, Vol. 20, No. 7, pp. 1050–1057.

Antonia, R. A., Danh, H. Q., and Prabhu, A.
- 1976. Bursts in turbulent shear flow. Phys. Fluids, Vol. 19, No. 11, pp. 1680–1686.

Antonia, R. A., Phan-Thien, N., and Satyaprakash, B. R.
- 1981. Autocorrelation and spectrum of dissipation fluctuation in a turbulent jet. Phys. Fluids, Vol. 24, No. 3, pp. 554–555.

Antonia, R. A., Prabhu, A., and Stephenson, S. E.
- 1975. Conditionally sampled measurements in a heated turbulent jet. J. Fluid Mech., Vol. 72, Pt. 3, pp. 455–480.

Antonia, R. A., Rajagopalan, S., Browne, L. W. B., and Chambers, A. J.
- 1982. Correlation of squared velocity and temperature derivatives in a turbulent plane jet. Phys. Fluids, Vol. 25, No. 7, pp. 1156–1158.

Antonia, R. A., Satyaprakash, B. R., and Chambers, A. J.
- 1982. Reynolds number dependence of velocity structure functions in turbulent shear flows. Phys. Fluids, Vol. 25, No. 1, pp. 29–37.

Antonia, R. A., Satyaprakash, B. R., and Hussain, A. K. M. F.
- 1982. Statistics of fine-scale velocity in turbulent plane and circular jet. J. Fluid Mech., Vol. 119, pp. 55–89.

Antonia, R. A. and Sreenivasan, K. R.
   1977. Lognormality of temperature dissipation in a turbulent jet. Phys. Fluids, Vol. 20, No. 11, pp. 1800–1804.
Baev, V. K., Golovichev, V. I., and Yasakov, V. A.
   1976. Two-dimensional turbulent flows of reacting gases. Novosibirsk, Nauka.
Baev, V. K. and Tretiyakov, P. K.
   1968. Characteristic combustion times of fuel-air mixtures. Fizika goreniya i vzryva (Physics of combustion and explosion), Vol. 4, No. 3, pp. 367–376.
   1972. Criterial description of combustion stability in a turbulent flow of a homogeneous mixture. Fizika goreniya i vzryva (Physics of combustion and explosion), Vol. 8, No. 1, pp. 46–51.
   1977. Use of integral characteristics of laminar flame for criterial description of turbulent flames. In: Investigation of combustion of gaseous fuels. Novosibirsk, Izd-vo SO AN SSSR, pp. 4–20.
Balescu, R.
   1975. Equilibrium and nonequilibrium statistical mechanics. New York, J. Wiley and Sons.
Ballal, D. R.
   1979. The structure of premixed turbulent flame. Proc. Roy. Soc., A367, No. 1730, pp. 353–380.
Ballal, D. R. and Lefebvre, A. H.
   1975. The structure and propagation of turbulent flames. Proc. Roy. Soc., A344, No. 1637, pp. 217–234.
Barenblatt, G. I., Zel'dovich, Ya. B., and Istratov, A. G.
   1962. On diffusion thermal stability of laminar flame. ZhPMTF, No. 4, pp. 21–26.
Barsoum, M. L., Kawall, J. G., and Keffer, J. F.
   1978. Spanwise structure of the plane turbulent wake. Phys. Fluids, Vol. 21, No. 2, pp. 157–161.
Bashir, J. and Uberoi, M. S.
   1975. Experiments on turbulent structure and heat transfer in a two-dimensional jet. Phys. Fluids, Vol. 18, No. 4, pp. 764–769.
Batchelor, G. K.
   1952. The effect of homogeneous turbulence on material lines and surfaces. Proc. Roy. Soc., A213, No. 1114, pp. 349–366.
   1953. The theory of homogeneous turbulence. Cambridge Univ. Press.
   1967. An introduction to fluid dynamics. Cambridge Univ. Press.
Batchelor, G. K. and Townsend, A. A.
   1949. The nature of turbulent motion at large wave numbers. Proc. Roy. Soc., A199, No. 1057, pp. 238–255.
   1956. Turbulent diffusion. In: Surveys in mechanics. Ed. by G. K. Batchelor and R. M. Davies. Cambridge Univ. Press, pp. 352–399.
Bateman, H. and Erdeylyi, A.
   1953a. Higher transcendental functions. Vol. 2. New York, McGraw-Hill.

1953b. Integral transfers. New York, McGraw-Hill.

Batt, R. G.
- 1977. Turbulent mixing of passive and chemically reacting species in a low speed shear layer. J. Fluid Mech., Vol. 82, Pt. 1, pp. 59–95.

Becker, H. A., Hottel, H. S., and Williams, G. S.
- 1967. The nozzle-fluid concentration field of the round, turbulent, free jet. J. Fluid Mech., Vol. 30, Pt. 2, pp. 285–303.

Beker, H. A., Rosensweig, R. E., and Gwozdz, J. R.
- 1966. Turbulent dispersion in a pipe flow. AIChE J., Vol. 12, No. 5, pp. 964–972.

Beguier, C., Dekeyser, I., and Launder, B. E.
- 1978. Ratio of scalar and velocity dissipation time scales in shear flow turbulence. Phys. Fluids, Vol. 21, No. 3, pp. 307–310.

Betchov, R.
- 1974. Nongaussian and irreversible event in isotropic turbulence. Phys. Fluids, Vol. 17, No. 8, pp. 1509–1512.
- 1975. Numerical simulation of isotropic turbulence. Phys. Fluids, Vol. 18, No. 10, pp. 1230–1236.
- 1976. On the nongaussian aspects of turbulence. Archiwum Mechaniki Stosowanej (Archives of Mechanics), Vol. 28, No. 5–6, pp. 837–845.

Betchov, R. and Larsen, P. S.
- 1981. A nongaussian model of turbulence (soccer ball integrals). Phys. Fluids, Vol. 24, No. 9, pp. 1602–1604.

Betchov, R. and Lorenzen, C.
- 1974. Phase relations in isotropic turbulence. Phys. Fluids, Vol. 17, No. 8, pp. 1503–1508.

Bezuglov, V. A.
- 1974. Measurement of velocity of concentration transport. In: Transactions of 20th scientific conference MFTI. Aerophysics and applied mathematics. Pt. 1. Dolgoprudnyi, MFTI, pp. 147–151.

Bilger, R. W.
- 1976. Turbulent jet diffusion flames. Progr. Energy and Combust. Sci., Vol. 1, No. 1, pp. 87–109.
- 1977. Reaction rates in diffusion flames. Combustion and Flame, Vol. 30, No. 3, pp. 277–284.
- 1978. Reply to comments of P. Bradshaw. Phys. Fluids, Vol. 21, No. 2, p. 304.
- 1980a. Perturbation analysis of turbulent nonpremixed combustion. Combust. Sci. and Tech., Vol. 22, pp. 251–261.
- 1980b. Turbulent flows with nonpremixed reactants. In: Turbulent Reacting Flows. Ed. by P. A. Libby and F. A. Williams. Topics in Applied Phys., Vol. 44, Berlin, Springer-Verlag, pp. 65–113.
- 1982. Molecular transport effects in turbulent diffusion flames at moderate Reynolds number. AIAA J., Vol. 20, No. 7, pp. 962–970.

Bilger, R. W., Antonia, R. A., and Sreenivasan, K. R.
  1976. Determination of intermittency from the probability density function of a passive scalar. Phys. Fluids, Vol. 19, No. 10, pp. 1471–1474.
Bilger, R. W. and Beck, R. F.
  1974. Further experiments on turbulent jet diffusion flame. Techn. Note F-67, The Univ. of Sydney.
Bill, R. G., Namer, I., and Talbot, L.
  1981. Flame propagation in grid-induced turbulence. Combustion and Flame, Vol. 43, No. 3, pp. 229–242.
Birch, A. D., Brown, D. R., Dodson, M. G., and Thomas, G. R.
  1978. The turbulent concentration field of a methane jet. J. Fluid Mech., Vol. 88, Pt. 3, pp. 431–450.
Borghi, R.
  1980. Models of turbulent combustion for numerical prediction. In: Prediction methods for turbulent flows. Ed. by W. Kollman. Washington D.C., Hemishpere Publishing Corp., pp. 423–458.
  1984. Modeling of turbulent homogeneous combustion. In: Structure of gas-phase flames. Pt. 1. Novosibirsk, ITPM SO AN SSSR, pp. 138–160.
Borisov, A. G., Kuznetsov, V. R., and Sekundov, A. N.
  1985. Large-scale motion in shear turbulent flows. In: Turbulent jet flows. Reports of the fifth all-union scientific conf. on theoretical and applied aspects of turbulent flows. Pt. 1. Tallin, Izd-vo AN ESSR, ITEF, pp. 63–68.
Boston, N. E. J. and Burling, R. W.
  1972. An investigation of high wave number temperature and velocity spectra in air. J. Fluid Mech., Vol. 55, Pt. 3, pp. 473–492.
Bowman, C. T.
  1973. Kinetics of nitric oxide formation in combustion processes. Fourteenth Int. Sympos. on Combustion. Pittsburgh, The Combust. Inst., pp. 729–738.
Bradbury, L. J. S.
  1965. The structure of self-preserving turbulent plane jet. J. Fluid Mech., Vol. 23, Pt. 1, pp. 31–64.
Bradley, D.
  1984. Problems of mathematical modeling of turbulent flames. In: Structure of gas-phase flames. Pt 1. Novosibirsk, ITPM SO AN SSSR, pp. 15–42.
Bradshaw, P.
  1978. Comments on "Determination of intermittency from the probability density function of a passive scalar". Phys. Fluids, Vol. 21, No. 2, p. 303.
Bray, K. N. C.
  1980. Turbulent flows with premixed reactants. In: Turbulent Reacting Flows. Ed. by P. A. Libby and F. A. Williams. Topics in Applied Physics, Vol. 44, Berlin, Springer-Verlag, pp. 115–183.

Bray, K. N. C. and Libby, P. A.
   1976. Interactions effect in turbulent premixed flames. Phys. Fluids, Vol. 19, No. 11, pp. 1687–1701.
Bray, K. N. C. and Moss, J. B.
   1977a. A closure model for turbulent premixed flame with sequential chemistry. Combustion and Flame, Vol. 30, No. 2, pp. 125–131.
   1977b. A unified statistical model for the premixed turbulent flame. Acta Astronautica, Vol. 4, No. 3–4, pp. 291–319.
Breidenthal, R. E.
   1981. Structure in turbulent mixing layers and wakes using a chemical reaction. J. Fluid Mech., Vol. 109, pp. 1–24.
Bretshnaider, S.
   1966. Properties of gases and liquids. Moscow, Khimiya.
Brechet, M. E., Meiron, D. I., Orszag, S. A., Nickel, B. G., Morf, R. H., and Frisch, U.
   1983. Small-scale structure of the Taylor-Green vortex. J. Fluid Mech., Vol. 130, pp. 411–452.
Broadwell, J. E. and Breidenthal, R. E.
   1982. A simple model of mixing and chemical reaction in a turbulent short layer. J. Fluid Mech., Vol. 125, pp. 397–410.
Browne, L. W. B., Antonia, R. A., and Shah, D. A.
   1987. Turbulent energy dissipation in a wake. J. Fluid Mech., Vol. 179, pp. 307–326.
Buriko, Yu. Ya. and Kuznetsov, V. R.
   1976. Effect of diffusion stratification on the combustion process of a homogeneous mixture in laminar and turbulent flows. Fizika goreniya i vzryva (Physics of combustion and explosion), Vol. 12, No. 3, pp. 390–397.
   1978. On the probable mechanism of formation of oxides of nitrogen in turbulent diffusion combustion. Fizika goreniya i vzryva (Physics of combustion and explosion), Vol. 14, No. 3, pp. 32–42.
   1980. Effect of adding air to the fuel on the formation of oxides of nitrogen in turbulent diffusion plume. Fizika goreniya i vzryva (Physics of combustion and explosion), Vol. 16, No. 4, pp. 60–67.
   1983a. Formation of oxides of nitrogen in nonequilibrium diffusion flame. Fizika goreniya i vzryva (Physics of combustion and explosion), Vol. 19, No. 2, pp. 71–81.
   1983b. Effect of turbulence on the formation of oxides of nitrogen in diffusion flame. In: Transactions of five scientific readings on cosmonautics. Theoretical problems of engine building. Moscow, IIET AN SSSR, pp. 124–133.
   1986. Effect of chemical reactions and mixing processes on the concentration of carbon monoxide and hydrocarbons in a turbulent diffusion flame. Fizika goreniya i vzryva (Physics of combustion and explosion), Vol. 22, No. 4, pp. 19–25.

Carter, J. E.
 1974. Solutions for laminar boundary layers with separation and reattachment. AIAA Paper, No. 583.
Caulliez, G.
 1980. Structure temporelle de la microturbulence dans la couche de surface atmospherique. J. Rech. Atmos., Vol. 15, No. 2, pp. 129–148.
Cercignani, C.
 1975. Theory and application of the Boltzman Equation. London, Scottish Academic Press.
Chambers, A. J. and Antonia, R. A.
 1984. Atmospheric estimates of power law exponents μ and μ_θ. Boundary-Layer Meteorol., Vol. 28, pp. 343–352.
Champagne, F. H.
 1978. The fine-scale structure of the turbulent velocity field. J. Fluid Mech., Vol. 86, Pt. 1, pp. 67–108.
Champagne, F. H., Freihe, C. A., La Rue, J. C., and Wingaard, J. C.
 1977. Flux measurements, flux estimation techniques and fine-scale turbulent measurements in the surface layer over land. J. Atmos. Sci., Vol. 34, No. 3, pp. 515–530.
Champagne, F. H., Pao, J. H., and Wygnanski, I. J.
 1976. On the two-dimensional mixing region. J. Fluid Mech., Vol. 74, Pt. 2, pp. 209–250.
Chang, P. M.
 1969. A simplified statistical model of turbulent chemically reacting shear flows. AIAA J., Vol. 7, No. 10, pp. 1982–1991.
 1970. Chemical reaction in turbulent flow field with uniform velocity gradient. Phys. Fluids, Vol. 13, No. 10, pp. 1153–1165.
 1972. Diffusion flame in homogeneous turbulent shear flows. Phys. Fluids, Vol. 15, No. 10, pp. 1735–1746.
 1976. A kinetic-theory approach to turbulent chemically reacting flows. Combustion Sci. and Tech., Vol. 13, pp. 123-153.
Chen, C. H. and Blackwelder, R. F.
 1978. Large-scale motion in a turbulent boundary layer: a study using temperature contamination. J. Fluid Mech., Vol. 89, Pt. 1, pp. 1–32.
Chen, W. Y.
 1971. Lognormality of small-scale structure of turbulence. Phys. Fluids, Vol. 14, No. 8, pp. 1639–1642.
Cheng, R. K. and Ng, T. T.
 1983. Velocity statistics in premixed turbulent flames. Combustion and Flame, Vol. 52, No. 2, pp. 185–202.
Chevray, R. and Tutu, N. K.
 1978. Intermittency and preferential transport of heat in a round jet. J. Fluid Mech., Vol. 88, Pt. 1, pp. 133–160.

Chorin, A. J.
   1981. Estimates of intermittency, spectra, and blow-up in developed turbulence. Comm. Pure and Appl. Math., Vol. 34, No. 6, pp. 853–866.
Chomiak, J.
   1970. A possible propagation mechanism of turbulent flames at high Reynolds numbers. Combustion and Flame, Vol. 15, No. 3, pp. 319–321.
   1979. Basic considerations in the turbulent flame propagation in premixed gases. Progr. Energy Combus. Sci., Vol. 5, pp. 207–221.
Corrsin, S.
   1943. Investigation of flow in an axially symmetric heated jet of air. Nat. Adv. Com. Aeronaut. Wartime Report, W-94.
   1951. The decay of isotropic temperature fluctuation in an isotropic turbulence. J. Aeronaut. Sci., Vol. 18, No. 6, pp. 417–423.
Corrsin, S. and Kistler, A. L.
   1955. The free-stream boundaries of turbulent flows. Nat. Adv. Com. Aeronaut. Report, No. 1244.
Curl, R. L.
   1963. Dispersed phase mixing: I. Theory and effects in simple reactors. Amer. Inst. Chem. Eng. J., Vol. 9, No. 2, pp. 175–181.
Damkohler, G.
   1940. Der Einfluss der Turbulenz auf die Flammengeschwindigkeit in Gasgemischen. Zs. Electrochemie, Vol. 6, No. 11, pp. 601–626.
Dibble, R. and Magre, P.
   1987. Finite chemical kinetics effects in a subsonic turbulent hydrogen flame. AIAA Paper, No. 0378.
Dopazo, C.
   1975. Probability density function PDF approach for a heated turbulent axisymmetric jet. Centerline evolution. Phys. Fluids, Vol. 18, No. 4, pp. 397–404.
   1976. A probabilistic approach to turbulent flame theory. Acta Astronaut, Vol. 3, No. 9–10, pp. 853–878.
   1977. On conditioned averages for intermittent turbulent flows. J. Fluid Mech., Vol. 81, Pt. 3, pp. 433–438.
   1979. Relaxation of initial probability density functions (PDFs) in the turbulent convection of scalar fields. Phys. Fluids, Vol. 22, No. 1, pp. 20–30.
Dopazo, C. and O'Brien, E. E.
   1976. Statistical treatment of nonisothermal chemical reactions in turbulence. Combustion Sci. and Tech., Vol. 13, pp. 99–122.
   1979. Intermittency in free turbulent shear flows. In: Turbulent shear flows. Ed. by F. Durst, B. E. Launder, F. W. Schmidt, and J. H. Whitelaw. Berlin, Springer-Verlag, pp. 6–23.

Doroshenko, V. E.
　1960. On the combustion process in the chamber of a gas turbine. In: Proceedings of the all-union conf. on theory of combustion, Vol. 2. Moscow, Izd-vo AN SSSR, pp. 262-270.

Doroshenko, V. E. and Nikitskii, A. I.
　1960. Investigation of the effect of mixture parameters on the characteristics of the turbulent combustion process. In: Combustion under low pressures and some problems of flame stabilization in one-phase and two-phase mixtures. Moscow, Izd-vo AN SSSR, 3-23.

Drake, M. C., Pitz, R. W., and Lapp, M.
　1984. Laser measurements of nonpremixed hydrogen-air flames for assessment of turbulent combustion models. AIAA Paper, No. 544.

Dubovkin, N. F.
　1961. Handbook of hydrocarbon fuels and their combustion products. Moscow, Gosenergoizdat.

Duhamel, P.
　1981. A detailed derivation of conditioned equations for intermittent turbulent flows. Letters in Heat and Mass Transfer, Vol. 8, pp. 491-502.

Ebrahimi, I., Günther, R., and Haberda, F.
　1977. Wahrscheinlichkeitsdichteverteilungen der Konzentrazion in isothermen Luft-Freistrahlen. Forsch. Ing. Wes., Vol. 43, No. 2, pp. 47-52.

Eckhaus, W.
　1961. Theory of flame front stability. J. Fluid Mech., Vol. 10, Pt. 1, pp. 80-100.

Everett, K. W. and Robins, A. G.
　1978. The development and structure of turbulent plane jet. J. Fluid Mech., Vol. 88, Pt. 3, pp. 563-583.

Fabris, G.
　1974. Conditionally sampled turbulent thermal and velocity fields in the wake of a warm cylinder and interaction with an equally cool wake. Ph.D. Thesis, Illinois Inst. of Technology.
　1979a. Turbulent temperature and thermal flux characteristics in the wake of a cylinder. In: Turbulent shear flows. Ed. by F. Durst, B. E. Launder, F. W. Schmidt, and J. H. Whitelaw. Berlin, Springer-Verlag, pp. 55-70.
　1979b. Conditional sampling study of the turbulent wake of cylinder. Part I. J. Fluid Mech., Vol. 94, Pt. 4, pp. 673-710.
　1983a. Third order conditional transport correlations in the two-dimensional turbulent wake. Phys. Fluids, Vol. 26, No. 2, pp. 422-427.
　1983b. Higher-order statistics of turbulent fluctuations in the plane wake. Phys. Fluids, Vol. 26, No. 6, pp. 1437-1445.

Fedoryuk, M. V.
　1983. Asymptotic methods for linear ordinary differential equations. Moscow, Nauka.

Ferziger, J. H.
  1985. Large eddy simulation: its role in turbulence research. Theor. Approaches Turbul., New York, e.a.
Fiedler, H. E.
  1974. Transport of heat across a plane turbulent mixing layer. Adv. Geophysics, Vol. 18A, pp. 93–109.
Fox, R. L.
  1971. Solution for turbulent correlations using multipoint distribution functions. Phys. Fluids, Vol. 14, No. 8, pp. 1806–1808.
  1975. Multipoint distribution functions hierarchy for compressible turbulent flow. Phys. Fluids, Vol. 18, No. 10, pp. 1245–1248.
Frank-Kamenetskii, D. A.
  1967. Diffusion and heat transfer in chemical kinetics. Moscow, Nauka.
Freihe, C. O., Van Atta, C. W., and Gibson, C. H.
  1972. Jet turbulence dissipation rate measurements and correlations. In: AGARD conf. proceedings No. 93 on turbulent shear flows. London, Techn. Ed. Reproduction.
Frenkiel, F. N. and Klebanoff, P. S.
  1975. On the lognormality of the small-scale structure of turbulence. Bound. Layer Meteorol., Vol. 8, No. 2, pp. 173–200.
  1979. Grid turbulence in air and water. Phys. Fluids, Vol. 22, No. 9, pp. 1606–1617.
Freymuth, P. and Uberoi, M. S.
  1971. Structure of temperature fluctuations in the turbulent wake behind heated cylinder. Phys. Fluids, Vol. 14, No. 12, pp. 2574–2580.
Freidman, A.
  1964. Partial differential equations of parabolic type. Englewood Cliffs, N.J., Prentice Hill.
Frisch, U., Sulem, Ph. L., and Nelkin, M.
  1978. A simple dynamical model of intermittent fully developed turbulence. J. Fluid Mech., Vol. 87, Pt. 4, pp. 719–737.
Frost, V. A.
  1960. Mathematical model of turbulent combustion. In: Transactions of the third all-union conf. on combustion theory. Moscow, Izd-vo AN SSSR, pp. 121–125.
  1967. Acceleration of molecular transport processes in turbulent flows. Dokl, AN SSSR, Vol. 176, No. 4, pp. 794–796.
  1973. Model of turbulent diffusion flame front. Izv. AN SSSR, Energetika i transp., No. 6, pp. 108–116.
  1977. Model of homogeneous turbulent flame plume. In: Combustion and explosion. Moscow, Nauka, pp. 361–365.
Gad-el-Hak, M. and Morton, J. B.
  1979. Experiments on the diffusion of smoke in isotropic turbulent flow. AIAA J., Vol. 17, No. 6, pp. 558–562.

Gagne, Y. and Hopfinger, E. J.
  1979. High order dissipation correlations and structure functions in an axisymmetric jet and plane channel flow. In: Second symposium on turbulent shear flows. London, Imperial College, pp. 11.7–11.15.
Gel'fand, I. M. and Shilov, G. E.
  1959. Generalized functions and the actions on them ("Generalized functions", Iss. 1), 2nd ed. Moscow, Nauka.
Genkins, P. E. and Goldschmidt, W. W.
  1976. Conditional (point averaged) temperature and velocity in a heated plane jet. Phys. Fluids, Vol. 19, No. 5, pp. 613–617.
Gibson, C. H.
  1968. Fine structure of scalar fields mixed by turbulence. I. Zero-gradient points and minimal gradient surfaces. Phys. Fluids, Vol. 11, No. 11, pp. 2305–2315.
Gibson, C. H., Friehe, C. A., and McConnell, S.
  1977. Structure of sheared turbulent fields. Phys. Fluids, Suppl., 1977, Vol. 20, pp. 156–167.
Gibson, C. H. and Masiello, P. J.
  1975. Observations of the variability of dissipation rates of turbulent velocity and temperature fields. In: Proc. Sympos. held at the Univ. of Calif., San Diego, 1971, Berlin, Springer-Verlag, pp. 428–452.
Gibson, C.H. and Schwartz, W. H.
  1963. The universal equilibrium spectra of turbulent velocity and scalar fields. J. Fluid Mech., Vol. 16, Pt. 3, pp. 365–384.
Gibson, C. H., Stegen, G. R., and Williams, R. B.
  1970. Statistics of the fine structure of turbulent velocity and temperature fields measured at high Reynolds number. J. Fluid Mech., Vol. 41, Pt. 1, pp. 153–167.
Gibson, C. H., Stegen, G. R., and McConnell, S.:
  1970. Measurements of the universal constant in Kolmogoroff's third hypothesis for high Reynolds number. Phys. Fluids, Vol. 13, No. 10, pp. 2448–2451.
Gibson, M. M.
  1963. Spectra of turbulence in a round jet. J. Fluid Mech., Vol. 15, Pt. 2, pp. 161–173.
Ginevskii, A. S., Ioselevich, V. A., Kolesnikov, A. V., Lapin, Yu. V., Pilipenko, V. N., and Sekundov, A. N.
  1978. Calculation methods of the turbulent boundary layer. Series: Rev. of science and tech. Mechanics of liquids and gases, 11. Moscow, VINITI, pp. 155–304.
Glaz, M.
  1981. Statistical behavior and coherent structures in two-dimensional inviscid turbulence. SIAM J. Appl. Math., Vol. 41, No. 3, pp. 459–479.

Gogos, D., Sadhal, S. S., Ayyaswamy, P. S., and Sundararajan, T.
   1986. Thin-flame theory for combustion of moving liquid drop: effects due to variable density. J. Fluid Mech., Vol. 171, pp. 121–144.
Golovanov, Yu. V.
   1977. Experimental investigation of statistical characteristics of turbulent fluctuations of contaminant concentration in a submerged jet. Cand. of physical and math. sciences thesis. Dolgoprudnyi, MFTI.
Golovanov, Yu. V. and Sherbina, Yu. A.
   1979. Distribution functions of concentration probability in submerged jets with defferent initial turbulence. Stored at VINITI, No. 4061-79, 24 pp.
Golubev, V. V., Yankovskii, V. M., Postnov, V. F., and Talantov, A. V.
   1973. Effects of pressure on the velocity of flame propagation in turbulent flow. Izv. vuzov, Aviats. tekhnika, No. 2, pp. 77–86.
Gol'dshtik, M. A. and Shtern, V. N.
   1977. Fluid dynamic stability and turbulence. Novosibirsk, Nauka.
Gorbatko, A. A., Kuznetsov, V. R., and Lipatov, P. A.
   1986. Application of the frontal model of combustion to the analysis of flame stabilization by blunt bodies. In: Combustion of heterogeneous and gaseous systems. Chernogolovka, Izd-vo OIKhF AN SSSR, pp. 35–39.
Grant, H. L., Stewart, R. W., and Moilliet, A.
   1962. Turbulence spectra from a tidal channel. J. Fluid Mech., Vol. 12, Pt. 2, pp. 241–268.
Gremyachkin, V. M. and Istratov, A. G.
   1972. On the stability of plane flame in a flow with velocity gradient. In: Combustion and Explosion. Moscow, Nauka, pp. 305–308.
Gromov, V. G., Larin, O. B., and Levin, V. A.
   1984. Application of the "immiscibility" model for the calculation of a turbulent near-wall jet of hydrogen in coflowing supersonic air stream. Khimicheskaya fizika, Vol. 3, No. 8, pp. 1190–1195.
Günther, R.
   1979. Strömungsturbulenz and verbrennung. Chemic. Ing. Techn., Vol. 51, No. 9, pp. 858–866.
Gurvich, A. S.
   1960a. Measurement of the asymmetry factor of the velocity difference distribution in the atmosphere layer above land. Dokl. AN SSSR, Vol. 134, No. 5, pp. 554–557.
   1960b. Experimental investigation of the frequency spectra and the functions of probability distribution of the vertical component of air velocity. Izv. AN SSSR, ser. geofiz., No. 7, pp. 1042–1055.
   1962. Fluctuation spectra of the vertical component of air velocity and their relation with micrometeorological conditions. In: Atmospheric Turbulence. Trudy in-ta fiziki atmosfery, AN SSSR, No. 4, pp. 101–136.

1966. On the probability distribution (PDF) of the squared velocity difference in turbulent flow. Izv. AN SSSR. Fizika atmosfery i okeana, Vol. 2, No. 10, pp. 1095–1098.

1967. On the probability distribution (PDF) of the squared temperature difference at two points of the turbulent flow. Dokl. AN SSSR, Vol. 172, No. 3, pp. 554–557.

Gurvich, A. S. and Yaglom, A. M.
1967. Breakdown of eddies and probability distributions (PDFs) for small-scale turbulence. Phys. Fluids, Suppl., Vol. 10, No. 9 (Pt. 2), pp. 59–65.

Gurvich, A. S. and Yurchenko, B. N.
1981. Frequency spectra of temperature fluctuations under turbulent convection. Izv. AN SSSR, Physics of the atmosphere and the ocean, Vol. 16, No. 8, pp. 854–857.

Gurvich, A. S. and Zubkovskii, S. L.
1963. On the experimental estimate of turbulence fluctuation energy dissipation. Izv. AN SSSR, Ser. geofiz., No. 2, pp. 1856–1858.
1965. Measurement of fourth and sixth moments of velocity gradient. Izv. AN SSSR. Fizika atmosfery i okeana, Vol. 2, No. 8, pp. 797–802.

Gutmark, E. and Wygnanski, I.
1976. The planar turbulent jet. J. Fluid Mech., Vol. 73, Pt. 3, pp. 465–495.

Hall, H. G.
1965. The boundary layer over an impulsively started flat plate. Proc. Roy. Soc., A310, No. 1502, pp. 401–414.

Hawthorne, W. R., Wedell, D. S., and Hottel, H. S.
1949. Mixing and combustion in turbulent gas jets. In: Third Int. sympos. on combustion, flame, and explosion phenomena. Baltimore, The Williams and Wilkins Co., pp. 267–300.

Hedley, T. B. and Keffer, J. F.
1947a. Turbulent/nonturbulent decisions in an intermittent flow. J. Fluid Mech., Vol. 64, Pt. 4, pp. 625–644.
1947b. Some turbulent/nonturbulent properties of the outer intermittent region of a boundary layer. J. Fluid Mech., Vol. 64, Pt. 4, pp. 645–678.

Helland, K. N., Van Atta, C. W., and Stegen, G. R.
1977. Spectral energy transfer in high Reynolds number turbulence. J. Fluid Mech., Vol. 79, Pt. 2, pp. 337–359.

Hill, J.
1976. Homogeneous turbulent mixing with chemical reaction. Annual Review of Fluid Mech., Vol. 8, pp. 135-161.

Hinze, J. O.
1959. Turbulence. An introduction to its mechanism and theory. New York, McGraw-Hill.

Hopf, E.
1952. Statistical hydromechanics and functional calculus. J. Rat. Mech. Anal., Vol. 1, No. 1, pp. 87–123.

Ibragimov, M. Kh., Petrishcheva, G. A., and Taranov, G. S.
  1968. Investigation of turbulent characteristics of a free round jet of incompressible gas. Inzh.-fiz. zhurnal (Journal of engineering physics), Vol. 14, No. 3, pp. 415–422.
Ievlev, V. M.
  1970. Approximate equations of turbulent motion of incompressible fluid. Izv. AN SSSR. MZhG, No. 1, pp. 91–103.
  1972. Equations for finite-dimensional probability distributions of fluctuating quantities in turbulent flow. Dokl. AN SSSR, Vol. 208, No. 5, pp. 1036–1038.
  1975. Turbulent flow of high-temperature continuum media. Moscow, Nauka.
Il'yashenko, S. M. and Talantov, A. V.
  1964. Theory and calculation of uniflow combustion chambers. Moscow, Mashinostroyeniye.
Istratov, A. G. and Librovich, V. B.
  1966a. On the effect of transport processes on the stability of plane flame front. PMM, Vol. 30, No. 3, pp. 451–466.
  1966b. Stability of flames. Science Review. Moscow, Izd-vo AN SSSR.
  1966c. On the stability of propagation of spherical flame. ZhPMTF, No. 1, pp. 67–78.
Janicka, J., Kolbe, W., and Kollman, W.
  1979. Closure of the transport equation for the probability density function (PDF) of turbulent scalar fields. J. Nonequilib. Thermodyn., Vol. 4, pp. 47–66.
Janicki, J. and Kollman, W.
  1979. Prediction model for PDF of turbulent flows. In: Second symposium on turbulent shear flows. London, Imperial College.
Jenkins, D. R., Yumlu, V. S., and Spalding, D. S.
  1966. Combustion of hydrogen and oxygen in steady-flow adiabatic stirred reactor. In: Eleventh Int. Sympos. on combustion. Pittsburgh, The Combustion Inst., pp. 779–787.
Jensen, D. E. and Jones, G. A.
  1978. Reaction rate coefficients for flame calculations. Combustion and Flame, Vol. 32, No. 1, pp. 1–33.
Kabanov, V. V. and Klubnikin, V. S.
  1985. Experimental determination of the resistance of spherical particle in a flow of argon plasma. Inzh.-fiz. zh., Vol. 48, No. 3, pp. 396–402.
Kaimal, J. C., Wingaard, J. C., Izumi, Y., and Coté, O. R.
  1972. Spectral characteristics of surface layer turbulence. Quart. J. Roy. Meteorol. Soc., 1972, Vol. 98, No. 417, pp. 563–589.
Kalghatgi, G. T. and Moss, J. B.
  1979. Quantitative schlieren measurements in a confined turbulent premixed flame. In: Second sympos. on turbulent shear flows. London, Imperial College, pp. 5.9–5.14.

Kalitkin, N. N.
  1978. Numerical methods. Moscow, Nauka.
Karlowitz, B., Denniston, D., and Wells, F.
  1951. Investigation of turbulent flames. J. Chem. Phys., Vol. 19, No. 4, pp. 541–552.
Karpov, V. P. and Severin, E. S.
  1977. Turbulent burn-out rates of gas mixtures for describing combustion in engines. In: Combustion of heterogeneous gas systems. Chernogolovka, OIKhF, AN SSSR, pp. 74–76.
  1980. Effect of the coefficients of molecular transport on the turbulent burn-out rate. Fizika goreniya i vzryva (Physics of combustion and explosion), Vol. 16, No. 1, pp. 45–51.
Kaskan, W. E.
  1958. The concentration of hydroxyl and of oxygen atoms in gases from lean hydrogen-air flames. Combustion and Flame, Vol. 3, No. 2, p. 286.
Kawall, J. G. and Keffer, J. E.
  1979. Interface statistics of a uniformly distorted heated turbulent wake. Phys. Fluids, Vol. 22, No. 1, pp. 31–39.
Keffer, J. E., Olsen, G. J., and Kawall, J. G.
  1977. Intermittency in a thermal mixing layer. J. Fluid Mech., Vol. 79, Pt. 3, pp. 595–607.
Keller, L. B. and Friedman, A. A.
  1924. Differentialgleichung fur die turbulente Bewegung einer kompressiblen Flussigkeit. In: Proc. First Intern. Congr. Appl. Mech., Delft, pp. 395–405.
Kent, J. H. and Bilger, R. W.
  1977. The prediction of turbulent diffusion flame fields and nitric oxide formation. In: Sixteenth Int. Sympos. on combustion. Baltimore, The Williams and Wilkins Co., pp. 1643–1656.
Kholmyanskii, M. Z.
  1970. Investigation of microfluctuations of the derivative of wind velocity at the earth's surface. Izv. AN SSSR, Fizika atmosfery i okeana (Physics of the atmosphere and the ocean), Vol. 6, No. 4, pp. 423–430.
  1972. Measurement of microturbulent fluctuations of the derivative of wind velocity at the earth's surface. Izv. AN SSSR, Fizika atmosfery i okeana (Physics of the atmosphere and the ocean), Vol. 8, No. 8, pp. 818–828.
Khramtsov, V. A.
  1960. Experimental investigation of the combustion of a homogeneous fuel-air mixture at low pressures. In: Combustion at low pressures and some problems of flame stabilization in single-phase and two-phase mixtures. Moscow, Izd-vo AN SSSR, pp. 43–53.

Kiselev, V. P.
    1977. Small-scale homogeneous turbulence in the Burgers problem. Trans. of 20th scientific conf. MFTI. Aerodynamics and applied math. Dolgoprudnyi, MFTI, 29–32.

Kislov, N. V.
    1980. Boundary-value problems for the equations of the mixed type in a rectangular region. Dokl. AN SSSR, Vol. 255, No. 1, pp. 26–30.

Kistler, A. L. and Vrebalovich, T.
    1966. Grid turbulence at large Reynolds numbers. J. Fluid Mech., Vol. 26, Pt. 1, pp. 37–47.

Kleinberg, J. M. and Steger, J. L.
    1974. On laminar boundary-layer separation. J. Fluid Mech., Vol. 53, Pt. 1, pp. 177–191

Klimov, A. M.
    1963. Laminar flame in turbulent flow. ZhPMTF, No. 3, pp. 49–58.
    1972. On the theory of arbitrary front of laminar flame. In: Combustion and explosion. Moscow, Nauka, pp. 299–304.
    1975. Flame propagation under strong turbulence. Dokl. AN SSSR, Vol. 221, No. 1, pp. 56–59.
    1977a. On the models of turbulent combustion. In: Combustion and explosion, Moscow, Nauka, pp. 349–356.
    1977b. On flame stabilization by recirculation zones. In: Combustion of heterogeneous and gas systems. Chernogolovka, OIKhF AN SSSR, pp. 81–84.

Klyachko, L. A.
    1958. Experimental investigation of combustion of a fuel drop. In: Combustion of two-phase systems. Moscow, Izd-vo AN SSSR, pp. 5–18.

Klyachko, L. A. and Strokin, V. N.
    1969. Turbulent diffusion combustion in a cylindrical tube. Inzh.-fiz. zhurnal, Vol. 17, No. 3, pp. 447–454.

Kokushkin, N. V.
    1958. Investigation of combustion of a homogeneous mixture in turbulent flow by monitoring temperature fluctuations. Izv. AN SSSR, OTN, No. 8, pp. 3–11.
    1960. Investigation of the structure of turbulent flame. In: Transactions of the third all-union conf. on theory of combustion. Moscow, Izd-vo AN SSSR, pp. 109–113.

Kollman, W. and Janicka, J.
    1982. The probability density functions (PDFs) of a passive scalar in turbulent shear flows. Phys. Fluids, Vol. 25, No. 10, pp. 1755–1769.

Kolmogorov, A. N.
    1935. La transformation de Laplace dans les espaces lineaires. Compt. Rend. Acad. Sci. (Paris), Vol. 200, pp. 1717–1718.
    1941. Local turbulence structure in incompressible viscous fluid at very high Reynolds numbers. Dokl. AN SSSR, Vol. 30, No. 4, pp. 299–303.

1962a. Refinement of the hypotheses concerning local turbulence structure in incompressible viscous fluid at high Reynolds numbers. Mécanique de la turbulence (Coll. Intern. du CNRS á Marseille), Paris Éd., CNRS, pp. 447–458

1962b. A refinement of previous hypotheses concerning the local structure of turbulence in a viscous incompressible fluid at high Reynolds number. J. Fluid Mech., Vol. 13, Pt. 1, pp. 82–85.

Kompaniets, V. Z., Ovsyannikov, A. A., and Polak, L. S.
1979. Chemical reactions in turbulent flows of gas and plasma. Moscow, Nauka.

Kotsovinos, N. E.
1977. Plane turbulent buoyant jets. Part II. Turbulent structure. J. Fluid Mech., Vol. 81, Pt. 1, pp. 46–52.

Kovasznay, L. S. G.
1956. Combustion in turbulent flow. Jet propulsion, Vol. 26, pp. 485–497.

Kovasznay, L. S. G., Kibens, V., and Blackwelder, R. F.
1970. Large-scale motion in the intermittent region of a turbulent boundary layer. J. Fluid Mech., Vol. 41, Pt. 2, pp. 283–325.

Kozachenko, L. S.
1960. Combustion of gasoline-air mixture in a turbulent flow. In: Transactions of the third all-union conf. on theory of combustion. Moscow, Izd-vo AN SSSR, 126–137.

Kraichnan, R. H.
1959. The structure of isotropic turbulence at very high Reynolds number. J. Fluid Mech., Vol. 5, Pt. 4, pp. 497-543.
1974. On Kolmogorov's inertial-range theories. J. Fluid Mech., Vol. 62, Pt. 2, pp. 305–330.
1975. Remarks on the turbulence theory. Adv. in Math., Vol. 16, pp. 305–331.

Krasitskii, V. P., Filimonov, M. L., and Frost, V. A.
1970. Mathematical description of turbulent combustion. In: Problems in the theory of combustion. Moscow, Nauka, pp. 7–16.

Kremer, H.
1966. Strömung und Mischung in frei brennenden diffusion Flammen. VDI. Berict., No. 95, pp. 55-69.

Kuo, A. Y. S. and Corrsin, S.
1971. Experiments on internal intermittency and fine structure distribution functions in fully developed turbulent fluid. J. Fluid Mech., Vol. 50, Pt. 2, pp. 285–319.

Kuo, Ying-Yan and O'Brien, E. E.
1981. Two-point probability density function (PDF) closure applied to a diffusive-reactive system. Phys. Fluids, Vol. 24, No. 2, pp. 194–201.

Kuzin, A. F. and Talantov, A. V.
- 1977. On the problem of the mechanism and characteristics of combustion in turbulent homogeneous mixture flow. In: Combustion and explosion. Moscow, Nauka, pp. 356–360.

Kuzin, A. F., Yankovskii, V. M., Appolonov, V. L., and Talantov, A. V.
- 1972. Effect of the initial temperature on the main characteristics of combustion in turbulent homogeneous mixture flow. In: Combustion and explosion. Moscow, Nauka, pp. 337–341.

Kuznetsov, V. R.
- 1967. On velocity difference probability density (PDF) at two points of a homogeneous, isotropic turbulent flow. PMM, Vol. 31, No. 6, pp. 1069–1972.
- 1969. Effect of temperature and concentration fluctuations on the mean reaction rate in turbulent flow. In: Transactions of the second all-union symposium on combustion and explosion. Chernogolovka, OIKhF AN SSSR, pp. 99–104.
- 1971. Passive contaminant concentration probability distribution (PDF) in submerged axisymmetric jet. Izv. AN SSSR MZhG, No. 2, pp. 161–164.
- 1972a. Passive contaminant concentration probability in turbulent flows with transverse shear. Izv. AN SSSR, MZhG, No. 5, pp. 86–91.
- 1972b. Effect of temperature and concentration fluctuations on ignition lag in turbulent flow. In: Combustion and explosion. Moscow, Nauka, pp. 342–346.
- 1975. Some specific features of flame front motion in turbulent flow of homogeneous combustible mixture. Fizika goreniya i vzryva, Vol. 11, No. 4, pp. 574–581.
- 1976a. Velocity difference probability distribution (PDF) in the inertial range of the turbulence spectrum. Izv. AN SSSR, MZhG, No. 3, pp. 32–41.
- 1976b. Flame propagation in turbulent flow of combustible mixture. Izv. AN SSSR, MZhG, No. 5, pp. 3–15.
- 1977a. Velocity of flame propagation in turbulent flow of homogeneous combustible mixture. In: Combustion and explosion. Moscow, Nauka, pp. 366–372.
- 1977b. Mixing to the molecular level and development of the chemical reaction in turbulent flow. Izv. AN SSSR, MZhG, No. 3, pp. 32–41.
- 1977c. Statistical characteristics of turbulence in the inertial range of the spectrum. In: Turbulent flows. Moscow, Nauka, pp. 123–129.
- 1979a. Estimate of correlation between pressure fluctuations and velocity divergence in variable density subsonic flows. Izv. AN SSSR, MZhG, No. 3, pp. 4–11.
- 1979b. Effect of turbulence on combustion processes. In: Transactions of the fifth lecture series devoted to the formulation of the scientific legacy and development of the ideas of A. F. Tsander. Theory and

construction of engines for flying machines. Moscow, IIET AN SSSR, pp. 66–75.
- 1980. Effect of flame instability on turbulent combustion of a homogeneous mixture. In: Combustion of gases and natural fuels. Chernogolovka, OIKhF AN SSSR, pp. 32–37.
- 1982a. Effect of turbulence on the formation of large superequilibrium concentrations of atoms and free radicals in diffusion flames. Izv. AN SSSR, NZhG, No. 6, pp. 3–9.
- 1982b. Boundary laws of turbulent flame propagation. Fizika goreniya i vzryva, Vol. 18, No. 2, pp. 52–60.
- 1983. Formation of carbon monoxide in turbulent diffusion combustion. Fizika goreniya i vzryva, Vol. 19, No. 4, pp. 42–45.

Kuznetsov, V. R. and Frost, V. A.
- 1973. Concentration probability distribution (PDF) and intermittency in turbulent jets. Izv. AN SSSR, MZhG, No. 2, pp. 58–64.

Kuznetsov, V. R., Lebedev, A. B., Sekundov, A. N., and Smirnova, I. P.
- 1977a. Calculation of turbulent diffusion combustion plume taking into account concentration fluctuations and Archimedean forces. Izv. AN SSSR, MZhG, No. 1, pp. 30–40.
- 1977b. Effect of concentration fluctuations on diffusion combustion. In: Combustion of heterogeneous and gas systems. Chernogolovka, OIKhF AN SSSR, pp. 57–60.
- 1980. Analysis of the possibilities of using different turbulence models for the description of diffusion combustion in jets. In: Combustion of gases and natural fuels. Chernogolovka, OIKhF AN SSSR, pp. 29–32.
- 1981. Investigation of quasi-one-dimensional turbulent combustion using the equations for the concentration probability density distribution functions (PDFs). Izv. AN SSSR, MZhG, No. 4, pp. 3–11.

Kuznetsov, V. R., Praskovskii, A. A., and Sabel'nikov, V. A.
- 1984a. Intermittency and the local structure of turbulent flow with transverse shear. Stored at VINITI, No. 4706-84, 130 pp.
- 1984b. Experimental investigation of intermittency and local structure of turbulent flow with transverse shear. In: Structure of gas-phase flames. Part II. Novosibirsk, ITPM SO AN SSSR, pp. 21–38.
- 1987. The variability of constants in Kolmogorov's second and third hypotheses. In: Proceedings of the Intern. Conf. on Fluid Mech. (Suppl.), Peking Univ. Press, Beijing, China, pp. 106–111.
- 1988. Local structure of turbulence in free flows with strong external intermittency. Izv. AN SSSR, MZhG (in press).

Kuznetsov, V. R. and Rashchupkin, V. I.
- 1977. Probability distribution (PDF) and conditional averaging in turbulent flows. Izv. AN SSSR, MZhG, No. 6, pp. 31–37.

Kuznetsov, V. R. and Sabel'nikov, V. A.
- 1977. Specific features of combustion of premixed gases in a strongly turbulent flow. Fizika goreniya i vzryva, Vol. 13, No. 4, pp. 499–511.

1981a. Intermittency and velocity probability distribution (PDF) in turbulent flows. Uspekhi mekhaniki (Progress in mechanics), Vol. 4, Iss. 3, pp. 81–134.

1981b. Intermittency and concentration probability distribution (PDF) in turbulent flows. Uspekhi mekhaniki (Progress in mechanics), Vol. 4, Iss. 2, pp. 123–166.

Ladyzhenskaya, O. A.
    1970. Mathematical problems of the dynamics of a viscous incompressible fluid. Moscow, Nauka.
    1972. On the dynamic system resulting from the Navier-Stokes equations. Dokl. AN SSSR, Vol. 205, No. 2, pp. 318–320.

Landau, L. D.
    1944. On the theory of slow combustion. ZhETF, Vol. 14, No. 6, pp. 240–244.

Landau, L. D. and Lifshits, E. M.
    1954. Continuum mechanics. Moscow, Gostekhizdat.

La Rue, J. C.
    1974. Detection of the turbulent-nonturbulent interface in slightly heated turbulent shear flows. Phys. Fluids, Vol. 17, No. 8, pp. 1513–1517.

LaRue, J. C. and Libby, P. A.
    1974. Temperature fluctuations in the plane turbulent wake. Phys. Fluids, Vol. 17, No. 11, pp. 1956–1967.
    1976. Statistical properties of the interface in the turbulent wake of a heated cylinder. Phys. Fluids, Vol. 19, No. 12, pp. 1864–1875.
    1978. Detailed similarity in the turbulent wake of a heated cylinder. Phys. Fluids, Vol. 21, No. 6, pp. 891–897.
    1981. Thermal mixing layer downstream of half-heated turbulent grid. Phys. Fluids, Vol. 24, No. 4, pp. 597–603.

LaRue, J. C., Libby, P. A., and Seshadri, D. V. R.
    1981. Further results on the thermal mixing layer downstream of a turbulent grid. Phys. Fluids, Vol. 24, No. 11, pp. 1927–1933.

Laufer, J.
    1954. The structure of turbulence in fully developed pipe flow. Nat. Adv. Com. Aeronaut., Rep. No. 1174.

Launder, B. E. and Spalding, D. B.
    1972. Mathematical models of turbulence. London, Academic Press.

Lavoie, G. A. and Schlader, A. F.
    1974. A scaling study of NO formation in turbulent diffusion flames of hydrogen burning in air. Combust. Sci. and Techn., Vol. 8, pp. 215–224.

Lawn, G. J.
    1971. The determination of the rate of dissipation in turbulent pipe flow. J. Fluid Mech., Vol. 48, Pt. 3, pp. 477–505.

Lee, J.
    1982. Development of mixing and isotropy in inviscid homogeneous turbulence. J. Fluid Mech., Vol. 120, pp. 155–183.

Leray, J.
  1934. Sur le mouvement d'un liquide visquex exsplissant l'espace. Acta Math., Vol. 63, pp. 198–248.
Lewis, B. and von Elbe, G.
  1961. Combustion, flames, and explosions of gases. New York, Academic Press.
Libby, P. A.
  1975. On the prediction of the intermittent turbulent flows. J. Fluid Mech., Vol. 68, Pt. 2, pp. 273–295.
  1976. Prediction of the intermittent turbulent wake of a heated cylinder. Phys. Fluids, Vol. 19, No. 4, pp. 494–501.
Libby, P. A. and Bray, K. N. C.
  1981. Countergradient diffusion in premixed turbulent flames. AIAA J., Vol. 19, No. 2, pp. 205–213.
Libby, P. A. and Williams, F. A.
  1976. Turbulent flows involving chemical reactions. Annual Rev. Fluid Mech., Vol. 8, pp. 351–376.
  1981. Some implications of recent theoretical studies in turbulent combustion. AIAA J., Vol. 19, No. 3, pp. 261–274.
Librovich, V. B. and Lisitsyn, V. I.
  1975. Interaction of flow fluctuations with the chemical reaction of a turbulent flame. Moscow, IPM, preprint No. 57.
Liew, S. K., Bray, K. N. C., and Moss, J. B.
  1984. A stretched laminar flamelet model of turbulent nonpremixed combustion. Combustion and Flame, Vol. 56, No. 2, pp. 199–213.
Lockwood, F. C. and Shah, N. G.
  1982. New method for the computation of probability density functions (PDFs) in turbulent flows. AIAA J., Vol. 20, No. 6, pp. 860–862.
Long, M. B. and Chu, B. T.
  1981. Mixing mechanism and structure of an axisymmetric turbulent mixing layer. AIAA J., Vol. 19, No. 9, pp. 1158–1163.
Long, M. B., Chu, B. T., and Chang, R. K.
  1981. Instantaneous two-dimensional gas concentration measurements by light scattering. AIAA J., Vol. 19, No. 9, pp. 1151–1157.
Lungren, T. S.
  1967. Distribution functions (PDFs) in the statistical theory of turbulence. Phys. Fluids, Vol. 10, No. 5, pp. 968–975.
  1969. Model equation for nonhomogeneous turbulence. Phys. Fluids, Vol. 12, No. 3, pp. 485–497.
  1975. A closure hypothesis for hierarchy of equations for turbulent probability distribution functions. Lect. Notes Phys., Vol. 12, pp. 70–100.
Lyubimov, B. Ya.
  1969. Lagrangian description of the dynamics of turbulent flow. Dokl. AN SSSR, Vol. 184, No. 5, pp. 1069–1071.

Lyubimov, B. Ya. and Ulinich, F. R.
  1970. Statistical equations of turbulent flow in Lagrangian variables. PMM, Vol. 34, No. 1, pp. 24–31.
Malkus, W. V.
  1956. Outline of a theory of turbulent shear flow. J. Fluid Mech., Vol. 1, Pt. 3, pp. 521–539.
Mandelbrot, B. B.
  1974. Intermittent turbulence in self-similar cascades: divergence of high moments and dimension of the carrier. J. Fluid Mech., Vol. 62, Pt. 2, pp. 331–358.
  1975. On the geometry of homogeneous turbulence, with stress on the fractal dimension of the iso-surfaces of scalars. J. Fluid Mech., Vol. 72, Pt. 2, pp. 401–416.
  1976. Intermittent turbulence and fractal dimension: kurtosis and the spectral exponent $5/3 + B$. In: Turbulence and Navier-Stokes Equations. Ed. by R. Temam, Lect. Notes in Math., Vol. 565, New York, Springer, pp. 121–145.
  1977. Fractals and turbulence: attractors and dispersion. Lect. Notes in Math., Vol. 615, New York, Springer, pp. 83–93.
Markstein, G. H.
  1975. Radiative energy transfer from turbulent diffusion flame. ASME Paper, No. 75-HT-7.
Marsden, J. E. and McCracken, M.
  1976. The Hopf bifurcation and its applications. New York, Springer-Verlag.
McBean, G. A., Stewart, R. W., and Miyake, M.
  1971. The turbulent energy budget near the surface. J. Geophys. Res., Vol. 76, No. 27, pp. 6540–6549.
Meiron, D. I., Baker, G. R., and Orszag, S. A.
  1982. Analytical structure of vortex sheet dynamics. Part I. Kelvin-Helmholtz instability. J. Fluid Mech., Vol. 114, pp. 283–298.
Meshcheryakov, E. A.
  1974. On the relation of the coefficients of turbulence diffusion of mean scalar value and its fluctuations in free jets. Sci. Records of TsAGI, Vol. 5, No. 1, pp. 113–118.
Meshcheryakov, E. A. and Sabel'nikov, V. A.
  1984a. Semiempirical model and computation of the intermittency factor in turbulent jet flows. Fizika goreniya i vzryva, Vol. 20, No. 4, pp. 45–52.
  1984b. Determination of the intermittency factor in turbulent jets. In: Structure of gas-phase flames. Part II. Novosibirsk, ITPM SO AN SSSR, pp. 3–20.
Meshkov, M. A.
  1976. Experimental investigation of the structure of concentration fluctuations in coflowing streams. In: Transaction of 21st scientific conf.

MFTI. Aerophysics and applied mathematics. Dolgoprudnyi, MFTI, pp. 22–24.

Meshkov, M. A. and Shcherbina, Yu. A.
- 1979. Experimental investigation of the statistical characteristics of the mixing field of coaxial jets with various initial conditions. Stored in VINITI, No. 4059–79, 59 pp.
- 1981. Development of plane turbulent mixing layer. Fifth all-union conf. on theoretical and applied mechanics. Annotations of papers, Alma-Ata, Nauka, pp. 254–255.

Mestayer, P.
- 1982. Local isotropy and anisotropy in a high Reynolds number turbulent boundary layer. J. Fluid Mech., Vol. 119, pp. 55–89.

Mikheev, A. A.
- 1949. Principles of heat transfer. Moscow. Gosenergoizdat.

Mirabel, A. P. and Monin, A. S.
- 1979. Two-dimensional turbulence. Uspekhi mekhaniki (Progress in mechanics), Vol. 2, Iss. 3, pp. 47–95.

Miyawaki, O., Tsujikawa, H., and Uraguchi, Y.
- 1974. Turbulent mixing in multinozzle injection tubular mixer. J. Chem. Eng., Japan, Vol. 7, No. 1, pp. 52–56.

Mobbs, F. R.
- 1968. Spreading and contraction of the boundaries of free turbulent flows. J. Fluid Mech., Vol. 33, Pt. 2, pp. 227–240.

Monin, A. S.
- 1967a. Equations for finite-dimensional turbulence field probability distributions (PDFs). Dokl. AN SSSR, Vol. 177, No. 5, pp. 1036–1038.
- 1967b. Equations for turbulent motion. PMM, Vol. 31, No. 6, pp. 1057–1068.
- 1978. On the nature of turbulence. UFN, Vol. 125, Iss. 1, pp. 97–122.

Monin, A. S. and Yaglom, A. M.
- 1963. On the laws of small-scale turbulent motion of liquids and gases. UMN, Vol. 18, Iss. 5 (113), pp. 93–114.
- 1965. Statistical fluid mechanics. Part I. Moscow, Nauka.
- 1967. Statistical fluid mechanics. Part II. Moscow, Nauka.

Moore, D. W.
- 1979. The spontaneous appearance of a singularity in the shape of an evolving vortex sheet. Proc. Roy. Soc., A365, No. 1720, pp. 105–119.

Morf, R. H., Orszag, S. A., and Frisch, U.
- 1980. Spontaneous singularity in three-dimensional, inviscid incompressible flow. Phys. Rev. Lett., Vol. 44, p. 572.

Mors, P. M. and Feshbach, H.
- 1953. Methods of theoretical physics. Part II. New York, McGraw-Hill.

Moss, J. B.
   1980. Simultaneous measurements of concentration and velocity in an open premixed turbulent flame. Combustion Sci. and Tech., Vol. 22, No. 3-4, pp. 119-129.

Moum, J. N., Kawall, J. G., and Keffer, J. F.
   1979. Structural features of the plane turbulent jet. Phys. Fluids, Vol. 22, No. 7, pp. 1240-1244.

Murlis, J., Tsai, H. M., and Bradshaw, P.
   1982. The structure of turbulent boundary layers at low Reynolds numbers. J. Fluid Mech., Vol. 122, pp. 13-56.

Nasmyth, P. M.
   1970. Oceanic turbulence. Ph.D. Dissertation. Inst. of Oceanography, Univ. of British Columbia, Canada.

Nedorub, S. A. and Shcherbina, Yu. A.
   1979. Analytical solution of the equation for single-point concentration probability distribution function (PDF) in turbulent flows. Stored in VINITI, No. 3406-79, 34 pp.

Nedorub, S. A., Frost, V. A., and Shcherbina, Yu. A.
   1979. Computation of turbulent diffusion and homogeneous turbulent flame on the basis of a statistical model. Stored in VINITI, No. 3405-79, 57 pp.

Neiland, V. Ya.
   1971. Flow beyond the boundary layer separation point in supersonic flow. Izv. AN SSSR, MZhG, No. 3, pp. 19-25.

Nelkin, M.
   1974. Turbulence, critical phenomena and intermittency. Phys. Rev., A9, No. 1, pp. 388-395.
   1975. Scaling theory of hydrodynamic turbulence. Phys. Rev., A11, No. 5, pp. 1737-1743.
   1981. Do the dissipation fluctuations in high Reynolds number turbulence define a universal exponent. Phys. Fluids, Vol. 24, No. 3, pp. 556-557.

Novikov, E. A.
   1965. On the correlation high-order constraints in a turbulent flow. Izv. AN SSSR. Fizika atmosfery i okeana (Physics of the atmosphere and the ocean), Vol. 1, No. 8, pp. 788-796.
   1966. Mathematical model of turbulent flow intermittency. Dokl. AN SSSR, Vol. 168, No. 6, pp. 1279-1282.
   1967. Kinetic equation for the vortex field. Dokl. AN SSSR, Vol. 177, No. 2, pp. 299-301.
   1969. Scaling similarity for random fields. Dokl. AN SSSR, Vol. 184, No. 5, pp. 1072-1075.
   1971. Intermittency and scaling similarity in the structure of turbulent flow. PMM, Vol. 35, No. 2, pp. 266-277.

Novikov, E. A. and Stewart, R. W.
- 1964. Intermittency of turbulence and energy dissipation fluctuations spectrum. Izv. AN SSSR, Ser. geogr. i. geofiz., No. 3, pp. 408–413.

O'Brien, E. E.
- 1978. Advection of scalars by a nonbuoyant plume. J. Fluid Mech., Vol. 89, No. 2, pp. 209–222.
- 1980a. Statistical methods in reacting turbulent flows. AIAA Paper, No. 137.
- 1980b. The probability density function (PDF) approach to reacting turbulent flows. In: Turbulent Reacting Flows. Topics in Applied Phys., Vol. 44. Ed. by P. A. Libby and F. A. Williams.

O'Brien, E. E. and Dopazo, C.
- 1978. Behavior of conditioned variables in free turbulent shear flows. In: Structure and mechanisms of turbulence. II. Berlin, Springer-Verlag, pp. 124–133.

Obukhov, A. M.
- 1941. On energy distribution in the turbulent flow spectrum. Dokl. AN SSSR, Vol. 32, No. 1, pp. 22–24.
- 1949. Structure of the temperature field in turbulent flow. Izv. AN SSSR, Ser. geogr. i geofiz., Vol. 13, No. 13, pp. 58–69.
- 1962. Some specific features of atmospheric turbulence. J. Fluid Mech., Vol. 13, Pt. 1, pp. 77–81.

Oler, J. W., Jenkins, P. E., and Goldschmidt, V. W.
- 1981. Interface related velocities in turbulent plane jets. Phys. Fluids, Vol. 24, No. 7, pp. 1235–1237.

Onsager, L.
- 1945. The distribution of energy in turbulence (abstr.) Phys. Rev., Vol. 68, No. 11–12, p. 286.
- 1949. Statistical hydrodynamics. Nuovo cimento, (9), 6, Suppl. No. 2, pp. 279–287.

Onufriev, A. T.
- 1977. On the model equation for probability density (PDF) in the semiempirical theory of turbulent transport. In: Turbulent Flows. Moscow, Nauka, pp. 110-116.

Ottino, J. M.
- 1982. Description of mixing with diffusion and reaction in terms of the concept of material surfaces. J. Fluid Mech., Vol. 114, pp. 83–103.

Paizis, S. T. and Schwarz, W. H.
- 1974. An investigation of the topography and motion of the turbulent interface. J. Fluid Mech., Vol. 63, Pt. 2, pp. 315–343.
- 1975. Entrainment rates in turbulent shear flows. J. Fluid Mech., Vol. 68, Pt. 2, pp. 297–308.

Palm-Leis, A. and Strehlow, R. A.
- 1969. On the propagation of turbulent flames. Combustion and Flame, Vol. 13, No. 2, pp. 111–129.

Paquin, J. E. and Pond, S.
    1971. The determination of the Kolmogorov constants for velocity, temperature, and humidity from second- and third-order structure functions. J. Fluid Mech., Vol. 50, Pt. 2, pp. 257–269.

Peters, N.
    1984. Laminar diffusion flamelet models in nonpremixed turbulent combustion. Progr. Energy and Combust. Sci., Vol. 10, No. 3, pp. 319–340.

Peters, N. and Williams, F. A.
    1981. Coherent structures in turbulent combustion. Lect. Not. Phys., Vol. 136, pp. 364–393.

Petersen, R. E. and Emmons, H. W.
    1961. Stability of laminar flames. Phys. Fluids, Vol. 1, No. 4, pp. 456–464.

Phillips, O. M.
    1955. The irrotational motion outside a free turbulent boundary. Proc. Cambr. Phil. Soc., Vol. 51, No. 1, pp. 220–229.
    1972. The entrainment interface. J. Fluid Mech., Vol. 51, Pt. 1, pp. 97–118.

Pond, S., Smith, S. D., Hamblin, P. F., and Burling, R. W.
    1966. Spectra of velocity and temperature fluctuations in the atmospheric boundary layer over sea. J. Atmos. Sci., Vol. 23, No. 4, pp. 376–386.

Pond, S. and Stewart, R. W.
    1965. Measurements of statistical characteristics of small-scale turbulent motions. Izv. AN SSSR. Fizika atmosfery i okeana (Physics of the atmosphere and the ocean), Vol. 1, No. 9, pp. 914–919.

Pond, S., Stewart, R. W., and Burling, R. W.
    1963. Turbulence spectra in the wind over waves. J. Atmos. Sci., Vol. 20, No. 3, pp. 319–324.

Pope, S. B.
    1976. The probability approach to the modelling of turbulent reacting flows. Combustion and Flame, Vol. 27, No. 3, pp. 299–312.
    1979a. The statistical theory of turbulent flames. Phil. Trans. Roy. Soc., A291, pp. 529–568.
    1979b. A rational method of determining probability distributions (PDFs) in turbulent reacting flows. J. Nonequil. Thermodyn., Vol. 4, p. 309.
    1980. Probability distributions (PDFs) of scalars in turbulent shear flow. In: Turbulent shear flows. II. Second Int. Sympos., London, 1979.
    1981a. A Monte Carlo method for the PDF equations of turbulent reactive flow. Combustion Sci. and Tech., Vol. 25, pp. 159–174.
    1981b. Transport equation for the joint probability density function (PDF) of velocity and scalars in turbulent flow. Phys. Fluids, Vol. 24, No. 4, pp. 588–596.
    1982. The application of PDF transport equations to turbulent reactive flows. J. Nonequil. Thermodyn., Vol. 7, pp. 1–14.
    1983. A Lagrangian two-time probability density function equation for inhomogeneous turbulent flows. Phys. Fluids, Vol. 26, No. 12, pp. 3448–3450.

Povinelli, L. A. and Fuchs, A. E.
  1962. The spectral theory of turbulent flame propagation. In: Eighth Int. Sympos. on combustion. Baltimore, The Williams and Wilkins Co., pp. 554–559.
Praskovskii, A. A.
  1982. Overshoots of turbulent fluctuations. Stored in VINITI, No. 4831–82, 81 pp.
  1983. Measurement of conditionally averaged characteristics of turbulence in the plane wake of a cylinder. ZhPMTF, No. 6, pp. 87–94.
Prudnikov, A. G., Volynskii, M. S., Sagaovich, V. N., et al.
  1971. Processes of mixture formation and combustion in jet engines. Moscow, Mashinostroyeniye.
Pullin, D. I.
  1981. The nonlinear behavior of a constant vorticity layer at a wall. J. Fluid Mech., Vol. 108, pp. 401–422.
Rabinovich, M. I.
  1978. Stochastic self-excitation in turbulence. UFN, Vol. 125, Iss. 1, pp. 123–168.
Rajagopalan, S. and Antonia, R. A.
  1980. Characteristics of a mixing layer of a two-dimensional turbulent jet. AIAA J., Vol. 18, No. 9, pp. 1052–1058.
Ramshaw, J. D.
  1980. Partial chemical equilibrium in fluid dynamics. Phys. Fluids, Vol. 23, No. 4, pp. 675–681.
Rasshchupkin, V. I. and Sekundov, A. N.
  1978. Experimental and theoretical investigation of temperature fluctuations in the wake behind a linear thermal source. Izv. AN SSSR, MZhG, No. 4, pp. 39–45.
Raushenbakh, B. V., Belyi, S. A., Bespalov, I. V., Borodachev, V. Ya., Volynskii, M. S., and Prudnikov, A. G.
  1964. Physical principles of the working process in the combustion chambers of jet engines. Moscow, Mashinostroyeniye.
Record, F. A. and Cramer, H. E.
  1966. Turbulent energy dissipation rates and exchange processes above a nonhomogeneous surface. Quart. J. Roy. Meteorol. Soc., Vol. 92, No. 394, pp. 519–532.
Reynolds, W. S.
  1976. Computation of turbulent flows. Annual Rev. of Fluid Mech., Vol. 8, pp. 183–208.
Ribner, H. S. and Tucker, M.
  1953. Spectrum of turbulence in a contracting stream. Nat. Adv. Com. Aeronaut. Report, No. 1113.
Richardson, L. F.
  1922. Weather prediction by numerical process. Cambridge Univ. Press.

1926. Atmospheric diffusion shown on a distance-neighbor graph. Proc. Roy. Soc., A110, No. 756, pp. 709–737.

Roache, P. J.
    1976. Computational fluid dynamics, Revised Printing, Albuquerque, Hermosa Publishers.

Rochko, A.
    1976. Structure of turbulent shear flows: a new look. AIAA Paper, No. 78.

Rodi, W.
    1980. Turbulence models for environmental problems. In: Prediction methods for turbulent flows. Ed. by W. Kollman. Hemisphere Publishing Corp., pp. 259–350.

Rose, H. A. and Sulem, P. L.
    1978. Fully developed turbulence and statistical mechanics. J. Phys., Vol. 39, pp. 441–484.

Sabel'nikov, V. A.
    1975. Some linear problems of the theory of homogeneous turbulence deformation. Trudy TsAGI, Iss. 1702.
    1979. On the problem of describing free turbulent flows taking into account the phenomenon of intermittency. Equations for conditionally averaged turbulence characteristics. Trudy TsAGI, Iss. 1998.
    1980a. Concentration probability distribution (PDF) in the turbulent diffusion plume. In: Combustion of gases and natural fuels. Chernogolovka, OIKhF AN SSSR, pp. 40–44.
    1980b. Equations for the velocity and concentration probability distributions (PDFs) in turbulent and nonturbulent regions of free flows. Sci. Records TsAGI, Vol. 11, No. 6, pp. 25–30.
    1981. Description of turbulent diffusion combustion with the aid of a joint velocity and concentration probability distribution. In: Reports at the seminar "Mechanics and physics of plasma and gas flows (aerodynamics of combustion of gases)". Riga, IPM AN SSSR and IPM Latv. SSR, pp. 42–43.
    1982a. On the use of single-point velocity probability distribution (PDF) for the description of turbulent flows. AhMTF, No. 5, pp. 66–74.
    1982b. Theoretical and numerical investigation of concentration probability distribution (PDF) in free turbulent flows. Fizika goreniya i vzryva (Physics of combustion and explosion), Vol. 18, No. 2, pp. 77–88.
    1982c. Passive contaminant concentration probability distribution in the mixing layer. Scientific Records of TsAGI, Vol. 13, No. 5, pp. 49–57.
    1982d. Computation of a plane turbulent jet using the equation for velocity probability distribution. In: Turbulent jet flows. Reports of the fourth all-union scientific conf. on theoretical and applied aspects of turbulent flows. Part II. Tallin, Izd-vo AN ESSR, ITEF, pp. 27–31.
    1983. Model equation for velocity and concentration probability distribution (PDF) in turbulent mixing and diffusion combustion of gases. Fiziki goreniya i vzryva, Vol. 19, No. 2, pp. 37–46.

1985a. Analysis of semiempirical hypotheses upon closure of the equation for concentration probability density (PDF). Stored in VINITI, No. 7212-85, 48 pp.

1985b. On the behavior of concentration probability density (PDF) in the region of large fluctuation amplitudes in jet turbulent flows. In: Turbulent jet flows. Reports of the fifth all-union scientific conf. on theoretical and applied aspects of turbulent flows. Part I. Tallin, Izd-vo AN ESSR, ITEF, pp. 75-79.

1985c. Semiempirical model for the computation of the intermittency factor, conditionally averaged velocities, and the second moments of turbulent flows. Sci. Records of TsAGI, Vol. 16, No. 5, pp. 48-59.

1986. Analysis of the models for concentration probability density in the theory of turbulent mixing and combustion. In: Transactions of the intern. school seminar " Processes of turbulent transport in reacting systems". Minsk, ITMO AN SSSR, pp. 19-29.

Saffman, P. G.
1978. Problems and progress in the theory of turbulence. Lecture Notes Phys., Vol. 76, pp. 279-306.
1981. Dynamics of vorticity. J. Fluid Mech., Vol. 106, pp. 49-58.

Sarofim, A. F. and Pohl, J. H.
1973. Kinetics of nitric oxide formation in premixed laminar flames. In: Fourteenth Int. Sympos. on Combustion. Pittsburgh, The Combustion Inst., pp. 739-753.

Schedwin, J., Stegen, G. R., and Gibson, C. H.
1974. Universal similarity at high grid Reynolds numbers. J. Fluid Mech., Vol. 65, Pt. 3, pp. 561-579.

Schlichting, H.
1960. Boundary layer theory: 4th ed. New York, McGraw-Hill.

Schon, J. P.
1977. Conditional sampling. In: Measurement of unsteady fluid phenomena. Ed. by B. E. Richards. New York, McGraw-Hill, pp. 291-325.

Sedov, L. I.
1977. Similarity and dimensional methods in mechanics. Moscow, Nauka.

Sekundov, A. N.
1971. Use of a differential equation for turbulent viscosity in the analysis of plane non-self-similar flows. Izv. AN SSSR, MZhG, No. 5, pp. 119-127.

Shchelkin, K. I.
1943. On the combustion in turbulent flow. ZhTF, Vol. 13, No. 9-10, pp. 520-530.

Shchelkin, K. I. and Troshin, Ya. K
1965. Combustion gas dynamics. Moscow, Izd-vo AN SSSR.

Shcherbina, Yu. A.
1982. Statistical characteristics of turbulent transport. Dolgoprudnyi, MFTI.

Shcherbina, Yu. A. and Mogilko, V. A.
    1985. Statistical characteristics of convective contaminant transport in a free jet. In: Turbulent jet flows. Reports of the fifth all-union scientific conf. on theoretical and applied aspects of turbulent flows. Part II. Tallin, Izd-vo AN ESSR, ITEF, pp. 69–74.

Shchetinkov, E. S.
    1965. Physics of combustion of gases. Moscow, Nauka.

Sheih, C. M., Tennekes, H., and Lumley, J. L.
    1971. Airborne hot-wire measurements of the small-scale structure of atmospheric turbulence. Phys. Fluids, Vol. 14, No. 2, pp. 201–215.

Shepard, I. G. and Moss, J. B.
    1981. Measurements of conditioned velocities in a turbulent premixed flame. AIAA Paper, No. 181.

Shraiber, A. A., Milyutin, V. N., and Yatsenko, V. P.
    1980. Hydromechanics of two-component flows with solid polydispersed substance. Kiev, Naukova dumka.

Shvab, V. A.
    1948. Relation between the temperature and velocity fields of a gas plume. In: Investigation of natural fuel combustion processes. Moscow, Gosenergoizdat, pp. 231–248.

Sigal, I. L.
    1977. Protection of the atmospheric basin when burning fuel. Moscow, Nauka.

Siggia, E. D.
    1981. Numerical study of small-scale intermittency in three-dimensional turbulence. J. Fluid Mech., Vol. 107, pp. 375–406.

Siggia, E. D. and Patterson, G. S.
    1978. Intermittency effects in a numerical simulation of stationary three-dimensional turbulence. J. Fluid Mech., Vol. 86, Pt. 3, pp. 567–592.

Sosinovich, V. A.
    1973. Finite-dimensional distribution functions in the statistical theory of turbulence. TMF, Vol. 17, No. 1, pp. 131–141.
    1974. Kinetic equations for homogeneous turbulence. Izv. AN SSSR, MZhG, No. 5, pp. 150–152.
    1981a. Closed equation for the structural function of the isotropic turbulent velocity field. Inzh.-fiz. zhurn. (Journal of engineering physics), Vol. 40, No. 6, pp. 980–992.
    1981b. Equation for the structural function of the turbulent stationary isotropic velocity field and its solution in the scale inertial range. Inzh. fiz. zhurnal, Vol. 41, No. 5, pp. 796–808.

Spalding, D. B.
    1971. Mixing and chemical reaction in steady confined turbulent flame. In: Thirteenth Int. Sympos. on Combustion. Pittsburgh, The Combustion Inst., 649–657.

1976. Mathematical models of turbulent flames: a review. Combust. Sci. and Technol., Vol. 13, pp. 3–25.

Sreenivasan, K. R.
1981. Evolution of the center-line probability density function (PDF) of temperature in a plane turbulent wake. Phys. Fluids, Vol. 24, No. 7, pp. 1232–1234.

Sreenivasan, K. R. and Antonia, R. A.
1978. Joint probability densities (PDFs) and quadrant contributions in a heated turbulent round jet. AIAA J., Vol. 16, No. 9, pp. 867–868.

Sreenivasan, K. R., Antonia, R. A., and Britz, D.
1979. Local isotropy and large structures in a heated turbulent jet. J. Fluid Mech., Vol. 94, Pt. 4, pp. 745–776.

Sreenivasan, K. R., Antonia, R. A., and Stephenson, S. E.
1978. Conditional measurements in a heated axisymmetric turbulent mixing layer. AIAA J., Vol. 16, No. 9, pp. 869–870.

Sreenivasan, K. R., Danh, H. Q., and Antonia, R. A.
1977. Temperature dissipation fluctuations in turbulent boundary layer. Phys. Fluids, Vol. 20, No. 10, pp. 1050–1057.

Starner, S. H. and Bilger, R. W.
1980a. LDA measurements in a turbulent diffusion flame with axial pressure gradient. Comb. Sci. and Technology, Vol. 21, pp. 259–276.
1980b. Measurements of velocity and concentration in turbulent diffusion flames with pressure gradients. AIAA Paper, No. 25.

Stewart, R. W.
1963. On the agreement of the available data on the spectrum and asymmetry of locally isotropic turbulence. Dokl. AN SSSR, Vol. 152, No. 3, pp. 324–328.

Stewart, R. W., Wilson, J. R., and Burling, R. W.
1970. Some statistical properties of small-scale turbulence in an atmospheric boundary layer. J. Fluid Mech., Vol. 41, Pt. 1, pp. 141–152.

Stewartson, C.
1951. The impulsive motion of a flat plate in a viscous fluid. J. Mech. and Appl. Math., Vol. 4, No. 2, pp. 182–198.
1974. Multistructural boundary layers on flat plates and related bodies. Adv. Appl. Mech., Vol. 14, pp. 145–239.

Sulem, C. and Sulem, Pl.
1983. The well-posedness of two-dimensional ideal flow. Journal de Mecanique théorique et appliquée Numéro Spécial, pp. 217–242.

Suzuki, T. Oba, M. Hirano, T., and Tsuji, H.
1979. An experimental study of turbulent premixed flame. Bulletin JSME, Vol. 22, No. 167, pp. 848–856.

Sychev, V. V.
1972. On laminar separation. Izv. AN SSSR, MZhG, No. 3, pp. 47–59.

Sychev, V. V., Ruban, A. I., Sychev, Vik. V., and Korolev, G. L.
1987. Asymptotic theory of detached flows. Moscow, Nauka.

Talantov, A. V.
    1975. Principles of the theory of combustion. Kasan, KAI.
Tavoularis, S. and Corrsin, S.
    1981a. Experiments in nearly homogeneous turbulent shear flow with uniform mean temperature gradient. Part 1. J. Fluid Mech., Vol. 104, pp. 311–347.
    1981b. Experiments in nearly homogeneous turbulent shear flow with uniform mean temperature gradient. Part 2. The fine structure. J. Fluid Mech., Vol. 104, pp. 349–367.
Taylor, G. I., and Green, A. E.
    1937. Mechanism of the production of small eddies from large ones. Proc. Roy. Soc., A158, No. 895, pp. 499–521.
Thomas, R. M.
    1973. Conditional sampling and other measurements in a plane turbulent wake. J. Fluid Mech., Vol. 57, Pt. 3, pp. 549–582.
Tikhomirov, V. G.
    1958. Principal characteristics of combustion of two-phase fuel-air mixture. In: Combustion of two-phase systems. Moscow, Izd-vo AN SSSR, pp. 19–25.
Tikhonov, A. N. and Arsenin, V. Ya.
    1974. Methods of solution of incorrect problems. Moscow, Nauka.
Townsend, A. A.
    1948. Local isotropy in the turbulent wake of a cylinder. Austral. J. Sci. Res., Vol. 1, No. 2, pp. 161–174.
    1949. The fully developed turbulent wake of a circular cylinder. Austral. J. Sci. Res., Vol. 2, pp. 451–468.
    1951. On the fine-scale structure of turbulence. Proc. Roy. Soc., A208, No. 1098, pp. 534–542.
    1956. The structure of turbulent shear flow. Cambridge Univ. Press.
Tung, T. S. and Adrian, R. J.
    1980. Higher-order estimates of conditional eddies in isotropic turbulence. Phys. Fluids, Vol. 23, No. 7, pp. 1469–1470.
Uberoi, M. S. and Freymuth, P.
    1969. Spectra of turbulence in wakes behind circular cylinders. Phys. Fluids, Vol. 12, No. 7, pp. 1359–1363.
    1970. Turbulence energy balance and spectra of the axisymmetric wake. Phys. Fluids, Vol. 13, No. 9, pp. 2205–2210.
Uberoi, M. S. and Singh, P. I.
    1975. Turbulent mixing in a two-dimensional jet. Phys. Fluids, Vol. 18, No. 7, pp. 764–770.
Ulinich, F. R.
    1968. Statistical dynamics of turbulent incompressible fluid. Dokl. AN SSSR, Vol. 183, No. 3, pp. 535–537.

Ulinich, F. R. and Lyubimov, B. Ya.
  1968. On the statistical theory of turbulence at high Reynolds numbers. ZhETF, Vol. 55, No. 3, pp. 951–965.
Van Atta, C. W. and Antonia, R. A.
  1980. Reynolds number dependence of skewness and flatness factors of turbulent velocity derivatives. Phys. Fluids, Vol. 23, No. 2, pp. 252–257.
Van Atta, C. W. and Chen, W. Y.
  1970. Structure functions of turbulence in the atmospheric boundary layer over the ocean. J. Fluid Mech., Vol. 44, Pt. 1, pp. 145–159.
Varma, A. K. and Beddini, R. A.
  1976. Second-order closure analysis of turbulent reacting flows. In: Proc. 1976 Heat Transfer and Fluid Mech. Inst., Davis, Calif., pp. 229–240.
Varshavskii, G. A.
  1945. Combustion of drops of liquid fuel (diffusion theory). TrNIINKAP, No. 6.
Venkataramani, K. S., Tutu, N. K., and Chevray, R.
  1975. Probability distributions (PDFs) in a round heated jet. Phys. Fluids, Vol. 18, No. 11, pp. 1413–1420.
Venkataramani, K. S. and Chevray, R.
  1978. Statistical features of heat transfer in grid generated turbulence: constant-gradient case. J. Fluid Mech., Vol. 86, Pt. 3, pp. 513–544.
Ventsel, A. D.
  1975. Course in the theory of random processes. Moscow, Nauka.
Verollet, E.
  1972. Etude d'une couche limite turbulente avec aspiration et chauffage á la paroi. Thése Docteur és Sciences, Univesité de Provence (also Rapport CEA-R-4872, CEN Saclay).
Vilyunov, V. N. and Dik, I. G.
  1975. On the frontal combustion in low-intensity large-scale turbulent flow. Fizika goreniya i vzryva (Physics of combustion and explosion), Vol. 11, No. 2, pp. 223–229.
  1976. On the statistical and phenomenological approach in describing turbulent flames. ZhPMTF, No. 5, pp. 61–68.
Vishik, M. I. and Fursikov, A. V.
  1980. Mathematical problems of statistical fluid mechanics. Moscow, Nauka.
Volodina, L. I. and Andreev, M. A.
  1958. Effect of heating the air on the process of flame stabilization by high-drag bodies in an open flow. In: Investigation of combustion processes. Moscow, Izd-vo AN SSSR, pp. 36–39.
Vulis, L. A.
  1960. On the problem of the role of temperature fluctuations in turbulent combustion. In: Transactions of the third all-union conf. on the theory of combustion. Moscow, Izd-vo AN SSSR, pp. 86–90.
Vulis, L. A., Ershin, Sh. A., and Yarin, L. P.
  1963. Fundamentals of the theory of gas plume. Leningrad, Energiya.

Wang, J. C. T.
    1983. On the numerical methods for the singular parabolic equations in fluid dynamics. J. Comput. Phys., Vol. 52, No. 3, pp. 464–479.

Weizsäcker, C. F. von
    1948. Das spectrum der Turbulenz bei grossen Reynolds'schen Zahlen. Zs. Phys., Vol. 124, No. 7–12, pp. 614–627.

Williams, F. A.
    1965. Combustion theory. Palo Alto, Addison-Wesley Publishing Co.

Williams, R. M. and Paulson, C. A.
    1977. Microscale temperature and velocity spectra in the atmosphere boundary. J. Fluid Mech., Vol. 83, Pt. 3, pp. 547–568.

Wingaard, J. C. and Coté, O. R.
    1971. The budgets of turbulent kinetic energy and temperature variance in the atmospheric surface layer. J. Atmos. Sci., Vol. 28, No. 2, pp. 190–201.

Wingaard, J. C. and Pao, Y. H.
    1975. Some measurements of fine structure at large Reynolds number turbulence. In: Proc. Sympos. held at the Univ. Calif., San Diego. Berlin, Springer-Verlag, pp. 384–400.

Wingaard, J. C., Pao, Y. H., and Wygnanski, G.
    1976. On the two-dimensional mixing region. J. Fluid Mech., Vol. 79, No. 2, pp. 209–250.

Wingaard, J. C. and Tennekes, H.
    1970. Measurements of the small-scale structure of turbulence at moderate Reynolds numbers. Phys. Fluids, Vol. 13, No. 8, pp. 1962–1969.

Wygnanski, I. and Fiedler, H.
    1969. Some measurements in the self-preserving jet. J. Fluid Mech., Vol. 38, Pt. 3, pp. 577–612.
    1970. The two-dimensional mixing region. J. Fluid Mech., Vol. 41, Pt. 2, pp. 327–361.

Yaglom, A. M.
    1966. On the effect of energy dissipation fluctuations on the form of turbulence characteristics in the inertial range. Dokl. AN SSSR, Vol. 166, No. 1, pp. 49–52.
    1981. Laws of fine-scale turbulence in the atmosphere and ocean (40 years of the theory of local turbulence). Izv. AN SSSR. Physics of atmosphere and ocean, Vol. 17, No. 2, pp. 1235–1257.

Yankovskii, V. M. and Talantov, A. V.
    1969. Effect of system size on the main characteristics of combustion in turbulent flow of homogeneous mixture. In: Transactions of the second all-union symposium on combustion and explosion. Chernogolovka, OIKhF AN SSSR, pp. 117–123.

Yoshida, A. and Günther, R.
    1980. Temperature and ionization measurements in turbulent premixed flames. AIAA Paper, No. 207.

Zel'dovich, Ya. B.
   1944. Theory of combustion and detonation of gases. Moscow, Izd-vo AN SSSR.
   1949. On the theory of combustion of nonpremixed gases. ZhTF, Vol. 19, No. 10, pp. 1199–1210.
   1966. On the stabilization of a distorted flame front. ZhPMTF, No. 1, pp. 102–104.
Zel'dovich, Ya. B. and Barenblatt, G. I.
   1959. Theory of flame propagation. Combustion and Flame, Vol. 3, No. 1, pp. 61–74.
Zel'dovich, Ya. B., Barenblatt, G. I., Librovich, V. B., and Makhviladze, G. M.
   1980. Mathematical theory of combustion and explosion. Moscow, Nauka.
Zel'dovich, Ya. B. and Frank-Kamenetskii, D. A.
   1938a. On the theory of steady flame propagation. Dokl. AN SSSR, Vol. 19, No. 9, pp. 693–695.
   1938b. Theory of thermal flame propagation. ZhFKh, Vol. 12, No. 1, pp. 100–105.
   1947. Turbulent and heterogeneous combustion. Moscow, MMI.
Zel'dovich, Ya. B., Sadovnikov, P. Ya., and Frank-Kamenetskii, D. A.
   1947. Oxidation of nitrogen during combustion. Moscow, Izd-vo AN SSSR.
Zhuk, V. I. and Ryzhov, O. S.
   1980. Free interaction and stability of the boundary layer in incompressible fluid. DAN SSSR, Vol. 253, No. 6, pp. 1326–1329.
Zimont, V. L.
   1977. On the calculation of turbulent combustion of partly premixed gases. In: Combustion of heterogeneous and gaseous systems. Chernogolovka, OIKhF AN SSSR, pp. 76–80.
   1979. On the theory of turbulent combustion of homogeneous combustible mixture of gases at high Reynolds numbers. Fizika goreniya i vzryva (Physics of combustion and explosion), Vol. 15, No. 3, pp. 23–32.
Zimont, V. L. and Meshcheryakov, E. A.
   1974. Calculation of turbulent combustion of submerged and coflowing jet taking into account the concentration fluctuations within the framework of the integral method. Fizika goreniya i vzryva, Vol. 10, No. 2, pp. 220–230.
Zimont, V. L., Meshcheryakov, E. A., and Sabel'nikov, V. A.
   1978. A simple model accounting for molecular mixing in turbulent combustion of nonpremixed gases. Fizika goreniya i vzryva, Vol. 14, No. 3, pp. 55–62.
   1981. On the calculation of turbulent diffusion combustion of nonpremixed gases taking into account concentration fluctuations and the effect of turbulence of coflowing streams. In: Theory and practice of burning of gases, Issue 7. Leningrad, Nedra, pp. 91–97.
   1983. Specific features of supersonic combustion of nonpremixed gases in channels. Fizika goreniya i vzryva, Vol. 19, No. 4, pp. 75–78.

Zimont, V. L. and Sabel'nikov, V. A.
- 1975a. On the equation of turbulent transport in the presence of molecular diffusion. Izv. AN SSSR. Fizika atmosfery i okeana (Physics of atmosphere and ocean), Vol. 11, No. 6, pp. 627–629.
- 1975b. Criterial description of the rate of combustion in the turbulent flow of a combustible mixture. In: Reports of the second all-union school conference on the theory of combustion. Moscow, IPM AN SSSR, pp. 52–54.

Zubkovskii, S. L.
- 1962. Frequency spectrum of the fluctuations of the horizontal component of air velocity in the air layer above land. Izv. AN SSSR, ser. geofiz., No. 10, pp. 1425–1433.

Zukoski, E. E. and Marble, F. E.
- 1955. The role of wake transition in the process of flame stabilization on bluff bodies. Combustion researches and reviews. In: Butterworth Scientific Publications. London, pp. 167–180.

INDEX

Absorption, 194
Adiabatic combustion, 208, 249
   temperature, 38
Airy formula, 91
Airy function, 87, 199
Analog-digital converter, 160
Anemometer wire, 156, 157
Angular brackets, 2
Archimedean forces, 24
Arrhyenius law, 38
"Artificial" averaging, 164
Asymmetry, 5
Atomized liquid fuel, 286
Atoms, 190
Axially symmetric flows, 4
Axisymmetric methane jet, 29
Axisymmetric wake, 165, 176

Back-flow region, 280
Bayes formula, 21, 50
Bayes theorem, 133, 142, 147
Bifurcation points, 184
Black-white mixing, 17, 53
Boundary layer, 2
   thickness, 8
Boundary-value problem, 92, 101, 117

Bounding isoscalar surfaces, 60
Bunsen burner, 38, 253
Buoyancy forces, 196, 199
Burgers model, 54
Burners, 189
Butterworth sinusoidal filter, 166

Calculated flowrate of nitrogen oxides, 229
Calibration curve, 159
Carbon dioxide, 208
Carbon monoxide, 84
Carbon particles, 205
Cartesian coordinate system, 1
Cartographic measurements, 14
Cascade disintegration, 62
Cascade "links," 133
Cauchy-Bunyakovskii inequality, 135
Cauchy equation, 76
Circular cylinder, 4, 6, 8, 63, 71, 84, 91, 154
Closure, 56
   hypotheses, 51, 61
Cold jet, 202
Combustion, 35, 267
   chamber, 189
   front, 39

Combustion (*Cont.*):
    of homogeneous mixtures, 233
    of hydrogen, 220
    of methane-air mixtures, 272, 280
    of nonpremixed gases, 194
    of propane, 224
    of propane-air mixtures, 272
    of suspended drops, 291
Concentration:
    gradients, 2
    PDF, 72, 89
    pulsation maximum, 29
Conical stabilizer, 36
Constant C, 151, 161
Constant μ, 152
Contortion of flame, 247
Conversion interval, 161
Curl system, 52
Curved flame, 282
Cylindrical channel, 201

Damköhler number, 211
Deformation, 247
Delta function, 29, 89
Density, 36, 191
Dependence, 196
Differing wave numbers, 61
Diffusion, 43
    combustion, 42, 189, 192, 201
    plume, 208
Discharge velocity, 36
Displacements, 175
Dissipation, 145, 196
    energy, 15
    spectrum, 160
Dorodnitsyn variable, 248

Earth's surface, 11
Eddies, 65, 79
Energy-carrying vortices, 183
Energy dissipation, 131, 146, 154, 167, 178, 186
Energy transfer, 61
Enthalpy, 194, 206
Equilibrium, 42, 67, 215
Ergodicity, 27
Ethane, 250
Ethylene, 264
Euler's equation, 2, 13
Evaporation, 285, 289

Experimental values of the constants C and μ, and ratios K_1 and K_2, 151, 152

Favre averaging, 197
Fine-scale:
    motions, 174
    vortices, 183
    turbulence structure, 159
Fine thirds law, 163
Finite scale structure, 163
Finite volume, 28
Flame:
    autoturbulization, 262
    contortion, 251
    front, 36, 192, 202, 206, 234, 286
    -generated turbulence, 260
    instability, 246, 253, 264
    length, 193
    propagation, 237, 251
    stabilizer, 265
    structure, 247
    temperature, 209
    tongue, 246
    tube, 302
Flameout, 192, 269, 272
Fluctuating jet boundary, 206
Fluctuating velocity, 1, 239
Fluctuations, 25
Fluid dynamics deformation, 247
Fluid dynamics stability, 2
Fluid particles (moles), 73
Fourier components, 1
Fourier method, 77
Fourier transform, 43, 44, 50, 251
Free flame, 191
Free flow, 13
Free jet, 196
Free turbulent flows, 57, 81, 93
Frozen spatial profiles, 31
Frozen temperature profiles, 98
Frozen turbulence, 158
Frontal combustion, 265
Fuel oxidation, 193
Fuel stream, 194

Gas dynamic effects in turbulent diffusion combustion, 194
Gases, 25
Gasoline-air mixture, 234
Gas turbine(s), 189, 233

Gaussian curve(s), 34, 77, 91
Gaussian function, 107
Global geometric configuration, 269
Gravitational acceleration, 196
Grid-induced turbulent flows, 25

Hamilton's operation, 2
Harmonic fluctuations, 22
Hausdorff-Bezikovich dimension, 128
Heat:
    equation, 38
    flux, 249, 252
    losses, 192
    release, 194, 269
Heaviside function, 27, 28
Heavy fluid (oxidant), 191
High-drag bodies, 239, 270
High-frequency filtering, 166
High turbulence intensity, 159
"Holes," 270
Homogeneous turbulence, 26, 48, 75
Hydrocarbon pyrolysis, 298
Hydrocarbons, 198, 215, 250, 297, 310
Hydrodynamic nonhomogeneity, 156
Hydrogen, 224, 296
Hydrogen-air mixtures, 250
Hydrogen plume, 197

Incoming flow, 5
Industrial furnaces, 189
Inert contaminant, 195, 204
Inertial range, 159
    of turbulence spectrum, 172
Integrals of function, 118
Integral turbulence scale, 14, 18, 48
Interaction between turbulent and nonturbulent fluids, 139
Intermittency factor, 27
Intermittency function, 166
Internal isoscalar surfaces, 123
Internal nodes, 118
Isoscalar surface(s), 20, 27, 47, 122, 125, 190
Isotherm, 236, 256
Isothermal jet, 198
Isotropic turbulence, 2, 31, 55, 67, 93
Isotropy, 150, 181

Jet axis, 197
Jet(s), 3, 21, 58, 131

Jet discharge, 5
Jet-type flows, 32
Jet width, 5

Karman-Hovars equation, 135
Kinematic molecular viscosity, 1
Kinematic viscosity, 130, 292
Kinetics of oxidation, 210
Kolmogorov:
    constant, 163
    frequency, 171
    scale, 18, 48, 261
    -Obukhov Law/Theory, 33, 55, 149, 173, 182
    scale, 126, 139, 157, 183
Krook model, 54
Kurtosis, 9, 89, 105, 109, 115, 131

Laminar:
    boundary layer, 13, 186
    combustion, 236
    flame, 235
        deformation, 242
    scale, 7
Langevin equation, 51
Laplace operator, 67
Large fluctuation amplitudes, 178
Large-scale motions, 33
Large-scale perturbations, 265
Large-scale vortices, 221
Laser anemometry, 38
Laser Doppler anemometer, 255
Laser Raman spectroscopy, 296
Layer thickness, 2
Lebesque measure, 15
Light fluid (combustion products), 191
Limit isoscalar surfaces, 125
Linear law, 22
Lognormal distribution, 11, 18
Lognormal law, 175
Longitudinal energy spectrum, 164
Longitudinal structure function, 163
Longitudinal velocity, 71, 94, 196
Low turbulence intensity, 159
Long-wave region, 255
Luminosity, 237

Mach number, 42, 194, 195
Mean reaction rate, 202

Mean velocities, 57
Methane-air mixture, 279, 280
Mechanism of turbulent combustion, 242
Mellin transform, 134, 142
Methane, 235
Microturbulent viscosity, 185
Mixing layer, 26, 30, 164, 181
Molecular:
    diffusion, 2, 16, 27, 123, 189, 242
    mixing, 98
    transfer, 34, 68
    transport, 2, 32, 37, 54, 89, 194, 210, 235, 239, 278
    viscosity, 33
Monatomic oxygen, 213
Multiple-wire sensor, 159

Narrow spatial zones, 61, 210
"Natural" averaging, 164
Navier-Stokes equations, 13, 25, 41, 42, 48, 51, 129, 144, 183
Nitrogen, 296
    oxide(s), 224, 227
    oxidation, 209
Nodes, 184
Nonburning jet, 191
"Nonsteady flame propagation," 244, 254
Nonturbulent fluid, 3, 9, 17, 21, 132
Nozzle, 4, 95, 209, 230
Nozzle edge, 26
Nozzle exit, 158
Numerical algorithm, 13

One-point velocity PDI, 140
Oscillating wire, 244
Oscillations, 156, 190
Oscillogram(s), 4, 36, 60, 170
Oxidant stream, 194
Oxidation, 193
Oxidation kinetics, 210
Oxygen, 296

Parabolic cylinder function, 107
Partially premixed gases, 285, 295
Passive contaminant, 24, 207
    concentrations PDF, 57
PDF, 30, 35, 38, 72, 83, 88, 120, 178
    equations for, 41
Peclet number, 25

Phase point, 15
Phase space, 59, 62
Plane:
    boundary, 170
    cylinder, 174
    flaws, 4
    layer, 246
    -parallel flows, 118
    reaction zone, 192
    of symmetry, 82
    wake(s), 157, 159, 181
Plate tip, 8
Platinum coated tungsten, 155
Plume, 196
Poisson's equation, 48
Polygons, 127
Pressure:
    fluctuations, 6
    forces, 54
    gradient, 191
Probability density fluctuations, 1
Propane, 199, 207, 235, 250
    -air mixture, 254, 279
    jet, 209
    plume, 208
Pyrolysis, 198

Radial coordinate, 196
Radiation, 193, 204, 207, 208, 227, 310
Radiation flux density, 206
Radiative heat transfer, 207
Radius curvature, 192
Radius vector, 1
Rapid vortex-free deformation of compressible fluid, 263
Ratios K_1 and K_2, 153
Reynolds number, 1, 2, 12, 15, 16, 25, 26, 34, 37, 59, 61, 80, 125, 139, 146, 167, 182, 194, 210, 254, 273
Reynolds stresses, 66, 264
Resistance thermometer, 4
Round nozzle, 193, 196

Scalar dissipation(s), 2, 6, 45, 56, 58, 78, 102, 112, 196, 218, 304
Scalar fields, 25
Scanning frequency, 160
Scatter, 150, 275
Schlichting, 2
Shear flows, 155, 161

Similarity hypothesis, 132, 138
Single-wire probe, 159
Skewness, 115
Small eddies, 65
Small fluctuations, 22
Small-scale pulsations, 34
    effect of viscosity on, 146
Small-scale turbulence, 50, 129, 247
"Smearing," 89
Smoothing, 169
Soot formation, 208
Soot particles, 207, 208
Spark ignition energy, 241
Spatial distribution, 3
    domain, 14
    grid scale, 183
    resolution, 156
    scale, 220
    wake, 158
Spherical flame, 245, 247
Sponge, 24
    pattern, 60
Stable laminar plume, 253
Stationary air, 29, 193
Stephan-Boltzmann constant, 207
Stoichiometric surface, 198
Stokes formula, 292
Structure functions, 158
Submerged:
    axisymmetric jet, 87
    diffusion flame, 191
    diffusion plume, 193
    free propane plume, 197
    heated plane jet, 121
    jet, 140
Suspended drops, 291

Taylor-Green vortex, 13
Taylor's hypothesis, 6, 20, 158
Taylor's microscale, 9
Thermal diffusivity, 36, 233, 253, 280
Thermal thickness, 36
Thermodynamic analysis, 194, 208
Thermodynamic equilibrium, 190, 193, 210
Thickness, 217
Thickness of reaction zone, 191
Thin boundary, 39
Thin channels, 14
Three-dimensional wake, 157, 181
Three-point PDF, 146
Three-point Gaussian correlation function, 54

Transition zone, 171
Transverse velocity, 71, 99, 196
Trimolecular reaction, 214
Turbulence:
    energy, 196
    isotropy, 6
    modeling, 183
    spectra, 149, 185, 265
Turbulent:
    combustion, 242
        of partially premixed gases, 285
    diffusion combustion, 189, 196
    diffusion flame, 223
    flows, 1, 3, 37, 86
    fluid(s), 5, 15, 18, 21, 23, 28, 98, 123, 126
    kinetic energy, 196
    ranges, 20
    viscosity, 196
Two-point probability density, 139
Two-point velocity, 21
Two-thirds law, 131, 161, 163

Universal equilibrium, 55

Velocity, 44, 66, 259
    difference PDF, 137
    field, 3, 5
    fluctuation(s), 21, 64, 73, 126
    gradient fluctuations, 9
    pulsation, 265
    shear, 201
Viscosity, 130, 139
    induced singularities, 34
Viscous:
    dynamics, 13
    fluids, 186
    forces, 13
    processes, 28
Volumetric combustion, 35
Vortex:
    field, 31
    flow, 20, 125
    formation, 61
    tubes, 13
Vortices, 183, 219
Vorticity vector, 2

Wake(s), 3, 21, 58, 64, 82, 112, 131
    burns, 266

Wake(s) (*Cont.*):
    of circular cylinder, 90, 129
    expansion, 266
Water vapor, 207, 208
Wave number, 19, 156
Wind tunnel, 158
Wind velocity, 11

Zel'dovich-Frank-Kamenetskii approximation, 271

MAY 1 0 1990